Chemistry of Dehydrogenation Reactions and Its Applications

The present book focuses on advancement in the application of heterogeneous catalytic materials for the dehydrogenative synthesis of valuable organic compounds from substrates such as alcohols and simple aliphatic compounds. Several heterogeneous transition metals-based catalytic materials are explored for the synthesis of valuable chemicals for industrial applications. The book provides insight into the application of state-of-the-art technology for energy utilization and clean chemical synthesis.

Features:

- Offers a wide overview of dehydrogenation catalytic chemistry catalyzed by transition metals and their compounds.
- Helps design novel and more benign and uncomplicated protocols for the synthesis of valuable chemicals from readily available raw materials.
- Provides deeper insight into the aspect of dehydrogenation reactions for clean chemical synthesis via a cascade process.
- Summarizes new mechanistic details of dehydrogenation reactions, experimental side development and applications of dehydrogenation techniques.
- Explores alternative solutions for the assimilation and transportation of clean energy in the form of hydrogen energy utilization.

This book is aimed at graduate students and researchers in chemical engineering, chemistry, catalysis, organic synthesis, pharmaceutical chemistry and petrochemistry.

Emerging Materials and Technologies

Series Editor: Boris I. Kharissov

The *Emerging Materials and Technologies* series is devoted to highlighting publications centered on emerging advanced materials and novel technologies. Attention is paid to those newly discovered or applied materials with potential to solve pressing societal problems and improve quality of life, corresponding to environmental protection, medicine, communications, energy, transportation, advanced manufacturing, and related areas.

The series takes into account that, under present strong demands for energy, material, and cost savings, as well as heavy contamination problems and worldwide pandemic conditions, the area of emerging materials and related scalable technologies is a highly interdisciplinary field, with the need for researchers, professionals, and academics across the spectrum of engineering and technological disciplines. The main objective of this book series is to attract more attention to these materials and technologies and invite conversation among the international R&D community.

Polymer Processing
Design, Printing and Applications of Multi-Dimensional Techniques
Abhijit Bandyopadhyay and Rahul Chatterjee

Nanomaterials for Energy Applications
Edited by L. Syam Sundar, Shaik Feroz, and Faramarz Djavanroodi

Wastewater Treatment with the Fenton Process
Principles and Applications
Dominika Bury, Piotr Marcinowski, Jan Bogacki, Michal Jakubczak, and Agnieszka Jastrzebska

Mechanical Behavior of Advanced Materials: Modeling and Simulation
Edited by Jia Li and Qihong Fang

Shape Memory Polymer Composites
Characterization and Modeling
Nilesh Tiwari and Kanif M. Markad

Impedance Spectroscopy and its Application in Biological Detection
Edited by Geeta Bhatt, Manoj Bhatt and Shantanu Bhattacharya

Nanofillers for Sustainable Applications
Edited by N.M Nurazzi, E. Bayraktar, M.N.F. Norrrahim, H.A. Aisyah, N. Abdullah, and M.R.M. Asyraf

Chemistry of Dehydrogenation Reactions and Its Applications
Edited by Syed Shahabuddin, Rama Gaur and Nandini Mukherjee

Biosorbents
Diversity, Bioprocessing, and Applications
Edited by Pramod Kumar Mahish, Dakeshwar Kumar Verma and Shailesh Kumar Jadhav

For more information about this series, please visit www.routledge.com/Emerging-Materials-and-Technologies/book-series/CRCEMT.

Chemistry of Dehydrogenation Reactions and Its Applications

Edited by Syed Shahabuddin, Rama Gaur and
Nandini Mukherjee

Boca Raton London New York

CRC Press is an imprint of the
Taylor & Francis Group, an **informa** business

First edition published 2024
by CRC Press
2385 NW Executive Center Drive, Suite 320, Boca Raton FL 33431

and by CRC Press
4 Park Square, Milton Park, Abingdon, Oxon, OX14 4RN

CRC Press is an imprint of Taylor & Francis Group, LLC

© 2024 selection and editorial matter, Syed Shahabuddin, Rama Gaur and Nandini Mukherjee; individual chapters, the contributors

Reasonable efforts have been made to publish reliable data and information, but the author and publisher cannot assume responsibility for the validity of all materials or the consequences of their use. The authors and publishers have attempted to trace the copyright holders of all material reproduced in this publication and apologize to copyright holders if permission to publish in this form has not been obtained. If any copyright material has not been acknowledged please write and let us know so we may rectify in any future reprint.

Except as permitted under U.S. Copyright Law, no part of this book may be reprinted, reproduced, transmitted, or utilized in any form by any electronic, mechanical, or other means, now known or hereafter invented, including photocopying, microfilming, and recording, or in any information storage or retrieval system, without written permission from the publishers.

For permission to photocopy or use material electronically from this work, access www.copyright.com or contact the Copyright Clearance Center, Inc. (CCC), 222 Rosewood Drive, Danvers, MA 01923, 978-750-8400. For works that are not available on CCC please contact mpkbookspermissions@tandf.co.uk

Trademark notice: Product or corporate names may be trademarks or registered trademarks and are used only for identification and explanation without intent to infringe.

ISBN: 978-1-032-34396-9 (hbk)
ISBN: 978-1-032-34397-6 (pbk)
ISBN: 978-1-003-32193-4 (ebk)

DOI: 10.1201/9781003321934

Typeset in Times
by Apex CoVantage, LLC

Contents

Preface xi
About the Editors xiii
Contributors xv

Chapter 1 Introduction to Dehydrogenation Reactions of Organic Compounds 1

Syed Shahabuddin, Nandini Mukherjee, and Rama Gaur

1.1 Introduction 1
1.2 Mechanism of Dehydrogenation Reactions 2
    1.2.1 Alkane to Alkene 2
    1.2.2 Alcohol to Carbonyl Derivatives 2
    1.2.3 Dehydrogenation to Yield Olefins with EWG at α-Position 4
    1.2.4 Ester and Nitrile to Activated Olefins 4
    1.2.5 Amide/Lactam to Activated Olefins 4
1.3 Heterogeneous and Homogeneous Catalysts for Dehydrogenation Reactions 6
1.4 Types of Reactors for Dehydrogenation Reactions 6
1.5 Dehydrogenation Methods and Reactions That Are Commercially Significant 7
    1.5.1 Dehydrogenation of Paraffins to Olefins 7
    1.5.2 Dehydrogenation of C2-C15 Alkanes to Alkenes 8
    1.5.3 Dehydrogenation of Ethylbenzene to Styrene 9
1.6 Recent Advances in Dehydrogenation Technology 10
1.7 Summary 10

Chapter 2 Transition Metal-Based Catalyst for Dehydrogenation Reactions of Organic Compounds 13

Atul Kumar

2.1 Introduction 13
2.2 N-Alkylation by Dehydrogenative Alcohol Activation 13
    2.2.1 N-Alkylation by Ruthenium Catalyst 15
    2.2.2 N-Alkylation by Iridium Catalyst 18
    2.2.3 N-Alkylation by Pd Catalyst 23
    2.2.4 N-Alkylation by Copper/Iron Catalyst 23
    2.2.5 Synthesis of Primary Amine from Alcohol and Ammonia/Ammonium Salt 25
    2.2.6 Enantioselective Substitution of Alcohols by Amine 27
    2.2.7 Reductive N-Alkylation with Alcohols 29
2.3 C-alkylation by Dehydrogenative Alcohol Activation 29
    2.3.1 α-Alkylation of Ketones and Its Derivative 30
    2.3.2 β-Alkylation of Secondary Alcohols 32
    2.3.3 α-Alkylation of Activated Nucleophile 34
    2.3.4 Asymmetric C-C Bond Formation by Alcohols Activation 36
    2.3.5 Versatile HA-sequence by Dehydrogenative Alcohol Coupling 37
2.4 Dehydrogenative Amine Activation 38
    2.4.1 Transamination 39
    2.4.2 Hydroimination 40

	2.5	Dehydrogenative Alkane Activation 41
	2.6	Net Dehydrogenative Oxidation Reactions 42
		2.6.1 Formation of Ester and Acid 42
		2.6.2 Formation of Amide 43
		2.6.3 Formation of Nitriles by Amine Oxidation 44
	2.7	Semi-Borrowing Hydrogen (SBH) Process 45
		2.7.1 Synthesis of Benzimidazoles 45
		2.7.2 Modified Fischer Indole Synthesis 46
		2.7.3 Synthesis of Quinazolines Derivatives 46
		2.7.4 Synthesis of Pyrroles Derivatives 47
	2.8	Conclusions 47

Chapter 3 Transition Metal Catalyst Free Dehydrogenative Organic Synthesis: Role of New Materials, Composites, and Nanomaterials 54

Prakash Chandra and Syed Shahabuddin

	3.1	Introduction 54
	3.2	Nanotechnology Catalysts for Hydrogenation Budge in Organic Synthesis 55
	3.3	Heterogenization of Homogeneous Catalysts for Dehydrogenation Reactions 58
	3.4	Anchoring Homogeneous Catalysts over Heterogeneous Support 58
	3.5	Direct Grafting of Metal Complexes 59
	3.6	Encapsulation of the Catalysts 60
	3.7	Ionic Liquid Assisted Organic Transformation 61
	3.8	Single and Double Atom Catalysts for Transfer Hydrogenation Reactions 62
	3.9	Conclusions 63
	3.10	Outlook 63

Chapter 4 Dehydrogenation Reaction of Aliphatic and Aromatic Alcohols 67

Vijay Bahadur and Chandni Pathak

	4.1	Objectives 67
	4.2	Dehydrogenation Reaction 67
	4.3	Aliphatic and Aromatic Alcohols 67
	4.4	Dehydrogenation of Alcohols 67
	4.5	Acceptorless Dehydrogenation of Alcohols 69
		4.5.1 Conversion of Alcohols into Carbonyl Compounds 69
		4.5.2 Conversion of Alcohols into Ester Compounds 70
		4.5.3 Conversion of Alcohols into Amide Compounds 73
		4.5.4 Conversion of Alcohols into Imines Compounds 76
		4.5.5 Conversion of Alcohols into Acylated Compounds 78
		4.5.6 Conversion of Alcohols into Acetals Compounds 78
		4.5.7 Conversion of Alcohols into Polyester and Lactones Compounds 78
		4.5.8 Direct Synthesis of Pyrrole from Alcohols 79
	4.6	Green Method for the Dehydrogenation of Alcohols 81
		4.6.1 Dehydrogenation of Alcohols with Nanoparticles 81
		4.6.2 Dehydrogenation of Alcohols with Photocatalyst 83
	4.7	Conclusions 84

Contents

Chapter 5 Dehydrogenation Reactions of Hydrocarbons: Alkane, Alkenes, and Aromatic Hydrocarbons 87

Chandni Pathak and Vijay Bahadur
5.1 Introduction 87
5.2 Non-Oxidative Dehydrogenation 88
    5.2.1 Platinum-Based Catalyst 88
    5.2.2 Chromium Oxide-Based Catalyst 89
    5.2.3 Vanadium Oxide-Based Catalyst 90
    5.2.4 Molybdenum Oxide-Based Catalysts 90
    5.2.5 Gallium Oxide-Based Catalyst 90
    5.2.6 Carbon-Based Catalyst 91
5.3 Oxidative Dehydrogenation 91
    5.3.1 Groups V and VI Transition Metal Oxides 91
    5.3.2 Ni-Based Catalyst Systems 91
    5.3.3 Lithium and Halide-Containing Catalysts 92
5.4 Dehydrogenation of Alkanes by Pincer Complexes 92
    5.4.1 Dehydrogenation of Alkane by Pincer Iridium Complexes 93
    5.4.2 Pincer-Ruthenium Complexes as Catalysts for Alkane Dehydrogenation 94
5.5 Dehydrogenation of Aromatic Hydrocarbons 95
    5.5.1 Catalytic Dehydrogenation of Aromatic Hydrocarbons Using Pd or Pt 96
    5.5.2 Dehydrogenation of Aromatic Hydrocarbons Using DDQ 96

Chapter 6 Dehydrogenation Reactions of Aliphatic and Aromatic Amines 101

Nandini Mukherjee and Sauvik Chatterjee
6.1 Introduction 101
6.2 Mechanistic Consideration 101
6.3 Dehydrogenation Reactions of Aliphatic Amines 102
    6.3.1 Ru-Catalyzed Dehydrogenation 102
    6.3.2 Mo-Catalyzed Dehydrogenation 104
    6.3.3 Ni-Catalyzed Dehydrogenation 104
    6.3.4 Ir-Based Catalyst for Dehydrogenation 104
6.4 Dehydrogenation Reactions of Aromatic Amines 105
6.5 Challenges and Future Prospects 110

Chapter 7 Dehydrogenation Reactions of Aliphatic and Aromatic Carboxylic Acids and Their Derivatives 113

Megha Balha
7.1 Introduction 113
7.2 Dehydrogenation Reactions of Aliphatic and Aromatic Carboxylic Acids 114
7.3 Conclusion 121

Chapter 8 Dehydrogenation Reactions of Heterocyclic Compounds and Their Derivatives 123

Prakash Chandra and Syed Shahabuddin
8.1 Introduction 123

8.2 Transition Metal-Catalyzed Synthesis of Heterocyclic Compounds 123
    8.2.1 Synthesis of Lactones by Heterogeneous TM Catalysts 124
    8.2.2 Benzofurans and Chromones from Ortho-Substituted Phenols 124
    8.2.3 Nitrogen-Containing Heterocycles by Heterogeneous TM Catalysts 125
    8.2.4 Indoles, Benzimidazoles, Quinazolinones and Pyrroles 127
8.3 Summary and Outlook 131

Chapter 9 Recent Advances in Dehydrogenative Technique for Hydrogen Energy Storage and Utilization 134

Prakash Chandra and Syed Shahabuddin

9.1 Introduction 134
9.2 Important Properties of LOHC 136
9.3 Mono- and Polyaromatic Systems for LOHC Applications 136
    9.3.1 Benzene-Cyclohexane System 136
    9.3.2 Toluene-Methylcyclohexane System 137
    9.3.3 Decalin-Naphthalene System 137
    9.3.4 Perhydrodibenzyltoluene–Dibenzyltoluene 139
9.4 Heterocyclic Compounds 140
    9.4.1 Carbazole Derivatives 140
    9.4.2 Pyridines and Quinolines 141
    9.4.3 Pyrroles and Indoles 141
9.5 Integration of LOHC Process 142
9.6 Reactor for LOHC 142
9.7 Theoretical and Computational Approach 144
9.8 Conclusions 144

Chapter 10 Dehydrogenation Reactions and Inspirations from Nature for the Synthesis of Building Blocks Leading to Valued Pharmaceutical Compounds 151

Pravin R. Bhansali, Vijayendran K. K. Praneeth, and Ronald E. Viola

10.1 Introduction 151
10.2 Dehydrogenation Reactions Found in Nature 152
10.3 Dehydrogenation Reactions Inspired by Nature 154
    10.3.1 Quinoline Derivatives 154
    10.3.2 Pyrrole Derivatives 158
    10.3.3 β-Carboline Derivatives 161
    10.3.4 Thienoquinolines Derivatives 162
    10.3.5 Benzimidazoles Derivatives 162
    10.3.6 Galantamine Derivatives 163
    10.3.7 Pyrazolone and Pyrazole Derivatives 163
10.4 Metabolic Oxidative Dehydrogenation Reactions 163
10.5 Miscellaneous Dehydrogenation Reactions 165
10.6 Conclusions and Perspectives 170

Chapter 11 Industrial Applications of Dehydrogenation Reactions: Process Design of Reactors 173

Ravi Tejasvi

11.1 Introduction 173

	11.2	Process Design of Reactors	174
		11.2.1 Ideal Reactors	175
		11.2.2 Non-Ideal Reactors	180
	11.3	Reactor Networking	180
	11.4	Additional Design Considerations for Real Reactors	182
	11.5	Conclusion	182

Chapter 12 Future Aspects of Dehydrogenative Reactions 184

Rama Gaur and Syed Shahabuddin

- 12.1 Introduction 184
- 12.2 Development of the Existing Catalytic Technologies via the Designing of More Selective, Active Catalysts 184
- 12.3 Developing Sustainable, Eco-Friendly, and Green Dehydrogenation Technologies 185
- 12.4 Selective Hydrogen Oxidation: Design and Development of a Novel Catalyst and Facile Process 185
- 12.5 Development of Membrane Separation Techniques for Removing Hydrogen from the Dehydrogenation Product 186
- 12.6 Green, Sustainable, Safe, and Environmentally Friendly Manufacturing Techniques 186
- 12.7 Development of New Catalysts Replacing Noxious Metals and Metal Oxides 187
- 12.8 Development of Reformed Heterogeneous Catalysts via Surface Modifications 187
- 12.9 Efficient Heterogenized-Homogeneous Catalyst Development for Modified Dehydrogenation Reactions 188
- 12.10 Development of Dehydrogenative Technologies for Hydrogen Energy Storage and Utilization 189

Chapter 13 Utilizing Ruthenium (Ru) Complexes in Dehydration Reactions of Saturated and Unsaturated Compounds 192

Khushbu G. Patel and Saami Ahmed

- 13.1 Introduction 192
- 13.2 Alcohol Dehydrogenation Reactions Based on Ru 192
  - 13.2.1. Aliphatic versus Aromatic Ligands 195
  - 13.2.2 Dehydrogenation of Formic Acid (FA) 201
  - 13.2.3 Dehydrogenation of C-N Bond by Ru Catalyst 201
  - 13.2.4 Ruthenium-Catalyzed Dehydrogenation of Alkene 203
- 13.3 Conclusion 204
- 13.4 Abbreviations 204

Chapter 14 Dehydrogenation Reactions Incorporating Membrane Catalysis 207

Hiralkumar Morker, Pratik Saha, Bharti Saini, and Anirban Dey

- 14.1 Introduction 207
- 14.2 The History of Membrane Catalysis 208
- 14.3 Major Merits of H_2-Absorptive Membrane Catalysts in Subsequent Reactions with the Expulsion of H_2 209
- 14.4 Dehydrogenation on Palladium Membranes 210
- 14.5 Dehydrogenation Using a Composite Membrane Catalyst 210

14.6 Dehydrogenation of Low Molecular Alkane, Alkene, and Alcohol 212
14.7 Dehydrogenation of Cyclohexane and Methylcyclohexane 213
14.8 Conclusion 214

Chapter 15 A Greener Dehydrogenation: Environmentally Benign Reactions 217

Ankita Saini, Monalisa Bourah and Sunil Kumar Saini
15.1 Introduction 217
    15.1.1 Oxidative Catalytic Dehydrogenation 217
    15.1.2 Oxygen-Based Oxidative Dehydrogenation 218
    15.1.3 Carbon Dioxide-Based Oxidative Dehydrogenation 219
    15.1.4 Nitrous Oxide-Based Oxidative Dehydrogenation 221
    15.1.5 Comparison of Catalytic ODH over DDH 221
15.2 Acceptorless Dehydrogenation 223
15.3 Nanoparticle-Based Catalyzed Dehydrogenation 224
15.4 Photocatalysis-Based Dehydrogenation 225
15.5 Water-Mediated Dehydrogenation Reactions 227
15.6 Conclusion and Future Prospects 230
15.7 Conflict of Interest 231

Chapter 16 Application of Pt- and Non Pt-Based Zeolitic Catalysts for the Dehydrogenation of Light Alkanes 238

Hardik Koshti and Rajib Bandyopadhyay
16.1 Introduction 238
16.2 Chemistry of Dehydrogenation 241
16.3 Non Pt-Based Zeolitic Catalyst 242
    16.3.1 Vanadium Oxide-Based Zeolitic Materials 242
    16.3.2 Chromium Oxide-Based Zeolitic Materials 244
16.4 Pt-Based Zeolitic Catalyst 249
    16.4.1 Sn Metal as Promoter 249
    16.4.2 Ce Metal as Promoter 250
    16.4.3 Gallium (III) Oxide as Promoter 251
    16.4.4 Alkaline Earth Metals as Promoter 251
    16.4.5 Transition Metals as Promoters 251
16.5 Conclusion 255

Chapter 17 Porous Inorganic Nanomaterials as Heterogeneous Catalysts for the Dehydrogenation of Paraffin 262

Sauvik Chatterjee and Nandini Mukherjee
17.1 Introduction 262
17.2 Principle of Dehydrogenation 263
17.3 Heterogeneous Catalysts for Paraffin Dehydrogenation 264
    17.3.1 Metal Oxide Catalyst 264
    17.3.2 Nanoporous Materials-Based Single Atom (SA) and Single Atom Alloy (SAA) Catalysts 267
    17.3.3 Metal Organic Frameworks (MOFs)-Based Catalyst 269
17.4 Outlook and Conclusion 271

Index 274

Preface

This text is intended for postgraduate and undergraduate researchers, students, and professionals in the research industries who are interested in learning about the chemistry of dehydrogenation reactions. It provides both basic theory and advanced applications of this topic. The authors of this book hope that it will also serve as introductory literature for those who are not familiar with the subject matter.

Chemistry plays a vital role in our daily lives, from the medicines we take to the materials we use. As industrialization advances, the demand for chemical reagents in various industries has skyrocketed. But what if we could use the existing reagents to develop useful chemicals with simple dehydrogenation reactions? This is where the chemistry of dehydrogenation reactions comes in. Dehydrogenation reactions play a vital role in various fields of chemistry, including organic synthesis, materials science, and catalysis. These reactions involve the removal of hydrogen from a molecule, resulting in the formation of a double or triple bond between the remaining atoms. In recent years, dehydrogenation reactions have attracted significant attention due to their potential for developing sustainable and energy-efficient processes.

The proposed book is a comprehensive source of information about the chemistry of dehydrogenation reactions and their applications in various domains of industry. It encompasses not only a detailed discussion of the different classes of organic compounds that undergo dehydrogenation reactions but also the role of transition metal and non-transition metal catalysts in these reactions.

This book has been designed for undergraduate and postgraduate students, academics, and industrial researchers, who want to learn about the fundamental principles of dehydrogenation reactions and their applications. It can serve as exploratory literature for those who are working towards the development of low-cost useful chemicals, pharmaceutical precursors, and hydrogen storage devices.

The content of this book is planned in such a manner that it will be useful and function as a guide for universities and industrial institutions, postgraduate and Ph.D. scholars, and researchers working in the areas of chemistry, biology, physics, polymer sciences, material sciences, nanoscience, and more.

We hope that this book will serve as an invaluable resource for those interested in the chemistry of dehydrogenation reactions and inspire further research in this exciting area of chemistry.

It is important to recognize that perfection is often an unattainable goal, especially when it comes to creating written works. Authors and editors alike may find themselves constantly striving for improvement even after multiple rounds of review and editing. Although the chapters and final draft of a book may have undergone comprehensive scrutiny, there is always room for refinement and modification to enhance the quality and quantity of the content.

Therefore, as readers of this book, your feedback is not only welcomed but also encouraged. We value the input of students, researchers, scientists, and professors who can offer valuable insights and suggestions regarding any errors, omissions, or important information that may have been overlooked. Your recommendations can help us improve and enhance the overall quality of this book.

About the Editors

Dr. Syed Shahabuddin obtained his M.Sc. in materials chemistry in 2011 from Jamia Millia Islamia, India. He was awarded a Ph.D. in polymer chemistry from the University of Malaya, Malaysia in September 2016. He has served as Assistant Manager in Samtel Avionics Limited for almost three years in the research and development of avionics grade displays. Formerly, he worked as Senior Research Fellow (senior lecture) at the Research Centre for Nano-Materials and Energy Technology (RCNMET), Sunway University, Malaysia for more than 2 years. Currently, he is working as an assistant professor at Pandit Deendayal Petroleum University, Gandhinagar and Gujrat. He has published more than 120 research articles in international journals of repute and presented 30 papers at international/national conferences. He has successfully completed three research projects as principal investigator (PI) and two research projects as co-investigator (Co-I), and he is currently Co-I for eight running projects with different institutes in Malaysia. He is a member of the Royal Society of Chemistry (MRSC) and a reviewer of many high impact journals. His current research focus is on the synthesis of nanomaterials, 2D-MXene, graphene, conducting polymer nanocomposites for water treatment, photocatalysis, supercapacitors, DSSCs, nanofluids for solar thermal applications, waste lubricant oil refining and phase change materials.

Dr. Rama Gaur is Assistant Professor, School of Technology, PDPU Gandhinagar, Gujarat. She received her Ph.D. in chemistry from IIT Roorkee in 2017. Dr. Gaur has expertise in the synthesis of nanoscale materials with interesting and unique morphologies through simple and economical chemical approaches; shape- and size-dependent optical, magnetic and electrochemical properties; applications in photocatalysis, optoelectronics, electrochemical sensing, electrocatalytic reduction/oxidation, solar energy conversion, energy storage devices, supercapacitors, water splitting and environmental remediation. Dr. Gaur has recently received a research DST-SERB-TARE Grant in collaboration with IIT Gandhinagar. Dr. Gaur has served as Assistant Professor at the National Institute of Technology, Hamirpur and the Indian Institute of Information Technology, Una (Himachal Pradesh). She has vast experience in teaching and research.

Dr. Nandini Mukherjee completed her bachelor's (B.Sc. Hons.) in chemistry from Scottish Church College, University of Calcutta in 2012 and master's in chemistry from Banaras Hindu University in 2014. She received her Ph.D. in bioinorganic and medicinal chemistry from the Indian Institute of Science, Bangalore in 2020. Currently, she is working as Assistant Professor in the Department of Chemistry, Pandit Deendayal Energy University (PDEU), Gandhinagar, Gujarat. Prior to joining PDEU, she also worked as a research associate in IISc Bangalore. She has more than 5 years of research experience in organic synthesis and coordination chemistry and has published her research in international peer-reviewed journals. Her current research interest focuses on the design and synthesis of molecules that have applications in the field of chemo-sensing of toxins and monitoring biological phenomena. She is also working to develop cost-effective nanomaterials for application in medicinal chemistry and the renewable energy sector.

Contributors

Saami Ahmed
District Institute of Education and Training
Ansari Road, Daryaganj
Delhi 110002, India

Vijay Bahadur
Department of Basic Sciences, Alliance University
Chandapura - Anekal Main Road
Anekal, Bengaluru 562106, India;
College of Pharmacy
University of Houston
Houston, TX 77204-5037, USA

Megha Balha
Department of Chemistry, School of Sciences
Gandhi Institute of Technology and Management (GITAM)
Bengaluru Campus, India

Rajib Bandyopadhyay
Department of Chemistry, School of Energy Technology
Pandit Deendayal Energy University, Knowledge Corridor Raysan
Gujarat 382426, India

Pravin R. Bhansali
Department of Science, Faculty of Science and Technology
Alliance University
Bangluru, Karnataka 562106, India

Monalisa Bourah
Department of Chemistry, Faculty of Physical Sciences
P.D.M. University
Bahadurgarh, Haryana 124507, India

Prakash Chandra
Department of Chemistry, School of Energy Technology
Pandit Deendayal Energy University, Knowledge Corridor
Raysan, Gandhinagar 382426, Gujarat, India

Sauvik Chatterjee
School of Materials Sciences, Indian Association for the Cultivation of Science
2A & B Raja S. C. Mallick Road, Jadavpur, Kolkata 700032, India

Anirban Dey
Department of Chemical Engineering, School of Technology
Pandit Deendayal Energy University
Gandhinagar 382007, Gujarat, India

Rama Gaur
Department of Chemistry, School of Energy Technology
Pandit Deendayal Energy University, Knowledge Corridor
Raysan, Gandhinagar 382426, Gujarat, India

Hardik Koshti
Department of Chemistry, School of Energy Technology
Pandit Deendayal Energy University, Knowledge Corridor Raysan
Gujarat 382426, India

Atul Kumar
Department of Chemistry
Birla Institute of Technology BIT-Mesra
Ranchi, India

Hiralkumar Morker
Department of Chemical Engineering, School of Technology
Pandit Deendayal Energy University
Gandhinagar 382007, Gujarat, India

Nandini Mukherjee
Department of Chemistry, School of Energy Technology
Pandit Deendayal Energy University, Knowledge Corridor
Raysan, Gandhinagar 382426, Gujarat, India

Khushbu G. Patel
Department of Chemistry, Rai School of Sciences
Rai University
SH144, Saroda, Gujarat 382260, India

Chandni Pathak
Department of Basic Sciences, Alliance University
Chandapura - Anekal Main Road, Anekal, Bengaluru 562106, India

Vijayendran K. K. Praneeth
Somaiya Vidyavihar University
Vidyavihar, Mumbai Maharashtra 400077, India

Pratik Saha
Department of Chemical Engineering, School of Technology
Pandit Deendayal Energy University
Gandhinagar 382007, Gujarat, India

Ankita Saini
Department of Chemistry, Faculty of Physical Sciences
P.D.M. University
Bahadurgarh, Haryana 124507, India

Bharti Saini
Department of Chemical Engineering, School of Technology
Pandit Deendayal Energy University, Knowledge Corridor Raysan
Gujarat 382426, India

Sunil Kumar Saini
Department of Zoology, Faculty of Life Sciences
P.D.M. University
Bahadurgarh, Haryana 124507, India

Syed Shahabuddin
Department of Chemistry, School of Energy Technology
Pandit Deendayal Energy University, Knowledge Corridor
Raysan, Gandhinagar 382426, Gujarat, India

Ravi Tejasvi
Department of Chemical Engineering, School of Technology
Pandit Deendayal Energy University, Knowledge Corridor
Raysan, Gandhinagar 382426, Gujarat, India

Ronald E. Viola
Biochemistry, University of Toledo
Toledo, OH 43606, USA

1 Introduction to Dehydrogenation Reactions of Organic Compounds

Syed Shahabuddin, Nandini Mukherjee, and Rama Gaur

1.1 INTRODUCTION

Dehydrogenation reactions include a wide variety of reactions where hydrogen is completely or partially eliminated from an organic compound to form a new compound (e.g., conversion of saturated into unsaturated compounds) [1]. Dehydrogenation reactions have applications in the production of hydrogen (H_2), alkenes, cycloalkanes, aromatic compounds, imines, and oxygenates such as carbonyl compounds [2]. Alkenes, imines, and oxygenates are crucial chemical intermediates for producing solvents, polymers, rubbers, detergents, insecticides, and pharmaceuticals [3].

Dehydrogenation is most important in the petrochemical industry especially in the refinery cracking process, where olefins are synthesized from saturated hydrocarbons [4]. Suitable hydrocarbon feedstocks (e.g., naphtha) are subjected to cracking in order to produce industrially important olefins such as ethylene, propylene, butene derivatives, and butadiene derivatives in large volumes. Traditionally, steam cracking and fluid catalytic cracking (FCC) of C2+ hydrocarbons (ethane, propane, etc.) have been extensively used to produce alkenes. However, as there is an increasing demand for a specific alkene to be produced, building new specific steam crackers or FCC units for each individual alkene is not a frugal choice. Dehydrogenation reaction, in contrast, provides a comparatively flexible, cost-effective method for producing single alkenes. The dehydrogenation of ethylbenzene to produce an important monomer 'styrene' has become a preferred commercial route in the polymer industry.

The benefits of building dehydrogenation reactors are the production not only of specific alkenes but also of aldehydes and ketones. Although the carbonyl compounds can be easily prepared via the oxidation of alkenes or alcohols, dehydrogenation routes may offer advantages such as a higher selectivity for products or the availability of a wider variety of feedstock [5]. Dehydrogenation reaction is also extremely useful in synthesizing α,β-unsaturated carbonyls as common organic building blocks in natural product and pharmaceutical synthesis. The introduction of a C=C group adjacent to an electron withdrawing group (EWG) provides greater synthetic opportunity that utilizes the resulting polarized double bond, since EWG is susceptible to regioselective functionalization by several methods [6, 7]. Acceptorless dehydrogenation of alcohols, especially biomass-derived carbohydrates, provides an atom-economical method to synthesize carbonyl derivatives and a low-temperature route for selective H_2 production [8–14]. It also has synthetic applications in tandem coupling reactions involving C-N and C-C bond formation for the synthesis of imines and amides and the β-functionalization of alcohols [15–24].

Another crucial significance of many dehydrogenation reactions is the generation of H_2 as a by-product. H_2, which is an alternative and highly in-demand clean fuel, is required to be produced and stored in a safe and efficient manner. Ammonia borane (AB or H_3N-BH_3), for instance, serves as one of the safest sources of H_2 (3 equivalents theoretically). Homogeneous (and heterogeneous) catalyst-driven dehydrogenation reaction of AB provides an efficient route for non-hydrolytic reversible H_2 generation [25].

Evidently, the vastness of dehydrogenation reactions originating from the infinite number of possible precursors and products leads to various procedural aspects to be considered systematically. The most important amongst these are the (a) catalyst type, (b) reactor type where the commercial production of organic derivatives and/or H_2 is concerned, (c) method of heat supply to the dehydrogenation reaction involved, (d) method of catalyst regeneration, (d) reaction conditions (P, T, solvent/diluent), and (f) yield of desired product and atom economy.

In the following section, these aspects are briefly introduced. In addition, where possible, the dehydrogenation reaction mechanism is discussed.

1.2 MECHANISM OF DEHYDROGENATION REACTIONS

The dehydrogenation mechanism from different starting materials primarily depends on three factors: the chemical nature of the starting material, the type of catalyst being used, and whether any acceptor molecule is present in the reaction medium. The following section provides insight on the mechanisms involved in the dehydrogenation of various substrates including alkanes, alcohols, esters, amides, carboxylic acids, and carbonyl compounds.

1.2.1 ALKANE TO ALKENE

Alkane to alkene conversion is the first example that comes to mind when dehydrogenation is mentioned; it is one of the most studied and commercially used dehydrogenation reactions known. This process involves alkanes (C_nH_{2n+2} with $n=2$ or more) where one (or more) H_2 is removed to form unsaturated hydrocarbons or olefins (Figure 1.1).

This reaction can be of two types: oxidative and non-oxidative. Although both methods involve the use of suitable metal-based catalysts, the mechanism is different. Non-oxidative dehydrogenation involves C-H bond activation directly by metal and is endothermic in nature (Figure 1.2).

Oxidative dehydrogenation of alkane involves C-H bond activation by O-species leading to hydrogen abstraction and is an exothermic reaction (Figure 1.3). Readers are directed to Chapters 5 and 17 for more details on this topic.

1.2.2 ALCOHOL TO CARBONYL DERIVATIVES

The dehydrogenation of aliphatic alcohol yields different products depending on the type of alcohol used, namely, primary, secondary, or tertiary. The reactions are usually endothermic and require the use of high-temperature/low-pressure conditions. Primary and secondary alcohols are converted into aldehyde and ketone compounds, respectively, which may further be functionalized or derivatized.

One of the most familiar examples of alcohol dehydrogenation is the Oppenauer-oxidation reaction of a secondary alcohol. However, this famous name reaction involves the use of aluminum tert-butoxide and acetone. Acetone accepts the eliminated hydrogen leading to the formation of isopropyl alcohol as a byproduct (Figure 1.4.a). A more promising approach in terms of atom economy is to opt for the transition metal catalysed acceptorless dehydrogenation of alcohols, which yields molecular hydrogen as a byproduct for further reaction/use as a fuel (Figure 1.4.b).

Alcohol dehydrogenation reaction is not limited to only carbonyl and H_2 generation. Depending on the acceptor molecule present and the reaction conditions, there is the possibility of the formation

$$C_nH_{2n+2} \xrightarrow{\text{Catalyst}} C_nH_{2n} + H_2$$

FIGURE 1.1 Dehydrogenation of alkanes to unsaturated hydrocarbons.

Introduction to Dehydrogenation Reactions of Organic Compounds

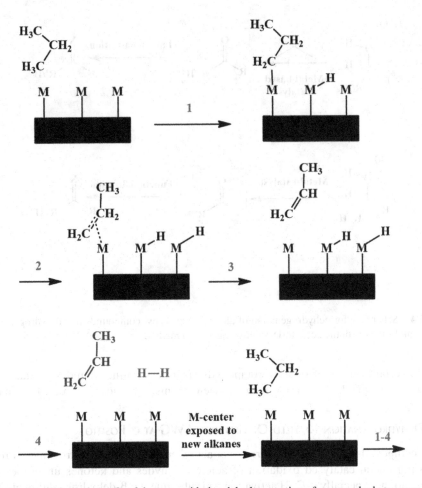

FIGURE 1.2 Simplified schematic of the non-oxidative dehydrogenation of propane by heterogeneous catalysis (M = Transition metal center).

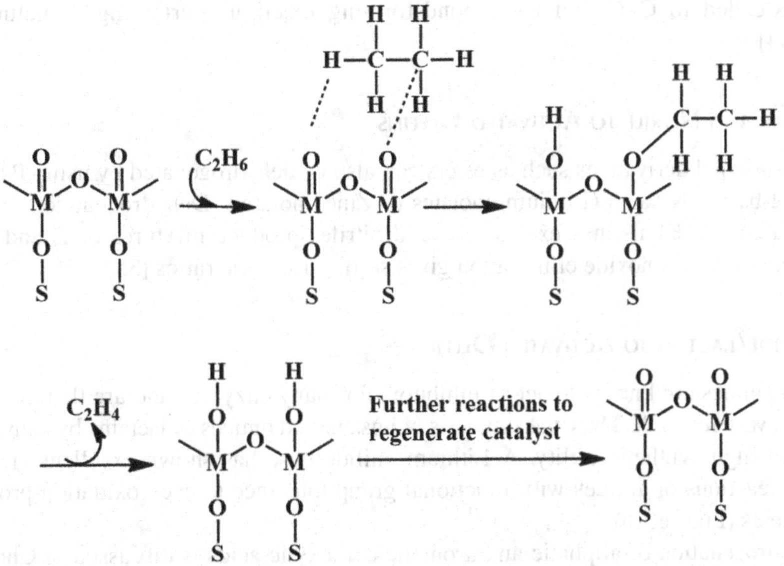

FIGURE 1.3 Simplified schematic of oxidative dehydrogenation of ethane on solid supported metal oxides (M = Metal center, S = Support).

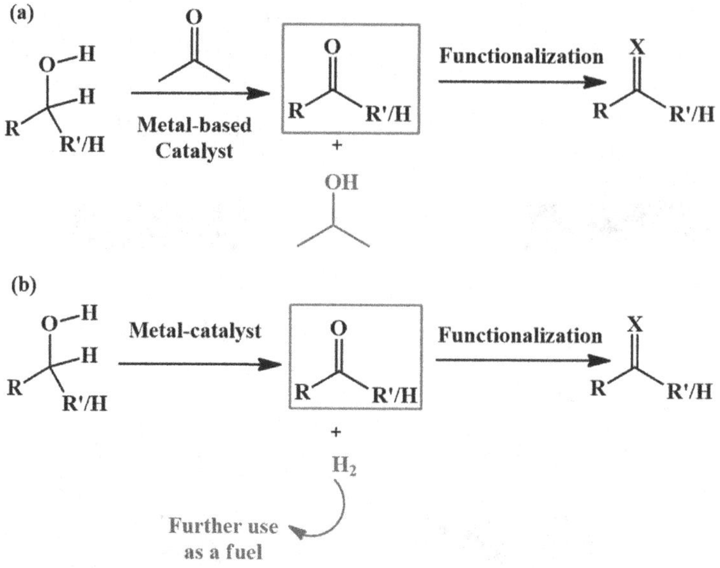

FIGURE 1.4 Schematic for dehydrogenation of alcohol to carbonyl compounds by (a) hydrogen transfer to an acceptor molecule or (b) the acceptorless dehydrogenation method.

of many important classes of organic compounds *viz*. esters, imines, amides, acetals, lactones, lactams, and even pyrroles. For a thorough overview on this, the reader is directed to Chapter 4.

1.2.3 Dehydrogenation to Yield Olefins with EWG at α-Position

Carbonyl compounds, albeit less explored, can act as precursors in dehydrogenation reactions. The transition metal catalysed oxidation of acidic aldehydes and ketones undergoes efficient dehydrogenation. Especially α-C–H activation and subsequent α,β-dehydrogenation of carbonyls have been reported by several groups [26–33]. In addition, esters, nitriles, amides, carboxylic acids, and electron-deficient heteroarenes compounds subjected to dehydrogenation reactions could be extended to C–C and C–X bond-forming reactions pertaining to natural product synthesis [34].

1.2.4 Ester and Nitrile to Activated Olefins

Less acidic carbonyl derivatives such as esters can also be dehydrogenated by using Pd or Ni catalyst and allyl-based oxidants (Lithium enolates or Zinc enolates). Dehydrogenation of ester leads to the formation of (E)-alkenes exclusively, and nitriles produce mixtures of E and Z isomers. Dehydrogenation by selenoxide elimination gives similar isomeric ratios [35].

1.2.5 Amide/Lactam to Activated Olefins

Unsaturated amides are known to act as inhibitors for many enzymes and are therefore important for drug discovery research. Thus, the synthesis of unsaturated amides or lactams by dehydrogenation reaction is of high synthetic utility. A Lithium anilide base has shown excellent yield in dehydrogenation reactions of amides with functional group tolerance to even oxidation-prone OH and NH-substituents (Figure 1.6).

The dehydrogenation of aliphatic and aromatic carboxylic acids is discussed in Chapter 7, and the dehydrogenation of heterocyclic compounds is examined at length in Chapter 8.

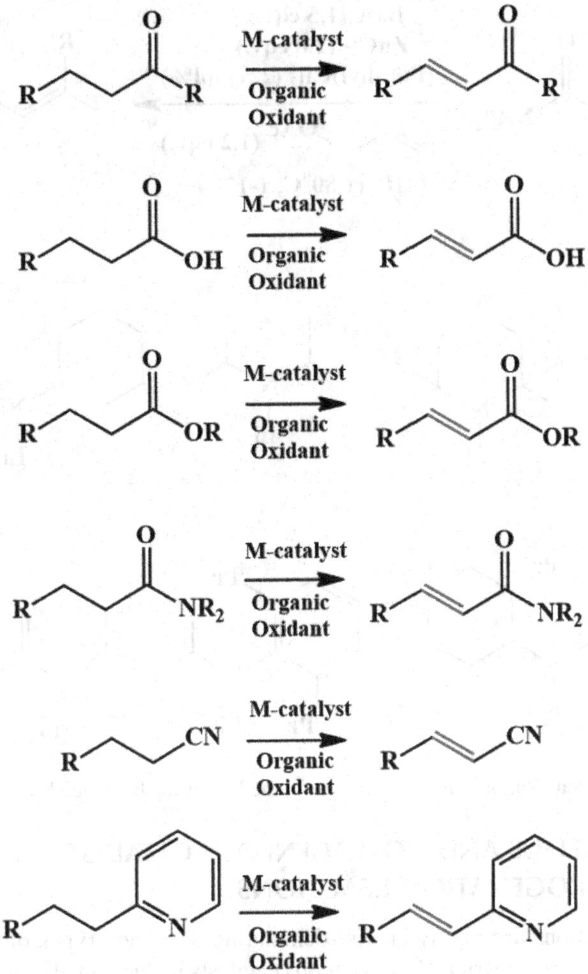

Introduction to Dehydrogenation Reactions of Organic Compounds

FIGURE 1.5 Dehydrogenation of carbonyl derivatives to activated olefin products catalysed by transition metal-based catalyst and organic oxidants.

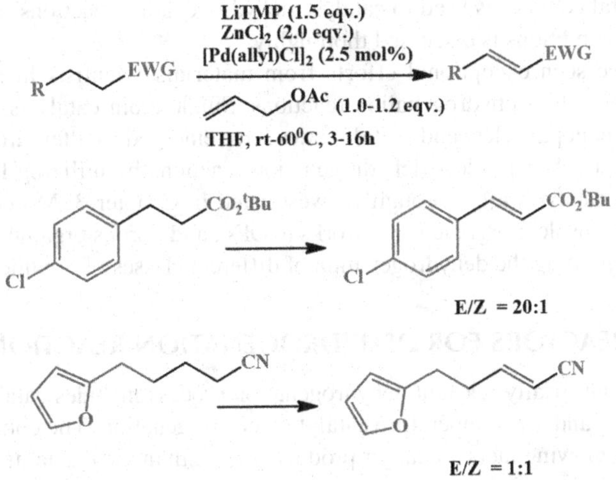

FIGURE 1.6 Dehydrogenation of ester/nitrile to activated olefins with different stereoselectivity.

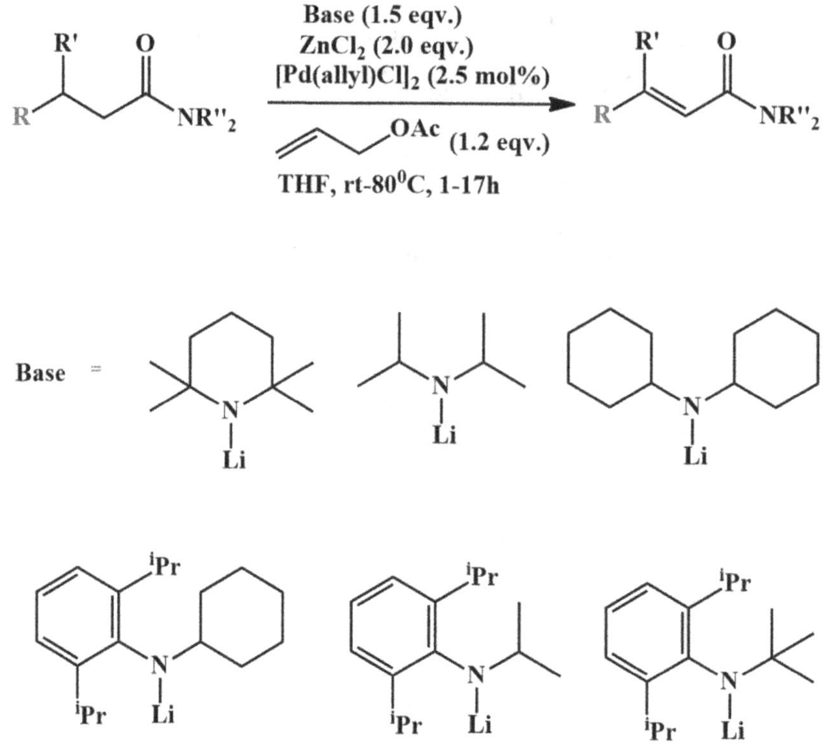

FIGURE 1.7 Dehydrogenation of amide to olefins and the Li-anilide bases leads to high yield.

1.3 HETEROGENEOUS AND HOMOGENEOUS CATALYSTS FOR DEHYDROGENATION REACTIONS

Dehydrogenation reactions are highly energy-demanding. Different types of catalysts are used to reduce the activation energy barrier. Homogeneous catalysts include 3d, 4d, and 5d-transition metal complexes of Ru, Ir, Pd, Rh, Cu, Ni, and Fe. To achieve efficient dehydrogenation and region/stereoselectivity, the transition metal catalyst must be rationally designed by modulating the metal-redox properties and metal-ligand cooperativity by using pincer ligands. The reader is directed to Chapter 2 where transition metal complexes used to catalyse dehydrogenation reactions in homogeneous and quasi-homogeneous conditions is discussed thoroughly.

Recent years have seen exceptional efforts from materials scientists in the development of heterogeneous catalysts for dehydrogenation reactions. Single atom catalysts (SAC), single atom alloys (SAA), metal nanoparticles, and metal oxides have emerged as alternatives to conventional homogeneous catalysts. Acceptorless dehydrogenation reactions by utilizing Pt, Pd, Au, Ag, Cu, Ru, Rh, Ni, and Co catalysts are thoroughly investigated in Chapter 3. Moreover, metal-organic frameworks (MOFs), covalent organic frameworks (COFs), and porous organic frameworks (POFs) are also discussed regarding the dehydrogenation of different classes of organic compounds.

1.4 TYPES OF REACTORS FOR DEHYDROGENATION REACTIONS

The most common industrially relevant dehydrogenation process includes catalytic cracking, thermochemical cracking, and low-temperature catalytic dehydrogenation. The common features in all of them include the following facts—a larger product volume than the reactants, the requirement of

Introduction to Dehydrogenation Reactions of Organic Compounds

a quick separation of produced H_2 from the other product chemicals generated, and the maintenance of sustainable reaction temperature/pressure [36]. The design of dehydrogenation process reactors is achieved by carefully considering the previous features in addition to the thermodynamics, kinetics, and phase of the reactants (gas-liquid, liquid-liquid, gas-solid, and liquid-solid). Additional factors such as safety and production cost are also considered by chemical engineers to sustain a commercially viable manufacturing process. A detailed discussion on batch reactors, mixed flow reactors, and plug flow reactors can be found in Chapter 11.

1.5 DEHYDROGENATION METHODS AND REACTIONS THAT ARE COMMERCIALLY SIGNIFICANT

Dehydrogenation reactions are reactions in which hydrogen is removed from the organic compound. These crucial groups of reactions are extremely helpful in the chemical industry. Dehydrogenation leads to the formation of unsaturated compounds from saturated compounds that can be used as intermediates during the production of different materials, polymers, and chemicals. Over the years, various methods of dehydrogenation reactions have been used and developed as this class of reactions is commercially significant. Dehydrogenation reactions can also be employed on a large scale in industrial processes.

One of the popular examples of a dehydrogenation reaction that is commercially used on a large scale is removing hydrogen from propane to form propylene, which also acts as an intermediate in the production of polypropylene. Polypropylene is a widely used thermosetting polymer. Some other examples of dehydrogenation reactions are the formation of butadiene from butane; benzene can be produced by the dehydrogenation of cyclohexane and the dehydrogenation of ethylbenzene to form styrene.

Thermal dehydrogenation, oxidative dehydrogenation, and catalytic hydrogenation are some methods of dehydrogenation that can be used industrially on a large scale. In thermal decomposition, the starting material is heated at high temperatures in the absence of oxygen. This is the simplest method of all. In oxidative dehydrogenation, an oxidising agent is used for the removal of hydrogen atoms. In a catalytic reaction, the requirement of activation energy is lowered by using a catalyst [37–41].

Hence, the development of new methodologies and dehydrogenation reactions are vital to synthesise new chemicals and materials. Dehydrogenation reactions are a useful class of reactions utilised in various chemical industries.

1.5.1 Dehydrogenation of Paraffins to Olefins

Dehydrogenation of paraffins to olefins is a beneficial class of reactions that is employed in the synthesis of different materials, chemicals, elastomers, and polymers. Formation of propylene from propane is an important dehydrogenation reaction that can be used as feedstock for the preparation of acrylonitrile, polypropylene, and several other chemicals [42].

Various catalysts, including supported metal catalysts, zeolites, and various mixed metal oxide catalysts, can be used for the dehydrogenation of propane. Chromia-alumina catalystsare popular and widely used catalysts for dehydrogenation of propane that can be functional and operated at temperatures between 500 to 600°C and pressures between 1–2 atm.

Dehydrogenation of butane (C_4H_{10}) to butane (C_3H_6) can also be used for the synthesis of various chemicals and materials including rubber and plastics. Dehydrogenation of butane can also be carried out in the presence of catalysts such as zeolites, meta-oxide, etc.

Hence, this class of dehydrogenation, where butane is converted into butene, is a useful process in chemical industries in the preparation of various chemicals and materials.

1.5.1.1 Reaction Mechanism

Dehydrogenation of C3 and C4 alkanes to alkenes includes the removal of hydrogen atoms from the alkane molecule, forming a carbon-carbon double bond. The reaction is as follows:

$$C_3H_8 \rightarrow C_3H_6 + H_2$$
$$C_4H_{10} \rightarrow C_4H_8 + H_2$$

Endothermic reactions, such as dehydrogenation, generally require high temperatures and the presence of an appropriate catalyst to perform effectively.

1.5.1.2 Catalysts

Mixed metal oxide catalysts, which are a combination of a transition metal oxide and an alkali metal oxide, are the most utilized catalysts for the dehydrogenation of C3 and C4 alkanes to alkenes. These catalysts enhance the dehydrogenation process by offering a surface for the adsorption and activation of alkane molecules and by allowing the removal of hydrogen atoms from the molecules.

As previously stated, chromium-alumina catalysts are also extensively utilised for the dehydrogenation of propane to propylene. Chromium-alumina catalyst is made of chromium oxide (Cr_2O_3) supported on alumina (Al_2O_3) and is commonly operated at temperatures between 500 and 600°C and pressures ranging from 1–2 atm.

1.5.1.3 Kinetics

The kinetics of dehydrogenation reactions are affected by a number of parameters, including the catalyst used, the temperature and pressure conditions, and the type of alkane molecule being dehydrogenated. In general, increasing the temperature improves the reaction rate because it offers more energy for the activation of the alkane molecule and the removal of the hydrogen atoms. However, higher temperature and more energy might cause catalyst deactivation and product degradation; therefore, the reaction rate and selectivity are trade-offs.

The rate of the reactions is affected by the alkane and catalyst concentrations, porosity, and the surface area of the catalyst. Since mixed metal oxide catalysts can supply both acidic and basic sites for the reaction, they have better activity and selectivity than other forms of catalysts.

So, chemical engineers are actively investigating ways to improve the kinetics of the dehydrogenation reaction by developing more efficient and environmentally friendly catalysts and reaction conditions.

1.5.2 Dehydrogenation of C2-C15 Alkanes to Alkenes

For the synthesis of a variety of polymers and chemicals, the C2-C15 alkane to alkene dehydrogenation reaction is economically one of the most important reactions. In the alkane, a carbon-carbon double bond is formed as a result of the elimination of hydrogen atoms, which leads to the formation of hydrogen gas and alkene as shown in the following.

$$C_nH_{2n+2} \rightarrow C_nH_{2n} + H_2$$

Dehydrogenation reactions require specific conditions such as high temperatures and suitable catalysts for the reaction to proceed; they are also endothermic. The yield and the selectively of the reaction are highly dependent on factors such as the type of catalyst and alkane, reaction conditions, etc.

1.5.2.1 Catalysts

Mixed metal oxides are widely employed catalysts for C2-C15 alkane to alkene dehydrogenation reactions. Mixed metal oxides comprise earth metal oxides or alkali metal oxides in combination

Introduction to Dehydrogenation Reactions of Organic Compounds

with transition metal oxides. The dihydrogen reaction on the catalyst is promoted by the adsorption and activation followed by the hydrogen atoms' removal on the surface of the catalysts.

Some commonly used catalysts for C2-C15 dehydrogenation are mentioned as follows:

i. **Chromia-alumina catalysts:** The chromia-alumina catalysts comprise Cr_2O_3 (Chromium oxide) reinforced on the AL_2O_3 (alumina) substrate. These are the most commonly employed catalysts for propane to propylene dehydrogenation. Their operating temperatures and pressures are between 500 and 600°C and 1 and 2 atm, respectively.

ii. **Nickel-based catalysts:** Most of the time, these catalysts are employed to convert butane and other higher alkanes into butenes and other higher alkenes. They are usually used at temperatures between 600 and 800°C and pressures between 1 and 3 atm.

iii. **Platinum-based catalysts:** Light alkanes such as ethane and propane are often dehydrogenated with these catalysts to synthesize ethylene and propylene, respectively. Their operating temperature and pressures are between 500 and 800°C and 1 and 2 atm, respectively.

1.5.2.2 Kinetics

Various factors such as the nature of the catalyst and alkane molecule, temperature, and pressure of reactions cause the alkane molecule to affect the kinetics of dehydrogenation reactions. Generally, as the temperature increases, the reaction rate also increases, which offers increased energy to activate the alkanes and promotes the dehydrogenation reaction. However, at high temperatures, catalysts get deactivated, and the product might become degraded. Hence, there is a compromise between the selectivity and reaction rate.

The rate of the reaction also depends on the amount of the catalyst, alkane concentration, porosity, and surface area of the catalyst. Mixed metal oxide catalysts are usually more active and selective than other catalysts as they can supply both acidic and basic sites for the reaction to happen.

In general, C2-C15 alkane to alkene dehydrogenation is an important reaction for the development of a variety of polymers and chemicals. In the field of chemical engineering, researchers are focused on the development of effective catalysts and reaction parameters.

1.5.3 Dehydrogenation of Ethylbenzene to Styrene

Styrene is a vital reagent used for the preparation of different types of plastics, resins, and polymers. In general, styrene is prepared by dehydrogenating ethylbenzene. The reaction proceeds via the formation of a carbon-carbon double bond and the elimination of two hydrogens [43]. The chemical equation for the reaction is expressed as follows.

$$C_6H_5CH_2CH_3 \rightarrow C_6H_5CH=CH_2 + H_2$$

At high temperatures, the reaction takes place in the gas phase with the favoured catalyst. The process is endothermic in nature, resulting in a high temperature requirement (500 to 600 °C). In addition, the reaction exhibits exothermic behavior leading to heat release in the process [44, 45].

1.5.3.1 Catalysts

The catalyst used during the preparation of styrene is composed of a mixed metal oxide, namely, potassium and iron oxide (K_2O and Fe_2O_3) reinforced on an alumina substrate. The catalyst facilitates the elimination of hydrogens and offers a surface for the activation and adsorption of the ethylbenzene.

K_2O present in the catalyst serves as a promoter, leading to an increase in selectivity and activity, whereas Fe_2O_3 offers the active site for the reaction to proceed. The alumina substrate offers a highly porous and large surface area, which enables effective reactant and product diffusion across the catalyst surface [44, 45].

1.5.3.2 Kinetics

The kinetics of the reaction are highly affected by various parameters, such as the pressure and temperature conditions, the catalyst used, and the type of ethylbenzene molecule being dehydrogenated. The activation of the ethylbenzene molecule and the elimination of the hydrogen atoms require more energy, so the reaction rate typically rises as the temperature rises.

Moreover, the rate of the reaction depends on the amount of catalyst and ethylbenzene used and on the catalysts' surface area and pore size. The catalyst Fe_2O_3-K_2O/Al_2O_3 offers high selectivity and activity because it provides basic and acidic sites for the reaction to take place.

1.6 RECENT ADVANCES IN DEHYDROGENATION TECHNOLOGY

Dehydrogenative reactions have proved to be of great importance in a number of important applications, such as catalysis, energy storage, polymer synthesis, materials science, etc. Overall, dehydrogenative reactions play a vital role in a variety of important applications and are essential for the development of new materials, catalysts, and energy storage systems. Tremendous growth has been observed in the applications of dehydrogenative reactions toward a better and sustainable future. The development or progress of any process demands safety, sustainability, and improved efficiency. Environmental safety entails minimizing the use of hazardous chemicals and minimal hazardous manufacturing waste. The use of green chemicals, catalysts, and eco-friendly methods not only aids in enhancing the effectiveness of the process but also makes it an environmentally friendly and sustainable approach. Similar perspectives are listed as follows:

- Development of the existing catalytic technologies via the designing of more selective, active catalysts
- Developing sustainable, eco-friendly, and green dehydrogenation technologies
- Selective hydrogen oxidation for the design and development of novel catalysts and facile processes
- Development of membrane separation techniques for removing hydrogen from the dehydrogenation product
- Development of oxidative dehydrogenation techniques
- Green, sustainable, safe, and environmentally friendly manufacturing techniques
- Development of new catalysts replacing noxious metals and metal oxides
- Development of reformed heterogeneous catalysts via surface modifications
- Efficient heterogenized-homogeneous catalysts developed for modified dehydrogenation reactions

A detailed discussion about the recent advances in dehydrogenation technology with a focus on these listed topics can be found in Chapter 12, Future Aspects of Dehydrogenative Reactions, of this book.

1.7 SUMMARY

Cross-coupling dehydrogenative reactions involve the elimination of an H_2 molecule that leads to the formation of a C-C or C-Heteroatom bond. The applications and chemistry of dehydrogenative reactions are vast. They are widely used for hydrocarbons, alcohols, amines, acids, and heterocyclic compounds to prepare molecules of commercial importance. Reports are also available on the use of dehydrogenative technologies for hydrogen storage and utilization and the preparation of valued pharmaceutical compounds. A lot of progress has been made in this area regarding understanding the mechanism of the reactions and their use for application in various fields. However, some challenges remain unanswered to date. These challenges of today are prospects for the future and provide directions for further exploration.

REFERENCES

1. Li, Chunyi, and Guowei Wang. "Dehydrogenation of light alkanes to mono-olefins." *Chemical Society Reviews* 50.7 (2021): 4359–4381.
2. Dobereiner, Graham E., and Robert H. Crabtree. "Dehydrogenation as a substrate-activating strategy in homogeneous transition-metal catalysis." *Chemical Reviews* 110.2 (2010): 681–703.
3. Bartholomew, C. H., and R. J. Farrauto. *Fundamentals of Industrial Catalytic Processes*. John Wiley & Sons, 2011.
4. Speight, James G. "Gasification processes." In *Handbook of Petroleum Refining*. CRC Press, 2016.
5. Huang, David, and Timothy R. Newhouse. "Dehydrogenative Pd and Ni catalysis for total synthesis." *Accounts of Chemical Research* 54.5 (2021): 1118–1130.
6. Chen, H., L. Liu, T. Huang, J. Chen, and T. Chen. "Direct dehydrogenation for the synthesis of α, β-unsaturated carbonyl compounds." *Advanced Synthesis & Catalysis* 362.16 (2020): 3332–3346.
7. Muzart, Jacques. "One-pot syntheses of α, β-unsaturated carbonyl compounds through palladium-mediated dehydrogenation of ketones, aldehydes, esters, lactones and amides." *European Journal of Organic Chemistry* 2010.20 (2010): 3779–3790.
8. Guoqi, Zhang. "Cobalt-catalyzed acceptorless alcohol dehydrogenation: Synthesis of imines from alcohols and amines." *Organic Letters* 15.3 (2013): 650–653.
9. Watson, Andrew J. A., and Jonathan M. J. Williams. "The give and take of alcohol activation." *Science* 329.5992 (2010): 635–636.
10. Hamid, Malai Haniti S. A., Paul A. Slatford, and Jonathan M. J. Williams. "Borrowing hydrogen in the activation of alcohols." *Advanced Synthesis & Catalysis* 349.10 (2007): 1555–1575.
11. Guillena, Gabriela, Diego J. Ramon, and Miguel Yus. "Hydrogen autotransfer in the N-alkylation of amines and related compounds using alcohols and amines as electrophiles." *Chemical Reviews* 110.3 (2010): 1611–1641.
12. Dobson, Alan, and Stephen D. Robinson. "Complexes of the platinum metals. 7. Homogeneous ruthenium and osmium catalysts for the dehydrogenation of primary and secondary alcohols." *Inorganic Chemistry* 16.1 (1977): 137–142.
13. Junge, Henrik, Björn Loges, and Matthias Beller. "Novel improved ruthenium catalysts for the generation of hydrogen from alcohols." *Chemical Communications* 5 (2007): 522–524.
14. Nielsen, Martin, et al. "Efficient hydrogen production from alcohols under mild reaction conditions." *Angewandte Chemie* 123.41 (2011): 9767–9771.
15. Andrushko, Natalia, et al. "Amination of aliphatic alcohols and diols with an iridium pincer catalyst." *ChemCatChem* 2.6 (2010): 640–643.
16. Patman, Ryan L., et al. "Carbonyl propargylation from the alcohol or aldehyde oxidation level employing 1, 3-enynes as surrogates to preformed allenylmetal reagents: A ruthenium-catalyzed C-C bond-forming transfer hydrogenation." *Angewandte Chemie* 120.28 (2008): 5298–5301.
17. Gnanaprakasam, Boopathy, Jing Zhang, and David Milstein. "Direct synthesis of imines from alcohols and amines with liberation of H_2." *Angewandte Chemie* 122.8 (2010): 1510–1513.
18. Maggi, Agnese, and Robert Madsen. "Dehydrogenative synthesis of imines from alcohols and amines catalyzed by a ruthenium N-heterocyclic carbene complex." *Organometallics* 31.1 (2012): 451–455.
19. Esteruelas, Miguel A., et al. "Direct access to pop-type osmium (II) and osmium (IV) complexes: Osmium a promising alternative to ruthenium for the synthesis of imines from alcohols and amines." *Organometallics* 30.9 (2011): 2468–2471.
20. Rigoli, Jared W., et al. "α, β-Unsaturated imines via Ru-catalyzed coupling of allylic alcohols and amines." *Organic & Biomolecular Chemistry* 10.9 (2012): 1746–1749.
21. Gunanathan, Chidambaram, Yehoshoa Ben-David, and David Milstein. "Direct synthesis of amides from alcohols and amines with liberation of H_2." *Science* 317.5839 (2007): 790–792.
22. Zhang, Jing, et al. "Facile conversion of alcohols into esters and dihydrogen catalyzed by new ruthenium complexes." *Journal of the American Chemical Society* 127.31 (2005): 10840–10841.
23. Kossoy, Elizaveta, et al. "Selective acceptorless conversion of primary alcohols to acetals and dihydrogen catalyzed by the ruthenium (II) complex Ru (PPh3) 2 (NCCH3) 2 (SO4)." *Advanced Synthesis & Catalysis* 354.2-3 (2012): 497–504.
24. Nixon, T. D., M. K. Whittlesey, and J. M. J. Williams. "Transition metal catalysed reactions of alcohols using borrowing hydrogen methodology." *Dalton Transactions* (2009): 753–762.
25. Zhang, Xingyue, et al. "Ruthenium-catalyzed ammonia borane dehydrogenation: Mechanism and utility." *Accounts of Chemical Research* 50.1 (2017): 86–95.
26. Stahl, S. S., and Tianning Diao. "Oxidation adjacent to CX bonds by dehydrogenation." In *Oxidation*. Elsevier Ltd, 2014, 178–212.

27. Turlik, Aneta, Yifeng Chen, and Timothy R. Newhouse. "Dehydrogenation adjacent to carbonyls using Palladium–allyl intermediates." *Synlett* 27.3 (2016): 331–336.
28. Iosub, Andrei V., and Shannon S. Stahl. "Palladium-catalyzed aerobic dehydrogenation of cyclic hydrocarbons for the synthesis of substituted aromatics and other unsaturated products." *ACS Catalysis* 6.12 (2016): 8201–8213.
29. Hirao, Toshikazu. "Synthetic strategy: Palladium-catalyzed dehydrogenation of carbonyl compounds." *The Journal of Organic Chemistry* 84.4 (2019): 1687–1692.
30. Gnaim, S., J. C. Vantourout, F. Serpier, P.-G. Echeverria, and P. S. Baran. "Carbonyl desaturation: Where does catalysis stand?" *ACS Catalysis* 11 (2021): 883–892.
31. Diao, T., and S. S. Stahl. "Synthesis of cyclic enones via direct palladium-catalyzed aerobic dehydrogenation of ketones." *Journal of the American Chemical Society* 133 (2011): 14566–14569.
32. Diao, T., T. J. Wadzinski, and S. S. Stahl. "Direct aerobic α,β-dehydrogenation of aldehydes and ketones with a Pd(TFA)2/4,5-diazafluorenone catalyst." *Chemical Science* 3 (2012): 887–891.
33. Diao, T., D. Pun, and S. S. Stahl. "Aerobic dehydrogenation of cyclohexanone to cyclohexenone catalyzed by Pd(DMSO)2(TFA)2: Evidence for ligand-controlled chemoselectivity." *Journal of the American Chemical Society* 135 (2013): 8205–8212.
34. Nicolaou, K. C., Paul G. Bulger, and David Sarlah. "Palladium-catalyzed cross-coupling reactions in total synthesis." *Angewandte Chemie International Edition* 44.29 (2005): 4442–4489.
35. Reich, H. J., and S. Wollowitz. "Preparation of α,β-unsaturated carbonyl compounds and nitriles by selenoxide elimination." *Organic Reactions* 44 (1993): 1–296.
36. Coker, A. K. "Process planning, scheduling, and flowsheet design." In *Ludwig's Applied Process Design for Chemical and Petrochemical Plants*, 3rd ed. Elsevier, 2007, 1–68.
37. Spivey, J. J. and Y. Wang. "Recent advances in dehydrogenation of light alkanes: A review." *ACS Catalysis* 7.5 (2017): 3523–3542.
38. Hanna, M. A., A. R. Al-Fatesh, M. A. Al-Hajji, and A. B. Al-Bassam. "Advances in the catalytic dehydrogenation of propane: A review." *Journal of Industrial and Engineering Chemistry* 20.4 (2014): 1217–1226.
39. Román-Martínez, M. C., M. A. Bañares, and M. J. Illán-Gómez. "Recent advances in the catalytic dehydrogenation of ethylbenzene to styrene." *Applied Catalysis A: General* 533 (2017): 33–48.
40. de Lera, R., A. B. García, and P. J. Pérez. "Oxidative dehydrogenation of light alkanes on multicomponent oxide catalysts: State of the art and future perspectives." *Chemical Reviews* 115.16 (2015): 8710–8753.
41. Moffat, J. B., and J. C. De Wilde. "Thermal dehydrogenation of hydrocarbons." In *Kirk-Othmer Encyclopedia of Chemical Technology*. John Wiley & Sons, Inc., 2000.
42. Sanfilippo, D., and P. N. Rylander. *Hydrogenation and Dehydrogenation*. Ullmann's Encyclopedia of Industrial Chemistry, 2000.
43. Resasco, D. E. "Dehydrogenation by heterogeneous catalysts." *Encyclopedia of Catalysis* (2000): 51–58.
44. Adesina, A. A., and M. Menendez. "Dehydrogenation of ethylbenzene to styrene over iron-oxide-based catalysts." *Industrial & Engineering Chemistry Research* 43.21 (2004): 6713–6722.
45. Suwannakarn, K., P. Praserthdam, and S. Assabumrungrat. "Kinetics of dehydrogenation of ethylbenzene over iron-oxide-based catalysts." *Industrial & Engineering Chemistry Research* 47.18 (2008): 6986–6993.

2 Transition Metal-Based Catalyst for Dehydrogenation Reactions of Organic Compounds

Atul Kumar

2.1 INTRODUCTION

Transition metal-based systems have emerged as homogeneous catalysts that have profound applications in organic synthesis. The unusual reactivity and selectivity via C-H/C-OH bond activation derive the reactions to more complex organic synthesis. Transition metal (TM) catalysts featuring the controlled fixation and release of hydrogen provide an alternative strategy to perform dehydrogenation and hydrogenation reactions within the same catalytic cycle in a single step. The process of bond formation between two substrates involving coupled dehydrogenation and a hydrogenation step without using any external hydrogen donor/acceptor source is termed hydrogen auto transfer (HA) or acceptor-less dehydrogenation coupling (ADC).[1-4] Traditional organic synthesis mostly utilizes reactive or activated reagents such as alkyl bromide (R-Br) or alkyl tosylate (R-OTs) as coupling reagents for cross-coupling or substitution reactions in bond formation reactions. Although these reactions feature high conversion, they lack selectivity in product formation and are non-atom economical as they generate multiple products.[5] Thus, organic synthesis through these reactive reagents on a bulk/industrial scale increases the large waste formation and leads to acute environmental problems. Replacing R-Br/R-OTs with less reactive R-H/R-OH as alkylating agents is an alternative green chemistry approach as it avoids the use of toxic reagents and is atom economical as it forms minimal byproduct.[6] However, C-H/C-OH bonds are unreactive and thus need activation components for substitution/cross-coupling reactions. TM catalysts provide an alternative mode of bond activation through oxidative dehydrogenation to overcome this intrinsic nonreactivity. Metal-substrate binding followed by bond construction steps in a one-pot synthesis with a single or two metals acting as a tandem catalyst are key steps involved in dehydrogenation reactions.[7] In these aspects, activation of the C-OH bond in alcohol, C-NH_2 bond in amine, and C-H bond in alkane via oxidative dehydrogenation provides a cheap alkylating agent source. In this chapter, we mainly focus on the activation of the substrate (alcohol, amine, and alkane) by a TM catalyst featuring oxidative dehydrogenation to construct a variety of C-C and C-N bonds in organic synthesis.

2.2 N-ALKYLATION BY DEHYDROGENATIVE ALCOHOL ACTIVATION

C-N bond construction is one of the most crucial synthetic steps followed to design a wide range of organic compounds. The formation of the C-N chemical bond can be achieved through addition, substitution, pericyclic, and cross-coupling reactions. The traditional approach utilizes the concept of N-alkylation of amine to construct a C-N bond through a cross-coupling reaction between amine and alkyl/aryl halide, tosylates, triflates, sulfonates, etc. However, this approach faces a severe drawback in governing the selectivity of the product as it often results in the formation of a mixture of different mono/di/tri/tetra-substituted alkylated products. In recent decades, significant progress had been made in C–N bond formation using a TM-catalyst via cross-coupling reactions.[8–10] The formation of the C-N bond is accomplished through the alkylation of amine (amination reaction) using activated alcohol. In general, the process involves three steps: i) the alcohol functional group is dehydrogenated (oxidized) to the corresponding aldehyde by a TM catalyst; ii) generated aldehyde reacts with amine and leads to the formation of hemiaminal that upon dehydration, gives imine intermediate, and iii) the imine is hydrogenated by an in situ-generated metal-hydride complex to give the final amine product (Figure 2.1). The reaction is highly atom economical as the generated H_2 during dehydrogenation of the starting material is reutilized in the hydrogenation of in situ formed imine intermediate within the same catalytic cycle and generates only water as a side product.

The scope of N-alkylation by alcohols ranges from the synthesis of simple amines to heterocyclic compounds such as quinolines, indoles, etc. When this reaction is used in tandem with multistep synthesis, the formation of more complex products can be achieved. Ruthenium- and iridium-based catalysts are generally used for these reactions. The reactions are usually performed at room temperature to moderate heating conditions in the presence of a mild base.[11]

The first report of the N-alkylation of alkyl/cycloalkyl amine via homogeneous catalysis was reported in 1981 by Grigg and co-workers using an Rh hydride complex [RhH(PPh$_3$)$_4$] (also known as Grigg's hydride Rh complex) (Figure 2.2).[12] Thereafter, several catalysts based on iridium, ruthenium, copper, iron, cobalt, manganese, and palladium were developed.

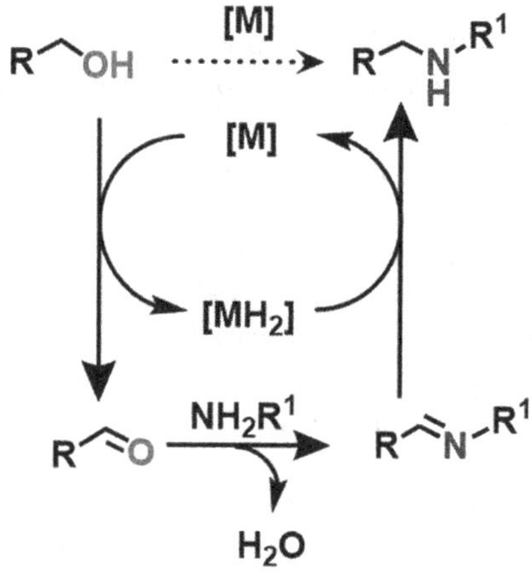

FIGURE 2.1 Schematic representation of metal-catalyzed dehydrogenative coupling of alcohol via a hydrogen auto-transfer (HA) process.

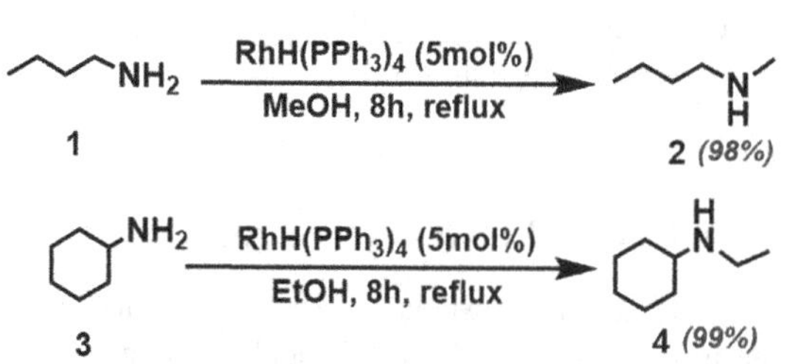

FIGURE 2.2 N-alkylation of alkyl amine using RhH(PPh$_3$)$_4$ catalyst.

2.2.1 N-Alkylation by Ruthenium Catalyst

Ruthenium complexes are typically effective catalysts for the N-alkylation of amine with alcohols through the HA process. The first ruthenium catalyst for N-alkylation reaction competent for aliphatic amines was reported in 1982 by Murahashi and co-workers using a [RuH$_2$(PPh$_3$)$_4$] catalyst.[13] This catalyst shows a similar catalytic activity as Grigg's hydride Rh complex as both catalysts are effective for alkyl substrates and remain ineffective for aryl substrates (Figure 2.3).

N-alkylation of arylamines/aminoarenes was first reported by Watanabe and co-workers using either a [RuCl$_2$(PPh$_3$)$_3$] or [RuCl$_3$L] (where L= monophosphine) complex.[14] Heterocyclic arylamine was also N-alkylated using these Ru catalysts. For example, [RuCl$_2$(PPh$_3$)$_3$] efficiently catalyzes the synthesis of 1-phenylpiperidine (**10**) by N, N'-dialkylation of aniline (**8**) with pentane-1,5-diol.[15] In another approach, the N-alkylation-induced cyclization of 2-(2-aminophenyl)ethan-1-ol (**11**) resulted in the synthesis of 1*H*-indole (**12**) in the presence of a [RuCl$_2$(PPh$_3$)$_3$] catalyst (Figure 2.4).[16]

However, the selectivity of monoalkylation vs dialkylation remains a major challenge in the N-alkylation of primary aryl/alkyl amines since the monoalkylation of primary amine results in the formation of more nucleophilic secondary amine, which tends to react further to form a corresponding dialkylated product. In this regard, conventional N-alkylation by aryl halide possesses serious drawbacks as it forms a mixture of different alkylated products. However, N-alkylation by alcohol activation has the advantage of a secondary amine, which is formed after the alkylation of primary amine becomes an inactive substrate for further alkylation. Since secondary amine upon further reaction with an aldehyde (resulting from alcohol activation) will generate an unstable iminium cation intermediate, the formation of this iminium cation is unfavorable under normal conditions, particularly in nonpolar solvents. Watanabe et al. has established the selective formation

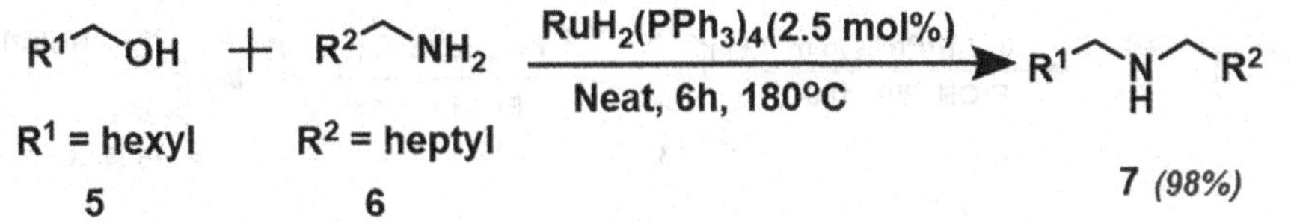

FIGURE 2.3 N-alkylation of alkyl amine using a RuH$_2$(PPh$_3$)$_4$ catalyst.

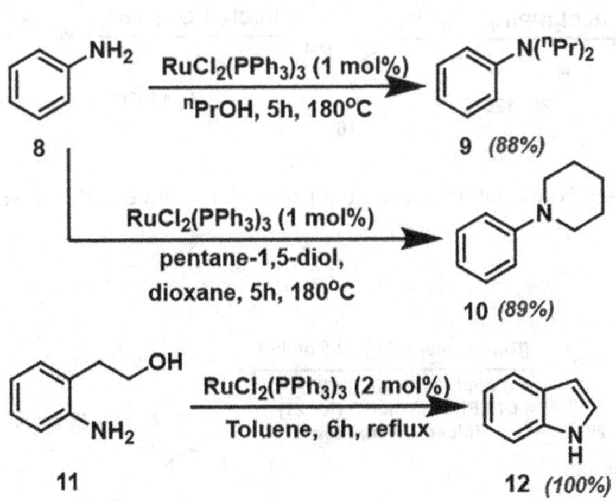

FIGURE 2.4 N-alkylation of aryl amine using a RuCl$_2$(PPh$_3$)$_3$ catalyst.

of either a *mono-* or *di-*alkylated product in an N-alkylation reaction that depends not only on the type of Ru-complex used for catalysis but also on experimental conditions. For example, the [RuCl$_2$(PPh$_3$)$_3$] catalyst promotes N, N'-dialkylation of 2-aminopyridine (**13**) with an excess of ethanol under a long reaction time.[13] Meanwhile, the [Ru(cod)(cot)] catalyst (cod=cyclooctadiene, cot=cyclooctatriene) selectivity of N-mono-ethylate 2-aminopyridine occurs in the presence of ethanol (Figure 2.5).[17]

N-alkylation selectivity has also been demonstrated by the reaction of ethylene glycol with secondary amine (**16**). In this case, the formation of either monosubstituted amino alcohol (**17**) or diamines (**18**) depends upon the ruthenium catalyst employed in the reaction. A phosphine-based ruthenium catalyst [RuH$_2$(PPh$_3$)$_4$] or [RuCl$_2$(PPh$_3$)$_3$] exclusively form diamines (**18**), while RuCl$_3$·H$_2$O forms the amino alcohol (**17**) (Figure 2.6).[18] However, the selectivity of diamine (**18**) formation using a [RuCl$_3$·nH$_2$O] catalyst decreases as PPh$_3$ is added to the reaction mixture. Similarly, [(PPh$_3$)$_2$Ru(CH$_3$CN)]BPh$_4$ is selective towards monoalkylation,[19] while a [h^5-CpRuCl(PR$_3$)$_2$] complex mostly gives a disubstituted product through the N-alkylation of aliphatic amine with methanol.[20]

The organometallic complex [Ru(p-cymene)Cl$_2$]$_2$ is another important catalyst that is most widely utilized in a variety of N-alkylation reactions. In general, the catalytic activity of [Ru(p-cymene)Cl$_2$]$_2$ is activated by bidentate phosphine such as dppf (bis(diphenylphosphino)ferrocene) or DPEphos.[21] A variety of alkyl/aryl substrates such as t-butylamine, aniline, 1-phenylethylamine, and 2-aminopyridine have been successfully alkylated by [Ru(p-cymene)Cl$_2$]$_2$ to give corresponding secondary amines with a high yield (Figure 2.7).

The medicinally important drug Piribedil (**26**) used for the treatment of Parkinson's disease is prepared by the N-alkylation of a cyclic amine-based piperidine substrate with piperonyl alcohol (Figure 2.8(i)). A [Ru(p-cymene)Cl$_2$]$_2$ and DPEphos have been found to be active catalysts for the

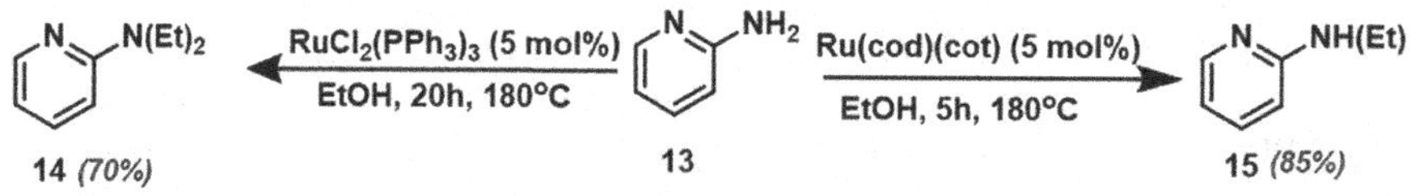

FIGURE 2.5 Mono vs di-N-alkylation of aryl amine using a ruthenium catalyst.

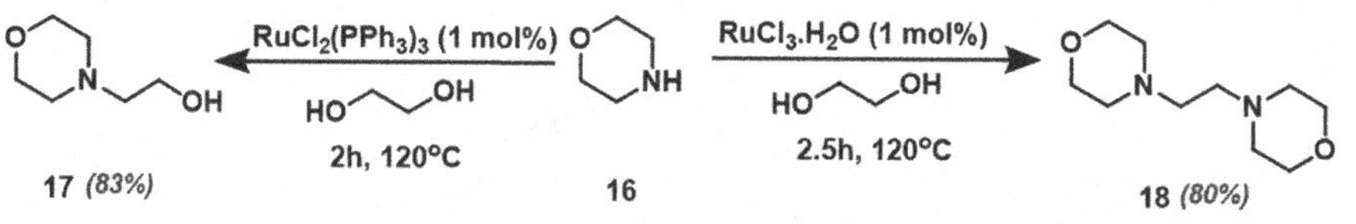

FIGURE 2.6 Ru-catalyzed N-alkylation selectivity of secondary amine in the presence of ethylene glycol.

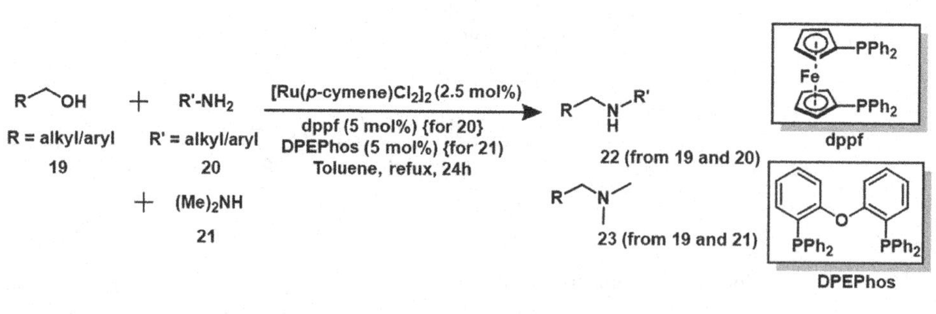

FIGURE 2.7 [Ru(p-cymene)Cl$_2$]$_2$ catalyzed N-alkylation alkyl/aryl amine with alcohol.

N-alkylation of even challenging substrate sulphonamide (**27**), which is an important reagent for drug development (Figure 2.8(ii)).[22]

The [Ru(*p*-cymene)Cl$_2$]$_2$ shows very high functional group tolerance as *para*- and *meta*-boronic ester substrate-bearing amines (**33**) groups are successfully N-alkylated without affecting boronic ester (Figure 2.9).[23]

A combination of an (arene)ruthenium (II) (**38**) complex and camphorsulfonic acid (CSA) promotes a highly regioselective cascade *N*- and *C(3)*-dialkylation of amines such as piperidine, pyrrolidine, and azepane with benzylic alcohol through the HA process.[24] The mechanistic detail involving the reaction of piperidine and benzyl alcohol suggests that CSA assists ammine activation and enables it for both *N*- and *C(3)*-dialkylation in the presence of an Ru-catalyst (**38**). However, the reaction often competes with its analog mono-N-alkylated product (Figure 2.10). Shvo's diruthenium complex {[Ph$_4$(η^5-C$_4$CO)]$_2$H]}Ru$_2$(CO)$_4$(μ-H) also promotes the N-alkylation of N-heterocyclic compounds such as indoles (Figure 2.11).[25] Williams and co-workers also synthesized benzimidazole via N-alkylation using a Ru(PPh$_3$)$_3$(CO)H$_2$ catalyst in the presence of Xantphos. In this case, piperidinium acetate acts as a base that converts imine (formed after the N-alkylation process) into an iminium ion and hence promotes nucleophilic attack by the adjacent amine group leading to the formation of cyclized product (**46**) (Figure 2.12).[26]

The primary alcohols are more reactive than the secondary alcohols for the N-alkylation reaction since the activation of primary alcohol produces the aldehyde, which is more electrophilic than the ketone that is formed after secondary alcohol activation. However, the activation of secondary alcohols (and primary alcohol) by [Ru$_3$(CO)$_{12}$] results in the successful N-alkylation of primary amines.

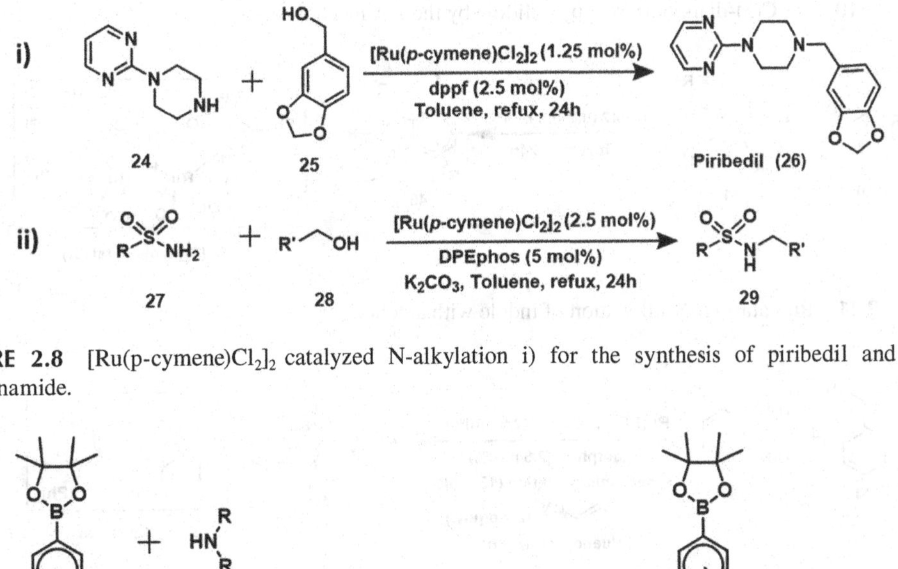

FIGURE 2.8 [Ru(p-cymene)Cl$_2$]$_2$ catalyzed N-alkylation i) for the synthesis of piribedil and ii) of sulphonamide.

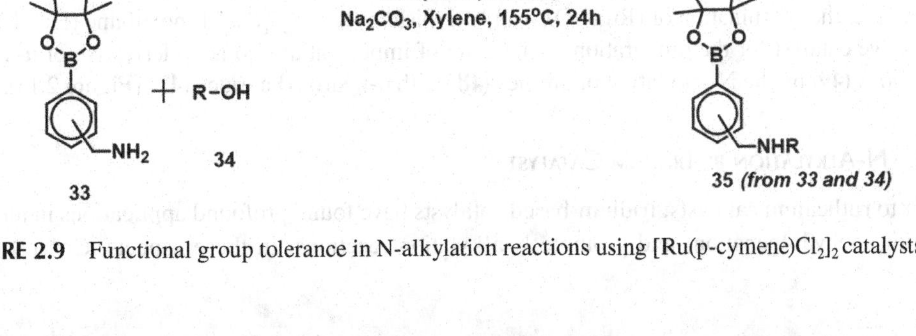

FIGURE 2.9 Functional group tolerance in N-alkylation reactions using [Ru(p-cymene)Cl$_2$]$_2$ catalysts.

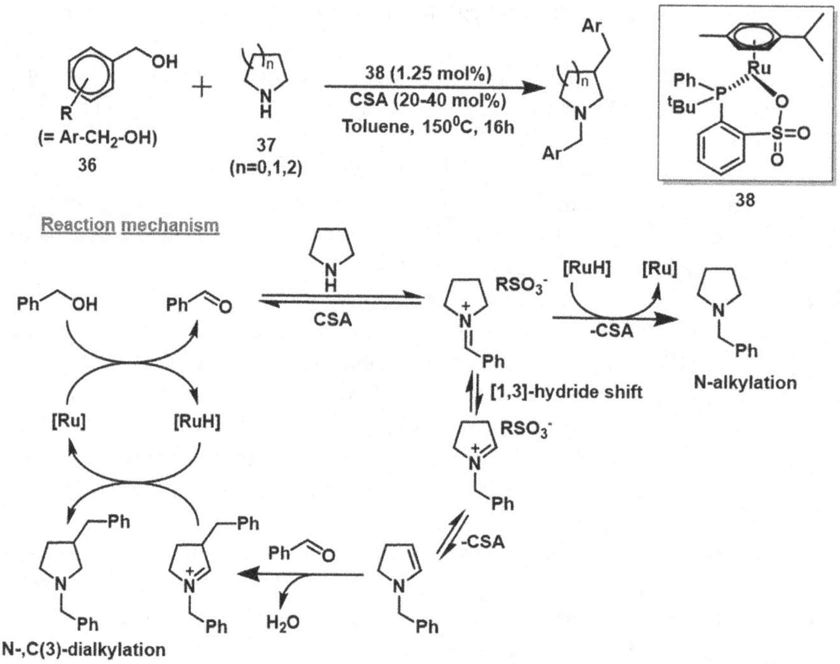

FIGURE 2.10 N-, C(3)-dialkylation of pyrrolidine by the HA mechanism.

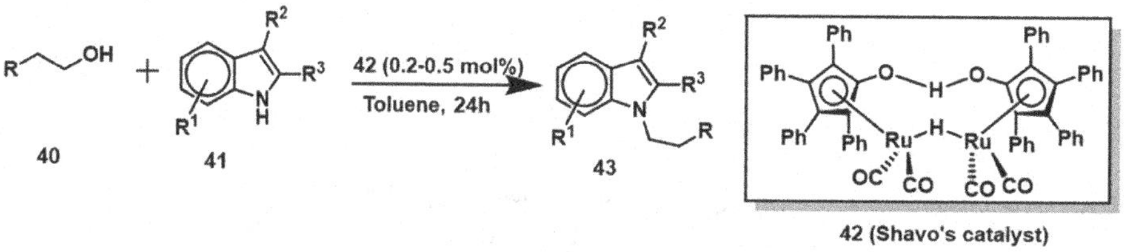

FIGURE 2.11 Ru-catalyzed N-alkylation of indole with alcohol.

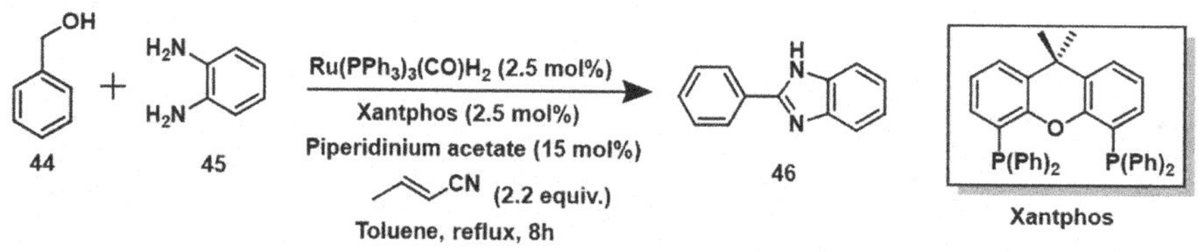

FIGURE 2.12 Ru-catalyzed N-alkylation for the synthesis of benzimidazole.

For example, the combination of [$Ru_3(CO)_{12}$] and 1,2-bis(dicyclohexylphosphino)ethane (DCPE) acts as an active catalyst for the preparation of the class of important amino acid derivative of α-amino acid amides (**49**) by the N-alkylation of amines (**48**) with α-hydroxyl amides (**47**) (Figure 2.13).[27]

2.2.2 N-Alkylation by Iridium Catalyst

Similar to ruthenium catalysts, iridium-based catalysts have found profound applications in alcohol activation (dehydrogenation) and promote N-alkylation reactions. In these aspects, Cp* ($\eta^5 - p$

TM-Based Catalyst for Dehydrogenation of Organic Compounds

entamethylcyclopentadienyl)-based Ir complexes were first utilized for N-alkylation reaction and show promising catalytic activity.

Fujita and Yamaguchi first reported Oppenauer-type oxidation of primary and secondary alcohols in the presence of catalyst [Cp*IrCl$_2$]$_2$ in acetone.[28] Later, [Cp*IrCl$_2$]$_2$ was utilized for the dehydrogenation of alcohols for N-alkylation reactions, where a wide range of substrates ranging from primary to secondary alcohol and amines (both primary and secondary) was employed.[29] Usually, a base is required in this case for catalyst activation. Although high catalyst loading (as high as 5 mol%) is required, the catalyst shows very high selectivity for the formation of a monoalkylated product. Even the catalyst shows the formation of secondary and tertiary amines by *di-* or *tri-*alkylation, respectively, by using quaternary ammonium salts under harsh conditions (Figure 2.14).[30]

Depending on the base used, [Cp*IrCl$_2$]$_2$ shows chemoselective alkylation of benzylic alcohols with 2'-aminoacetophenone (**58**) under microwave conditions. N-alkylation was observed with a mild base K$_2$CO$_3$, while a stronger base such as KOH gives a C-alkylated product (Figure 2.15).[31]

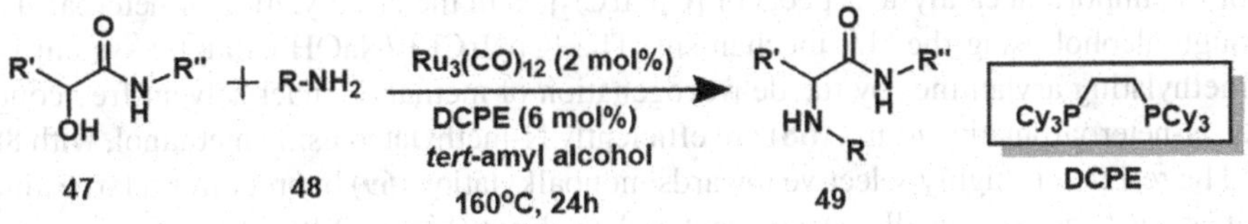

FIGURE 2.13 Ru$_3$(CO)$_{12}$-catalyzed N-alkylation of amines with α-hydroxyl amides.

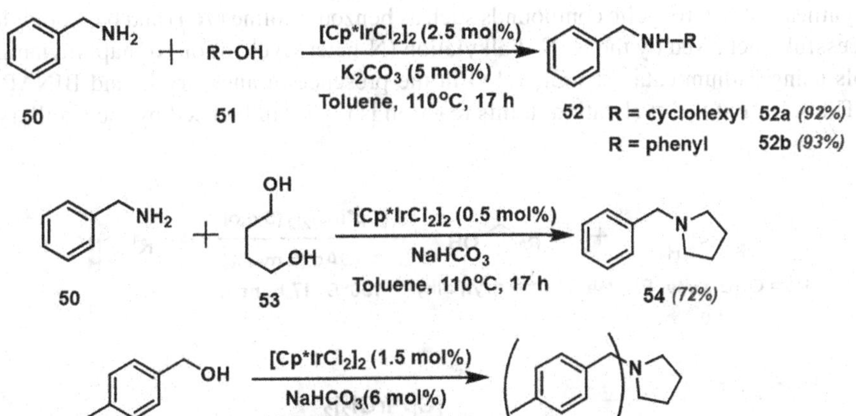

FIGURE 2.14 [Cp*IrCl$_2$]$_2$ catalyzed N-alkylation reactions by alcohol dehydrogenation.

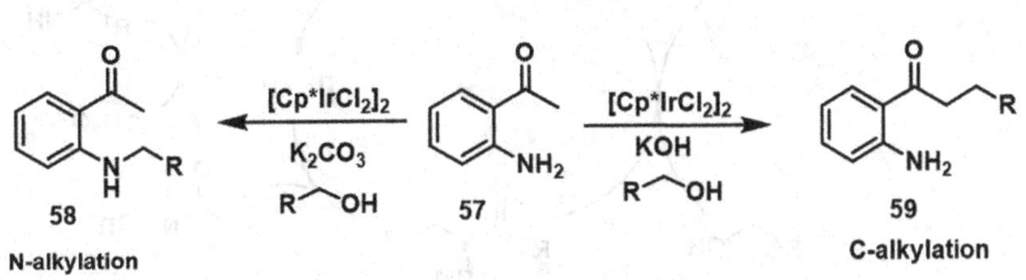

FIGURE 2.15 [Cp*IrCl$_2$]$_2$ catalyzed selective N-/C-alkylation of 2'-aminoacetophenone with benzyl alcohols.

[Cp*IrCl$_2$]$_2$ in the presence of NaOAc acts as an effective catalyst for the N-alkylation of amides and carbamate (**60**) by an alcohol dehydrogenation process under solvent-free conditions.[32] In the first step of the catalytic cycle, alcohol coordinates to the Ir-complex and forms an iridium alkoxy complex that is subsequently dehydrogenated to its corresponding aldehyde. Then, the catalytic cycle follows Schiff base condensation to form imine, which is hydrogenated by an in situ-generated iridium-hydride complex to give N-alkylated amide products (Figure 2.16).

[Cp*IrCl$_2$]$_2$ also catalyzes the N-alkylation of sulphonamides (**62**) with good to excellent yield under reflux conditions in toluene/p-xylene in the presence of ^{t}BuOK.[33] The reaction mechanism suggests the formation of unsaturated sulfonylimido-bridged unsaturated diiridium complex [(Cp*Ir)$_2$(μ-NTs)$_2$] as an active catalyst that promotes the dehydrogenative coupling of alcohol (Figure 2.17).

The N-alkylation of N-monosubstituted ureas (**65**) is also carried out in the presence of [Cp*IrCl$_2$]$_2$ to form N, N'-dialkyl ureas/N, N'-diaryl ureas (**67**) with a 70–93% yield.[34] The reaction is highly regioselective as no trace of N1-alkylated or N3-dialkylated products was observed in this case (Figure 2.18).

One of the important catalytic aspects of [Cp*IrCl$_2$]$_2$ is in the N-alkylation of heteroaryl amines (**68**) through alcohol using the HA mechanism. The [Cp*IrCl$_2$]$_2$/NaOH catalytic system is effective for methylating arylamines by the dehydrogenation of methanol under solvent-free conditions. Similarly, N-heteroaromatic amines (**68**) are efficiently N-methylated using methanol, with 88–95% yields.[35] The reaction is highly selective towards monoalkylation (**69**) but requires a large amount of methanol as it acts as both an alkylating agent and a solvent (Figure 2.19).

Anionic P, N-ligand based iridium(I) complex **72** or **73** was also used for the N-alkylation of N-heterocyclic amines (**70**) with benzyl alcohol (**71**). In this case, Ir complex **72** is more stable than **73** and thus exhibits higher catalytic activity than **73** (Figure 2.20).[36]

The synthesis of heterocyclic compounds such as benzoquinoline (**76a**) and benzoindoles (**76b**) was successfully achieved by the N, N-dialkylation (N-heterocyclization) of naphthylamines (**75**) with diols using iridium catalyst IrCl$_3$·3H$_2$O in the presence of ancillary ligand BINAP (Figure 2.21).[37] The selectivity of cyclization in this reaction is highly influenced by the ancillary ligand,

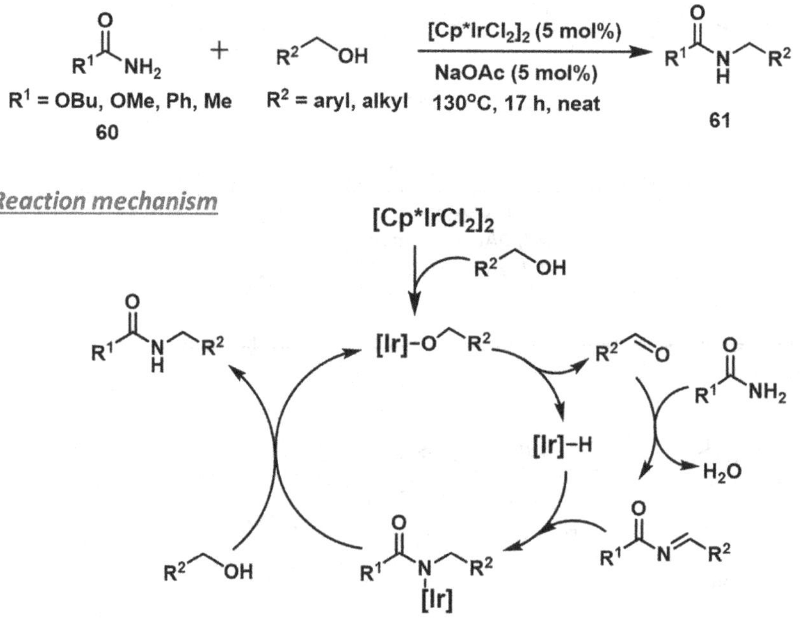

FIGURE 2.16 N-alkylation of amide by iridium-catalyzed dehydrogenative coupling of alcohol.

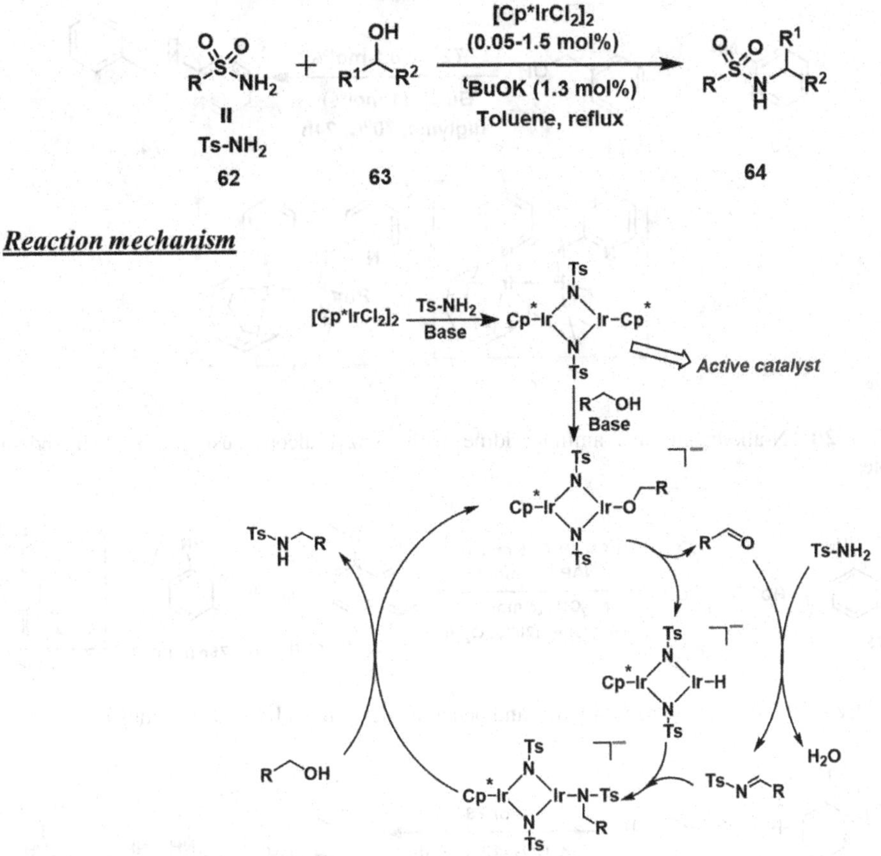

TM-Based Catalyst for Dehydrogenation of Organic Compounds

FIGURE 2.17 [Cp*IrCl$_2$]$_2$ catalyzed N-alkylation of sulphonamides and its reaction mechanism.

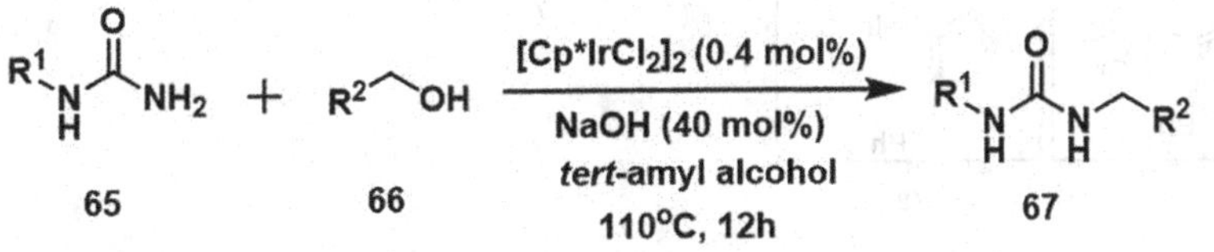

FIGURE 2.18 Regioselective mono-N-alkylation of N-monosubstituted ureas using the HA process.

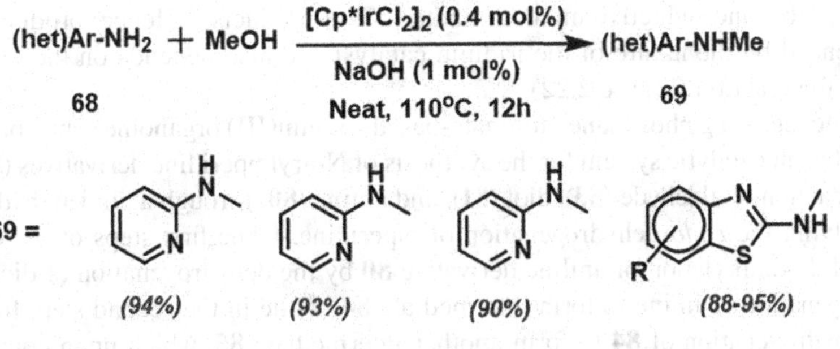

FIGURE 2.19 Regioselective mono-N-alkylation of heterocyclic amines.

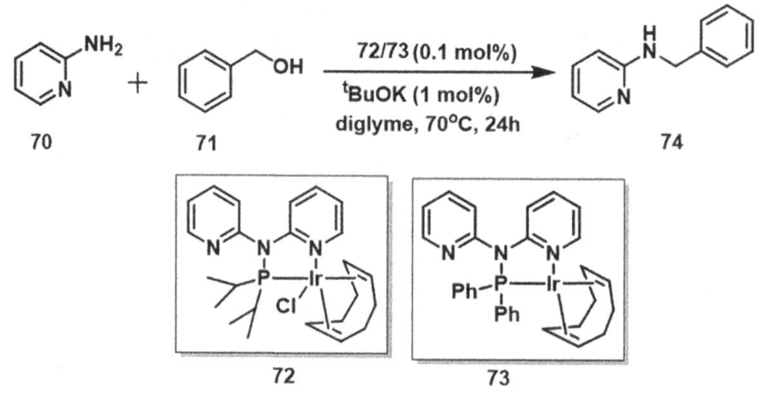

FIGURE 2.20 N-alkylation of 2-aminopyridine with benzyl alcohol using a P, N-ligand-supported Ir-complex.

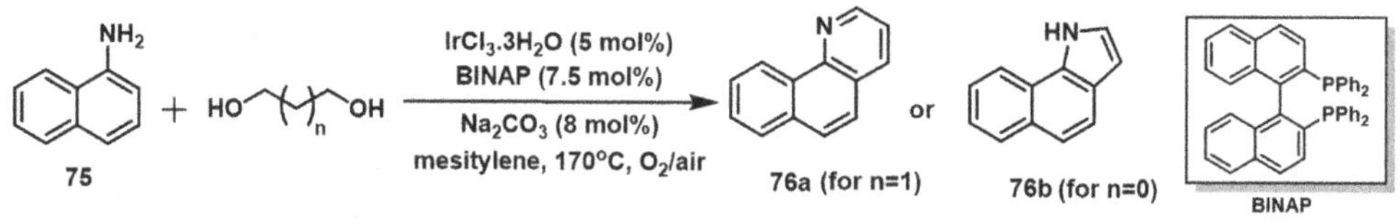

FIGURE 2.21 Synthesis of benzoquinoline and benzoindoles using a $IrCl_3 \cdot 3H_2O$ catalyst.

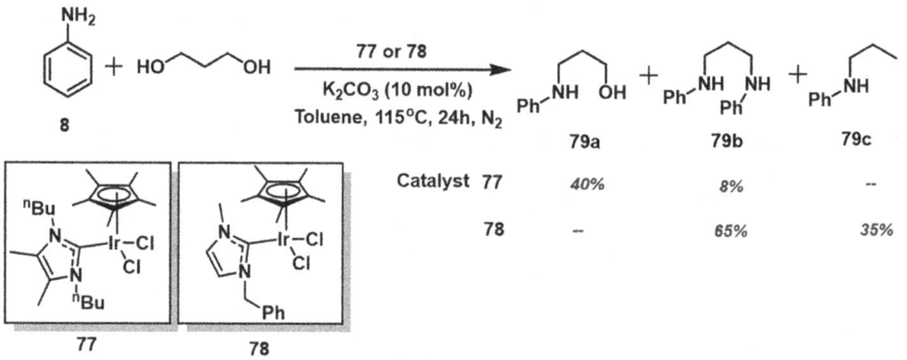

FIGURE 2.22 N-alkylation selectivity of aniline with diol using an iridium-NHC complex.

as under similar reaction conditions, N-alkylation aniline (2 equiv.) (**8**) with propane-1,3-diol using Ir–NHC catalyst **77** or **78** results in the formation of a mixture of mono-aminated (**79a**), di-aminated (**79b**), and reductive mono-aminated (**79c**) products.[38] Hence, product selectivity is highly influenced by the nature of the iridium catalyst and also depends on the external ligands employed in the reaction (Figure 2.22).

Chelate ring featuring phosphane-sulfonate-based iridium(III) organometallic complex (**82**) was used as an efficient catalytic system for the synthesis of N-arylpiperidine derivatives (**86**) by a three-component reaction of aldehyde (**83**), diol (**81**), and amine (**80**) through a tandem hydrogen transfer process involving the *endo*-dehydrogenation of piperidine.[39] The first steps of the catalytic cycle involve the N, N-dialkylation of aniline derivative **80** by the dehydrogenation of diols **81** followed by the hydrogenation of imine to form intermediate **84**, while in the second step, Ir(III) catalyzes the *endo*-dehydrogenation of **84** to form another intermediate **85**, which upon condensation with aldehyde **83**, formed final N-arylpiperidine derivatives **86** (Figure 2.23).

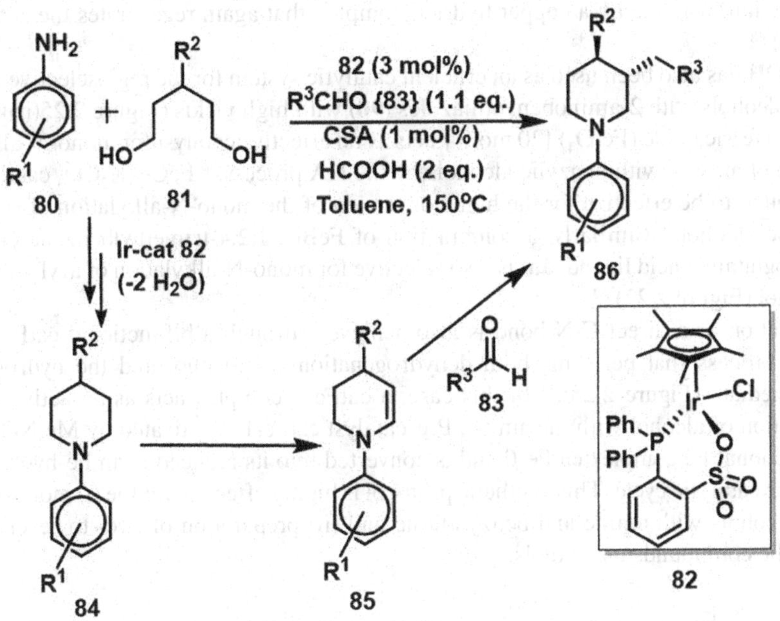

FIGURE 2.23 Tandem hydrogen transfer for the synthesis of N-arylpiperidine derivatives.

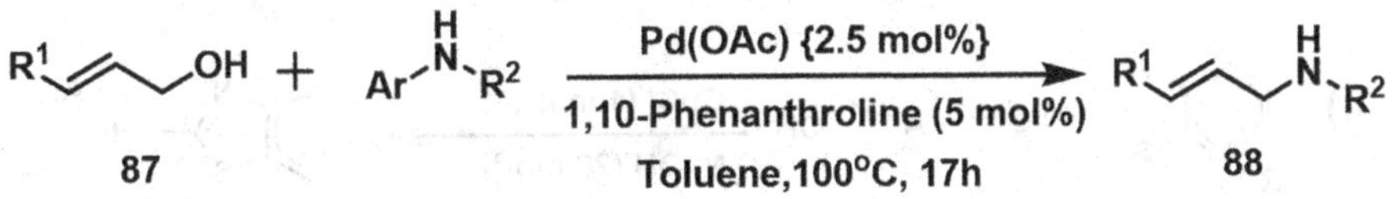

FIGURE 2.24 N-alkylation of aryl amine with allylic alcohol using a Pd(OAc)$_2$ catalyst.

2.2.3 N-Alkylation by Pd Catalyst

Compared to ruthenium or iridium, palladium-based catalysts have been less explored for the N-alkylation of amine with alcohol. In this aspect, Pd(OAc)$_2$ acts as a versatile catalyst for the N-monosubstitution of primary alcohol with poor nucleophilic amines (aromatic/heteroaromatic amines, carboxamides, and phosphazenes).[40] For example, Pd(OAc)$_2$ in the presence of 1,10-phenanthroline can effectively alkylate aryl amine by allylic alcohol (**87**), affording the formation of allylic amines (**88**) (Figure 2.24).[41] Pd(OAc)$_2$/K$_2$CO$_3$ in aerobic condition[42] and PdCl$_2$/dppe/LiOH {dppe = 1,2-Bis(diphenylphosphino)ethane}[43] catalytic systems have been found to be effective for the N-alkylation of amines/amide with both primary and secondary alcohols.

2.2.4 N-Alkylation by Copper/Iron Catalyst

Compared to 4d/5d-TMs such as ruthenium and iridium, 3d-TM is less explored as a catalyst for N-alkylation reaction. However, significant progress had been made in the development of Cu- and Fe-based catalysts due to their promising features of being cheaper and more readily available than their heavier metal congeners. In this aspect, Cu(OAc)$_2$ acts as an efficient catalyst of the N-(mono) alkylation of the less nucleophilic aromatic and heteroaromatic amine and sulfonamides (**62**), phosphinamides, carboxamides, and phosphazenes with primary alcohols (Figure 2.25(i)).[44–47] Cu(OAc)$_2$ requires a strong base ^{t}BuOK for the formation of catalytic active species Cu(OAc)(NH-Ph), and the catalytic cycle consists of a sequential three-step process: i) dehydrogenation of alcohol and formation of a copper hydride complex, ii) Schiff base condensation of aldehyde and amine to form imine, and iii) reduction of imine by a copper hydride complex that again regenerates the active catalyst (Figure 2.26).[48]

CuCl/NaOH has also been used as an efficient catalytic system for the regioselective substitution of benzylic alcohols with 2-aminobenzothiazoles (**90**) with high yields (Figure 2.25(ii)).[49]

Iron-based ferric oxide (Fe_3O_4) {20 mol%} acts as an effective catalyst for mono-N-alkylation for a wide range of amines with benzylic alcohols by the HA process.[50] $FeCl_2/K_2CO_3$ catalyst systems have been found to be effective for the high conversion of the mono-N-alkylation of sulfonamides with benzylic alcohol.[51] Similarly, a combination of $FeBr_3$, 1,2,4-trimethylbenzene (1,2,4-TMB), and DL-pyroglutamic acid ligand (**aa**) is also selective for mono-N-alkylation of aryl amine (**8**) with benzyl alcohol (Figure 2.27).[52]

The formation of a direct C-N bond is also achieved through a bifunctional catFe-0 complex via the HA process that performs both dehydrogenations of alcohol and the hydrogenation of imine intermediate (Figure 2.28).[53] In this case, a catFe-0 complex acts as an active catalyst for the N-alkylation of alcohol with the amine. Pre-catalyst catFe-1 is activated by Me_3NO to form an active bifunctional Fe complex catFe-0 and is converted into its reduced form Fe-hydride complex catFe-H in the catalytic cycle. This synthetic protocol is highly effective for the (i) mono-substitution of various alcohols with amine and benzylamine and (ii) preparation of five- to seven-membered N-heterocyclic compounds using diols.

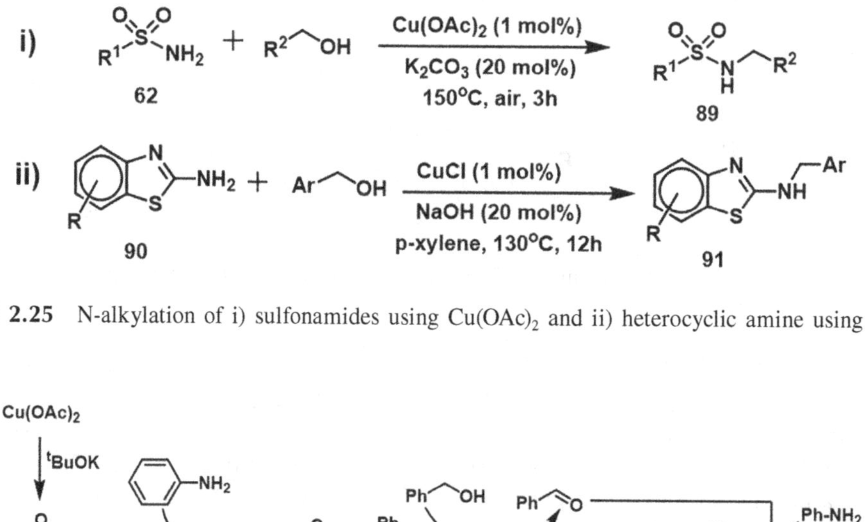

FIGURE 2.25 N-alkylation of i) sulfonamides using $Cu(OAc)_2$ and ii) heterocyclic amine using a CuCl catalyst.

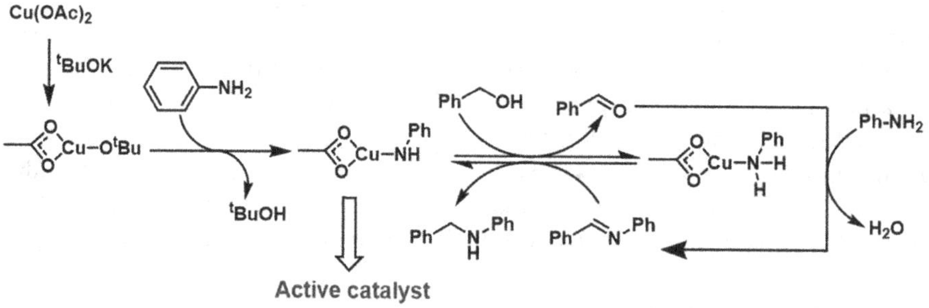

FIGURE 2.26 Possible catalytic cycle for Cu-catalyzed N-alkylation by the HA process.

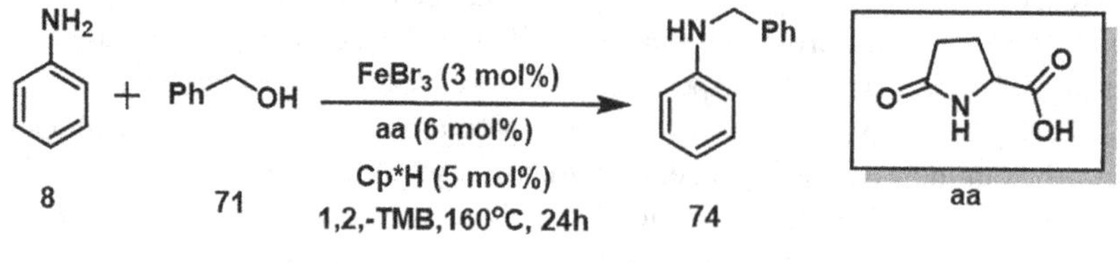

FIGURE 2.27 Iron-catalyzed N-alkylation of aniline with benzyl alcohol.

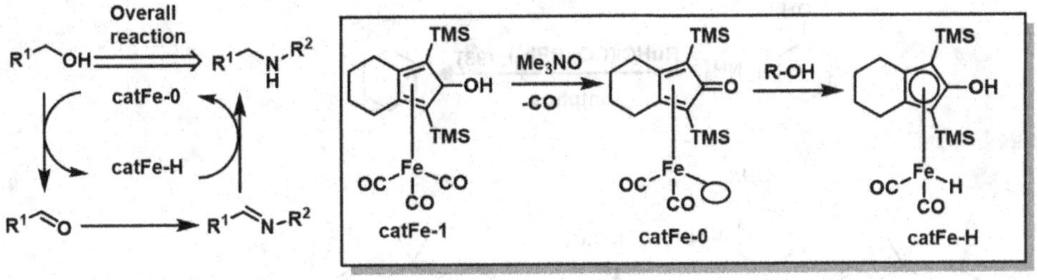

FIGURE 2.28 Iron-catalyzed N-alkylation of aniline with benzyl alcohol.

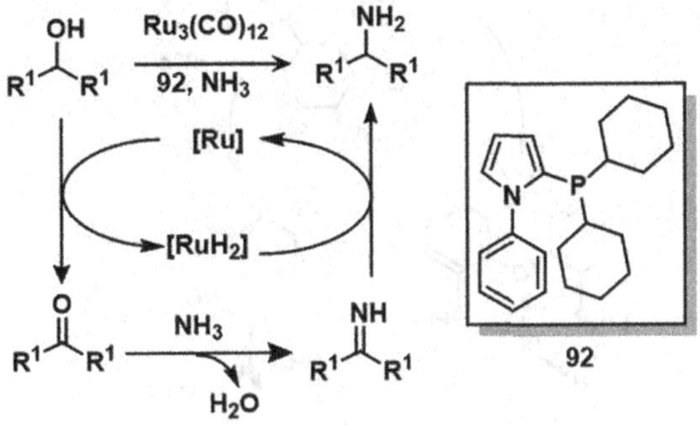

FIGURE 2.29 Ru-catalyzed amination of secondary alcohols.

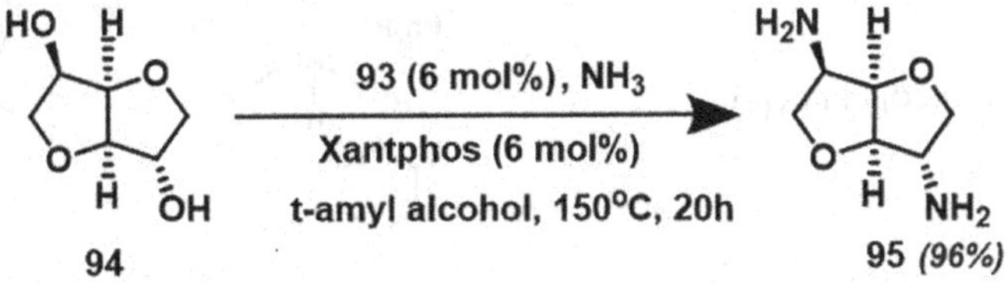

FIGURE 2.30 Ru-catalyzed di-amination of isosorbide by ammonia.

2.2.5 Synthesis of Primary Amine from Alcohol and Ammonia/Ammonium Salt

Primary amines are important precursors that are widely used for the synthesis of a wide range of organic derivatives. The traditional approach for the synthesis of amines includes the reduction of functional groups such as nitro and nitrile. These methods of amine preparation are still widely used in laboratory synthesis, but the high cost of nitro/nitrile derivative is a serious drawback to the bulk synthesis of amines on an industrial scale. In these aspects, the HA process provides an alternative approach to synthesizing primary amine using very cheap precursors of alcohols and ammonia.[54] Beller[55] and Vogt[56] at the same and first time reported the amination of secondary alcohol with liquid ammonia using catalyst $Ru_3(CO)_{12}$ in the presence of pyrrole phosphine ligand in an autoclave at 140–150°C (Figure 2.29). In another approach, $RuHCl(CO)(PPh_3)_3$/Xantphos catalyzes the diamination of isosorbide using ammonia in tert-amyl alcohol at 150°C (Figure 2.30).[57] This catalytic system was found to be effective for the synthesis of amine from corresponding primary/secondary alcohol, diols, and hydroxyl-based esters with a high yield. Mechanistic investigation of amine substitution in cyclohexanol by $RuHCl(CO)(PPh_3)_3$/Xantphos catalyst indicates the formation of inactive dihydrido ruthenium complex (**93d**) through a reaction of Ru-xantphos complex with alcohol in the presence of a base. Inactive dihydrido Ru-complex (**93d**) gets reactivated (**93e**) by in situ-generated imine intermediate

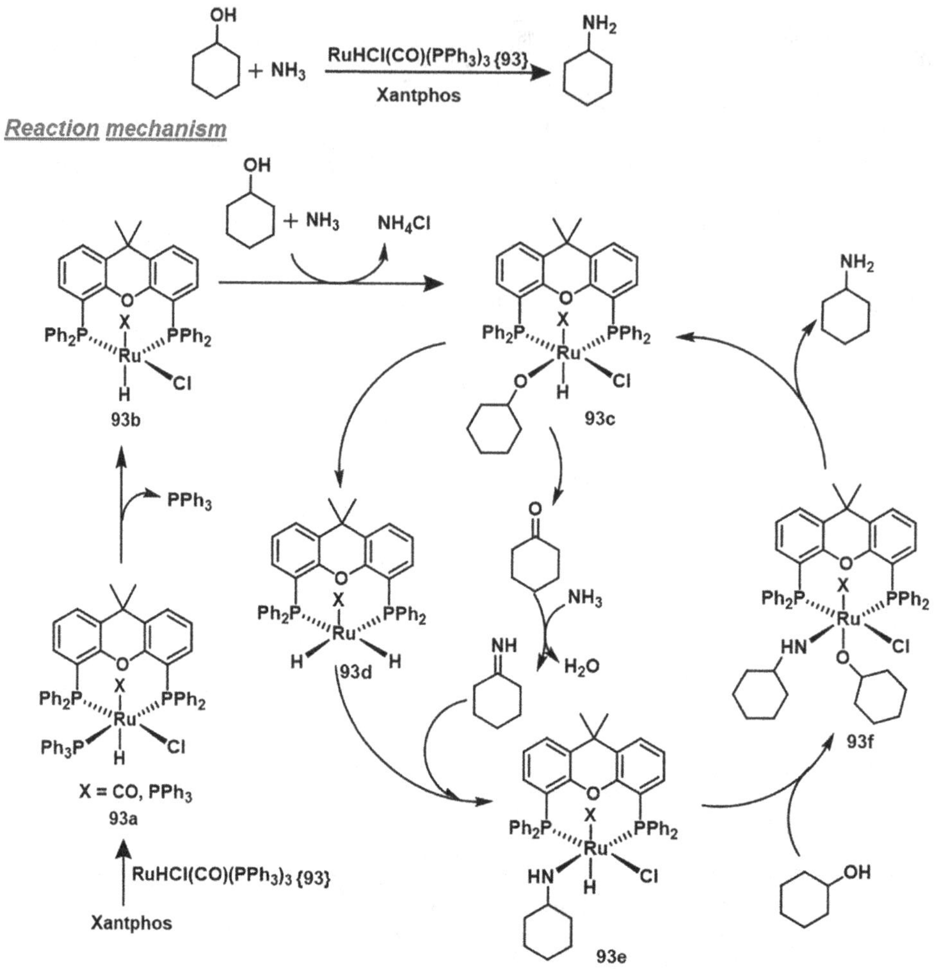

FIGURE 2.31 Reaction mechanism for the amination of alcohols with NH_3 by a $RuHCl(CO)(PPh_3)_3$ catalyst in the presence of Xantphos.

resulting from the Schiff base condensation of formed ketone (cyclohexanone) and ammonia. The catalytic cycle closely resembles the HA mechanism comprising both the dehydrogenation (of alcohol) and hydrogenation (of imine intermediate) steps (Figure 2.31).[58]

The modified analog of Milstein's acridine-based PNP-Ru(II) pincer complex **96** was found to be a highly efficient catalyst for the amination of primary alcohols to generate corresponding amines (Figure 2.32).[54] In this case, the high catalytic activity and selectivity of the product do not depend on metal-ligand cooperation as investigated from reaction optimization and density functional theory calculation on the behavior of substrates, intermediate, and active catalyst forms.[59]

The formation of higher-order amine from alcohol was successfully realized with water-soluble Cp*-based Ir-amine complex $[Cp*Ir(NH_3)_3]I_2$ (**98**).[60] The catalyst found its versatility in the multi-alkylation of liquid ammonia with primary or secondary alcohol affording a verity of secondary and tertiary amines. For example, $[Cp*Ir(NH_3)_3]I_2$ complex effectively catalyzes the amination reaction of ammonia with 1,5,9-nonanetriol (**97**) to generate N-heterocyclic compound as quinolizidine (**99**) with an 85% yield in a one-pot synthesis (Figure 2.33).

Commercially available ammonium salts are the chief source of the starting materials used as alkylating agents in the preparation of higher-order ammines. For example, $[Cp*IrCl_2]_2$ efficiently catalyzes the trialkylation of ammonium acetate with an excess of benzyl alcohol at 160°C under a

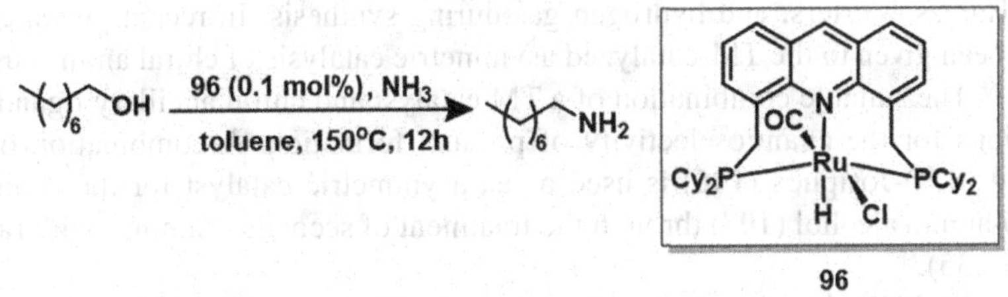

FIGURE 2.32 Acridine-based PNP-Ru (II) catalyst for the amination of primary alcohol.

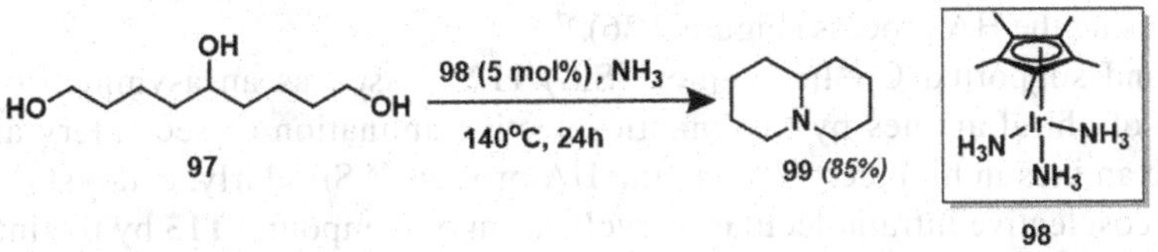

FIGURE 2.33 Synthesis of quinolizidine by amination reaction using [Cp*Ir(NH$_3$)$_3$]I$_2$ catalyst.

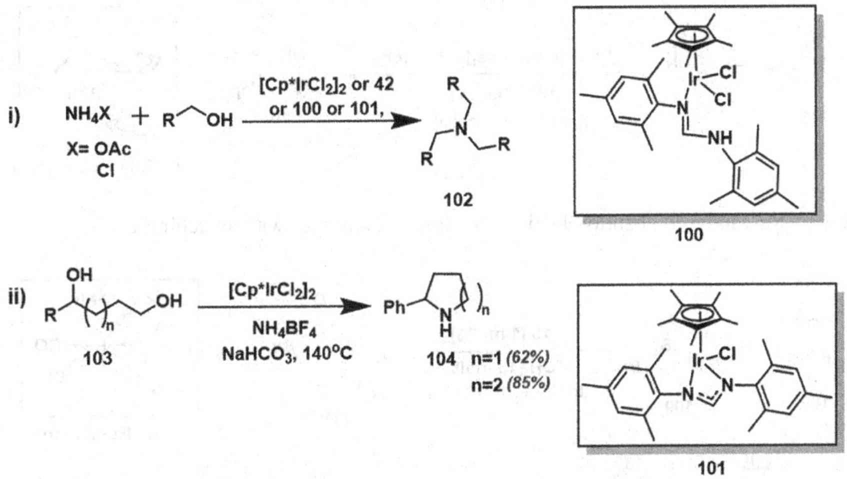

FIGURE 2.34 Metal-catalyzed multi-alkylation of alcohols with ammonium salt.

microwave condition, affording the formation of tertiary amines (Figure 2.34(i)).[61] [Cp*IrCl$_2$]$_2$ also catalyzes a one-pot synthesis of five/six-membered cyclic amine from multialkylation of NH$_4$BF$_4$ with diols (Figure 2.34(ii)).[62]

The formation of tertiary amine using the alkylation of a primary amine with NH$_4$Cl/NH$_4$OAc was also achieved using [Cp*IrCl$_2$(amidine)] (**100**) and Ru-based Shvo's catalyst with an excellent yield of up to 99%.[63] The high catalytic efficiency of **100** is attributed to the presence of the NH group in amidine ligand in Ir-complex (**100**), which generates a highly active precatalyst **101** during the catalytic cycle (Figure 2.34(i)).

2.2.6 Enantioselective Substitution of Alcohols by Amine

The synthesis of chiral amines has special importance due to their vast demand in the pharmaceutical and chemical industries. Traditionally, chiral amines were synthesized by asymmetric hydrogenation and reductive amination of ketimines and ketone, respectively. However, all of these reactions are not atom economical as they require various external reductants such as silanes, formic acid, Hantzsch esters, and hydrogen gas during synthesis. In recent years, considerable attention had been given to the TM-catalyzed asymmetric catalysis of chiral amines using the HA mechanism.[64,65] The suitable combination of a TM catalyst and chiral ancillary ligand has important implications for the enantioselectivity of product formation. A combination of [RuCl$_2$(p-cymene)]$_2$ and (S,R)-Josiphos (**106**) is used as an asymmetric catalyst for the enantioselective synthesis of β-amino alcohol (**107**) through the treatment of secondary amines with racemic diols (**105**) (Figure 2.35).[66]

Ruthenium (II) PNP-pincer complex (Ru-Macho) (**110**) acts as an efficient catalyst for the formation of α-chiral *tert*-butanesulfinamides (**111**) with high ee > 95:5 dr (89% yield) by the enantioselective N-alkylation of chiral *tert*-butanesulfinamides (**109**) with racemic secondary alcohol (**108**) using the HA process (Figure 2.36).[67]

Chiral ligand supported Cp*Ir-complex (S,S)-**112** is used as an asymmetric catalyst for the synthesis of chiral amines by the enantioselective amination of secondary alcohol (**108**) with aromatic amines in high ee>97% via the HA process.[68] Similarly, catalyst (S,S)-**112** also catalyzes stereoselective intramolecular N-cyclization of compound **115** by the intramolecular amination reaction, affording the formation of chiral methyl substituted quinoline (**116**) (Figure 2.37).

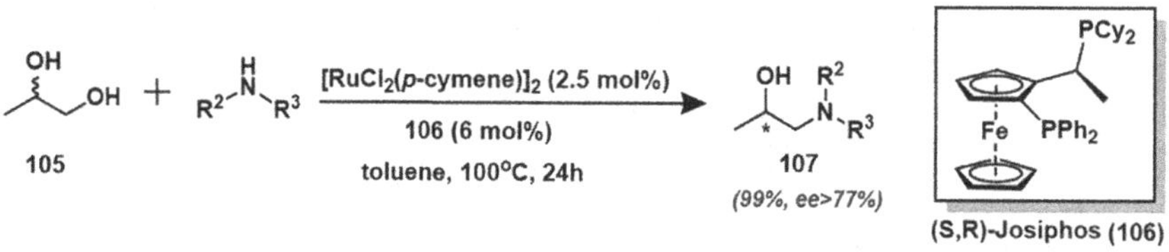

FIGURE 2.35 Ru-catalyzed enantioselective substitution of amines with an achiral diol.

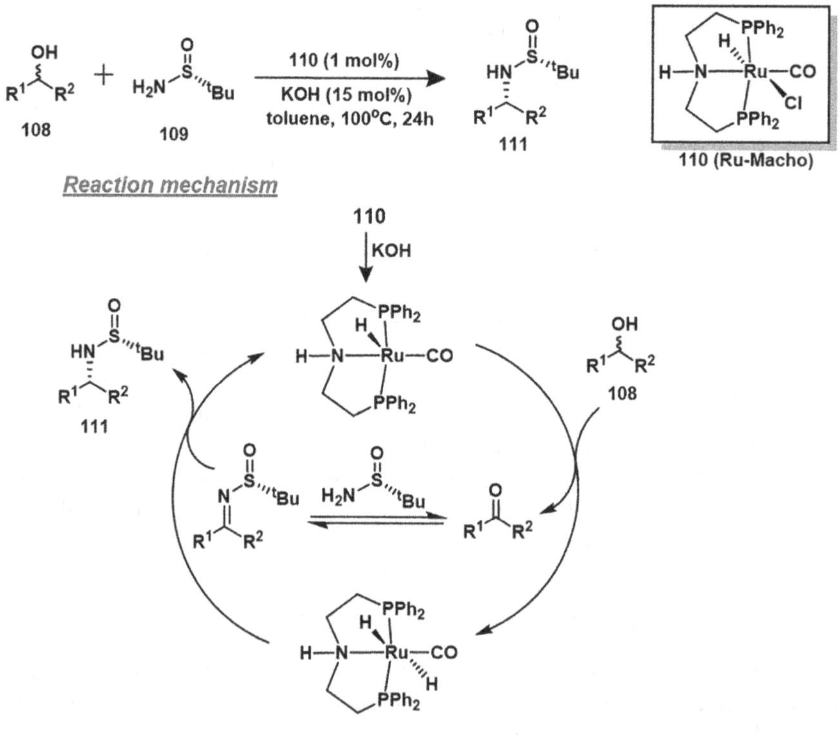

FIGURE 2.36 Ru-catalyzed asymmetric amination of alcohol by Ellman's sulfinamides.

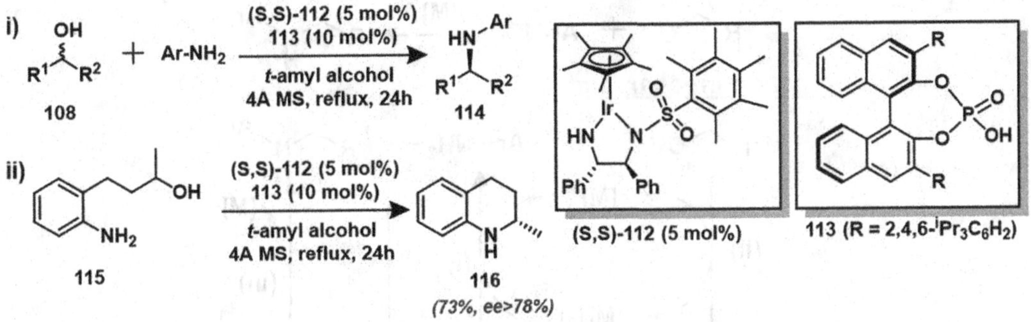

FIGURE 2.37 Enantioselective N-alkylation by chiral Ir-complex.

2.2.7 Reductive N-Alkylation with Alcohols

Nitroarenes are one of the cheap starting materials available that are used on a commercial scale for the production of a wide range of industrially important analogs through the reduction of the nitro group. In our previous discussion, we emphasized C-N bond formation by amines and alcohol via the TM-catalyze HA process. In recent years, a synthetic protocol has been developed where the N-alkylation of alcohols can be accomplished with in situ reductions of nitroarenes through the HA strategy.[69] As depicted in Figure 2.38, the reduction of alcohol (by a TM catalyst) serves as both a hydrogen source for nitro reduction and an imine intermediate. The catalytic cycle comprises a couple of hydrogen transfer steps in a cascade fashion: i) reduction of nitro to amine, ii) alcohol reduction, and iii) reduction of imine intermediate, which is formed by Schiff base condensation from in situ-generated aldehyde and amine. This synthetic strategy requires a large amount of alcohol but is still used as a simple and versatile approach for the synthesis of bulk amounts of secondary amines using inexpensive nitroarenes and alcohols on an industrial scale.

$Ru(CO)(H)_2(PPh_3)_3$ in the presence of N-heterocyclic carbene (NHC) acts as an efficient catalyst for the preparation of tertiary amines by the reductive amination of nitroarenes with an excessive amount of primary alcohols (7.5 equiv.) at 150°C under an inert atmosphere through the HA mechanism (Figure 2.39).[69] Similarly, a [Ru(p-cymene)Cl$_2$]/dppb/K$_2$CO$_3$ (dppb = diphenyl phosphobutane) catalytic combination affords the exclusive formation of secondary amines by the reductive amination of nitroarenes with primary alcohols at a high temperature (130°C) under an inert atmosphere.[70] However, the reductive N-alkylation of benzonitrile substrate using RuCl$_3$/PPh$_3$/K$_2$CO$_3$ requires harsher conditions to synthesize the same secondary amine products (Figure 2.40).[70]

The synthesis of diarylamines (**117**) with excellent yield was achieved from the reaction of cyclohexanone and nitroarenes using Pd(OAc)$_2$/Xantphos or Pd(OAc)$_2$/XPhos as a catalyst system (Figure 2.41).[71] The catalytic cycle involves three consequent steps: i) Pd-catalyzed dehydrogenation of cyclohexanone to form cyclic enone and PdH$_2$-complex, ii) hydrogenation of nitrobenzene by an in situ-generated PdH$_2$-complex to form corresponding amine, and iii) aldol condensation of in situ-generated cyclic enone and amine intermediates to form imine, which upon subsequent reduction with a metal catalyst, generates diarylamines as a final product.

2.3 C-ALKYLATION BY DEHYDROGENATIVE ALCOHOL ACTIVATION

The carbon-carbon bond formation is one of the most challenging and versatile reactions used for the synthesis of a wide range of organic compounds. In this aspect, α-alkylation of enolate via aldol condensation is widely used to construct a C-C bond in the presence of either an acid or base or stoichiometric reagents. In recent decades, the development of hydrogen transfer complexes has brought these reactions into the realms of homogeneous TM catalysts. Dehydrogenative coupling of alcohol catalyzed by TM provides an alternative mode of aldol condemnation for the construction

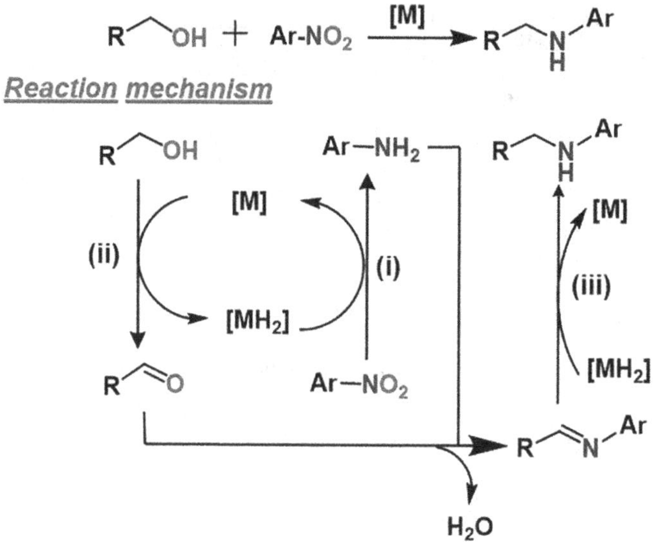

FIGURE 2.38 Metal-catalyzed amination of nitroarenes with alcohols.

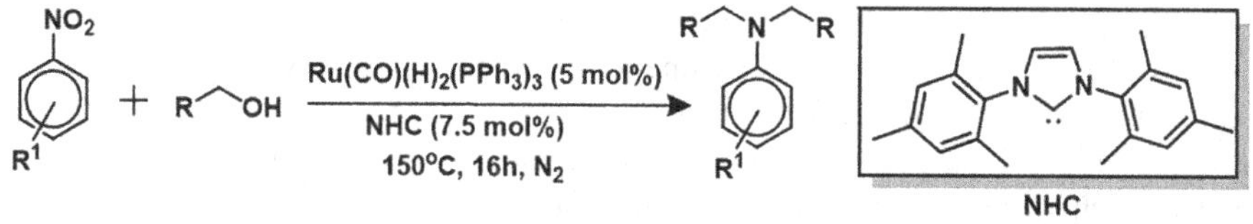

FIGURE 2.39 Ruthenium-catalyzed amination of nitrobenzene derivatives with alcohols.

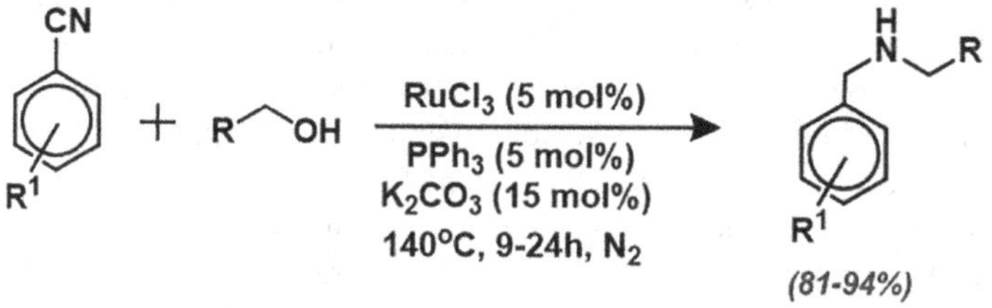

FIGURE 2.40 Ruthenium-catalyzed N-alkylation of benzonitrile derivatives with alcohols.

of C-C bonds.[72] There are two modes of aldol-based coupling by an alcohol activation strategy that can form a C-C bond: i) by alcohol-alcohol coupling or β-alkylation of secondary alcohols and ii) alcohol-ketone coupling or α-alkylation of ketones. As depicted in Figure 2.42, the α-alkylation of ketones often generates either ketone or α, β-unsaturated ketone, while alcohol-alcohol coupling mostly generates alcohol as a product. TM-catalyzed C-alkylation reactions provide a versatile approach to construct C-C bonds as the scope of substrate selection as alcohol derivatives are high compared to traditional aldol condensation reactions using ketone. In addition, the reactions are atom economical and generate minimal waste, and in most cases, water is the only by-product

2.3.1 α-Alkylation of Ketones and Its Derivative

Dehydrogenative coupling of alcohol with ketone will furnish alkylated ketone, α,β-unsaturated ketone, N-heterocycle, or alcohol derivatives depending on the substrate used (starting with ketone or functionalized alcohol).

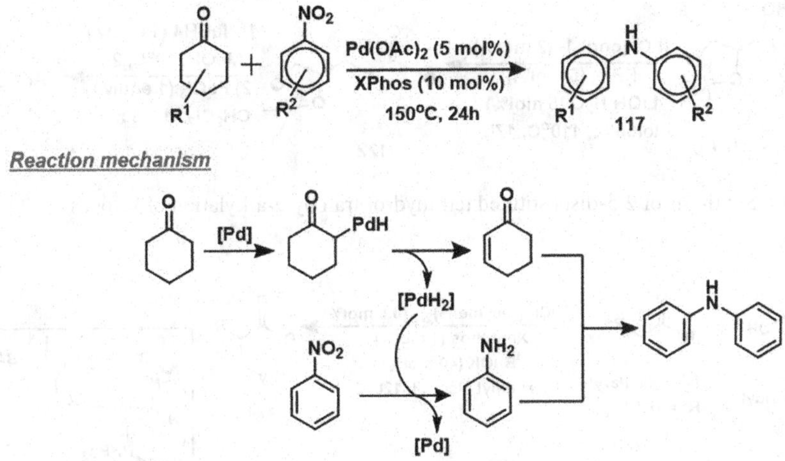

TM-Based Catalyst for Dehydrogenation of Organic Compounds

FIGURE 2.41 Pd-catalyzed reductive amination for the synthesis of diarylamine derivatives.

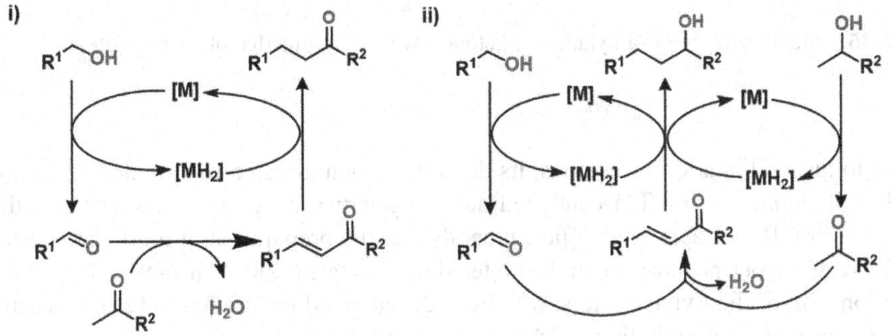

FIGURE 2.42 Schematic representation for the i) α-alkylation of ketones and b) β-alkylation of secondary alcohols.

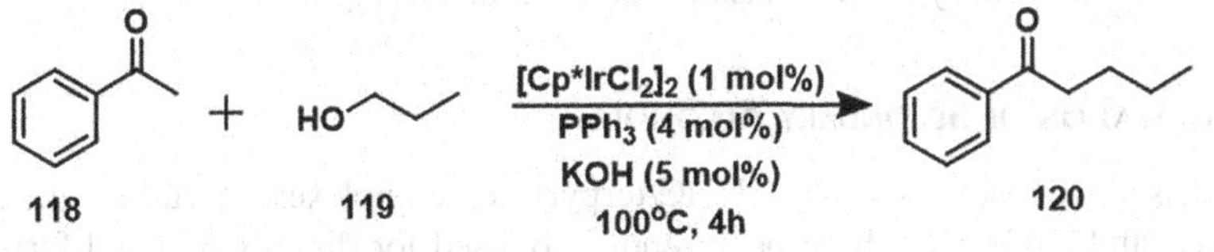

FIGURE 2.43 Iridium catalyzed α-alkylation of ketones.

Using the HA mechanism, [Cp*IrCl$_2$]$_2$/PPh$_3$/KOH catalytic system efficiently catalyzes the α-alkylation of acetophenone (**118**) with 1-propanol (**119**) (Figure 2.43).[73] Similarly, [IrCl(cod)]$_2$/PPh$_3$/LiOH.H$_2$O efficiently catalyzes the C-alkylation of substituted acetophenones with solketal (**121**) in toluene at 110C. In this case, the synthesis of 2.5-disubstituted tetrahydrofuran (**123**) is achieved by a reduction of **122** followed by subsequent cyclization promoted by FeCl$_3$ (Figure 2.44).[74] In addition, [RuCl$_2$(p-cymene)]$_2$] in the presence of Xantphos and ^{t}BuOK acts as a catalyst for the α-alkylation of substituted ketone with pyridyl methanol.[75] The formation of branched ketone was efficiently achieved using NHC-phosphine-Ir complex by dehydrogenative oxidation of methanol followed by aldol condensation with ketone.[76] In a similar manner, Rhodium (III) complex {[Cp*RhCl$_2$]$_2$} is also used for the alkylation of methanol with a ketone (Figure 2.45).[77]

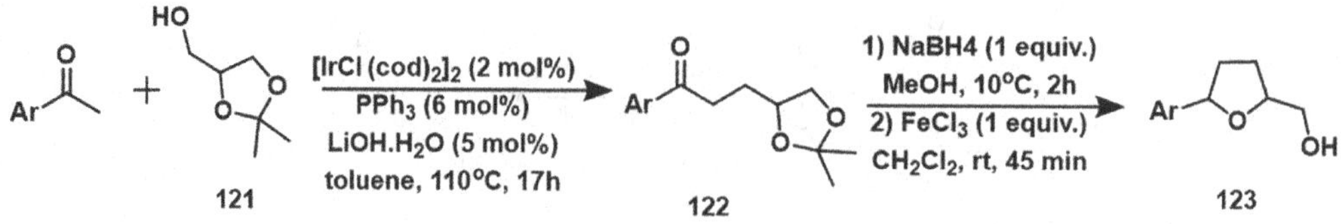

FIGURE 2.44 Synthesis of 2.5-disubstituted tetrahydrofuran by α-alkylation of ketones.

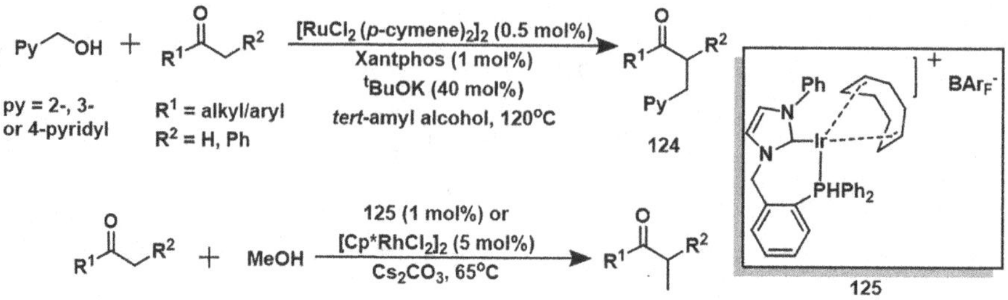

FIGURE 2.45 Metal catalyzed α-alkylation of ketones by (pyridyl)/methanol.

Similar to the α-alkylation of ketones, its derivative such as ester and amide can also be C(α)-alkylated with alcohols using TM-catalyzed dehydrogenative coupling. Using [{IrCl(cod)}$_2$]/PPh$_3$ in the presence of ^{t}BuOK acts as an efficient catalyst in the preparation of α-alkylated esters (**127**) through the reaction of a primary alcohol with tert-butyl acetate (**126**) with an 89% yield.[78] Similarly, the formation of α, α'-dialkylated esters, which are di-tert-butyl tridecanoate (**129**), was achieved by replacing primary alcohol with diols (**128**) {Figure 2.46}.

Iridium hydride complex (**131**) in the presence of a base (^{t}BuOK/^{t}BuOLi) acts as an efficient catalyst for the α-alkylation of inactivated esters (**130**) and lactone derivative (cyclic ester) {**133**} with primary alcohol.[79] With an iridium-based pincer complex (**136**) as the catalyst, the α-alkylation of acetamide (**135**) with primary alcohol has also been reported (Figure 2.47).[80]

2.3.2 β-Alkylation of Secondary Alcohols

Several TM-based complexes such as NHC/terpyridine-complexes of ruthenium and iridium, Ru-alkylidenes, and [IrCp*Cl$_2$]$_2$ have been reportedly used for direct C-C bond formation by the coupling of a primary alcohol with a secondary alcohol (Figure 2.48).[81–83]

NHC-complexes of Ru(II) or Ir(III) in the presence of a strong base such as KOH are the most selective in forming alcohol.[83] β-alkylation catalyzed by [RuCl$_2$(dmso)$_4$] in the presence of KOH results in the exclusive formation of alcohol product (**141**) by alcohol-alcohol coupling in 1,4-dioxane solvent (Figure 2.49).[84] Studies suggest that 1,4-dioxane acts as a good hydrogen acceptor in the catalytic cycle, which governs the selective formation of the alcohol product. However, the [{Ru(cod)C$_{12}$}$_n$]/PTA (PTA=1,3,5-triaza-7 phosphaadamantane) catalyst in the presence of a^{t}BuOK base gives ketone (**142**) as a major product by the coupling of 1-phenylethanol (**140**) and primary alcohols.[85]

β-alkylation of alcohol has greater prospects in the synthesis of a wide range of alcohol derivatives for various applications. For example, important biofuel material n-butanol is efficiently synthesized with 94% selectivity by the TM-catalyzed β-alkylation of alcohols from ethanol-ethanol coupling (Figure 2.50).[86] The self-coupling of w —arylalkanols (**143**) catalyzed by [IrCp*Cl$_2$]$_2$ provides a useful strategy to synthesize industrially important α,w-diarylalkanes (**144**).[87] The possible

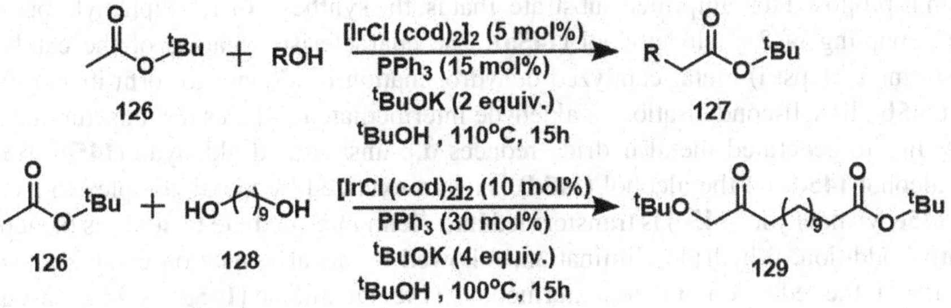

TM-Based Catalyst for Dehydrogenation of Organic Compounds

FIGURE 2.46 Iridium-catalyzed α-alkylation of esters.

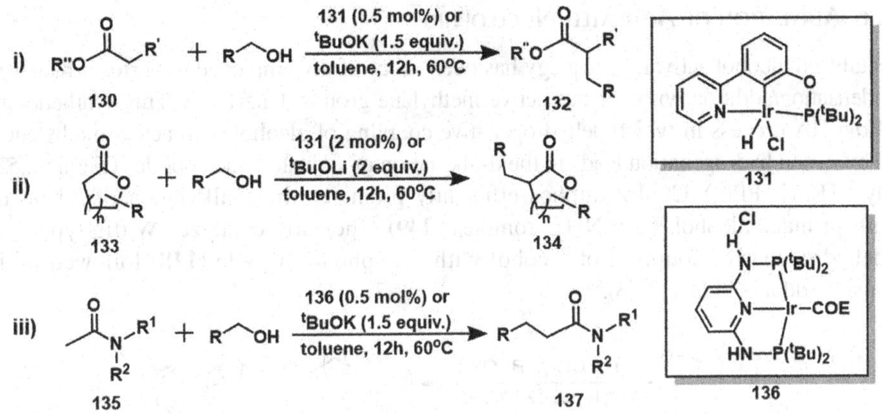

FIGURE 2.47 Iridium-pincer complex catalyzed α-alkylation of esters and amide.

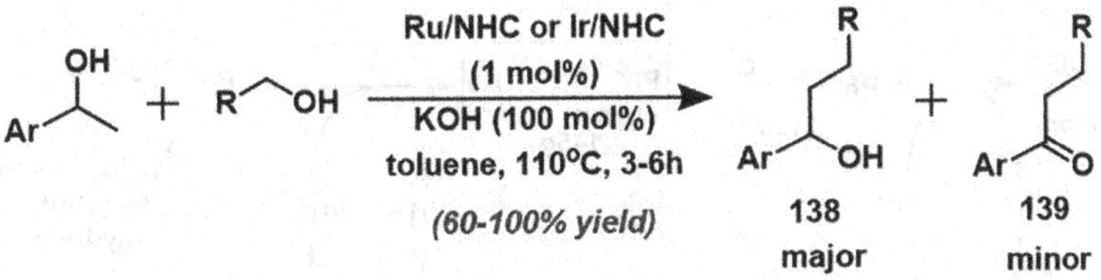

FIGURE 2.48 Formation of possible products by metal-catalyzed alcohol-alcohol coupling.

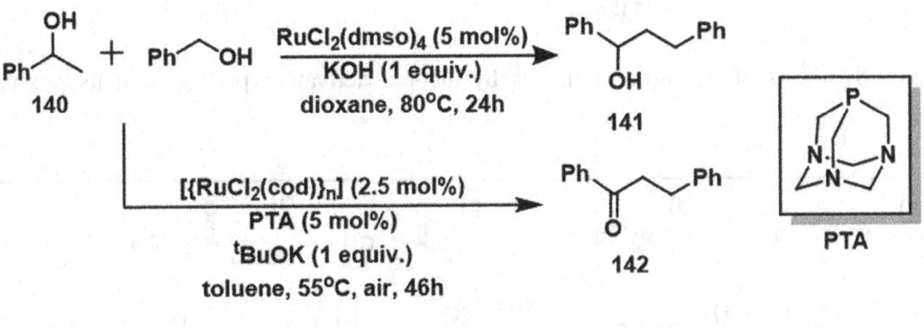

FIGURE 2.49 Metal-dependent product selectivity of alcohol-alcohol coupling.

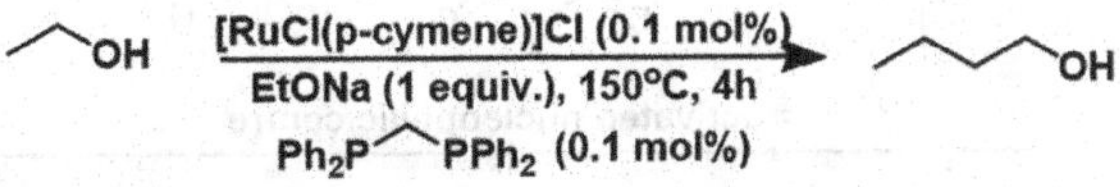

FIGURE 2.50 Ru-catalyzed self-coupling of alcohols by the HA process.

mechanism is proposed for simplified substrate, that is, the synthesis of 1,3-diphenylpropane (**145g**) by the self-coupling of 2-phenylethanol (**145a**). The characteristic feature of the catalytic cycle involves six main steps: i) metal-catalyzed dehydrogenation of alcohol to form its corresponding aldehyde (**145b**); ii) self-condensation of aldehyde intermediate produces α,β-unsaturated aldehyde (**145c**); iii) in situ-generated metal hydride reduces α,β-unsaturated aldehyde (**145c**) to its corresponding alcohol **145d**; iv) the alcohol (**145d**) is again oxidized by metal-complex to its saturated aldehyde **145e**; v) aldehyde (**145e**) is transformed into olefin intermediate by a series of steps involving oxidative addition, β-hydride elimination, followed by decarbonylation catalyzed by a metal complex; and vi) the reduction of olefin intermediate (**145e**) to alkane (**145g**) by an in situ-generated metal hydride complex (Figure 2.51).

2.3.3 α-Alkylation of Activated Nucleophile

A TM-catalyzed alcohol activation strategy has been successfully employed to perform Knoevenagel-type condensation/addition with various active methylene groups (Chart 2.1). This synthetic protocol involved the HA process in which dehydrogenative coupling of alcohol with active methylene group followed by in situ hydrogenation leads to the α-alkylation of activated nucleophiles (Figure 2.52).

Catalyst [RuH$_2$(PPh$_3$)$_3$(CO)]/Xantphos efficiently promotes the α-alkylation of β-keto nitriles (**146**) with primary alcohol.[88] Ru-NHC complex (**149**) efficiently catalyzes Wittig-type reactions by the dehydrogenative coupling of alcohol with phosphonium ylide (**148**) followed by in situ reductions of adduct (Figure 2.53).[89]

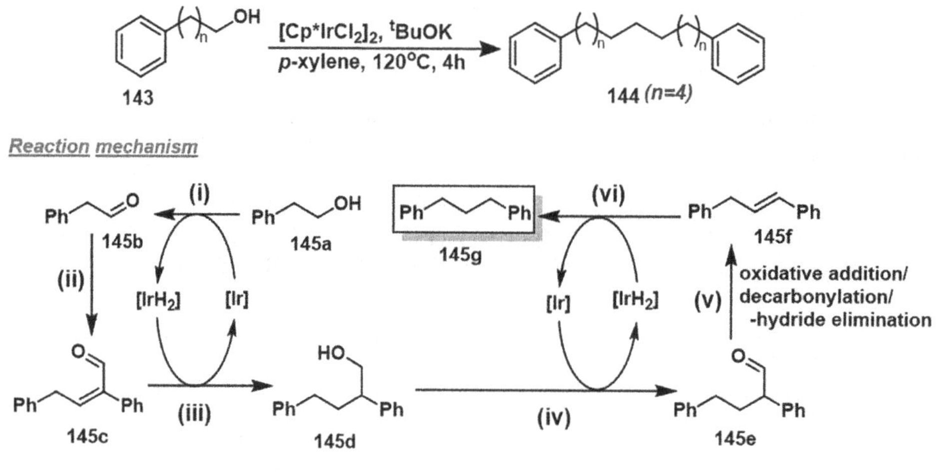

FIGURE 2.51 Synthesis of α, ω-diarylalkanes by alcohol activation process and its possible reaction mechanism.

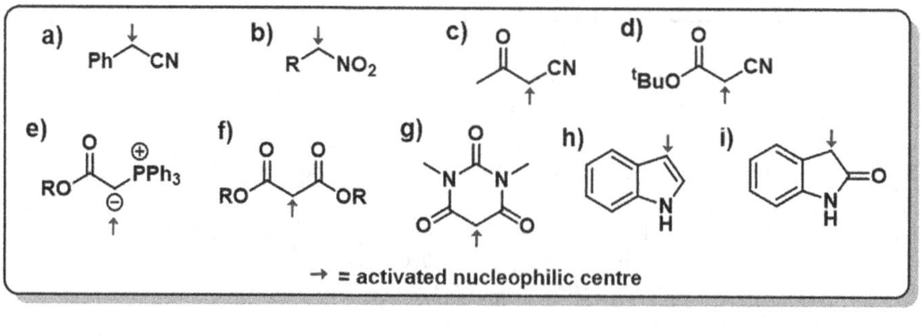

CHART 2.1 Representative examples of activated nucleophiles for alcohol coupling by HA.

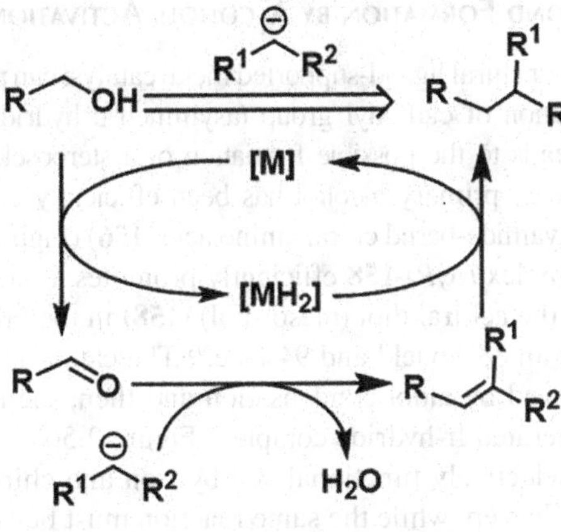

TM-Based Catalyst for Dehydrogenation of Organic Compounds 35

FIGURE 2.52 General mechanism of the metal-catalyzed α-alkylation of the activated nucleophile with alcohols.

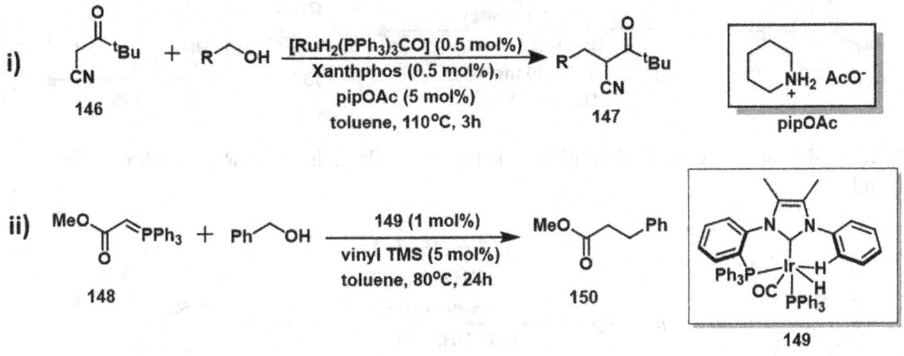

FIGURE 2.53 α-alkylation of i) β-keto nitriles and ii) ylide by activated alcohols.

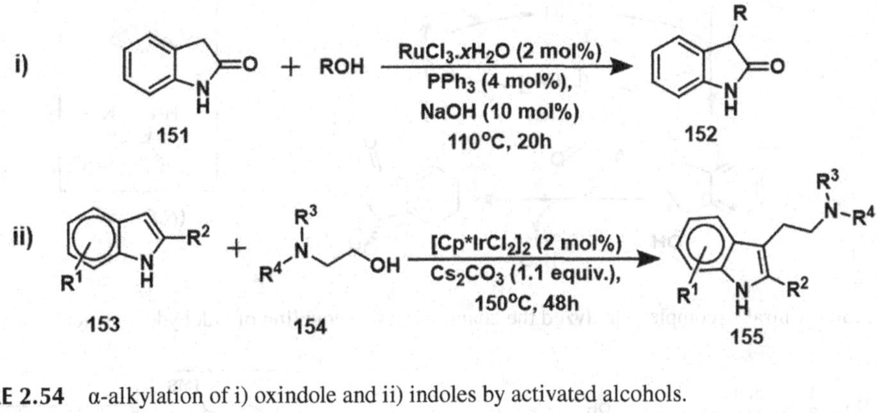

FIGURE 2.54 α-alkylation of i) oxindole and ii) indoles by activated alcohols.

Alcohols also serve as an alkylating agent for N-heterocycles or for the functional group attached to it using a TM catalyst via the HA process. Using this process, monoalkylation at 3-position in oxindole (**151**) has been successfully achieved by alkylation with primary/secondary alcohol using a $RuCl_3 \cdot xH_2O$ catalyst.[90] Meanwhile, catalyst $[Cp*IrCl_2]_2$ selectively performs a 3-alkylation of indoles (**153**) with functionalized primary alcohol (**154**) (Figure 2.54).[91]

2.3.4 Asymmetric C-C Bond Formation by Alcohols Activation

Use of an ancillary chiral ligand or chiral ligand-supported metal catalyst can promote tandem α-alkylation and the asymmetric hydrogenation of carbonyl group (asymmetric hydride transfer by metal-hydride complex to carbonyl), which leads to the possible formation of a stereoselective product. Asymmetric β-alkylation of acetophenone with primary alcohol has been efficiently catalyzed by [{Ru(p-cymene)Cl$_2$}$_2$] in the presence of hydroxyamide-based chiral amino acid (**156**) (Figure 2.55).[92]

A chiral iridium-based complex (*R,R*)-**158** efficiently promotes a one-pot asymmetric coupling of an aromatic aldehyde with the achiral diol (meso-diol) (**158**) in the presence of 2-propanol as an external hydrogen donor with an 88% yield and 94% ee.[93] The catalytic cycle features the dissymmetric oxidation of diol followed by aldol condensation and then, the reduction of formed enone intermediate by an in situ-generated Ir-hydride complex (Figure 2.56).

Allylic alcohol is enantioselectively functionalized by iron and chiralamine-mediated tandem asymmetric catalysis in a single step, while the same reaction must be performed in multiple steps in the classical synthetic approach (Figure 2.57).[94] Here, two catalytic processes (dual catalysis) operate in tandem in which the Fe-complex first catalyzes the dehydrogenative oxidation of allylic

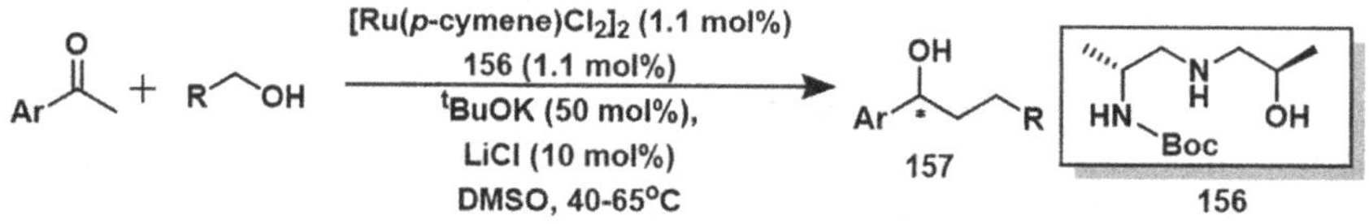

FIGURE 2.55 Enantioselective β-alkylation of ketone with alcohols by an Ru-complex in the presence of chiral ligand.

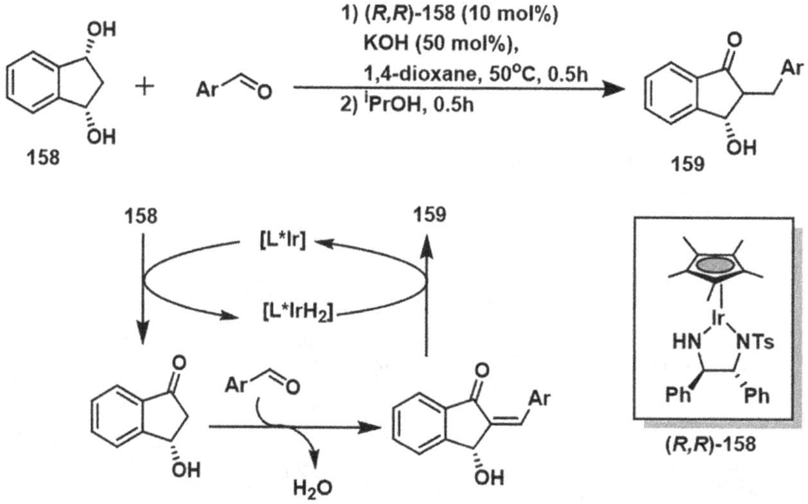

FIGURE 2.56 Chiral Ir-complex catalyzed the enantioselective coupling of aldehyde and alcohols.

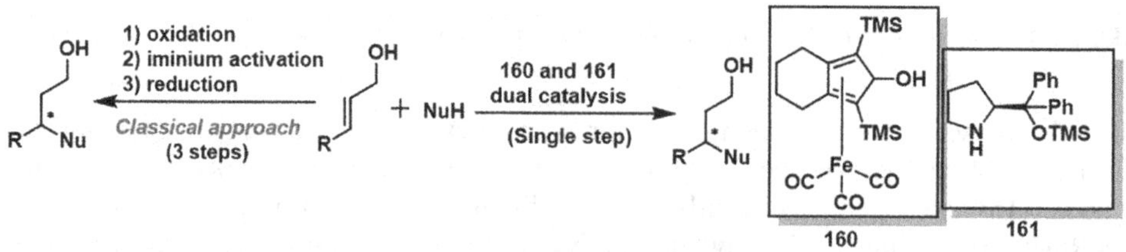

FIGURE 2.57 Dual catalysis vs the classical approach for the functionalization of allylic alcohols.

alcohol followed by iminium activation through chiral amine. Finally, enantioselective hydride transfer by an in situ-generated Fe-hydride complex leads to the formation of β-chiral alcohols.

2.3.5 Versatile HA-sequence by Dehydrogenative Alcohol Coupling

A TM-catalyzed alcohol activation process has been successfully employed in various reactions such as crotylations, carbonyl allylations, propargylations, and vinylations. Ruthenium- and iridium-based metal complexes are mostly used for these reactions. These reactions shared a similar mechanism where in situ-generated aldehyde by the metal activation of primary alcohol reacts with unsaturated substrates leading to the formation of a new C-C bond. The overall catalytic cycle is a redox neutral process as the hydrogen generated during alcohol activation is reintroduced to hydrogenate the unsaturated bonds (Figure 2.58).

Phosphine-supported iridium complex **163** efficiently catalyzes the coupling of benzyl alcohol with 3-methylbuta-1,2-diene (**162**) with a 90% yield (Figure 2.59).[95] [RuHCl(CO)(PPh$_3$)$_3$] in the presence of chelating phosphine ligand dppf promotes the propargylations reactions.[96] Formation of the enantioselective product (**170**) was observed by an allylation reaction involving allylic alcohol and allyl acetate with [Ir(COD)Cl]$_2$ catalysts in the presence of a chiral phosphine chelating ligand and *m*-nitro benzoic acid.[97] Here, the binding of chiral ligand **168** with an iridium complex (in the presence of *m*-nitro benzoic acid) promotes asymmetric carbonyl attack and hence introduces chiral induction into the final product (Figure 2.60). Hydroacylation reaction (alkene-alcohol/imine coupling) has also been achieved by chelate-assisted hydroamination reaction. The process involves the formation of a chelate imine complex through the reaction of aldehyde (in situ-generated) with a 2-aminopicoline derivative and thus promotes coupling with a terminal alkene. Hydrolysis after the

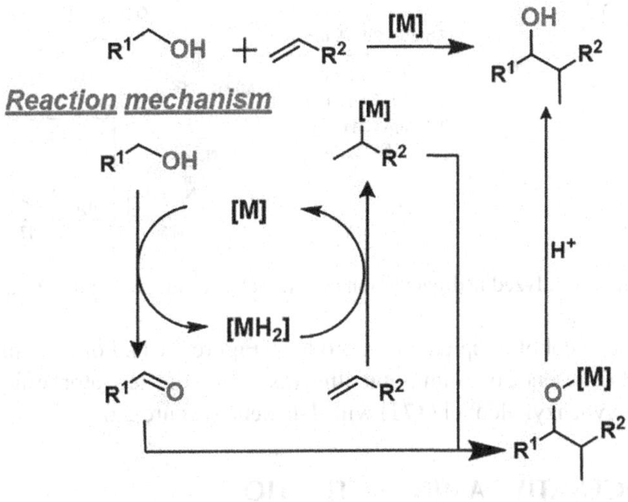

FIGURE 2.58 Metal-catalyzed olefin-alcohol coupling.

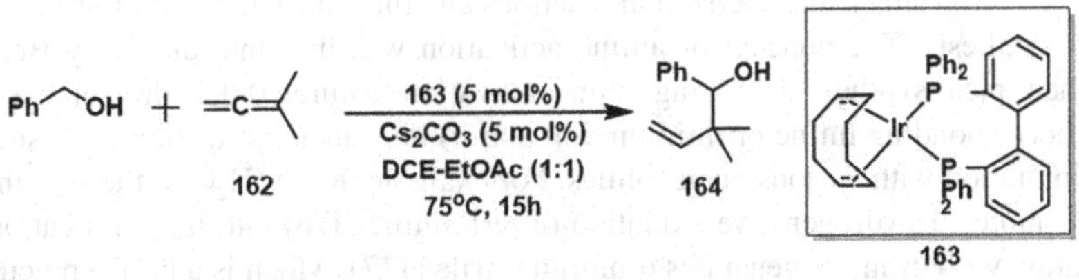

FIGURE 2.59 Iridium catalyzed alcohol-diene coupling.

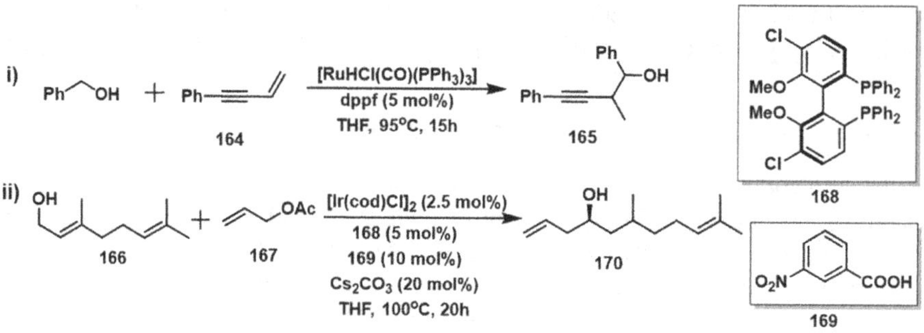

FIGURE 2.60 Metal-catalyzed alcohol dehydrogenation for i) propargylation and ii) allylation reactions.

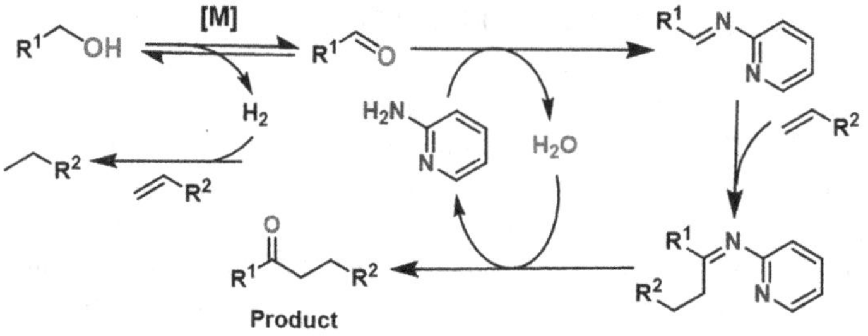

FIGURE 2.61 Chelate-assisted hydroacylation of alcohol and alkene in the presence of a coupling reagent.

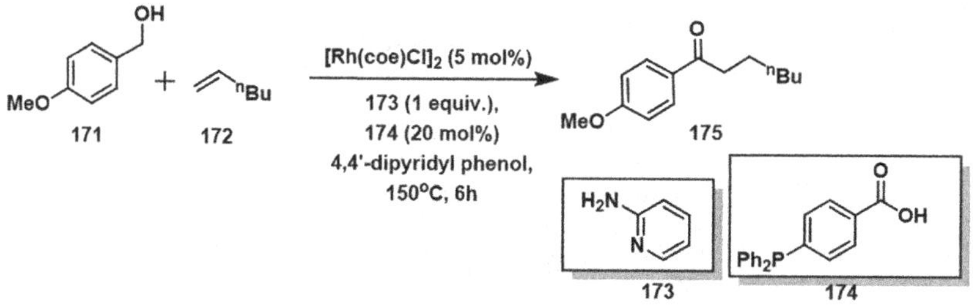

FIGURE 2.62 Rhodium catalyzed hydroacylation of primachartry alcohol and terminal alkene.

coupling step gives an alcohol-coupled olefin product (Figure 2.61). For example, RhCl$_3$.H$_2$O/PPh$_3$ catalytic system in the presence of 2-aminopicoline (as a coupling activator) efficiently promotes the coupling of 4-methoxybenzyl alcohol (**171**) with 1-hexene (Figure 2.62).[98]

2.4 DEHYDROGENATIVE AMINE ACTIVATION

Amine activation analogous to alcohol activation is a powerful strategy to generate imines and iminium ions as reactive functional species that are used in a variety of organic transformations including nucleophilic attack and cyclization reactions and thus introduces new C-C and C-N bonds in single pot synthesis. The concept of amine activation was first introduced by Beller and co-workers.[99] The typical synthesis involving amine activation features i) dehydrogenative amine oxidation to its corresponding imine or iminium ion and ii) the functionalization of in situ-generated imine or iminium ion with various nucleophiles. For example, RuCl$_3$.H$_2$O in the presence of O$_2$ as an oxidant promotes dehydrogenative oxidation of tert-amine (**176**) into iminium cation that upon functionalization with cyanide, generates α-amino nitrile (**177**), which is a useful precursor for the synthesis of α-amino acid (Figure 2.63).[100] Amine activation through the dehydrogenation process

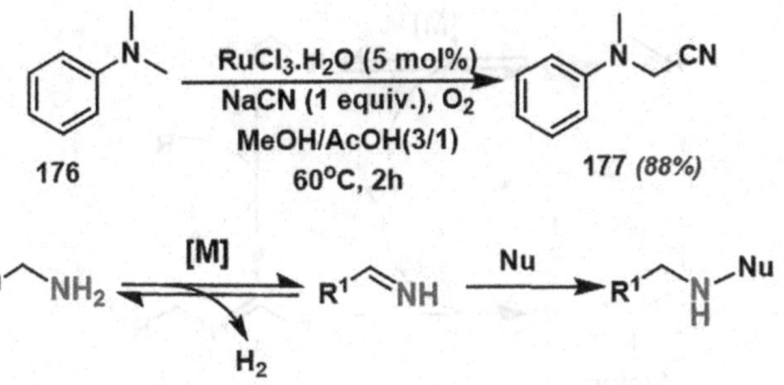

FIGURE 2.63 Ruthenium-catalyzed amine dehydrogenation for the synthesis of α-amino nitrile.

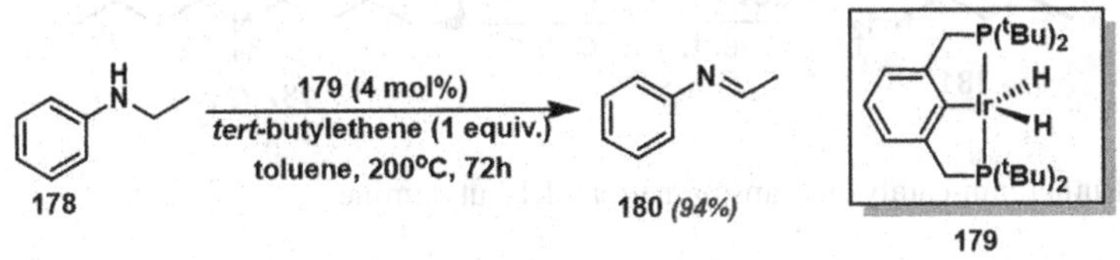

FIGURE 2.64 Iridium-catalyzed dehydrogenation of secondary amine.

is less prevalent in organic synthesis than dehydrogenative alcohol activation. The two factors that greatly impact dehydrogenative amine functionalization are the i) slower β-hydride elimination of amine/amido-metal complex compared to a metal-alkoxy complex and ii) enhanced nucleophilic character of amine in a basic medium (primary amines even possess high nucleophilic character in a neutral medium) promotes self-coupling with formed electrophilic imines.

PCP-iridium pincer complex (**179**) is used as a homogeneous catalyst for the dehydrogenation of secondary amine (**178**) to its corresponding imine (**180**) (Figure 2.64).[101] In this case, the formed imine product inhibits the progress of the reaction in the forward direction. Therefore, these reactions are usually conducted in high dilution to overcome the inhibition caused by a product formation. Two possible pathways have been suggested for the formation of an imine product: i) direct dehydrogenation of the C-N bond of secondary amine and ii) dehydrogenation of the α,β C-C bond followed by isomerization. However, mechanistic and theoretical studies suggest the formation of imine products via C-N bond dehydrogenation.

Several metal complexes exhibit catalytic activity toward the racemization of chiral amine. In these cases, TM promotes the racemization process through the amine activation process involving rapid dehydrogenation and hydrogenation via a transient imine intermediate. Rhodium-based metal complexes such as [Rh(cod)Cl]$_2$ [RhCl(PPh$_3$)$_3$] and [Rh(cod)$_2$-BF$_4$] in the presence of PCy$_3$ have shown good catalytic activity for the racemization of primary amine.[102]

2.4.1 Transamination

Transamination or amine exchange reactions via the TM-catalyzed amine dehydrogenation process provide a superior approach for the single pot synthesis of highly substituted amines without using any typical activating agents. In the dehydrogenative amine activation process, the generated imine is electrophilic in nature, and there is a good chance that it can react with an easily available nucleophile present in the reaction mixture, which is another equivalent of amine. The typical reaction mechanism of a TM-catalyzed transamination reaction is represented in Figure 2.65. Self-coupling of 1-butylamine (**181**) by transamination reaction has been efficiently catalyzed by a RuCl$_2$(PPh$_3$)$_3$

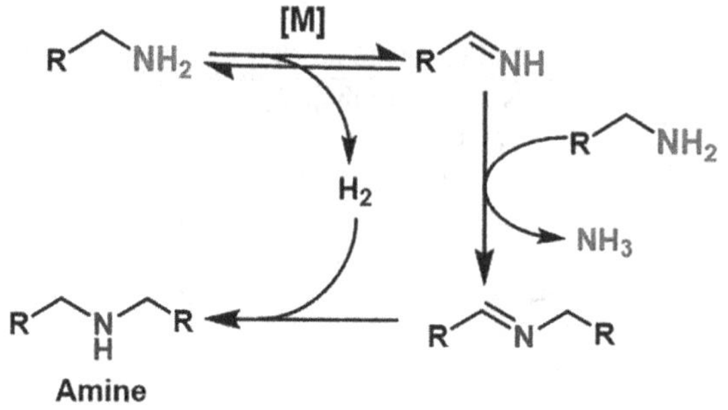

FIGURE 2.65 Metal-catalyzed transamination reaction.

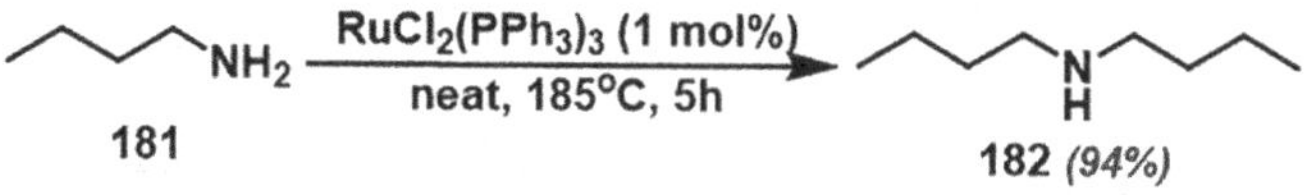

FIGURE 2.66 Ruthenium-catalyzed transamination of 1-butylamine

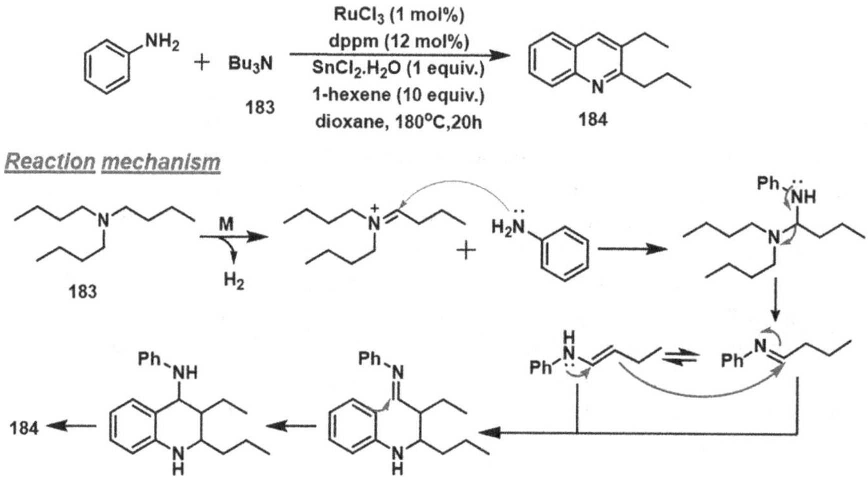

FIGURE 2.67 Ruthenium-catalyzed hetro-transamination reaction for the synthesis of quinoline.

catalyst (Figure 2.66).[103] With a judicial selection of ammine, it is also possible to achieve heterocouple transaminated products.

The synthesis of N-heterocyclic compounds has been realized with easily available starting materials using the TM-catalyzed transamination process. For example, a RuCl$_3$/dppm {dppm= bis(diphenylphosphino)methane} catalytic combination is used for the synthesis of quinoline derivative by the coupling of aniline and tributylamine (**183**) through the transamination process.[104] The catalytic cycle features the dehydrogenation of amine by a metal catalyst followed by the dimerization of heteroannulated intermediates (Figure 2.67).

2.4.2 Hydroimination

The coupling of imine and unsaturated hydrocarbon represents a hydroimination reaction. In general, alkyl-transferring reagents often promote hydroimination reactions. For example, RhCl(PPh$_3$)$_3$

drives the coupling of benzylamine with terminal alkene in the presence of alkyl transferring agent 2-amino-3-picoline (**185**).[105] The typical reaction mechanism involves two alkyl transfers catalyzed by 2-amino-3-picoline to form ketamine that upon subsequent hydrolysis, generates ketone. However, hydroimination is a less efficient process, since the maximum theoretical yield is only 50%. Investigation of the catalytic cycle suggests that only half equivalent of both amine and alkene is used to form product, while another half equivalent of amine is used in the formation of ketamine intermediate, and another half equivalent of alkene accepts hydrogen formed during the first step and thus acts as a sacrificial reagent (Figure 2.68).

2.5 DEHYDROGENATIVE ALKANE ACTIVATION

Alkane activation is by far the more difficult process compared to the previously discussed alcohol or amine bond activation. The intrinsic non-reactivity of the C-H bond in alkanes makes the dehydrogenation process extremely endogenic. In recent decades, striking development has been made by Goldman and Brookhart in the field of alkane metathesis by a TM catalyst that brings these reactions in a single pot, three-step Fig in a homogeneous manner.[106,107] The catalytic cycle of alkene metathesis comprises three steps: i) dehydrogenation of alkane to an alkene by a TM catalyst; ii) alkene metathesis; and iii) hydrogenation of alkene formed after the metathesis step (Figure 2.69). The first step of alkane dehydrogenation is an endothermic process, while alkene metathesis is a neutral process. However, the hydrogenation of final alkenes in step (iii) is an exothermic process (to the same extent as the endothermic step i)) and thus drives the reactions in the forward direction. Usually, two independent catalysts are found to be more effective for alkane metathesis. For

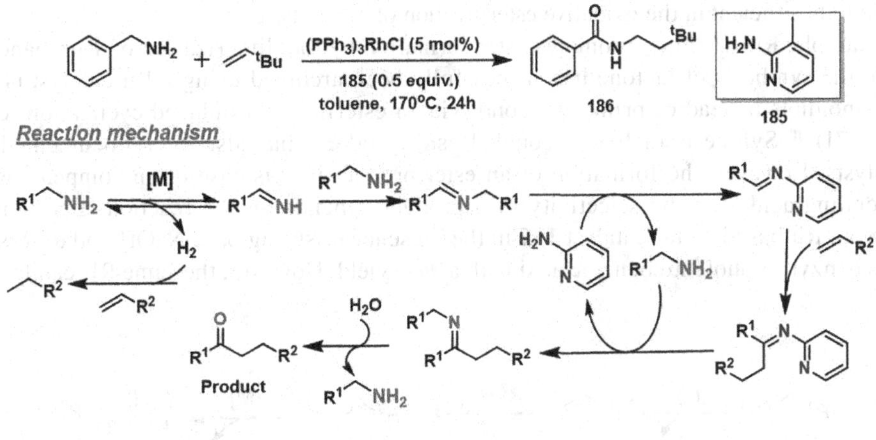

FIGURE 2.68 Metal-catalyzed hydroimination reactions.

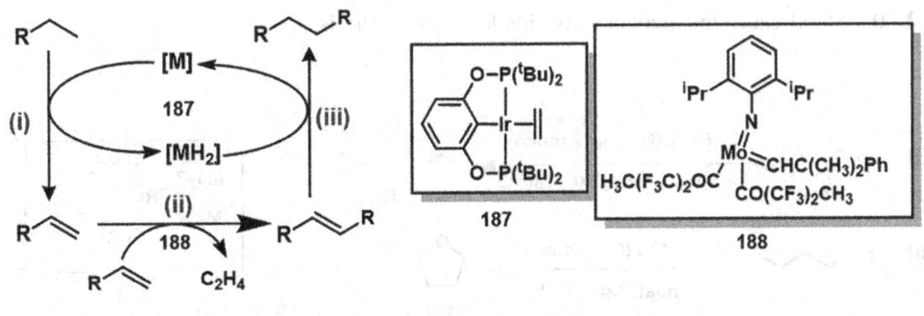

FIGURE 2.69 Metal-catalyzed alkane metathesis process. The (i), (ii), and (iii) shows the reaction pathway and the steps involved in the reaction. Following steps are involved: (i) dehydrogenation, (ii) alkylation, and (iii) hydrogenation.

example, Goldman's Ir-based catalyst **187** is mostly used for the dehydrogenation of alkanes to form a small concentration of corresponding alkenes, while Schrock Mo-based catalyst **188** performs the metathesis step and converts alkene into its corresponding lower or higher octane number.[108] Heating, reflux, or photolytic experimental conditions are generally employed during the reaction to sweep out the hydrogen from the solvent that is generated during the alkane dehydrogenation step. For example, acceptor-less dehydrogenative coupling of n-hexane at 125°C by the **187/188** catalytic system efficiently converts it into the mixture of C_2-C_{15+} alkanes.

2.6 NET DEHYDROGENATIVE OXIDATION REACTIONS

The reactions discussed so far address the HA process where hydrogen generated during the dehydrogenation of substrates returns to intermediates and forms the final product. However, in some cases, it was observed that the final hydrogenation does not proceed, which results in a net oxidation reaction. Thus, the process involving dehydrogenative functionalization results in the formation of a more oxidized product than starting precursors. For example, dehydrogenation of alcohol followed by alcohol and amine coupling will generate ester and amide, respectively.

2.6.1 Formation of Ester and Acid

The traditional approach of synthesizing ester through the Tishchenko process involves the dimerization of aldehyde by an alkali/lanthanoid metal oxide catalyst. In these aspects, the TM-catalyzed alcohol activation strategy provides an alternative method to synthesize ester by alcohol instead of aldehyde by the Tishchenko reaction (Figure 2.70). Ruthenium- and iridium-based metal complexes are found to be efficient in the oxidative esterification of alcohols.

For example, $RuH_2(PPh_3)_4$ promotes ester formation by the dimerization of 1-propanol.[109] In addition, the synthesis of lactone has successfully been archived using a Ru-catalyst (**190**) by 1,4-propanoldiols instead of primary alcohol via an esterification-induced cyclization reaction (Figure 2.71).[110] Synthesis of hetero-coupled ester products has also been documented using Ru-catalysts. However, the formation of an ester product in this case often competes with its corresponding acid, and the selectivity of a product depends on the reaction conditions. For example, an Rh-based metal catalyst **192** in the presence of strong base NaOH and excess water converts benzyl alcohol into benzoic acid with a 94% yield. However, the same Rh-catalyst (**192**)

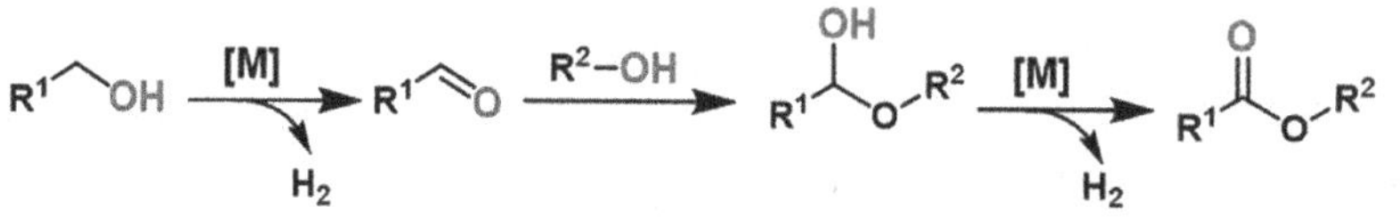

FIGURE 2.70 Metal-catalyzed alcohol activation for ester synthesis.

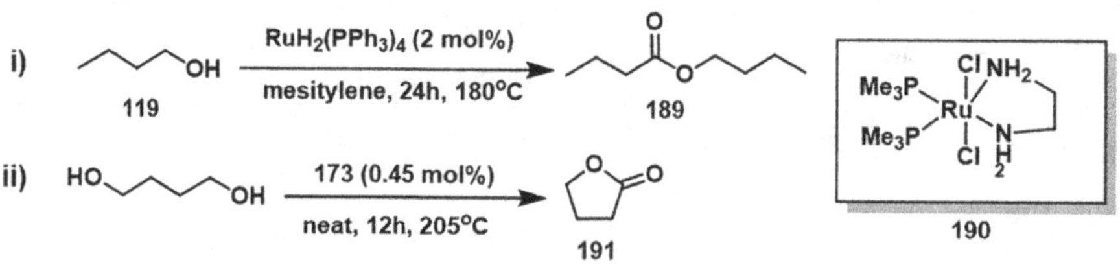

FIGURE 2.71 Ruthenium-catalyzed dehydrogenative esterification of (i) primary monoalcohol and (ii) 1,4-propanoldiols.

in the presence of mild base K_2CO_3 and excess methanol transform substituted benzyl alcohol into a methyl ester analog with an 86% yield (Figure 2.72).[111]

Ester and hemiacetal are always in dynamic equilibrium, and their relative presence depends on the experimental conditions. It has been observed that the presence of a base is required for ester formation, while under neutral conditions, the hemiacetal form persists (Figure 2.73).[112]

2.6.2 Formation of Amide

The formation of amide through the dehydrogenative oxidation of alcohol is a net oxidative reaction but often competes with the N-alkylation process. Moreover, the wide array of metal complexes is active for N-alkylation reactions, but only a few metal complexes exhibit selectivity for amide bond formation. The formation of an amide bond between alcohol and amine was first realized with a $[RuH_2(PPh_3)_4]$ catalyst in the presence of a hydrogen acceptor.[113]

For example, the synthesis of a five- to six-membered cyclic lactam ring has been achieved through the dehydrogenative intramolecular cyclization of amino-alcohol derivatives **196** in the presence of a metal catalyst (Figure 2.75).[113] The presence of water is usually required for amide formation as it inhibits the formation of other products through the N-alkylation process. Since the formation of both amide and N-alkylated product proceed via hemiaminal intermediate, the presence of water in the reaction mixture favours its irreversible oxidation (by the dehydrogenation of hemiaminal intermediate) and hence shifts the equilibrium towards the formation of amide.

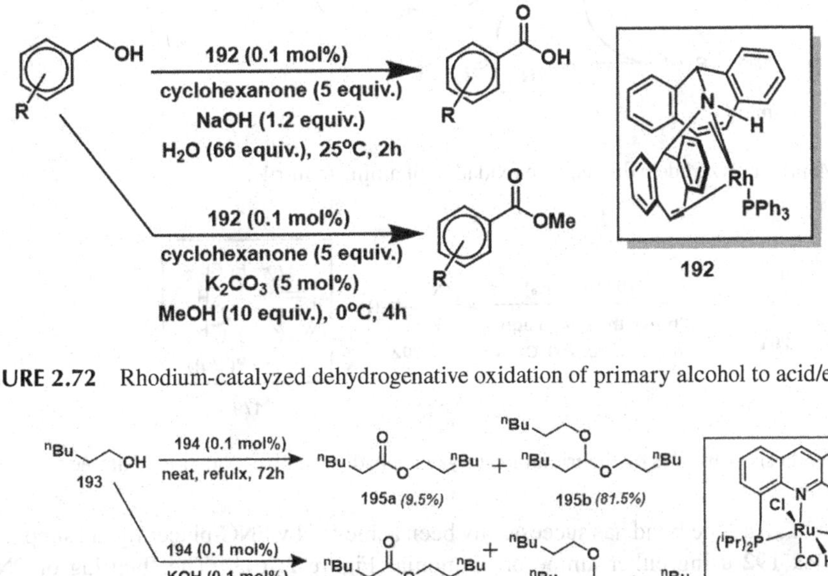

FIGURE 2.72 Rhodium-catalyzed dehydrogenative oxidation of primary alcohol to acid/ester.

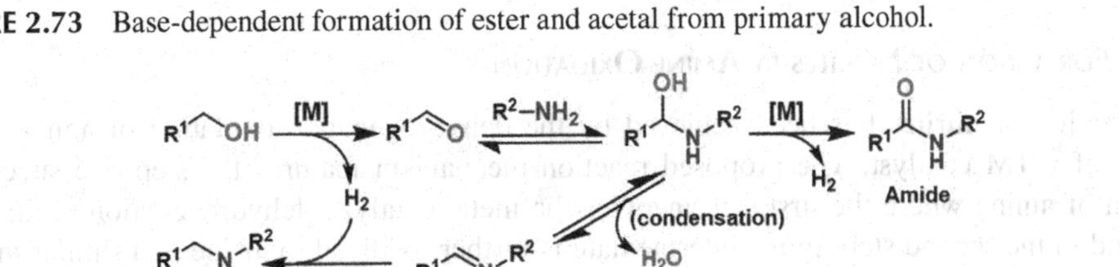

FIGURE 2.73 Base-dependent formation of ester and acetal from primary alcohol.

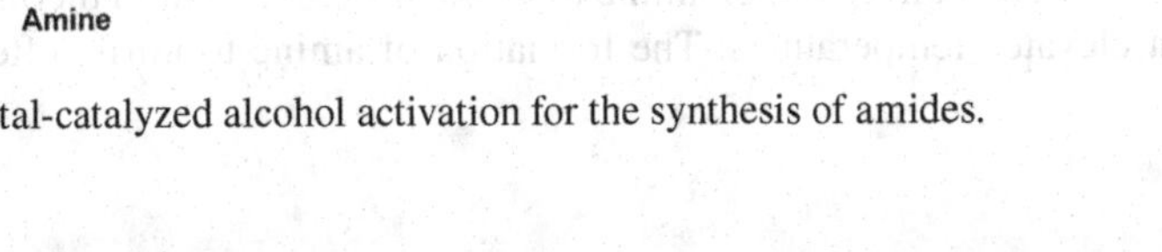

FIGURE 2.74 Metal-catalyzed alcohol activation for the synthesis of amides.

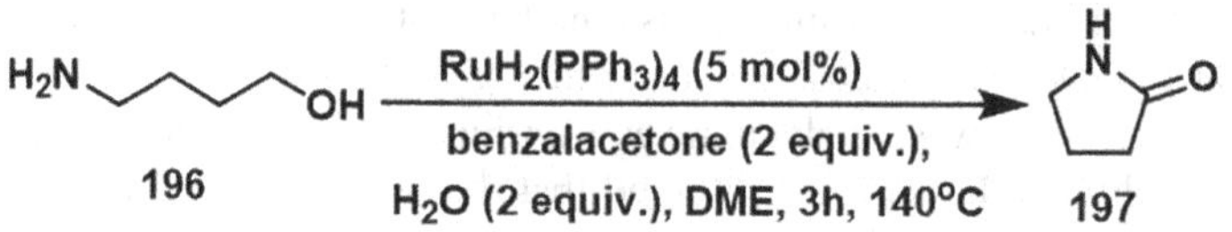

FIGURE 2.75 Ruthenium-catalyzed amidation-induced cyclization of amino alcohol.

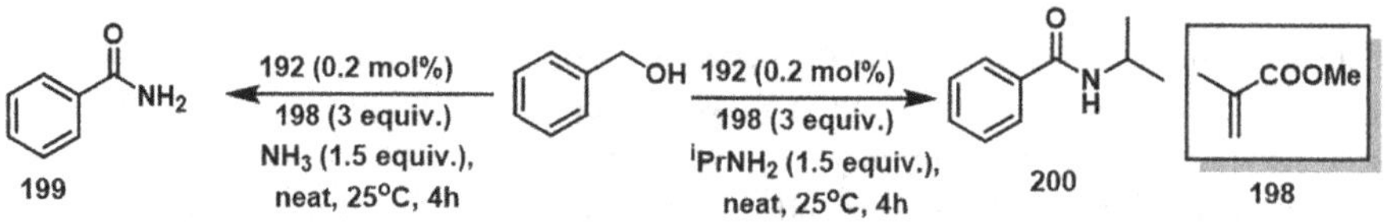

FIGURE 2.76 Rhodium-catalyzed dehydrogenative coupling of alcohols with ammonia or amine for amide synthesis.

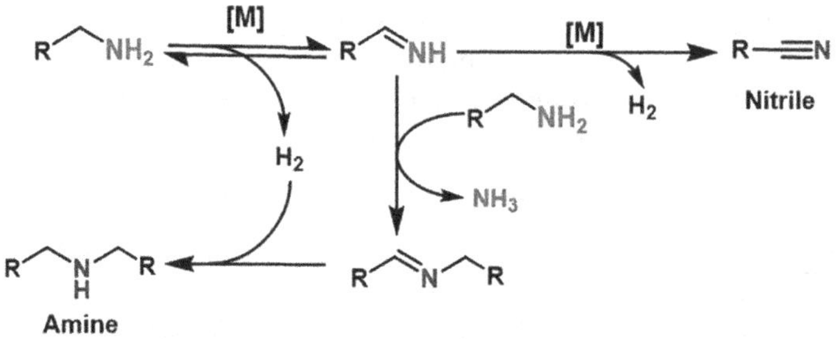

FIGURE 2.77 Metal-catalyzed dehydrogenative oxidation of amine to nitrile.

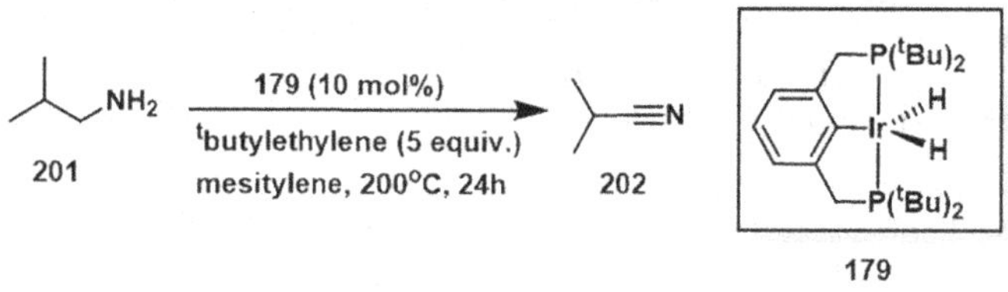

FIGURE 2.78 Synthesis of nitriles by the iridium-catalyzed dehydrogenative oxidation of amine

The formation of an amide bond has successfully been achieved by PNC-pincer ligand supported ruthenium complex **192** using either amine or ammonia (Figure 2.76).[111] The shuttling of PNC-ligand between its aromatized and non-aromatized forms during the reaction pathways enhances the chance for a net oxidative process, that is, the formation of amide.

2.6.3 Formation of Nitriles by Amine Oxidation

The formation of nitriles has been achieved by the dehydrogenative oxidation of amine in the presence of a TM catalyst. The proposed reaction mechanism features the stepwise successive oxidation of amine where the first step involves the metal catalyze-dehydrogenation of amine to imine, and in the second step, imine intermediate is further oxidized to nitrile in a similar manner as amine (Figure 2.77). The conversion of amine to nitrile possesses a high thermodynamic barrier and thus occurs at elevated temperatures. The formation of amine to nitrile often competes with

TM-Based Catalyst for Dehydrogenation of Organic Compounds

N-alkylation through the transamination process. However, the presence of an additional hydrogen acceptor favours the formation of nitrile from amine as it reduces the in situ-generated metal hydride during the reaction. Goldman first reported the double dehydrogenation of amine **201** to form nitrile **202** by PCP-pincer Ir complex (**179**) as a catalyst in the presence of hydrogen acceptor *tert*-butyl ethylene (Figure 2.79).[114]

2.7 SEMI-BORROWING HYDROGEN (SBH) PROCESS

The semi-borrowing hydrogen (SBH) or semi-hydrogen auto-transfer (SHA) process involves partial hydrogen transfer during the catalytic cycle. In the SBH process, the hydrogen that is generated by the reduction of alcohol (catalyzed by a TM-catalyst) is not reintroduced further in the catalytic cycle and thus requires an external hydrogen acceptor (usually unsaturated molecules) for its dissipation (Figure 2.79).[115] Although the SBH process is not as atom economical as the HA process, in recent years, it has been established as an efficient synthetic protocol for the preparation of various N-heterocyclic compounds including indoles, benzimidazole, pyrroles, quinazolines, etc. by using alcohol derivatives with high to excellent yield. In this process, the hydrogen acceptor source acts as a sacrificial reagent, is required in an equivalent amount as the starting material, and cannot be recovered in its original form after the completion of the reaction.

2.7.1 Synthesis of Benzimidazoles

Using the SBH strategy, Ru(PPh$_3$)$_3$(CO)(H)$_2$/Xantphos in the presence of 2.2 equivalent of crotonitrile (sacrificial reagent) acts as an efficient catalyst for the synthesis of benzimidazoles through the reactions of o-aminoaniline (**45**) with primary alcohol (Figure 2.80).[116]

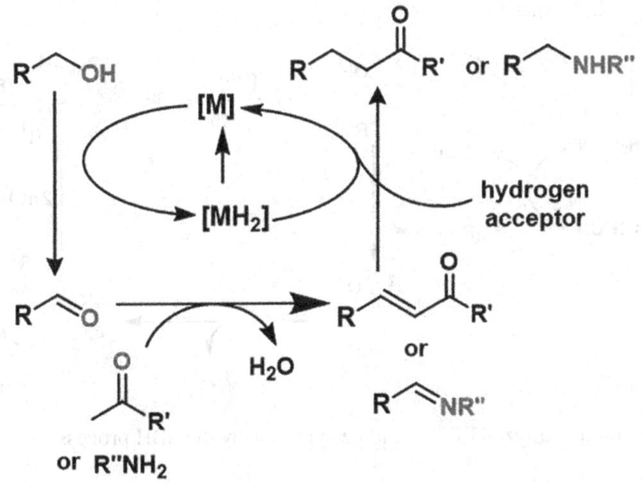

FIGURE 2.79 Semi-HA process for the construction of new C-C/C-N bonds.

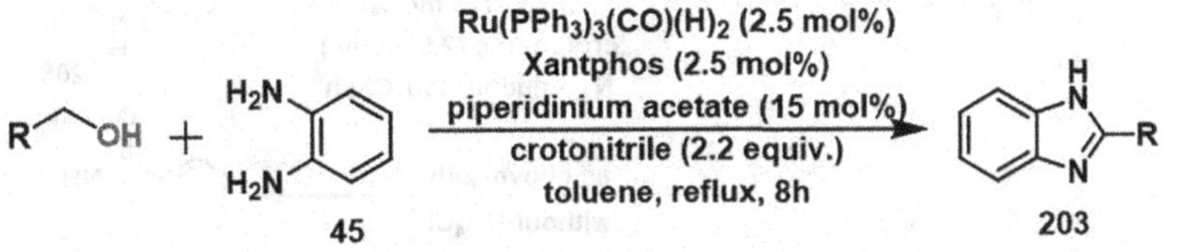

FIGURE 2.80 Synthesis of benzimidazoles by the SBH process.

2.7.2 Modified Fischer Indole Synthesis

An Ru$_3$(CO)$_{12}$/BIPHEP ((2,2'-bis(diphenylphosphanyl)-1,1'-biphenyl)) catalytic combination has been documented for indole synthesis through an alcohol activation strategy.[117] The process involves the dehydrogenative oxidation of primary or secondary alcohols (**204**) followed by condensation with arylhydrazines (**205**). Lewis's acids ZnCl$_2$ then catalyze further steps that resulted in the formation of an indole product (**206**). The sacrificial reagent crotonitrile significantly improves the yield and purity of the product, suggesting an efficient SBH process involved in this modified version of Fisher indole synthesis (Figure 2.81).

2.7.3 Synthesis of Quinazolines Derivatives

The synthesis of N-heterocyclic quinazolines derivatives (**208/209**) has also been achieved by the SBH process using the Ru-catalyzed reaction of a primary alcohol with o-aminobenzamides (**207**).[118] Interestingly, the presence of ammonium chloride governs the formation of sole product **208**. However, in the absence of NH$_4$Cl, it exclusively forms dehydrogenated product **209**. This reaction is so efficient that it does not require any purification by column chromatography, and the final product can be purified through just a simple recrystallization method (Figure 2.82).

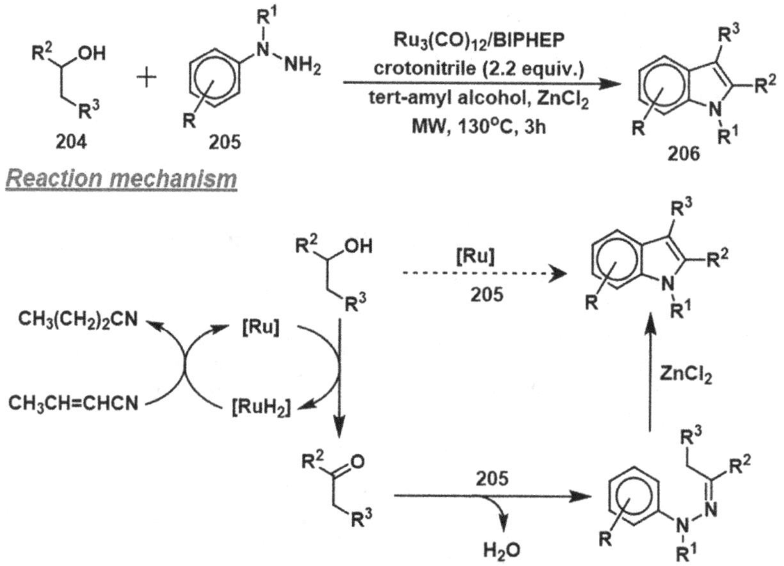

FIGURE 2.81 Ruthenium-catalyzed Fischer indole synthesis by the SBH process.

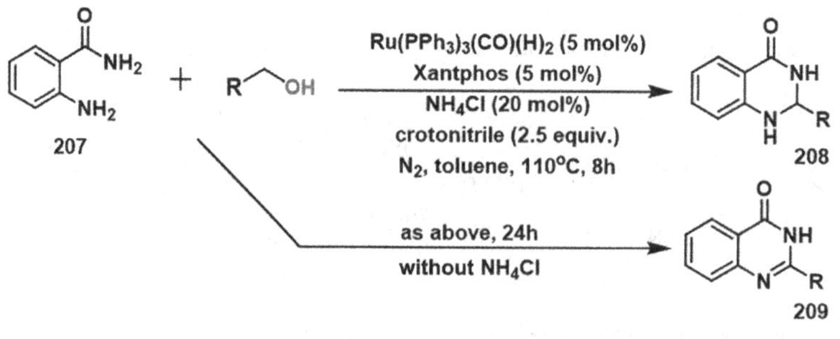

FIGURE 2.82 Synthesis of quinazolines derivatives by the SBH process.

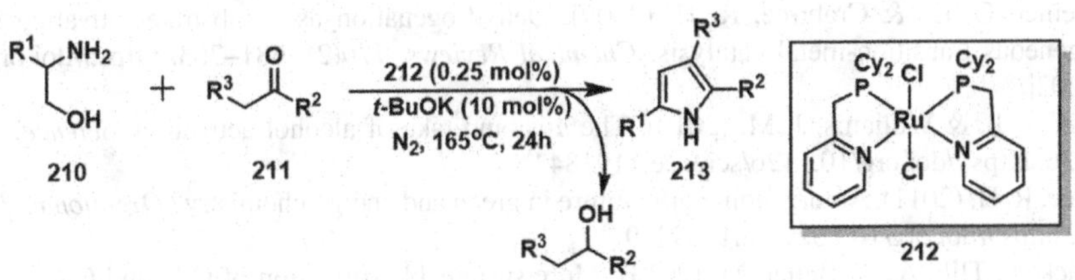

FIGURE 2.83 Synthesis of pyrroles by the SBH process.

2.7.4 Synthesis of Pyrroles Derivatives

Pyrroles synthesis has been achieved by ruthenium-catalyst **212** through reactions of α-amino alcohols (**210**) with ketones (**211**) (Figure 2.83).[119] The catalytic process involves the metal-catalyzed dehydrogenative coupling of α-amino alcohols with ketone followed by cyclization. The process involves the SBH mechanism as an additional equivalent of the ketone that serves as a sacrificial agent (hydrogen acceptor).

2.8 CONCLUSIONS

Accordingly, we have discussed homogeneous TM-catalyzed dehydrogenation reactions as a substrate-activating protocol to construct wide arrays of C-C and C-N bonds by the hydrogen autotransfer or semi-hydrogen borrowing process. These processes are highly atom economical and have greener prospects than the traditional synthesis of C-C/C-N bonds by activated substrates R-X (X = Cl, Br, OTs, etc.). The oxidative activation of alcohols and amines in organic synthesis features multiple advantages such as the substrate scope is high, single pot synthesis is possible, and minimal waste is produced. In general, the dehydrogenation reactions possess a high activation barrier and thus require an elevated temperature to operate. These challenges can be overcome by using metals with suitable ligand systems so that the process can be achieved under mild conditions. In addition, the use of chiral ancillary ligands with achiral/chiral metal complexes promotes asymmetric induction and results in enantioselective synthesis. The single pot synthesis of N-heterocyclic compounds can easily be accomplished by activating alcohols or amines through TM catalysts. Hence, the development of suitable and efficient TM catalysts for the dehydrogenation of organic substrates has important implications in core-organic synthesis.

REFERENCES

1. Alonso, F., Foubelo, F., González-Gómez, J. C., Martínez, R., Ramón, D. J., Riente, P., & Yus, M. (2009). Efficiency in chemistry: From hydrogen autotransfer to multicomponent catalysis. *Molecular Diversity*, *14*(3), 411–424. https://doi.org/10.1007/s11030-009-9195-z
2. Guillena, G., Ramón, D. J., & Yus, M. (2007). Alcohols as electrophiles in C-C bond-forming reactions: The hydrogen autotransfer process. *Angewandte Chemie International Edition*, *46*(14), 2358–2364. https://doi.org/10.1002/anie.200603794
3. Guillena, G., Ramón, D. J., & Yus, M. (2009). Hydrogen autotransfer in the *n*-alkylation of amines and related compounds using alcohols and amines as electrophiles. *Chemical Reviews*, *110*(3), 1611–1641. https://doi.org/10.1021/cr9002159
4. Jaiswal, G., Subaramanian, M., Sahoo, M. K., & Balaraman, E. (2019). A reusable cobalt catalyst for reversible acceptorless dehydrogenation and hydrogenation of n-heterocycles. *ChemCatChem*, *11*(10), 2449–2457. https://doi.org/10.1002/cctc.201900367
5. Yeung, C. S., & Dong, V. M. (2011). Catalytic dehydrogenative cross-coupling: Forming carbon–carbon bonds by oxidizing two carbon–hydrogen bonds. *Chemical Reviews*, *111*(3), 1215–1292. https://doi.org/10.1021/cr100280d
6. Anastas, P. T., & Warner, J. C. (2014). *Green Chemistry: Theory and Practice*. Oxford University Press.

7. Dobereiner, G. E., & Crabtree, R. H. (2009). Dehydrogenation as a substrate-activating strategy in homogeneous transition-metal catalysis. *Chemical Reviews*, *110*(2), 681–703. https://doi.org/10.1021/cr900202j
8. Watson, A. J., & Williams, J. M. (2010). The give and take of alcohol activation. *Science*, *329*(5992), 635–636. https://doi.org/10.1126/science.1191843
9. Crabtree, R. H. (2011). An organometallic future in green and energy chemistry? *Organometallics*, *30*(1), 17–19. https://doi.org/10.1021/om1009439
10. Schranck, J., Tlili, A., & Beller, M. (2013). More sustainable formation of C-N and C-C bonds for the synthesis of N-heterocycles. *Angewandte Chemie International Edition*, *52*(30), 7642–7644. https://doi.org/10.1002/anie.201303015
11. Grigg, R., Mitchell, T. R., Sutthivaiyakit, S., & Tongpenyai, N. (1981). Transition metal-catalyzed N-alkylation of amines by alcohols. *Journal of the Chemical Society, Chemical Communications*, *12*, 611. https://doi.org/10.1039/c39810000611
12. Almeida, M. L., Beller, M., Wang, G.-Z., & Bäckvall, J.-E. (1996). Ruthenium(II)-catalyzed Oppenauer-type oxidation of secondary alcohols. *Chemistry: A European Journal*, *2*(12), 1533–1536. https://doi.org/10.1002/chem.19960021210
13. Murahashi, S.-I., Kondo, K., & Hakata, T. (1982). Ruthenium catalyzed synthesis of secondary or tertiary amines from amines and alcohols. *Tetrahedron Letters*, *23*(2), 229–232. https://doi.org/10.1016/s0040-4039(00)86792-1
14. Watanabe, Y., Tsuji, Y., & Ohsugi, Y. (1981). The ruthenium catalyzed N-alkylation and N-heterocyclization of aniline using alcohols and aldehydes. *Tetrahedron Letters*, *22*(28), 2667–2670. https://doi.org/10.1016/s0040-4039(01)92965-x
15. Watanabe, Y., Tsuji, Y., Ige, H., Ohsugi, Y., & Ohta, T. (1984). Ruthenium-catalyzed N-alkylation and N-benzylation of aminoarenes with alcohols. *The Journal of Organic Chemistry*, *49*(18), 3359–3363. https://doi.org/10.1021/jo00192a021
16. Tsuji, Y., Huh, K. T., Ohsugi, Y., & Watanabe, Y. (1985). Ruthenium complex catalyzed N-heterocyclization. syntheses of n-substituted piperidines, morpholines, and piperazines from amines and 1,5-diols. *The Journal of Organic Chemistry*, *50*(9), 1365–1370. https://doi.org/10.1021/jo00209a004
17. Watanabe, Y., Morisaki, Y., Kondo, T., & Mitsudo, T.-A. (1996). Ruthenium complex-controlled catalytic *N*-mono- or *N,N*-dialkylation of heteroaromatic amines with alcohols. *The Journal of Organic Chemistry*, *61*(13), 4214–4218. https://doi.org/10.1021/jo9516289
18. Marsella, J. A. (1987). Homogeneously catalyzed synthesis of .beta.-amino alcohols and vicinal diamines from ethylene glycol and 1,2-propanediol. *The Journal of Organic Chemistry*, *52*(3), 467–468. https://doi.org/10.1021/jo00379a035
19. Naskar, S., & Bhattacharjee, M. (2007). Selective N-monoalkylation of anilines catalyzed by a cationic ruthenium(ii) compound. *Tetrahedron Letters*, *48*(19), 3367–3370. https://doi.org/10.1016/j.tetlet.2007.03.075
20. Del Zotto, A., Baratta, W., Sandri, M., Verardo, G., & Rigo, P. (2004). Cyclopentadienyl RU ii complexes as highly efficient catalysts for the *N*-methylation of alkylamines by methanol. *European Journal of Inorganic Chemistry*, *2004*(3), 524–529. https://doi.org/10.1002/ejic.200300518
21. Hamid, M. H., Allen, C. L., Lamb, G. W., Maxwell, A. C., Maytum, H. C., Watson, A. J., & Williams, J. M. (2009). Ruthenium-catalyzed *N*-alkylation of amines and sulfonamides using borrowing hydrogen methodology. *Journal of the American Chemical Society*, *131*(5), 1766–1774. https://doi.org/10.1021/ja807323a
22. Watson, A. J., Maxwell, A. C., & Williams, J. M. (2011). Borrowing hydrogen methodology for amine synthesis under solvent-free microwave conditions. *The Journal of Organic Chemistry*, *76*(7), 2328–2331. https://doi.org/10.1021/jo102521a
23. Ma, W. M., James, T. D., & Williams, J. M. (2013). Synthesis of amines with pendant boronic esters by borrowing hydrogen catalysis. *Organic Letters*, *15*(18), 4850–4853. https://doi.org/10.1021/ol402271a
24. Sundararaju, B., Tang, Z., Achard, M., Sharma, G. V., Toupet, L., & Bruneau, C. (2010). Ruthenium-catalyzed Cascade N- and c(3)-dialkylation of cyclic amines with alcohols involving hydrogen auto-transfer processes. *Advanced Synthesis & Catalysis*, *352*(18), 3141–3146. https://doi.org/10.1002/adsc.201000546
25. Bähn, S., Imm, S., Mevius, K., Neubert, L., Tillack, A., Williams, J. M. J., & Beller, M. (2010). Selective ruthenium-catalyzed N-alkylation of indoles by using alcohols. *Chemistry: A European Journal*, *16*(12), 3590–3593. https://doi.org/10.1002/chem.200903144
26. Blacker, A. J., Farah, M. M., Hall, M. I., Marsden, S. P., Saidi, O., & Williams, J. M. (2009). Synthesis of benzazoles by hydrogen-transfer catalysis. *Organic Letters*, *11*(9), 2039–2042. https://doi.org/10.1021/ol900557u

27. Zhang, M., Imm, S., Bähn, S., Neumann, H., & Beller, M. (2011). Synthesis of α-amino acid amides: Ruthenium-catalyzed amination of α-hydroxy amides. *Angewandte Chemie International Edition*, *50*(47), 11197–11201. https://doi.org/10.1002/anie.201104309
28. Fujita, K.-I., Enoki, Y., & Yamaguchi, R. (2008). CP*IR-catalyzed N-alkylation of amines with alcohols. A versatile and atom economical method for the synthesis of amines. *Tetrahedron*, *64*(8), 1943–1954. https://doi.org/10.1016/j.tet.2007.11.083
29. Fujita, K.-I., Li, Z., Ozeki, N., & Yamaguchi, R. (2003). N-alkylation of amines with alcohols catalyzed by a CP*IR complex. *Tetrahedron Letters*, *44*(13), 2687–2690. https://doi.org/10.1016/s0040-4039(03)00371-x
30. Yamaguchi, R., Kawagoe, S., Asai, C., & Fujita, K.-I. (2007). Selective synthesis of secondary and tertiary amines by cp*iridium-catalyzed multialkylation of ammonium salts with alcohols. *Organic Letters*, *10*(2), 181–184. https://doi.org/10.1021/ol702522k
31. Bhat, S., & Sridharan, V. (2012). Iridium catalyzed chemoselective alkylation of 2′-aminoacetophenone with primary benzyl type alcohols under microwave conditions. *Chemical Communications*, *48*(39), 4701. https://doi.org/10.1039/c2cc31055d
32. Fujita, K.-I., Komatsubara, A., & Yamaguchi, R. (2009). N-alkylation of carbamates and amides with alcohols catalyzed by a CP*IR complex. *Tetrahedron*, *65*(18), 3624–3628. https://doi.org/10.1016/j.tet.2009.03.002
33. Zhu, M., Fujita, K.-I., & Yamaguchi, R. (2010). Simple and versatile catalytic system for N-alkylation of sulfonamides with various alcohols. *Organic Letters*, *12*(6), 1336–1339. https://doi.org/10.1021/ol1002434
34. Li, F., Sun, C., Shan, H., Zou, X., & Xie, J. (2013). From regioselective condensation to regioselective N-alkylation: A novel and environmentally benign strategy for the synthesis of *N*,*N*′-alkyl aryl ureas and *N*,*N*′-dialkyl ureas. *ChemCatChem*, *5*(6), 1543–1552. https://doi.org/10.1002/cctc.201200648
35. Li, F., Xie, J., Shan, H., Sun, C., & Chen, L. (2012). General and efficient method for direct N-monomethylation of aromatic primary amines with methanol. *RSC Advances*, *2*(23), 8645–8652. https://doi.org/10.1039/c2ra21487c
36. Blank, B., Michlik, S., & Kempe, R. (2009). Selective Iridium-catalyzed alkylation of (hetero)aromatic amines and diamines with alcohols under mild reaction conditions. *Chemistry: A European Journal*, *15*(15), 3790–3799. https://doi.org/10.1002/chem.200802318
37. Aramoto, H., Obora, Y., & Ishii, Y. (2008). *N*-heterocyclization of naphthylamines with 1,2- and 1,3-diols catalyzed by an iridium chloride/BINAP system. *The Journal of Organic Chemistry*, *74*(2), 628–633. https://doi.org/10.1021/jo801966u
38. Liu, S., Rebros, M., Stephens, G., & Marr, A. C. (2009). Adding value to renewables: A one pot process combining microbial cells and hydrogen transfer catalysis to utilise waste glycerol from biodiesel production. *Chemical Communications*, *17*, 2308–2310. https://doi.org/10.1039/b820657k
39. Yuan, K., Jiang, F., Sahli, Z., Achard, M., Roisnel, T., & Bruneau, C. (2012). Iridium-catalyzed oxidant-free dehydrogenative C-H bond functionalization: Selective preparation of N-Arylpiperidines through tandem hydrogen transfers. *Angewandte Chemie International Edition*, *51*(35), 8876–8880. https://doi.org/10.1002/anie.201204582
40. Ramón, D., Martínez-Asencio, A., & Yus, M. (2011). Palladium(II) acetate as catalyst for the N-alkylation of aromatic amines, sulfonamides, and related nitrogenated compounds with alcohols by a hydrogen autotransfer process. *Synthesis*, *2011*(22), 3730–3740. https://doi.org/10.1055/s-0030-1260238
41. Banerjee, D., Jagadeesh, R. V., Junge, K., Junge, H., & Beller, M. (2012). An efficient and convenient palladium catalyst system for the synthesis of amines from allylic alcohols. *ChemSusChem*, *5*(10), 2039–2044. https://doi.org/10.1002/cssc.201200247
42. Yu, X., Jiang, L., Li, Q., Xie, Y., & Xu, Q. (2012). Palladium-catalyzed *N*-alkylation of amides and amines with alcohols employing the aerobic relay race methodology. *Chinese Journal of Chemistry*, *30*(10), 2322–2332. https://doi.org/10.1002/cjoc.201200462
43. Dang, T. T., Ramalingam, B., Shan, S. P., & Seayad, A. M. (2013). An efficient palladium-catalyzed N-alkylation of amines using primary and secondary alcohols. *ACS Catalysis*, *3*(11), 2536–2540. https://doi.org/10.1021/cs400799n
44. Shi, F., Tse, M. K., Cui, X., Gördes, D., Michalik, D., Thurow, K., Deng, Y., & Beller, M. (2009). Copper-catalyzed alkylation of sulfonamides with alcohols. *Angewandte Chemie International Edition*, *48*(32), 5912–5915. https://doi.org/10.1002/anie.200901510
45. Cui, X., Shi, F., Tse, M. K., Gördes, D., Thurow, K., Beller, M., & Deng, Y. (2009). Copper-catalyzed *N*-alkylation of sulfonamides with benzylic alcohols: Catalysis and mechanistic studies. *Advanced Synthesis & Catalysis*, *351*(17), 2949–2958. https://doi.org/10.1002/adsc.200900490

46. Martínez-Asencio, A., Ramón, D. J., & Yus, M. (2010). N-alkylation of poor nucleophilic amine and sulfonamide derivatives with alcohols by a hydrogen autotransfer process catalyzed by copper(ii) acetate. *Tetrahedron Letters, 51*(2), 325–327. https://doi.org/10.1016/j.tetlet.2009.11.009
47. Martínez-Asencio, A., Ramón, D. J., & Yus, M. (2011). N-alkylation of poor nucleophilic amines and derivatives with alcohols by a hydrogen autotransfer process catalyzed by copper(ii) acetate: Scope and mechanistic considerations. *Tetrahedron, 67*(17), 3140–3149. https://doi.org/10.1016/j.tet.2011.02.075
48. Zhao, G.-ming, Liu, H.-ling, Zhang, D.-dan, Huang, X.-ri, & Yang, X. (2014). DFT study on mechanism of N-alkylation of amino derivatives with primary alcohols catalyzed by copper(ii) acetate. *ACS Catalysis, 4*(7), 2231–2240. https://doi.org/10.1021/cs5004678
49. Li, F., Shan, H., Kang, Q., & Chen, L. (2011). Regioselective N-alkylation of 2-aminobenzothiazoles with benzylic alcohols. *Chemical Communications, 47*(17), 5058–5060. https://doi.org/10.1039/c1cc10604j
50. Martínez, R., Ramón, D. J., & Yus, M. (2009). Selective N-monoalkylation of aromatic amines with benzylic alcohols by a hydrogen autotransfer process catalyzed by unmodified magnetite. *Organic & Biomolecular Chemistry, 7*(10), 2176–2181. https://doi.org/10.1039/b901929d
51. Cui, X., Shi, F., Zhang, Y., & Deng, Y. (2010). Fe(II)-catalyzed N-alkylation of sulfonamides with benzylic alcohols. *Tetrahedron Letters, 51*(15), 2048–2051. https://doi.org/10.1016/j.tetlet.2010.02.056
52. Zhao, Y., Foo, S. W., & Saito, S. (2011). Iron/amino acid catalyzed direct N-alkylation of amines with alcohols. *Angewandte Chemie International Edition, 50*(13), 3006–3009. https://doi.org/10.1002/anie.201006660
53. Yan, T., Feringa, B. L., & Barta, K. (2014). Iron catalyzed direct alkylation of amines with alcohols. *Nature Communications, 5*(1), 5602. https://doi.org/10.1038/ncomms6602
54. Gunanathan, C., & Milstein, D. (2008). Selective synthesis of primary amines directly from alcohols and ammonia. *Angewandte Chemie International Edition, 47*(45), 8661–8664. https://doi.org/10.1002/anie.200803229
55. Imm, S., Bähn, S., Neubert, L., Neumann, H., & Beller, M. (2010). An efficient and general synthesis of primary amines by ruthenium-catalyzed amination of secondary alcohols with ammonia. *Angewandte Chemie International Edition, 49*(44), 8126–8129. https://doi.org/10.1002/anie.201002576
56. Pingen, D., Müller, C., & Vogt, D. (2010). Direct amination of secondary alcohols using ammonia. *Angewandte Chemie International Edition, 49*(44), 8130–8133. https://doi.org/10.1002/anie.201002583
57. Imm, S., Bähn, S., Zhang, M., Neubert, L., Neumann, H., Klasovsky, F., Pfeffer, J., Haas, T., & Beller, M. (2011). Improved ruthenium-catalyzed amination of alcohols with ammonia: Synthesis of diamines and amino esters. *Angewandte Chemie International Edition, 50*(33), 7599–7603. https://doi.org/10.1002/anie.201103199
58. Pingen, D., Lutz, M., & Vogt, D. (2014). Mechanistic study on the ruthenium-catalyzed direct amination of alcohols. *Organometallics, 33*(7), 1623–1629. https://doi.org/10.1021/om4011998
59. Ye, X., Plessow, P. N., Brinks, M. K., Schelwies, M., Schaub, T., Rominger, F., Paciello, R., Limbach, M., & Hofmann, P. (2014). Alcohol amination with ammonia catalyzed by an acridine-based ruthenium pincer complex: A mechanistic study. *Journal of the American Chemical Society, 136*(16), 5923–5929. https://doi.org/10.1021/ja409368a
60. Kawahara, R., Fujita, K.-ichi, & Yamaguchi, R. (2010). Multialkylation of aqueous ammonia with alcohols catalyzed by water-soluble cp*ir–ammine complexes. *Journal of the American Chemical Society, 132*(43), 15108–15111. https://doi.org/10.1021/ja107274w
61. Zhang, W., Dong, X., & Zhao, W. (2011). Microwave-assisted solventless reaction of Iridium-catalyzed alkylation of amines with alcohols in the absence of base. *Organic Letters, 13*(19), 5386–5389. https://doi.org/10.1021/ol202281h
62. Yamaguchi, R., Kawagoe, S., Asai, C., & Fujita, K.-ichi. (2007). Selective synthesis of secondary and tertiary amines by cp*iridium-catalyzed multialkylation of ammonium salts with alcohols. *Organic Letters, 10*(2), 181–184. https://doi.org/10.1021/ol702522k
63. Segarra, C., Mas-Marzá, E., Mata, J. A., & Peris, E. (2011). Shvo's catalyst and [ircp*cl2(amidine)] effectively catalyze the formation of tertiary amines from the reaction of primary alcohols and ammonium salts. *Advanced Synthesis & Catalysis, 353*(11–12), 2078–2084. https://doi.org/10.1002/adsc.201100135
64. Hollmann, D. (2014). Advances in asymmetric borrowing hydrogen catalysis. *ChemSusChem, 7*(9), 2411–2413. https://doi.org/10.1002/cssc.201402320
65. Nordstrøm, L. U., & Madsen, R. (2007). Iridium catalyzed synthesis of piperazines from diols. *Chemical Communications, 47*, 5034. https://doi.org/10.1039/b712685a
66. Eka Putra, A., Oe, Y., & Ohta, T. (2013). Ruthenium-catalyzed enantioselective synthesis of β-amino alcohols from 1,2-diols by "borrowing hydrogen". *European Journal of Organic Chemistry, 2013*(27), 6146–6151. https://doi.org/10.1002/ejoc.201300692

67. Miao, L., DiMaggio, S. C., Shu, H., & Trudell, M. L. (2009). Enantioselective syntheses of both enantiomers of noranabasamine. *Organic Letters, 11*(7), 1579–1582. https://doi.org/10.1021/ol9002288
68. Oldenhuis, N. J., Dong, V. M., & Guan, Z. (2014). From racemic alcohols to enantiopure amines: RU-catalyzed diastereoselective amination. *Journal of the American Chemical Society, 136*(36), 12548–12551. https://doi.org/10.1021/ja5058482
69. Feng, C., Liu, Y., Peng, S., Shuai, Q., Deng, G., & Li, C.-J. (2010). Ruthenium-catalyzed tertiary amine formation from nitroarenes and alcohols. *Organic Letters, 12*(21), 4888–4891. https://doi.org/10.1021/ol1020527
70. Cui, X., Zhang, Y., Shi, F., & Deng, Y. (2011). Ruthenium-catalyzed nitro and nitrile compounds coupling with alcohols: Alternative route for N-substituted amine synthesis. *Chemistry: A European Journal, 17*(9), 2587–2591. https://doi.org/10.1002/chem.201003095
71. Pérez, J. M., Cano, R., Yus, M., & Ramón, D. J. (2012). Straightforward synthesis of aromatic imines from alcohols and amines or nitroarenes using an impregnated copper catalyst. *European Journal of Organic Chemistry, 2012*(24), 4548–4554. https://doi.org/10.1002/ejoc.201200319
72. Rana, J., Babu, R., Subaramanian, M., & Balaraman, E. (2018). Ni-catalyzed dehydrogenative coupling of primary and secondary alcohols with methyl-n-heteroaromatics. *Organic Chemistry Frontiers, 5*(22), 3250–3255. https://doi.org/10.1039/c8qo00764k
73. Onodera, G., Nishibayashi, Y., & Uemura, S. (2006). IR- and Ru-catalyzed sequential reactions: Asymmetric α-alkylative reduction of ketones with alcohols. *Angewandte Chemie International Edition, 45*(23), 3819–3822. https://doi.org/10.1002/anie.200600677
74. Rueping, M., & Phapale, V. B. (2012). Effective synthesis of 2,5-disubstituted tetrahydrofurans from glycerol by catalytic alkylation of ketones. *Green Chemistry, 14*(1), 55–57. https://doi.org/10.1039/c1gc15764g
75. Yan, F.-X., Zhang, M., Wang, X.-T., Xie, F., Chen, M.-M., & Jiang, H. (2014). Efficient ruthenium-catalyzed α-alkylation of ketones using pyridyl methanols. *Tetrahedron, 70*(6), 1193–1198. https://doi.org/10.1016/j.tet.2013.12.065
76. Quan, X., Kerdphon, S., & Andersson, P. G. (2015). C-C coupling of ketones with methanol catalyzed by a N-heterocyclic carbene-phosphine Iridium Complex. *Chemistry: A European Journal, 21*(9), 3576–3579. https://doi.org/10.1002/chem.201405990
77. Chan, L. K., Poole, D. L., Shen, D., Healy, M. P., & Donohoe, T. J. (2013). Rhodium-catalyzed ketone methylation using methanol under mild conditions: Formation of α-branched products. *Angewandte Chemie International Edition, 53*(3), 761–765. https://doi.org/10.1002/anie.201307950
78. Iuchi, Y., Obora, Y., & Ishii, Y. (2010). Iridium-catalyzed α-alkylation of acetates with primary alcohols and diols. *Journal of the American Chemical Society, 132*(8), 2536–2537. https://doi.org/10.1021/ja9106989
79. Guo, L., Ma, X., Fang, H., Jia, X., & Huang, Z. (2015). A general and mild catalytic α-alkylation of unactivated esters using alcohols. *Angewandte Chemie International Edition, 54*(13), 4023–4027. https://doi.org/10.1002/anie.201410293
80. Guo, L., Liu, Y., Yao, W., Leng, X., & Huang, Z. (2013). Iridium-catalyzed selective α-alkylation of unactivated amides with primary alcohols. *Organic Letters, 15*(5), 1144–1147. https://doi.org/10.1021/ol400360g
81. Gnanamgari, D., Leung, C. H., Schley, N. D., Hilton, S. T., & Crabtree, R. H. (2008). Alcohol cross-coupling reactions catalyzed by Ru and Ir terpyridine complexes. *Organic & Biomolecular Chemistry, 6*(23), 4442–4445. https://doi.org/10.1039/b815547j
82. Fujita, K.-ichi, Asai, C., Yamaguchi, T., Hanasaka, F., & Yamaguchi, R. (2005). Direct β-alkylation of secondary alcohols with primary alcohols catalyzed by a CP*IR complex. *Organic Letters, 7*(18), 4017–4019. https://doi.org/10.1021/ol051517o
83. Gnanamgari, D., Sauer, E. L., Schley, N. D., Butler, C., Incarvito, C. D., & Crabtree, R. H. (2008). Iridium and ruthenium complexes with chelating N-heterocyclic carbenes: Efficient catalysts for transfer hydrogenation, β-alkylation of alcohols, and N-alkylation of amines. *Organometallics, 28*(1), 321–325. https://doi.org/10.1021/om800821q
84. Martínez, R., Brand, G. J., Ramón, D. J., & Yus, M. (2005). [ru(dmso)4]cl2 catalyzes the α-alkylation of ketones by alcohols. *Tetrahedron Letters, 46*(21), 3683–3686. https://doi.org/10.1016/j.tetlet.2005.03.158
85. Jumde, V. R., Gonsalvi, L., Guerriero, A., Peruzzini, M., & Taddei, M. (2015). A ruthenium-based catalytic system for a mild borrowing-hydrogen process. *European Journal of Organic Chemistry, 2015*(8), 1829–1833. https://doi.org/10.1002/ejoc.201403636
86. Dowson, G. R., Haddow, M. F., Lee, J., Wingad, R. L., & Wass, D. F. (2013). Catalytic conversion of ethanol into an advanced biofuel: Unprecedented selectivity for *N*-butanol. *Angewandte Chemie International Edition, 52*(34), 9005–9008. https://doi.org/10.1002/anie.201303723

87. Obora, Y., Anno, Y., Okamoto, R., Matsu-ura, T., & Ishii, Y. (2011). Iridium-catalyzed reactions of ω-Arylalkanols to α,ω-diarylalkanes. *Angewandte Chemie International Edition, 50*(37), 8618–8622. https://doi.org/10.1002/anie.201104452
88. Ledger, A. E., Slatford, P. A., Lowe, J. P., Mahon, M. F., Whittlesey, M. K., & Williams, J. M. (2009). Ruthenium xantphos complexes in hydrogen transfer processes: Reactivity and mechanistic studies. *Dalton Transactions, 4*, 716–722. https://doi.org/10.1039/b813543f
89. Cami-Kobeci, G., & Williams, J. M. (2004). Conversion of alcohols into N-alkyl anilines via an indirect aza-wittig reaction. *Chemical Communications, 9*, 1072. https://doi.org/10.1039/b402020k
90. Jensen, T., & Madsen, R. (2009). Ruthenium-catalyzed alkylation of oxindole with alcohols. *The Journal of Organic Chemistry, 74*(10), 3990–3992. https://doi.org/10.1021/jo900341w
91. Bartolucci, S., Mari, M., Bedini, A., Piersanti, G., & Spadoni, G. (2015). Iridium-catalyzed direct synthesis of tryptamine derivatives from indoles: Exploiting N-protected β-amino alcohols as alkylating agents. *The Journal of Organic Chemistry, 80*(6), 3217–3222. https://doi.org/10.1021/acs.joc.5b00195
92. Kovalenko, O. O., Lundberg, H., Hübner, D., & Adolfsson, H. (2014). Tandem α-alkylation/asymmetric transfer hydrogenation of acetophenones with primary alcohols. *European Journal of Organic Chemistry, 2014*(30), 6639–6642. https://doi.org/10.1002/ejoc.201403032
93. Suzuki, T., Ishizaka, Y., Ghozati, K., Zhou, D.-Y., Asano, K., & Sasai, H. (2013). Enantioselective multicatalytic synthesis of α-benzyl-β-hydroxyindan-1-ones. *Synthesis, 45*(15), 2134–2136. https://doi.org/10.1055/s-0033-1338479
94. Quintard, A., Constantieux, T., & Rodriguez, J. (2013). An iron/amine-catalyzed cascade process for the enantioselective functionalization of allylic alcohols. *Angewandte Chemie International Edition, 52*(49), 12883–12887. https://doi.org/10.1002/anie.201307295
95. Bower, J. F., Skucas, E., Patman, R. L., & Krische, M. J. (2007). Catalytic C–C coupling via transfer hydrogenation: Reverse prenylation, crotylation, and allylation from the alcohol or aldehyde oxidation level. *Journal of the American Chemical Society, 129*(49), 15134–15135. https://doi.org/10.1021/ja077389b
96. Patman, R. L., Williams, V. M., Bower, J. F., & Krische, M. J. (2008). Carbonyl propargylation from the alcohol or aldehyde oxidation level employing 1,3-enynes as surrogates to preformed Allenylmetal Reagents: A ruthenium-catalyzed C-C bond-forming transfer hydrogenation. *Angewandte Chemie International Edition, 47*(28), 5220–5223. https://doi.org/10.1002/anie.200801359
97. Kim, I. S., Ngai, M.-Y., & Krische, M. J. (2008). Enantioselective iridium-catalyzed carbonyl allylation from the alcohol or aldehyde oxidation level via transfer hydrogenative coupling of allyl acetate: Departure from chirally modified allyl metal reagents in carbonyl addition. *Journal of the American Chemical Society, 130*(44), 14891–14899. https://doi.org/10.1021/ja805722e
98. Chang, D.-H., Lee, D.-Y., Hong, B.-S., Choi, J.-H., & Jun, C.-H. (2003). A new solvent system for recycling catalysts for chelation-assisted hydroacylation of olefins with primary alcohols. *Journal of the American Chemical Society, 126*(2), 424–425. https://doi.org/10.1021/ja038071w
99. Krüger, K., Tillack, A., & Beller, M. (2009). Recent innovative strategies for the synthesis of amines: From C-N bond formation to C-N bond activation. *ChemSusChem, 2*(8), 715–717. https://doi.org/10.1002/cssc.200900121
100. Murahashi, S.-I., Komiya, N., Terai, H., & Nakae, T. (2003). Aerobic ruthenium-catalyzed oxidative cyanation of tertiary amines with sodium cyanide. *Journal of the American Chemical Society, 125*(50), 15312–15313. https://doi.org/10.1021/ja0390303
101. Gu, X.-Q., Chen, W., Morales-Morales, D., & Jensen, C. M. (2002). Dehydrogenation of secondary amines to imines catalyzed by an iridium PCP pincer complex: Initial aliphatic or direct amino dehydrogenation? *Journal of Molecular Catalysis A: Chemical, 189*(1), 119–124. https://doi.org/10.1016/s1381-1169(02)00200-5
102. Hateley, M. J., Schichl, D. A., Kreuzfeld, H.-J., & Beller, M. (2000). Rhodium-catalyzed racemisation of N-acyl α-amino acids. *Tetrahedron Letters, 41*(20), 3821–3824. https://doi.org/10.1016/s0040-4039(00)00539-6
103. Bui-The-Khai, Concilio, C., & Porzi, G. (1981). A facile synthesis of symmetrical secondary amines from primary amines promoted by the homogeneous catalyst rucl2(ph3p)3. *Journal of Organometallic Chemistry, 208*(2), 249–251. https://doi.org/10.1016/s0022-328x(00)82680-7
104. Cho, C. S., Oh, B. H., Kim, J. S., Kim, T.-J., & Shim, S. C. (2000). Synthesis of quinolines via ruthenium-catalyzed amine exchange reaction between anilines and Trialkylamines. *Chemical Communications, 19*, 1885–1886. https://doi.org/10.1039/b005966h
105. Jun, C.-H., Chung, K.-Y., & Hong, J.-B. (2001). C–H and C–C bond activation of primary amines through dehydrogenation and transimination. *Organic Letters, 3*(5), 785–787. https://doi.org/10.1021/ol015563+

106. Xu, W.-wei, Rosini, G. P., Krogh-Jespersen, K., Goldman, A. S., Gupta, M., Jensen, C. M., & Kaska, W. C. (1997). Thermochemical alkane dehydrogenation catalyzed in solution without the use of a hydrogen acceptor. *Chemical Communications, 23*, 2273–2274. https://doi.org/10.1039/a705105k
107. Zhu, K., Achord, P. D., Zhang, X., Krogh-Jespersen, K., & Goldman, A. S. (2004). Highly effective pincer-ligated iridium catalysts for alkane dehydrogenation. DFT calculations of relevant thermodynamic, kinetic, and spectroscopic properties. *Journal of the American Chemical Society, 126*(40), 13044–13053. https://doi.org/10.1021/ja047356l
108. Goldman, A. S., Roy, A. H., Huang, Z., Ahuja, R., Schinski, W., & Brookhart, M. (2006). Catalytic alkane metathesis by tandem alkane dehydrogenation-olefin metathesis. *Science, 312*(5771), 257–261. https://doi.org/10.1126/science.1123787
109. Murahashi, S., Naota, T., Ito, K., Maeda, Y., & Taki, H. (1987). Ruthenium-catalyzed oxidative transformation of alcohols and aldehydes to esters and lactones. *The Journal of Organic Chemistry, 52*(19), 4319–4327. https://doi.org/10.1021/jo00228a032
110. Zhao, J., & Hartwig, J. F. (2005). Acceptorless, neat, ruthenium-catalyzed dehydrogenative cyclization of diols to lactones. *Organometallics, 24*(10), 2441–2446. https://doi.org/10.1021/om048983m
111. Zweifel, T., Naubron, J.-V., & Grützmacher, H. (2009). Catalyzed dehydrogenative coupling of primary alcohols with water, methanol, or amines. *Angewandte Chemie International Edition, 48*(3), 559–563. https://doi.org/10.1002/anie.200804757
112. Gunanathan, C., Shimon, L. J., & Milstein, D. (2009). Direct conversion of alcohols to acetals and H2 catalyzed by an acridine-based ruthenium pincer complex. *Journal of the American Chemical Society, 131*(9), 3146–3147. https://doi.org/10.1021/ja808893g
113. Naota, T., & Murahashi, S.-I. (1991). Ruthenium-catalyzed transformations of amino alcohols to lactams. *Synlett, 1991*(09), 693–694. https://doi.org/10.1055/s-1991-20840
114. Bernskoetter, W. H., & Brookhart, M. (2008). Kinetics and mechanism of Iridium-catalyzed dehydrogenation of primary amines to nitriles. *Organometallics, 27*(9), 2036–2045. https://doi.org/10.1021/om701148t
115. Watson, A. J., Maxwell, A. C., & Williams, J. M. (2009). Ruthenium-catalyzed oxidation of alcohols into amides. *Organic Letters, 11*(12), 2667–2670. https://doi.org/10.1021/ol900723v
116. Blacker, A. J., Farah, M. M., Hall, M. I., Marsden, S. P., Saidi, O., & Williams, J. M. (2009). Synthesis of benzazoles by hydrogen-transfer catalysis. *Organic Letters, 11*(9), 2039–2042. https://doi.org/10.1021/ol900557u
117. Porcheddu, A., Mura, M. G., De Luca, L., Pizzetti, M., & Taddei, M. (2012). From alcohols to indoles: A tandem RU catalyzed hydrogen-transfer Fischer indole synthesis. *Organic Letters, 14*(23), 6112–6115. https://doi.org/10.1021/ol3030956
118. Watson, A. J., Maxwell, A. C., & Williams, J. M. (2012). Ruthenium-catalyzed oxidative synthesis of heterocycles from alcohols. *Organic and Biomolecular Chemistry, 10*(2), 240–243. https://doi.org/10.1039/c1ob06516e
119. Iida, K., Miura, T., Ando, J., & Saito, S. (2013). The dual role of ruthenium and alkali base catalysts in enabling a conceptually new shortcut to *N*-unsubstituted pyrroles through unmasked α-amino aldehydes. *Organic Letters, 15*(7), 1436–1439. https://doi.org/10.1021/ol4001262

3 Transition Metal Catalyst Free Dehydrogenative Organic Synthesis
Role of New Materials, Composites, and Nanomaterials

Prakash Chandra and Syed Shahabuddin

3.1 INTRODUCTION

The acceptorless dehydrogenation (AD) process, which involves the liberation of H_2, has recently drawn considerable attention for the production of environmentally friendly and sustainable chemicals. The use of oxidizing agents or of sacrificial acceptors are avoided in AD processes. As a result, the hydrogen produced during these processes has the potential to be used as a source of energy (Turner 2004; Maeda and Domen 2016). Bond activation and formation archetypally happen at the same time in "one-pot" AD processes, which use one or a maximum of two catalysts. Alkanes, alcohols, and amines are the most often utilized substrates for dehydrogenation to produce alkene, carbonyl compounds (aldehydes, ketones, carboxylic acids, or esters), and reactive imines as reactive intermediates, respectively. Moreover, dehydrogenative N-alkylation of amines and amides utilizing alcohols as alkylating agents has been thoroughly studied to produce water as the sole by-product. This tactic is more atom-economical and ecologically benign. Alcohols are also simpler to handle and store and are more widely accessible than more hazardous organohalide chemicals. Each less reactive substrate is dehydrogenated to generate alkenes, carbonyl compounds, and imines that exhibit superior reactivity. These more reactive intermediates are then further processed in tandem using a "one-pot" method (Dobereiner and Crabtree 2010).

AD reactions are responsible for providing facile access to the diversity of products that alcohol dehydrogenation reactions have valued. The efficient production of multivarious organic compounds through carbon-oxygen, carbon-nitrogen, carbon-sulfur, and carbon-carbon bond formation reactions has been made possible by the effective promotion of AD of alcohols by transition metal (TM)-based homogeneous catalysts (Borthakur, Sau, and Kundu 2022; Liu et al. 2022b; Wang et al. 2021; Subaramanian, Sivakumar, and Balaraman 2021; Siddiki, Toyao, and Shimizu 2018; Choi et al. 2011). Nevertheless, AD processes primarily utilize homogeneous TM catalysts, and there are few examples of heterogeneous catalysts being used in these reactions (Polukeev et al. 2022; Fanara et al. 2021; Polukeev, Abdelaziz, and Wendt 2022; Tocqueville et al. 2022; Qin et al. 2022; Wang et al. 2022; Das et al. 2022). The complex multistep synthetic process and challenging product separation processes required throughout the reactions are the limits of using homogeneous catalysts. Using these homogeneous catalysts in the industrial scale manufacture of commodity chemicals is particularly challenging. For the synthesis of valorized products, AD reactions require highly active, stable, and recyclable heterogeneous catalysts with an easy-to-follow synthetic methodology (Wang and Astruc 2015; Hofmann and Hultzsch 2021). To address the issue, several TMs such as Cu

(Mitsudome et al. 2008), Co (Shimizu et al. 2013), Ni (Chen et al. 2017), Mn (Filonenko et al. 2018; Das et al. 2022), Pt (Kon, Siddiki, and Shimizu 2013), Pd (Nicolau, Tarantino, and Hammond 2019), Ru (Kim, Park, and Park 2006), Au (Fang et al. 2011), Ag (Bayat et al. 2015), and Re have been considered for AD reactions (Sun et al. 2021; Bera, Kabadwal, and Banerjee 2021; Kaźmierczak et al. 2021; Filonenko et al. 2018).

3.2 NANOTECHNOLOGY CATALYSTS FOR HYDROGENATION BUDGE IN ORGANIC SYNTHESIS

Heterogeneous catalysis is quite prevalent in the chemical industry. Approximately 90% of chemical manufacturing relies on the chemical industry, which is 35% of the gross GDP of the world (Armor 2011). Traditional heterogeneous catalysts were created using straightforward techniques with little control over the solid's shape or makeup. This method has the effect of producing not only small metal particles with a wide range of sizes and shapes but also the accompanying array of surface groups. Non-uniform active sites present on the catalyst surface perform non-uniform catalytic activity, which further affects the product selectivity, that is, sterio- and enantioselectivity. Novel nanotechnologies have been developed to well-defined characteristics and applications in catalytic processes. The fruitful mutual interaction between nanotechnology and the surface chemistry of a catalyst is in a very preliminary stage but has led to many exciting developments and can upend chemical manufacturing. Keeping these things in mind, the present chapter focuses on the synthesis of supported TMs such as Mo, V, Cu, Zn, Ni, Co, W, etc. on inorganic nanomaterials such as SiO_2, ZrO_2, TiO_2, Al_2O_3, and CeO_2 micro- or nanospheres that can be utilized in catalytic reactions such as oxidation, epoxidation, esterification, hydroamination, olefin metathesis, oxidative coupling reactions, condensation reactions, etc. Several of these synthetic approaches, including the colloidal and reverse micelle method, the sol-gel solid growth method, and the surface protected etching method, use poly (vinyl pyrrolidone) as a protective agent while utilizing NaOH solution as an etching agent to protect the silica surfaces from etching. The atomic layer deposition (ALD) method includes organic templates grafting on the solid support, making of dumb-bell-, core@shell-, and yolk@shell-related nanoarchitectures, and silylation and other surface derivatization, tethering, and immobilization procedures. A diverse array of heterogeneous TM catalysts prepared by nanotechnology methods have been investigated for dehydrgenation, transfer hydrogenation, and AD reactions.

Diversity of alcohol dehydrogenatively oxidized to carboxylic acid with more than a 90% yield has been performed by manganese ferrite (MnFe-S-Ag) nanoparticles functionalized with 3,3'-thiodipropanoic acid. The MnFe-S-Ag catalyst was prepared via a series of steps involving the application of $FeCl_3.6H_2O$ and $MnCl_2.4H_2O$ as precursors, 3,3'-thiodipropanoic acid as the functionalizing agent, and silver nitrate as silver precursors (Yazdani and Heydari 2020). Mo_2N nanobelts (60-nm wide and 0.5–7.2-μm long) efficiently facilitate the dehydrogenative oxidation of a broad variety of primary/secondary aliphatic, aromatic, heterocyclic, and benzylic alcohols to the corresponding aldehyde/ketones (with up to 16–94% yield). These nanobelts were synthesized via the reaction of α-MoO_3 precursor with ammonia at 850 °C (Li et al. 2014). A 0.6 wt% Au/TiO_2 as photocatalysts promotes the dehydrogenation of linear and branched aliphatic alcohols (with up to 85% 3-phnylpropanal) to form aldehydes and molecular hydrogen via a photon-mediated process of UV photon illumination. The catalytic system effectively promoted the dehydrogenation of a variety of aliphatic alcohols with yields as high as 85% for 3-phenylpropanol. In addition, this catalytic system bears C-Cl and C-C bonds (Shibata et al. 2017). Mono- and bi-metallic copper- and silver-based nanoparticles dispersed over high surface area graphite materials via an impregnation technique have been investigated for the dehydrogenation of alcohols using a continuous flow reactor. Kinetic experiments clearly demonstrate that copper nanoparticles outperform silver nanoparticles for alcohol dehydrgenation with 100% selectivity for acetaldehyde. Both copper- and silver-based monometallic systems quickly undergo deactivation. These mono- and multimetallic catalysts exhibit deactivation at 250 °C because the active sites become obstructed due to coke deposition (Conesa et al. 2019). Rhenium nanocrystalline particles (Re NPs ≈2 nm size) composed of NH_4ReO_4 were prepared via refluxing at 180 °C in only an alcohol medium. During the process, the alcohol became dehydrogenated to form the corresponding carbonyl (aldehyde or ketone) compound. Moreover, molecular hydrogen was generated as the byproduct. These Re nanoparticles successfully help γ-C-H bond activation during the alcohol dehydrogenation process (Yi et al. 2014).

Dehydrogenation of propane to propylene has recently received considerable interest because it provides a cost-effective and effective technique that bridges the gap between supply and demand in the propylene market. Pt-based catalyst has been effectively used for the fast and efficient dehydrogenation of propane to propylene. The low dispersion and poor sintering of the Pt nanoparticles over the surface of the support that applies a severe reaction condition are the major shortcomings associated with these catalytic systems. Pt/Sn-Beta catalysts have been effectively used for the dehydrogenation of propane, which displays high activity, selectivity, and stability during the reaction. The physicochemical characterizations clearly demonstrate that Pt clusters are restricted at the Sn single-site under the zeolitic framework, which permits the uniform dispersion of Pt clusters over the surface of the Sn-beta zeolite. The high activity of Pt/Sn-Beta catalysts at high temperature was mainly because of the strong collaboration between the Pt cluster and Sn-zeolite. The Pt-Sn-beta catalysts exhibit more than 50% propane conversion and more than 99% selectivity for propylene. Moreover, this catalytic system exhibits sluggish deactivation kinetics with a rate constant of 0.006 h^{-1} and a turnover frequency (TOF) as high as 114 s^{-1}. Amongst the various Pt-Sn-Beta compositions, the Pt-$Sn_{2.00}$/Sn-Beta catalyst's highest activity for the propane dehydrogenation occurs at 570 °C (see Figure 3.1) (Xu et al. 2020).

Ag-doped (1 wt%) Zr-containing molecular sieves ZrBEA, ZrMCM-41, and ZrO_2 reinforced on silica effectually encourage the synthesis of 1,3-butadiene to form EtOH. The catalytic systems efficiently encourage the EtOH to butadiene conversion, and the reactivity of these catalytic

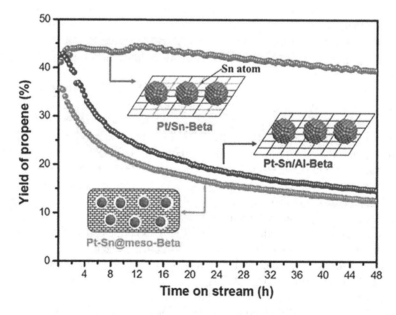

FIGURE 3.1 Pt/Sn-beta-based catalysts effectively promote the dehydrogenation of propane to propylene.
Source: Reproduced with permission (Xu et al. 2020).

systems is as follows: Ag/ZrO$_2$/SiO$_2$ < Ag/ZrMCM-41 < Ag/ZrBEA. The physicochemical characterizations of these catalytic systems using multiple physicochemical techniques clearly demonstrate that the catalytic performance was influenced by the Lewis acid sites that existed over the surface of the catalyst. Ag/ZrBEA (Si/Zr = 100), where EtOH conversion was 48% and butadiene selectivity was 56%, was achieved (Sushkevich, Ivanova, and Taarning 2015). For the ethanol dehydrogenation necessary to produce 1,3-butadiene, a Ta(V)-doped Si-BEA zeolite framework system integrating Ag, Cu, and Zn was employed as the heterogeneous catalytic material. The support's acid-base characteristics were crucial for the 1,3-butadiene synthesis. In the following order, the catalytic system efficiently helped the transformation of EtOH into butadiene: TaSi-BEA, ZnTaSi-BEA, AgTaSi-BEA, and Cu TaSi-BEA. At T = 598 K with WHSV = 0.5 h^{-1}, the best catalytic system promoted a 73% butadiene yield and 88% ethanol conversion (Kyriienko et al. 2017).

By doping metal (M=Ag, Cu, Ni) over oxide catalysts (MOx=MgO, ZrO$_2$, Nb$_2$O$_5$, TiO$_2$, Al$_2$O$_3$), EtOH dehydration/dehydrogenation to 1,3-butadiene was fruitfully accomplished. Through the insipient wetness impregnation of silica gel with a metal nitrate solution in water, all of the catalytic materials were created. Studies on the reaction kinetics clearly demonstrate that the Ag/ZrO$_2$/SiO$_2$ catalyst has good selectivity for butadiene and ethanol conversions (2–90%) (Sushkevich et al. 2014). An Au/MgO-SiO2 catalyst was created using two distinct techniques: (a) wet impregnation of MgO over a silica surface; and (b) deposition-precipitation of gold nanoparticles over MgO-SiO$_2$. Moreover, the resultant Au/Mg-SiO$_2$ was employed as an excellent catalytic material for the transformation of EtOH into 1,3-butadiene, butenes, ethylene, and diethyl ether with great selectivity. The physicochemical analysis of these materials demonstrates that the activity and selectivity of Au/MgOSiO$_2$ are a result of an appropriate balance between acid-base sites, which is noteworthy for the structure-activity relationship. Gold nanoparticles controlled the redox properties, whereas magnesium silicate hydrate was responsible for the acid-base properties in the catalytic system. Compared to production-based 1,3-butadiene manufacturing, the aforementioned catalytic device helped convert ethanol into 1,3-butadiene while reducing greenhouse gas emissions by 155% (Shylesh et al. 2016). At an ethanol flow rate of 2.45 g1,3-BD gcat1 h1, Zn(II) and Ta(V) catalysts supported on TUD-1 (ZnTa-TUD-1)-based mesoporous materials efficiently promoted the dehydrogenation of ethanol with up to a 94% conversion and selectivity for 1,3-butadiene as high as 73%. For 60 hours on stream, the catalytic system successfully encouraged the dehydrogenation and dehydration of ethanol (Pomalaza et al. 2018). 1,3-butadine was produced from

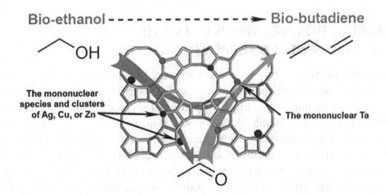

FIGURE 3.2 Deposition of silver nanoparticles over molecular sieves composed of zirconium boosts the dehydrogenation of butadiene generated from biological sources.

Source: **Reproduced with permission (Kyriienko et al. 2017).**

ethanol using a silver/magnesium-silica catalyst. For the 1,3-butadiene synthesis in the catalytic system, Ag and the type of silica supply were both investigated. The experimental results unmistakably show that for the best 1,3-butadiene selectivity, a suitable ratio of active sites was essential. The selectivity for 1,3-butadiene was increased in the catalytic system by 1–2% silver loadings with a Mg/Si ratio of 2. Moreover, the 1,3-butadien generation, ethanol production, and aldol condensation stages were optimally promoted by silver species and basic sites made up of two Mg-O units and basic OH groups, respectively. The 1,3-butadiene yield was increased by the catalyst surface's aforementioned distinguishing characteristics (Janssens et al. 2015). Using a Pt nanocluster over the metal oxides and a reduction with H2 at 500 °C, 2-octanol was reduced in the liquid phase. The acid-base properties of the oxide support, such as -Al_2O_3 amphoteric oxides with outstanding catalytic performance, were important for the dehydrogenation of 2-octanol's catalytic performance. The effect of the TOF on a variety of Pt/Al_2O_3 catalysts made up of various Pt particle shapes is evident from kinetic performance. The catalytic performance unequivocally demonstrates that the corroborative interaction between the metal/support interface and the low-coordinated Pt sites (Pt nanoparticles of 1.4 nm) successfully promoted the AD of 2-octanol. Other TMs, such as Co, Ni, Cu, Ru, Rh, Pd, Ag, Re, Ir, and Au, were also examined for the AD reaction, but they fell short of the Pt-based catalyst system's catalytic performance. A Lewis acid ($Al^{\delta+}$) and Lewis base (Al-O-) pair distributed over the alumina surface produced an alkoxide on the $Al^{\delta+}$ site by the Al-O proton abstraction, while Pt0 assisted the dissociation of the C-H bond to produce Pt-H and ketone and released H_2 gas through the protolysis of the Pt-H bond by the adjacent proton. In situ infrared (IR) analysis supports these findings. Studies on the mechanisms show that Pt is present across the acid-base bifunctional support (Al2O3) (see Figure 24) (Kon, Hakim Siddiki, and Shimizu 2013).

3.3 HETEROGENIZATION OF HOMOGENEOUS CATALYSTS FOR DEHYDROGENATION REACTIONS

Researchers from all over the world have given close attention to the heterogenization of homogeneous TM catalysts because such a system might combine the most beneficial characteristics of both homogeneous and heterogeneous systems. Together with the attained high activity and selectivity, the catalyst's ability to be recycled from the final product would be made easier. Recyclability of the often expensive catalysts is a key priority in industrial applications; hence, the problem of immobilizing catalysts operating in homogeneous phases is being thoroughly researched. By using the following techniques, homogeneous catalysts can be heterogenized.

3.4 ANCHORING HOMOGENEOUS CATALYSTS OVER HETEROGENEOUS SUPPORT

Anchoring the homogeneous catalyst on solid support by indirectly grafting, that is, attaching on heterogeneous support (silica supports as MCM-41, SBA-15, silica microsphers, etc.) by employing one or more ligands, for example, silylation of surface Si-OH using a (3-aminopropyl) trimethoxysilane–tetramethoxysilane (APTMS) reaction, is quickly gaining popularity as the most effective method for modifying the surface reactivity of such materials or for adjusting other surface characteristics including polarity, hydrophobicity, or hydrophilicity. By grafting [3-(2-aminoethyl) aminopropyl] triethoxysilane over a $Fe_3O_4@SiO_2$ surface and then reacting palladium acetate, a NHPd(0)@MNP-based catalytic system was created. The tandem process of dehydrogenation followed by Knoevenagel condensation with up to 9% yield for benzalmalononitriles is promoted by the bifunctional NHPd(0)@MNP-based catalytic system. The six cycles of the recyclable, heterogeneous catalytic material produced consistent results (Yuan et al. 2020).

3.5 DIRECT GRAFTING OF METAL COMPLEXES

The direct grafting of metal complexes is another method employed, but it drastically changes the ligands' environment around the metal centre, which results in a complete loss in identity of the metal centre and the generation of a new heterogenized metal complex quite different from the original metal complex.

By using IR, high field solid-state NMR, EXAFS, and elemental analysis, the interactions of Ga(i-Bu)$_3$ (i-Bu = CH$_2$CH(CH$_3$)$_2$) with the dehydrated and partially dehydroxylated surfaces of alumina (Al$_2$O$_3$-500) and silica (SiO$_2$-700) were investigated. By selectively protonolyzing at specific surface hydroxyl groups during grafting onto Al$_2$O$_3$-500, mononuclear [(AlO)Ga(i-Bu)$_2$L] sites—the main surface organometallic entities—are formed. In contrast, grafting silica results in a dinuclear species [(SiO)$_2$Ga$_2$(i-Bu)$_3$] by a process that combines protonolysis and the transfer of an isobutyl group to Si. By examining the WT-EXAFS, more proof of the difference in nuclearity was found. Compared to their dinuclear silica-supported counterparts, the mononuclear alumina-supported Ga sites exhibit significantly greater activity in the dehydrogenation of propane. Al-O-Ga bonds may be necessary for the propane dehydrogenation reaction in order to encourage heterolytic C-H bond activation. The effect of the catalyst diluent is important under the reaction circumstances and must be carefully examined to attribute reactivity appropriately, according to comparisons with benchmark catalysts and similar systems (see Figure 3.3) (Szeto et al. 2018).

Non-oxidative propane dehydrogenation was performed using isolated Fe^{II} and iron oxide particles, and metallic nanoparticles were used on silica. It was discovered that the open cyclopentadienide iron complex, bis(2,4-dimethyl-1,3-pentadienide) iron(II) or Fe(oCp)$_2$, was grafted onto silica to create an isolated Fe^{II} species, which was the most selective catalyst. The first grafting of iron proceeds by one surface hydroxyl Si-OH reacting with Fe(oCp)$_2$ to liberate one diene ligand (oCpH), resulting in a SiO$_2$-bound Fe^{II}(oCp) species, 1-FeoCp, according to several physicochemical characterization techniques. The residual diene ligand is lost during a subsequent reaction with H$_2$ at 400 °C, which also causes the production of 1-C, a nanoscale iron oxide cluster. At 650 °C, these Fe oxide clusters disperse, resulting in the formation of 1-Fe^{II}, an isolated, ligand-free Fe^{II} on silica that is catalytically active and extremely selective (99%) for the conversion of propane into propene. By using in situ XANES, there is no indication of metallic Fe under reaction circumstances. By grafting Fe[N(SiMe$_3$)$_2$]$_2$ onto silica, 2-FeN*, and reducing it to 650 °C in hydrogen, metallic Fe nanoparticles, 2-NP-Fe^0, were created independently for comparison. The Fe NPs had low selectivity (14%) for propene but were very active for converting propane. Low activity can be seen in independently created Fe oxide clusters on silica. These discoveries reveal that isolated Fe^{II} sites are responsible for the selective propane dehydrogenation (see Figure 3.4) (Hu et al. 2015).

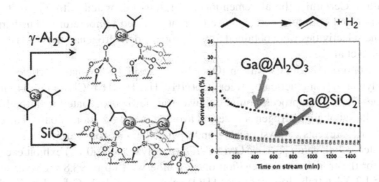

FIGURE 3.3 Direct grafting of gallium over alumina surface for propane dehydrogenation reaction.

Source: Reproduced with permission (Szeto et al. 2018).

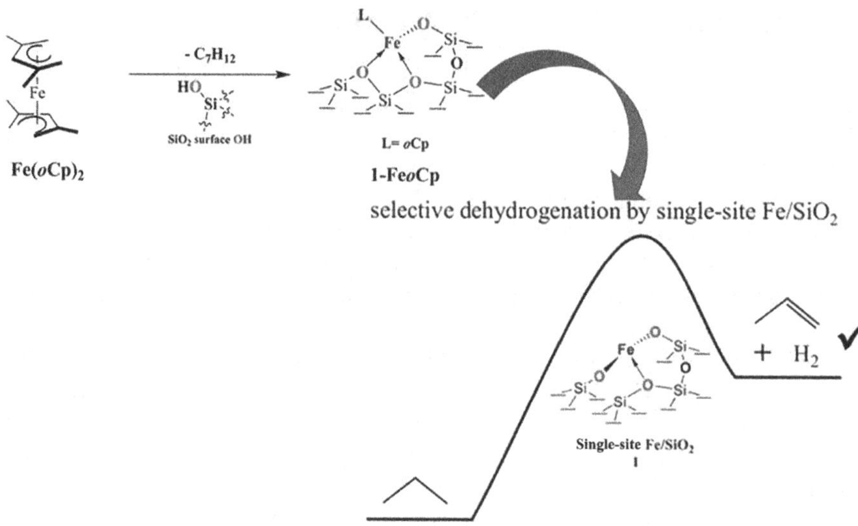

FIGURE 3.4 Single-site Fe^{II} iron oxide particles grafting over a silica surface of the propane dehydrogenation reactions.

Source: Reproduced with permission (Hu et al. 2015).

3.6 ENCAPSULATION OF THE CATALYSTS

Temperature has a remarkable influence on catalytic performance. At very high temperatures, catalyst are prone to sintering. Therefore, encapsulation of nanoparticles is an efficient method to stabilize them to utilize them under harsh reaction conditions such as performing the reactions at a high temperature under a highly corrosive environment. Magnetic nanoparticles (MagNPs) that can facilitate catalysis by acting as heating agents have recently emerged as a fascinating technique to perform challenging reactions. Nevertheless, the foremost restraint is the limited durability of the catalysts functioning under harsh conditions. Customarily, in temperatures greater than 500 °C, a noteworthy sintering of MagNPs occurs. Cobalt-based magnetic nanocatalysts [FeCo@C or composite catalyst FeCo@C:Co@C (2:1)] encapsulated under a carbon matrix have been used as efficient catalysts for propane dehydrogenation. These materials exhibit excellent thermal stability at high temperature. Certainly, the aforementioned catalysts decorated with Ni or Pt–Sn displayed decent constancy and optimum performance even at elevated temperatures. Furthermore, steady conversions and selectivities were obtained for the propane dehydrogenation (PDH) (see Figure 3.5) (Martínez-Prieto et al. 2020).

Ordered mesoporous N-doped carbon (OMNC) encapsulates Co nanoparticles (NPs) synthesized via a polymerization reaction between [Co(NH$_2$CH$_2$CH$_2$NH$_2$)$_2$]Cl$_2$ and CCl4 via a hard template technique. During the process, quinoline was transfer hydrogenated with the help of formic acid as a facile liquid phase hydrogen source and avoided the application of base. The encapsulation of a carbon matrix prevents catalyst poising under an acidic environment in the reaction medium. The encapsulated cobalt catalysts Co@OMNC-700 (pyrolyzed at 700 °C) exhibit excellent catalyst performance for the transfer hydrogenation of quinoline with up to 98.8% conversion and >99% selectivity for 1,2,3,4-tetrahydroquinoline (THQ) selectivity at 140 °C for 4 h. Furthermore, the catalyst was recycled for 5 cycles with a significant dip in catalytic performance (see Figure 3.6) (Li et al. 2018).

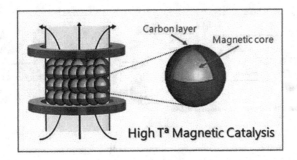

FIGURE 3.5 Cobalt-based magnetic nanoparticles as the catalytic materials for the propane dehydrogenation reaction.

Source: Reproduced with permission (Martínez-Prieto et al. 2020).

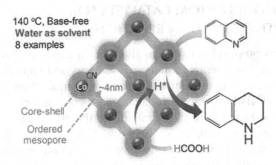

FIGURE 3.6 Transfer hydrogenation of quinoline with formic acid as a hydrogen source and a cobalt-based nanocatalyst.

Source: Reproduced with permission (Li et al. 2018).

3.7 IONIC LIQUID ASSISTED ORGANIC TRANSFORMATION

Immobilization is achieved by anchoring the homogeneous catalyst in a nonvolatile and hydrophilic solvent film deposited on a solid surface. Reactants and products are either in the gas phase or in two immiscible solvents. These are the so-called 'supported liquid (aqueous) phase catalysts' (SLPCs). Cations and the anions components in the ionic liquids along with their hygroscopicity hold the key in altering their performance in catalysis. Ionic liquid mixtures have also been successfully applied to raise the catalyst's efficiency. Rh-NPs that are water-soluble are created by combining imidazole-based ionic liquids, $(NH_4)_3RhCl_6$, trimethylbenzene (TMB), and $NABH_4$ as a reducing agent. With good to exceptional yields, the resultant catalytic system effectively promotes the AD of a variety of alcohols to the corresponding carbonyl compounds (up to 95% yields) (see Figure 3.7) (Yao, Zhao, and Lee 2017).

An ionic copolymer (DIM-AN) made of dicationic imidazole ionic liquid and acrylonitrile was used in a two-step procedure to create the PMo@DIM-AN$_{400/800}$ catalyst, which involved carbonization at 400 °C under aerobic circumstances and 800 °C in Ar. The produced catalytic material efficiently promotes the dehydrogenation of a wide range of alcohols without the use of oxygen. Without changing the crystal shape, the obtained catalytic material was employed for the BnOH dehydrogenation for five consecutive cycles. With 100% selectivity for benzaldehyde, the catalytic system effectively promoted total conversion of the alcohols. The catalytic system also encouraged the conversion of a variety of alcohols to their corresponding carbonyl molecules (Leng et al. 2017).

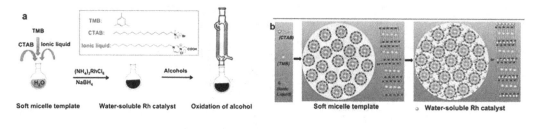

FIGURE 3.7 (a-b) Soft micelle approach for the synthesis of Rh-nanoparticles; (c-d) HRTEM picture of Rh-nanoparticles.

Source: Replicated with permission (Yao, Zhao, and Lee 2017).

3.8 SINGLE AND DOUBLE ATOM CATALYSTS FOR TRANSFER HYDROGENATION REACTIONS

The focus of the present proposal is to design single atom catalysts (SACs) and nanocatalysts with the help of recent advancements in materials sciences and nanotechnologies. SACs provide complete metal dispersion to maximize the metal utilization; therefore, they have emerged as a novel class of materials for catalytic organic transformation and have great potential for industrial catalysis in the near future. SACs maximize the efficiency of the metal atom used. With well-defined and uniform single atom dispersion, SACs definitely have great potential for achieving high activity and selectivity for the reduction of an olefinic double bond. Currently, the focus of the contemporary research is towards the development of a scalable and economically viable approach for the synthesis of highly active, selective, and stable technology for the synthesis of SACs. The size of the metal particles is the detrimental factor for the performance of the heterogeneous catalysts because the specific catalytic activity increases with the low-coordinated metal atoms that often function as the catalytically active sites. The specific catalytic performance per metal atom is enhanced by desizing the metal NPs. Appropriate support is necessary to prevent the aggregation of these small clusters with an increased shelf-life (Polukeev et al. 2022).

The dehydrogenative oxidation of ethanol to acetaldehyde and alcohols was efficiently enhanced by NiCu single atom alloy (SAA) NPs and silica. The NiCu SAA's catalytic activity and selectivity were on par with those of monometallic copper and bimetallic alloys such as PtCu and PdCu SAAs. In the flow reactor, the catalytic reaction was carried out under reasonable reaction conditions. The ethanolic O-H bond cleavage employing ethoxy intermediates over metal catalysts is obviously facilitated by silica as the support, according to mechanistic investigations using in-situ DRIFTS tests. The kinetic tests unequivocally demonstrate that Ni in Cu helps to decrease the rate-limiting stage of C-H bond activation compared to monometallic copper catalysts. The comparable Pt and Pd atoms, however, were less successful in reducing the catalytic performance under the same reaction circumstances (Shan et al. 2018).

Dehydrogenation of alcohols to the corresponding carbonyl compound is facilitated by iron- and cobalt-based SACs and FeCo-based double atom catalysts (Fe-Co-DACs). The FeCo-DAC, which is 100% metal-dispersed and represented by the $FeCoN_6(OH)$ group, was created by heating a mixture of $FeCl_2·4H_2O$, $CoCl_2·6H_2O$, and 1,10-phenanthroline to 700 °C in a nitrogen environment. Experimental tests utilizing o-xylene as the solvent at 140 °C in an inert argon atmosphere show 99% benzyl alcohol conversion with 99% yield for benzaldehyde. In addition, a variety of primary, secondary, and substituted benzylic alcohols were efficiently converted by the catalytic material with good to outstanding yield (see Figure 3.8) (Liu et al. 2022a).

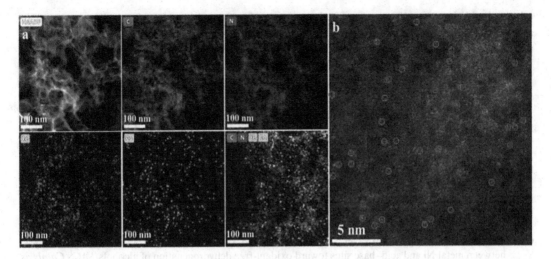

FIGURE 3.8 (a) HAADF image and EDAX image of FeCo-DAC; (b) The red circle represents the spherical abberation correction of the HAADF-STEM image.

Source: Reproduced with permission (Liu et al. 2022a).

3.9 CONCLUSIONS

Accordingly, carbon-carbon and carbon-heteroatom bond creation by TM-based heterogeneous catalysts efficiently promoted a variety of dehydrogenation and hydrogen bond processes without introducing oxidant-useful chemicals. The primary benefit of the AD method is the atom-economic route for the synthesis of important compounds from inexpensive and easily accessible substrate molecules. Furthermore, heterogeneous catalysts made of noble and non-noble metals offer an alternate method for the reagent- or additive-free synthesis of complex organic compounds. Acid/base oxides with multifunctional metal loading are essential components in the production of chemicals via the environmentally friendly route. The higher alcohols were synthesized, and the Knoevenagel condensation processes occurred as a result of the oxidant-free dehydrogenative techniques' effective promotion of C-C bond formation. Moreover, hydrogen is a renewable, cost-effective source of energy. For environmentally friendly, cost-effective, and one-pot tandem synthesis, a wide range of carbonyl compounds, including aldehydes, ketones, carboxylic acids, esters, and a variety of heterocyclic compounds, including pyrrolidine, pyrazine, indoles, quinolines, quinoxalines, benzimidazole, benzothiazole, etc., can be produced. Moreover, the dehydration and dehydrogenation cascade mechanism effectively promotes the synthesis of 1, 2-butadine and olefin.

3.10 OUTLOOK

Despite considerable progressions in TM-catalyzed AD reactions, there are still a number of questions that need to be resolved in the ongoing field research. It is crucial to solve the fundamental information gap about the molecular pathway for the AD of organic molecules at the interface between the metal and support. These mechanistic laws help create new heterogeneous catalysts for extremely effectual AD reactions that are selective. For the enantioselective heterocyclic synthesis for the pharmaceutical industry, effective heterogenous TM catalysts are highly anticipated. Moreover, the process's hydrogen can be utilized to create liquid organic hydrogen carriers (LOHCs), which is a potential first step towards clean, sustainable, and green energy storage. In addition, the hydrogen produced by the AD reaction may be employed for the hydrogenation of other valuable heterocyclic compounds that serve as significant pharmacophores. The synthesis of a variety of valuable medications can also be researched using the novel synthetic methods incorporating AD reactions.

REFERENCES

Armor, John N. 2011. "A history of industrial catalysis." *Catalysis Today* 163 (1):3–9.

Bayat, Ahmad, Mehdi Shakourian-Fard, Nona Ehyaei, and Mohammad Mahmoodi Hashemi. 2015. "Silver nanoparticles supported on silica-coated ferrite as magnetic and reusable catalysts for oxidant-free alcohol dehydrogenation." *RSC Advances* 5 (29):22503–22509.

Bera, Sourajit, Lalit Mohan Kabadwal, and Debasis Banerjee. 2021. "Recent advances in transition metal-catalyzed (1,n) annulation using (de)-hydrogenative coupling with alcohols." *Chemical Communications* 57 (77):9807–9819. https://doi.org/10.1039/D1CC03404A.

Borthakur, Ishani, Anirban Sau, and Sabuj Kundu. 2022. "Cobalt-catalyzed dehydrogenative functionalization of alcohols: Progress and future prospect." *Coordination Chemistry Reviews* 451:214257.

Chen, Hao, Shan He, Ming Xu, Min Wei, David G. Evans, and Xue Duan. 2017. "Promoted synergic catalysis between metal Ni and acid–base sites toward oxidant-free dehydrogenation of alcohols." *ACS Catalysis* 7 (4):2735–2743.

Choi, Jongwook, Amy H. Roy MacArthur, Maurice Brookhart, and Alan S. Goldman. 2011. "Dehydrogenation and related reactions catalyzed by iridium pincer complexes." *Chemical Reviews* 111 (3):1761–1779.

Conesa, J. M., M. V. Morales, C. López-Olmos, I. Rodríguez-Ramos, and A. Guerrero-Ruiz. 2019. "Comparative study of Cu, Ag and Ag-Cu catalysts over graphite in the ethanol dehydrogenation reaction: Catalytic activity, deactivation and regeneration." *Applied Catalysis A: General* 576:54–64. https://doi.org/10.1016/j.apcata.2019.02.031.

Das, Kuhali, Satyadeep Waiba, Akash Jana, and Biplab Maji. 2022. "Manganese-catalyzed hydrogenation, dehydrogenation, and hydroelementation reactions." *Chemical Society Reviews* 51 (11):4386–4464.

Dobereiner, Graham E., and Robert H. Crabtree. 2010. "Dehydrogenation as a substrate-activating strategy in homogeneous transition-metal catalysis." *Chemical Reviews* 110 (2):681–703.

Fanara, Paul M., Vipulan Vigneswaran, Parami S. Gunasekera, Samantha N. MacMillan, and David C. Lacy. 2021. "Reversible photoisomerization in a Ru cis-dihydride catalyst accessed through atypical metal–ligand cooperative H2 activation: Photoenhanced acceptorless alcohol dehydrogenation." *Organometallics* 41 (1):93–98.

Fang, Wenhao, Jiashu Chen, Qinghong Zhang, Weiping Deng, and Ye Wang. 2011. "Hydrotalcite-supported gold catalyst for the oxidant-free dehydrogenation of benzyl alcohol: Studies on support and gold size effects." *Chemistry: A European Journal* 17 (4):1247–1256.

Filonenko, Georgy A., Robbert van Putten, Emiel J. M. Hensen, and Evgeny A. Pidko. 2018. "Catalytic (de)hydrogenation promoted by non-precious metals—Co, Fe and Mn: Recent advances in an emerging field." *Chemical Society Reviews* 47 (4):1459–1483. https://doi.org/10.1039/C7CS00334J.

Hofmann, Natalie, and Kai C. Hultzsch. 2021. "Borrowing hydrogen and acceptorless dehydrogenative coupling in the multicomponent synthesis of N-heterocycles: A comparison between base and noble metal catalysis." *European Journal of Organic Chemistry* 2021 (46):6206–6223. https://doi.org/10.1002/ejoc.202100695.

Hu, Bo, Neil M. Schweitzer, Guanghui Zhang, Steven J. Kraft, David J. Childers, Michael P. Lanci, Jeffrey T. Miller, and Adam S. Hock. 2015. "Isolated FeII on silica as a selective propane dehydrogenation catalyst." *ACS Catalysis* 5 (6):3494–3503. https://doi.org/10.1021/acscatal.5b00248.

Janssens, Wout, Ekaterina V. Makshina, Pieter Vanelderen, Filip De Clippel, Kristof Houthoofd, Stef Kerkhofs, Johan A. Martens, Pierre A. Jacobs, and Bert F. Sels. 2015. "Ternary Ag/MgO-SiO2 catalysts for the conversion of ethanol into butadiene." *ChemSusChem* 8 (6):994–1008. https://doi.org/10.1002/cssc.201402894.

Kaźmierczak, Kamila, Aliyu Salisu, Catherine Pinel, Michèle Besson, Carine Michel, and Noémie Perret. 2021. "Activity of heterogeneous supported Cu and Ru catalysts in acceptor-less alcohol dehydrogenation." *Catalysis Communications* 148:106179. https://doi.org/10.1016/j.catcom.2020.106179.

Kim, Won-Hee, In Soo Park, and Jaiwook Park. 2006. "Acceptor-free alcohol dehydrogenation by recyclable ruthenium catalyst." *Organic Letters* 8 (12):2543–2545.

Kon, Kenichi, S. M. A. Hakim Siddiki, and Ken-ichi Shimizu. 2013. "Size- and support-dependent Pt nanocluster catalysis for oxidant-free dehydrogenation of alcohols." *Journal of Catalysis* 304:63–71. https://doi.org/10.1016/j.jcat.2013.04.003.

Kyriienko, Pavlo I., Olga V. Larina, Sergiy O. Soloviev, Svitlana M. Orlyk, Christophe Calers, and Stanislaw Dzwigaj. 2017. "Ethanol conversion into 1,3-butadiene by the lebedev method over MTaSiBEA zeolites (M = Ag, Cu, Zn)." *ACS Sustainable Chemistry & Engineering* 5 (3):2075–2083. https://doi.org/10.1021/acssuschemeng.6b01728.

Leng, Yan, Jingjing Li, Chenjun Zhang, Pingping Jiang, Yue Li, Yuchen Jiang, and Shengyu Du. 2017. "N-Doped carbon encapsulated molybdenum carbide as an efficient catalyst for oxidant-free dehydrogenation of alcohols." *Journal of Materials Chemistry A* 5 (33):17580–17588. https://doi.org/10.1039/C7TA04763K.

Li, Guoqiang, Huanhuan Yang, Haifu Zhang, Zhiyuan Qi, Minda Chen, Wei Hu, Lihong Tian, Renfeng Nie, and Wenyu Huang. 2018. "Encapsulation of nonprecious metal into ordered mesoporous N-doped carbon for efficient quinoline transfer hydrogenation with formic acid." *ACS Catalysis* 8 (9):8396–8405. https://doi.org/10.1021/acscatal.8b01404.

Li, Zhongcheng, Chunhui Chen, Ensheng Zhan, Na Ta, and Wenjie Shen. 2014. "Mo2N nanobelts for dehydrogenation of aromatic alcohols." *Catalysis Communications* 51:58–62. https://doi.org/10.1016/j.catcom.2014.03.029.

Liu, Ce, Teng Li, Xingchao Dai, Jian Zhao, Dongcheng He, Guomin Li, Bin Wang, and Xinjiang Cui. 2022a. "Catalytic activity enhancement on alcohol dehydrogenation via directing reaction pathways from single- to double-atom catalysis." *Journal of the American Chemical Society* 144 (11):4913–4924. https://doi.org/10.1021/jacs.1c12705.

Liu, Ce, Teng Li, Xingchao Dai, Jian Zhao, Dongcheng He, Guomin Li, Bin Wang, and Xinjiang Cui. 2022b. "Catalytic activity enhancement on alcohol dehydrogenation via directing reaction pathways from single-to double-atom catalysis." *Journal of the American Chemical Society* 144 (11):4913–4924.

Maeda, Kazuhiko, and Kazunari Domen. 2016. "Development of novel photocatalyst and cocatalyst materials for water splitting under visible light." *Bulletin of the Chemical Society of Japan* 89 (6):627–648.

Martínez-Prieto, Luis M., Julien Marbaix, Juan M. Asensio, Christian Cerezo-Navarrete, Pier-Francesco Fazzini, Katerina Soulantica, Bruno Chaudret, and Avelino Corma. 2020. "Ultrastable magnetic nanoparticles encapsulated in carbon for magnetically induced catalysis." *ACS Applied Nano Materials* 3 (7):7076–7087. https://doi.org/10.1021/acsanm.0c01392.

Mitsudome, Takato, Yusuke Mikami, Kaori Ebata, Tomoo Mizugaki, Koichiro Jitsukawa, and Kiyotomi Kaneda. 2008. "Copper nanoparticles on hydrotalcite as a heterogeneous catalyst for oxidant-free dehydrogenation of alcohols." *Chemical Communications* 39:4804–4806. https://doi.org/10.1039/B809012B.

Nicolau, Guillermo, Giulia Tarantino, and Ceri Hammond. 2019. "Acceptorless alcohol dehydrogenation catalysed by Pd/C." *ChemSusChem* 12 (22):4953–4961.

Polukeev, Alexey V., Omar Y. Abdelaziz, and Ola F. Wendt. 2022. "Combined experimental and computational study of the mechanism of acceptorless alcohol dehydrogenation by POCOP iridium pincer complexes." *Organometallics* 41 (7):859–873.

Polukeev, Alexey V., Reine Wallenberg, Jens Uhlig, Christian P. Hulteberg, and Ola F. Wendt. 2022. "Iridium-catalyzed dehydrogenation in a continuous flow reactor for practical on-board hydrogen generation from liquid organic hydrogen carriers." *ChemSusChem* 15 (8):e202200085.

Pomalaza, G., G. Vofo, M. Capron, and F. Dumeignil. 2018. "ZnTa-TUD-1 as an easily prepared, highly efficient catalyst for the selective conversion of ethanol to 1,3-butadiene." *Green Chemistry* 20 (14):3203–3209. https://doi.org/10.1039/C8GC01211C.

Qin, Yuhuan, Mingming Hao, Zhengxin Ding, and Zhaohui Li. 2022. "Pt@MIL-101(Fe) for efficient visible light initiated coproduction of benzimidazoles and hydrogen from the reaction between o-Phenylenediamines and alcohols." *Journal of Catalysis* 410:156–163. https://doi.org/10.1016/j.jcat.2022.04.023.

Shan, Junjun, Jilei Liu, Mengwei Li, Sylvia Lustig, Sungsik Lee, and Maria Flytzani-Stephanopoulos. 2018. "NiCu single atom alloys catalyze the CH bond activation in the selective non- oxidative ethanol dehydrogenation reaction." *Applied Catalysis B: Environmental* 226:534–543. https://doi.org/10.1016/j.apcatb.2017.12.059.

Shibata, Masaki, Ryoko Nagata, Susumu Saito, and Hiroshi Naka. 2017. "Dehydrogenation of primary aliphatic alcohols by Au/TiO$_2$ photocatalysts." *Chemistry Letters* 46 (4):580–582.

Shimizu, Ken-ichi, Kenichi Kon, Mayumi Seto, Katsuya Shimura, Hiroshi Yamazaki, and Junko N. Kondo. 2013. "Heterogeneous cobalt catalysts for the acceptorless dehydrogenation of alcohols." *Green Chemistry* 15 (2):418–424. https://doi.org/10.1039/C2GC36555C.

Shylesh, Sankaranarayanapillai, Amit A. Gokhale, Corinne D. Scown, Daeyoup Kim, Christopher R. Ho, and Alexis T. Bell. 2016. "From sugars to wheels: The conversion of ethanol to 1,3-butadiene over metal-promoted magnesia-silicate catalysts." *ChemSusChem* 9 (12):1462–1472. https://doi.org/10.1002/cssc.201600195.

Siddiki, S. M. A. Hakim, Takashi Toyao, and Ken-ichi Shimizu. 2018. "Acceptorless dehydrogenative coupling reactions with alcohols over heterogeneous catalysts." *Green Chemistry* 20 (13):2933–2952.

Subaramanian, Murugan, Ganesan Sivakumar, and Ekambaram Balaraman. 2021. "First-row transition-metal catalyzed acceptorless dehydrogenation and related reactions: A personal account." *The Chemical Record* 21 (12):3839–3871. https://doi.org/10.1002/tcr.202100165.

Sun, Kangkang, Hongbin Shan, Guo-Ping Lu, Chun Cai, and Matthias Beller. 2021. "Synthesis of N-heterocycles via oxidant-free dehydrocyclization of alcohols using heterogeneous catalysts." *Angewandte Chemie International Edition* 60 (48):25188–25202. https://doi.org/10.1002/anie.202104979.

Sushkevich, Vitaly L., Irina I. Ivanova, Vitaly V. Ordomsky, and Esben Taarning. 2014. "Design of a metal-promoted oxide catalyst for the selective synthesis of butadiene from ethanol." *ChemSusChem* 7 (9):2527–2536. https://doi.org/10.1002/cssc.201402346.

Sushkevich, Vitaly L., Irina I. Ivanova, and Esben Taarning. 2015. "Ethanol conversion into butadiene over Zr-containing molecular sieves doped with silver." *Green Chemistry* 17 (4):2552–2559. https://doi.org/10.1039/C4GC02202E.

Szeto, Kai C., Zachary R. Jones, Nicolas Merle, César Rios, Alessandro Gallo, Frederic Le Quemener, Laurent Delevoye, Régis M. Gauvin, Susannah L. Scott, and Mostafa Taoufik. 2018. "A strong support effect in selective propane dehydrogenation catalyzed by Ga(i-Bu)3 grafted onto γ-alumina and silica." *ACS Catalysis* 8 (8):7566–7577. https://doi.org/10.1021/acscatal.8b00936.

Tocqueville, D., Crisanti, F., Guerrero, J., Nubret, E., Robert, M., Milstein, D., and von Wolff, N. (2022). "Electrification of a Milstein-type catalyst for alcohol reformation. RSC advances." *Chemical Science*, 13 (44):13220–13224.

Turner, John A. 2004. "Sustainable hydrogen production." *Science* 305 (5686):972–974.

Wang, Dong, and Didier Astruc. 2015. "The golden age of transfer hydrogenation." *Chemical Reviews* 115 (13):6621–6686.

Wang, Qian, Yihao Xia, Zhijian Chen, Yifan Wang, Fanrui Cheng, Lei Qin, and Zhiping Zheng. 2022. "Hydrogen production via aqueous-phase reforming of ethanol catalyzed by ruthenium alkylidene complexes." *Organometallics* 41 (8):914–919.

Wang, Tao, Jin Sha, Maarten Sabbe, Philippe Sautet, Marc Pera-Titus, and Carine Michel. 2021. "Identification of active catalysts for the acceptorless dehydrogenation of alcohols to carbonyls." *Nature Communications* 12 (1):1–7.

Xu, Zhikang, Yuanyuan Yue, Xiaojun Bao, Zailai Xie, and Haibo Zhu. 2020. "Propane dehydrogenation over Pt clusters localized at the sn single-site in zeolite framework." *ACS Catalysis* 10 (1):818–828. https://doi.org/10.1021/acscatal.9b03527.

Yao, Lin, Jun Zhao, and Jong-Min Lee. 2017. "Small size Rh nanoparticles in micelle nanostructure by ionic liquid/CTAB for acceptorless dehydrogenation of alcohols only in pure water." *ACS Sustainable Chemistry & Engineering* 5 (3):2056–2060. https://doi.org/10.1021/acssuschemeng.6b02994.

Yazdani, Elahe, and Akbar Heydari. 2020. "Acceptorless dehydrogenative oxidation of primary alcohols to carboxylic acids and reduction of nitroarenes via hydrogen borrowing catalyzed by a novel nano-magnetic silver catalyst." *Journal of Organometallic Chemistry* 924:121453. https://doi.org/10.1016/j.jorganchem.2020.121453.

Yi, Jing, Jeffrey T. Miller, Dmitry Y. Zemlyanov, Ruihong Zhang, Paul J. Dietrich, Fabio H. Ribeiro, Sergey Suslov, and Mahdi M. Abu-Omar. 2014. "A reusable unsupported rhenium nanocrystalline catalyst for acceptorless dehydrogenation of alcohols through γ-C–H activation." *Angewandte Chemie International Edition* 53 (3):833–836. https://doi.org/10.1002/anie.201307665.

Yuan, Xiaofeng, Zijuan Wan, Jinfeng Ning, Qiang Zhang, and Jun Luo. 2020. "One-pot oxidant-free dehydrogenation-Knoevenagel tandem reaction catalyzed by a recyclable magnetic base-metal bifunctional catalyst." *Applied Organometallic Chemistry* 34 (11):e5897. https://doi.org/10.1002/aoc.5897.

4 Dehydrogenation Reaction of Aliphatic and Aromatic Alcohols

Vijay Bahadur and Chandni Pathak

4.1 OBJECTIVES

After studying this unit, students should be able to understand the following:

- Dehydrogenation of alcohols
- Acceptorless dehydrogenation of alcohols
- Conversion into different functionality through acceptorless dehydrogenation
- Acceptorless dehydrogenation of alcohols through a nano-catalyst
- Acceptorless dehydrogenation of alcohols through a photo-catalyst

4.2 DEHYDROGENATION REACTION

Removal of a hydrogen molecule from feedstock is the specific meaning of a dehydrogenation reaction. It is a type of elimination reaction. The major application of dehydrogenation reactions is in the petrochemicals industry, which converts non-reactive alkane into unsaturated hydrocarbon and aromatic compounds. It is a type of endothermic reaction favoured by high temperatures. The dehydrogenation product of aliphatic alcohol will transform into carbonyl compounds, depending on the type of alcohol. This chapter discusses recent progress in the dehydrogenation of alcohols.

4.3 ALIPHATIC AND AROMATIC ALCOHOLS

Alchemists referred to alcohols as "spirits of wine" and started to speak of "alcohol of wine" and then simply "alcohol."

Organic compounds having hydroxyl (–OH) groups are known as alcohols. Alcohols are very familiar and valuable compounds in nature, industry, and everyday things. The general formula for simple acyclic alcohol is $C_nH_{2n+1}OH$, where n=1, 2, 3, etc. Aromatic compounds contain a hydroxy group on a side chain in which the –OH group is attached to a sp3 hybridized carbon atom next to an aromatic ring.

In allylic and benzylic alcohol, the –OH group is glued to an sp3 hybridized carbon next to the carbon-carbon double bond and aromatic ring. In vinylic alcohol, the –OH group is directly bonded to a carbon-carbon double bond. Allylic and benzylic alcohols may be primary, secondary, or tertiary.

4.4 DEHYDROGENATION OF ALCOHOLS

Hydrogen transfer in the oxidation of alcohol in organic chemistry has been studied since the 1930s. An early breakthrough in the chemistry of catalyzed hydrogen borrowing from alcohol is

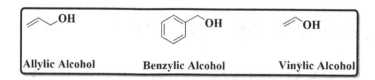

FIGURE 4.1 Different types of alcohols.

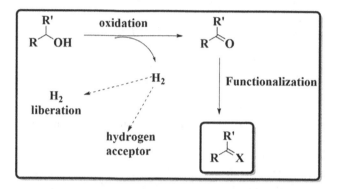

FIGURE 4.2 Different routes for oxidation of alcohols.

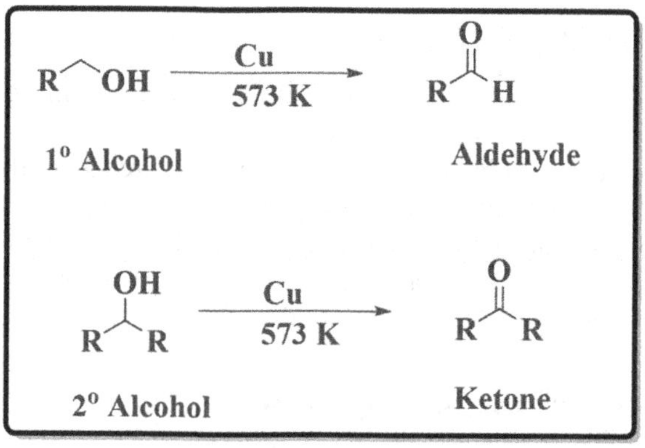

FIGURE 4.3 Oxidation of alcohols to aldehyde and ketone using a Cu catalyst.

Oppenauer's description of the aluminium tert-butoxide oxidation of secondary alcohols in the presence of acetone.[1]

Dehydrogenation of alcohol has received great attention in the last two decades. In this reaction, hydrogen gas is liberated as a by-product, and alcohols get converted into different functional groups. Released hydrogen gas can be used as fuel or reagents for further reactions.

This type of conversion has been stated as "hydrogen borrowing methodology," the "hydrogen auto-transfer process," or simply "hydrogen transfer."

Dehydrogenation of alcohols is one of the best examples of atom economy and sustainable development in green chemistry.

Conventionally, dehydrogenation of alcohol has been performed by passing alcohol vapour over copper heated at 573 K.

4.5 ACCEPTORLESS DEHYDROGENATION OF ALCOHOLS

The acceptorless dehydrogenation of alcohols has received extensive consideration for the synthesis of aldehyde, ketone, imines, amides, esters, carboxylic acids, acetals, heterocycles, and aldol products.[2] The only by-product of reactions is hydrogen gas and most importantly, does not involve any stoichiometric and hazardous oxidants. Conventionally, dehydrogenation reactions of organic compounds have been accomplished using stoichiometric quantities of inorganic oxidants, co-catalysts, and a variety of additives that lead to the generation of plentiful, often contaminated excess. Contemporary developments in catalysis by transition metal complexes have resulted in acceptorless dehydrogenation reactions that release a hydrogen molecule in which dehydrogenation is trailed by in situ consumption of the produced hydrogen equivalents, and no remaining hydrogen gas is liberated or liberated H_2 gas is vigorously expelled from the reaction mixture and collected for potential use elsewhere. The highly used catalysts are based on ruthenium and iridium.[3]

4.5.1 CONVERSION OF ALCOHOLS INTO CARBONYL COMPOUNDS

In 2002, L. A. Hulshof introduced **Robinson catalyst** [Ru (OCOCF$_3$)$_2$(CO)(PPh$_3$)$_2$] for the selective dehydrogenation of secondary alcohols into corresponding ketone in a short period of time without using solvents.

A catalyst can be recycled thrice without substantial loss of its activity. Primary alcohols undergo decarbonylation and aldol condensation while treated with Robinson catalyst.[4]

David Milstein in 2011 developed an electron rich PNP and PNN ruthenium hydrido borohydride pincer complexes for the acceptorless dehydrogenation of primary alcohols to ester, which are secondary alcohols to the ketone. The same catalyst was used to synthesize lactone through dehydrogenative cyclization of a diol with the evolution of hydrogen gas.[5]

For the development of alternative unlimited energy based on renewable resources, the results of a "hydrogen-economy" have been proposed with the dehydrogenation of alcohols. This is the first example of acceptorless dehydrogenation under mild and neutral conditions. The high turnover number for isopropyl and ethanol to produce hydrogen was observed at below 100 °C. Conventionally, an excess of base, more than 150 °C, and light have been required to produce

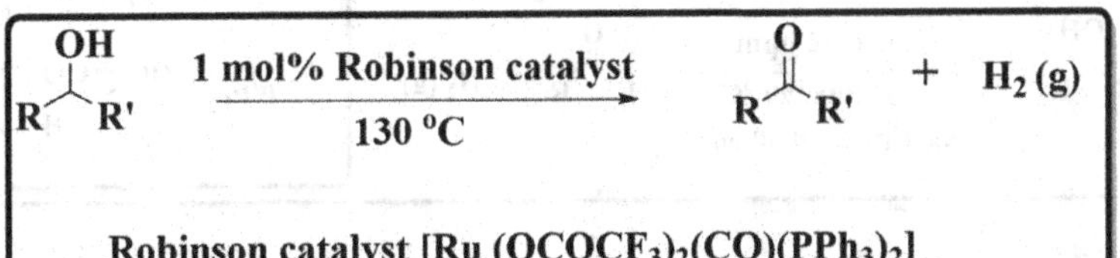

FIGURE 4.4 Conversion of secondary alcohols to ketone.

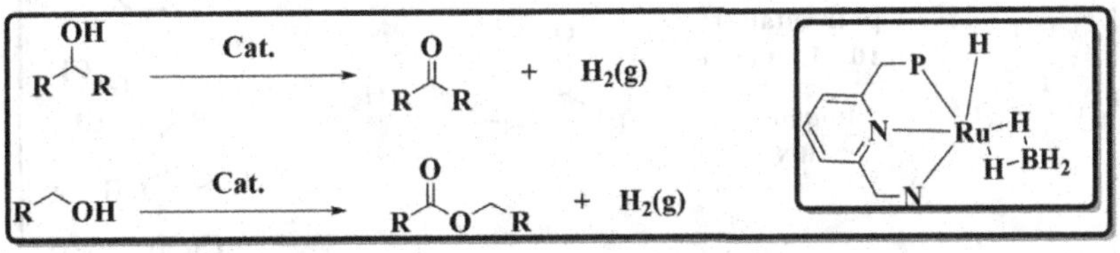

FIGURE 4.5 Conversion of secondary alcohols to ketone and primary alcohols to esters.

hydrogen. This ruthenium-based catalyst is the first example of acceptorless dehydrogenation under mild and neutral conditions as reported by Matthias Beller in 2011.[6]

In 2007, Ryohei Yamaguchi and colleagues designed and synthesized a new iridium-based catalyst for the acceptorless dehydrogenation of secondary alcohols into corresponding ketone in neutral conditions with a high turnover number (TON). They used hydroxypyridine ligands for the synthesis of catalysts.[7]

In 2011, Ryohei Yamaguchi developed an iridium-based C, N-chelated ligand complex for the acceptorless dehydrogenation of primary alcohols to aldehyde and secondary alcohols to ketone. Iridium-based catalyst is well-versed with a wide range of primary and secondary alcohols.[8]

Iridium catalyst, which is also used for the dehydrogenation of benzylic alcohols, gives benzaldehyde and hydrogen gas as the sole products.

Ryohei Yamaguchi in 2012 developed a new iridium-based catalyst for the dehydrogenation of alcohols in an aqueous medium. Bipyridine-based water-soluble Cp*Ir catalyst was first used for dehydrogenating primary and secondary alcohols into carbonyl compounds in a water system.[9]

4.5.2 CONVERSION OF ALCOHOLS INTO ESTER COMPOUNDS

David Milstein and his team in 2005 designed a new Ru (II) hydride catalyst for the acceptorless dehydrogenation of primary alcohols into ester in benign and neutral conditions. The synthesized catalyst was ligated with electron-rich PNP and PNN ligands.[10]

Synthesized electron-rich ruthenium complex is well-versed with primary alcohols and used to directly convert primary alcohols into ester.

Dmitry G. Gusev in 2012 developed a ruthenium-based air-stable $RuCl_2(PPh_3)$ [$PyCH_2NHC_2H_4PPh_2$] catalyst for the acceptorless dehydrogenation of ethanol and form ethyl acetate and hydrogen gas with a TON of up to 1700. This versatile catalyst is used for the dehydrogenation of alcohol and reduction of the C=X polar bond even using less than 50 ppm ruthenium catalyst.[11]

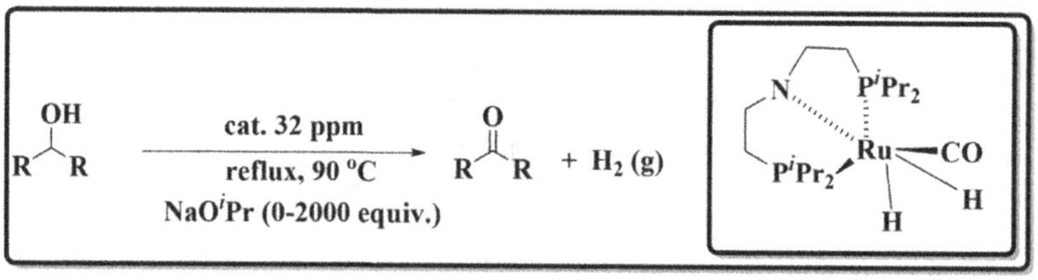

FIGURE 4.6 Conversion of secondary alcohols to ketone under 100 °C.

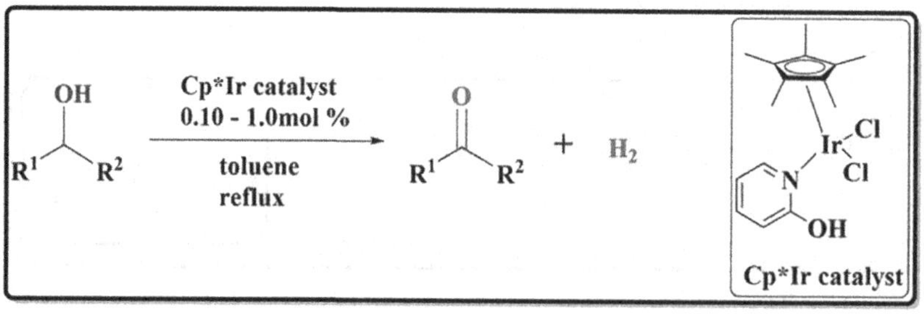

FIGURE 4.7 Conversion of secondary alcohols into ketone by Ir catalyst.

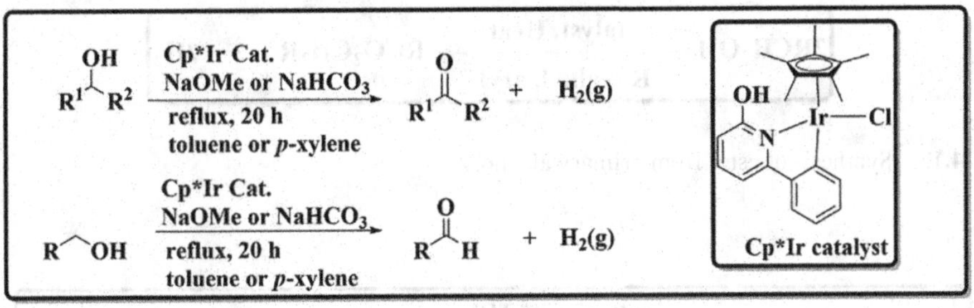

Dehydrogenation Reaction of Aliphatic and Aromatic Alcohols

FIGURE 4.8 Conversion of alcohols into carbonyl compounds using Ir catalyst.

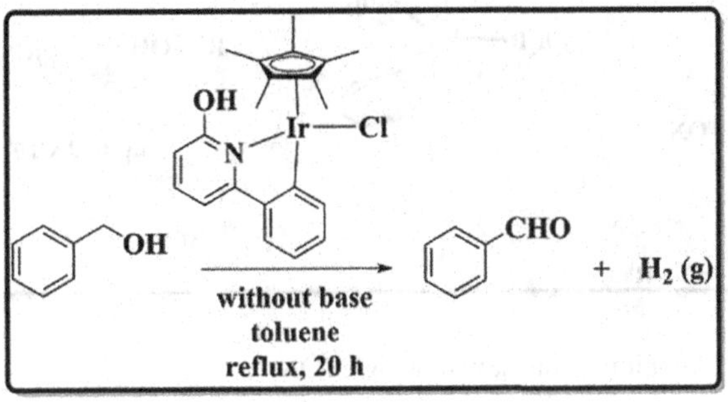

FIGURE 4.9 Conversion of aromatic alcohols into aldehyde with Ir catalyst.

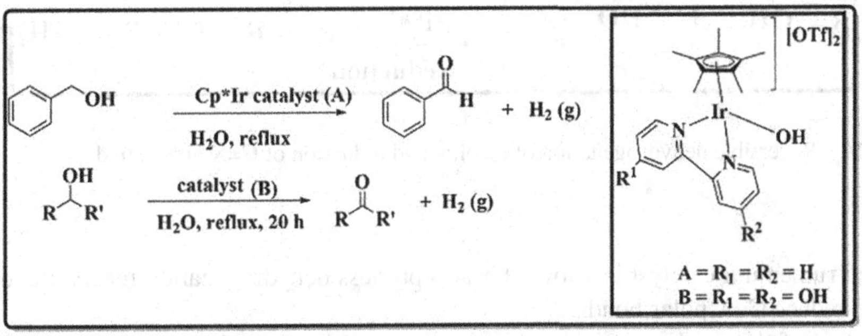

FIGURE 4.10 Conversion of alcohols into carbonyl compounds in an aqueous medium.

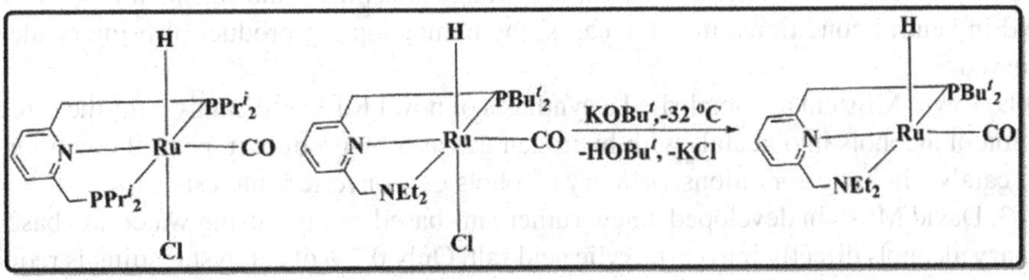

FIGURE 4.11 Synthesis of electron-rich PNP- and PNN-based Ru catalyst.

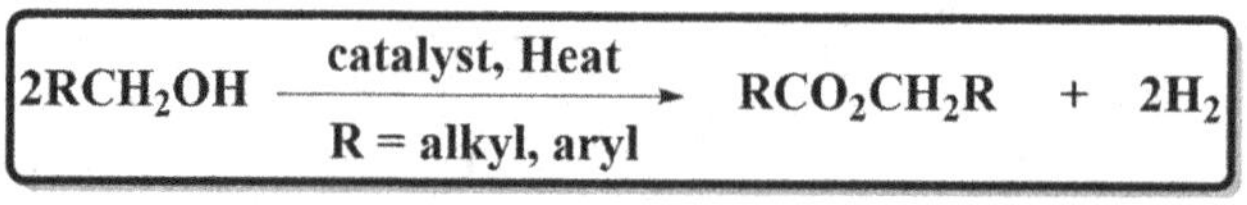

$$2RCH_2OH \xrightarrow[R = alkyl, aryl]{catalyst, Heat} RCO_2CH_2R + 2H_2$$

FIGURE 4.12 Synthesis of ester from primary alcohols.

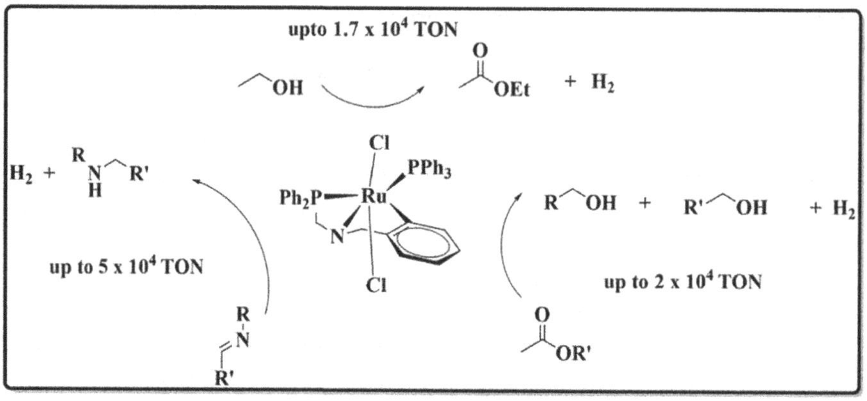

FIGURE 4.13 Versatile ruthenium catalyst with a high TON.

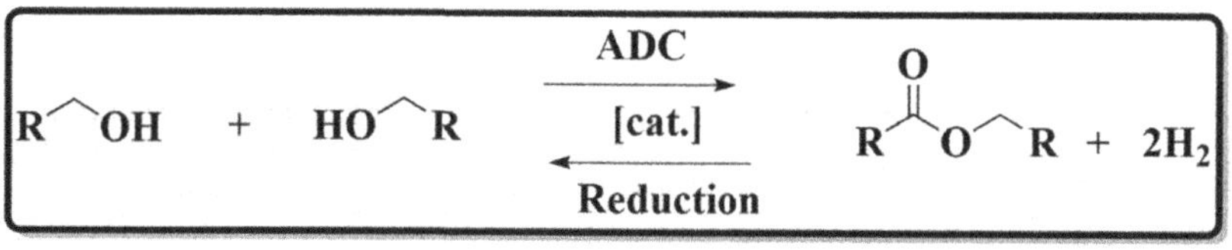

$$R\frown OH + HO\frown R \underset{Reduction}{\overset{ADC\ [cat.]}{\rightleftarrows}} R\overset{O}{\underset{\|}{C}}O\frown R + 2H_2$$

FIGURE 4.14 Reversible dehydrogenation of alcohol and reduction of C=X polar bond.

Reported ruthenium catalyst is known for acceptorless dehydrogenation reaction and is reversibly used to reduce C=X polar bonds.

The loading of the catalyst and amount of base are unique for the reduction of different C=X polar bonds such as carbonyl, ester, and imines as shown in the following figure.

In 2012, David Milstein developed an electron-rich bipyridyl-based ruthenium PNN pincer catalyst for the cross-dehydrogenative coupling of alcohols into mixed ester with the evolution of hydrogen gas. The protocol is well-tolerated with a wide range of primary and secondary alcohols having high yield in neutral conditions; in some cases, the homocoupling product of primary alcohols is also observed.[12]

In 2009, David Milstein accomplished a synthesis of novel Ru-based catalyst for the direct dehydrogenation of alcohols into acetals with hydrogen gas as a by-product in neutral conditions. With the same catalyst in basic conditions, primary alcohols get converted into ester.[13]

In 2013, David Milstein developed a new ruthenium-based catalyst using water and base to convert primary alcohols directly into carboxylic acid salt. Only 0.2% of catalyst loading is required for the transformation described previously. This is a cheaper, cleaner, and safer process for the direct conversion of alcohols into acid at laboratory and industrial levels.[14]

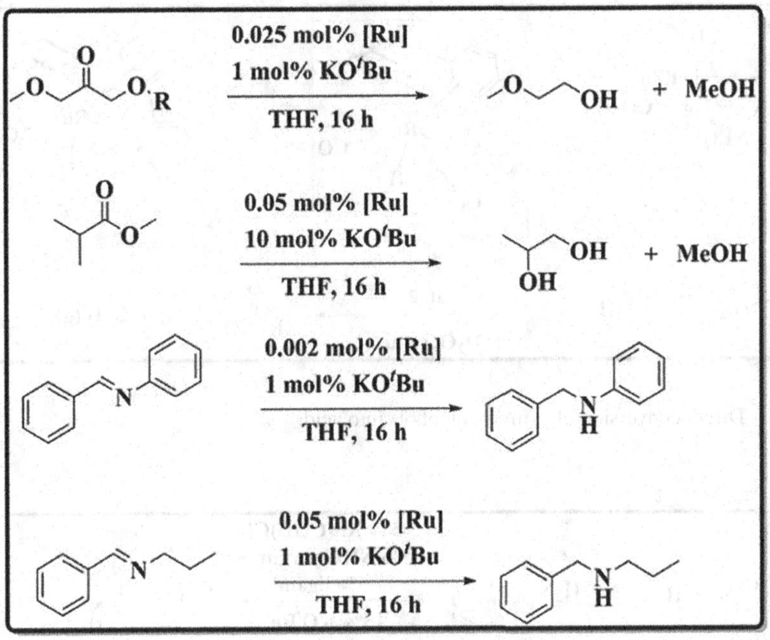

Dehydrogenation Reaction of Aliphatic and Aromatic Alcohols

FIGURE 4.15 Reduction of C=X polar bond with Ru-based catalyst.

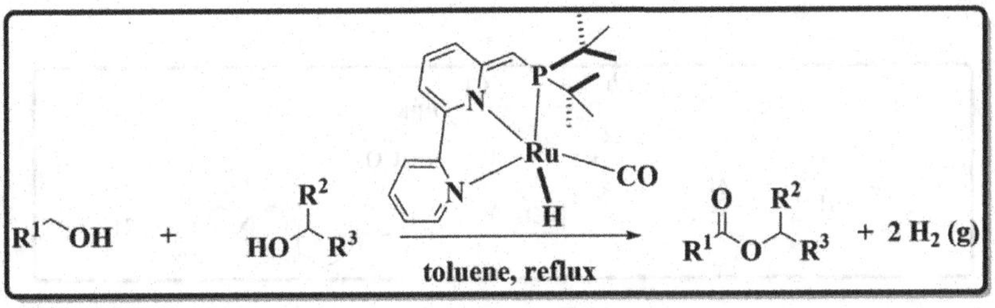

FIGURE 4.16 Synthesis of ester through cross-dehydrogenative coupling of alcohols.

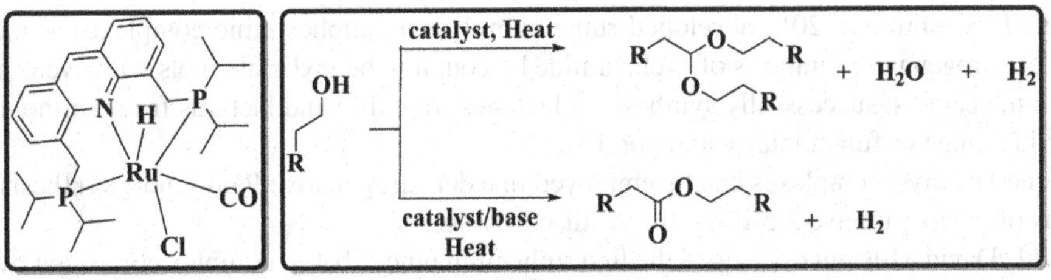

FIGURE 4.17 Conversion of alcohol into acetals and ester.

4.5.3 Conversion of Alcohols into Amide Compounds

Robert Madsen in 2008 developed a new catalyst with ruthenium, *N*-heterocyclic carbene and phosphine ligand, for the synthesis of amide by treating primary alcohol and amine with the evolution of hydrogen gas.[15]

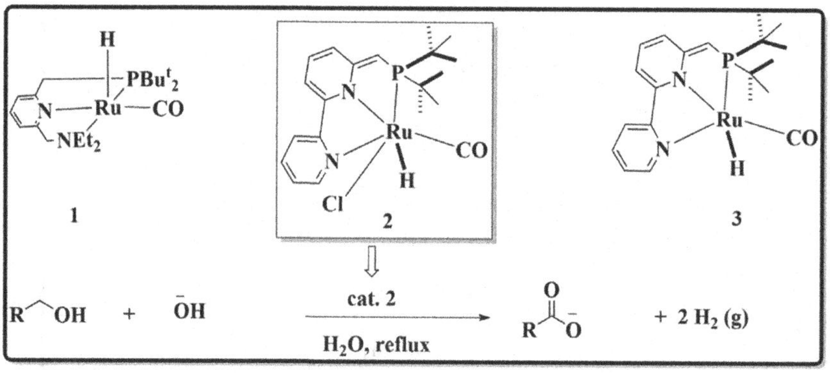

FIGURE 4.18 Direct conversion of primary alcohols into acid.

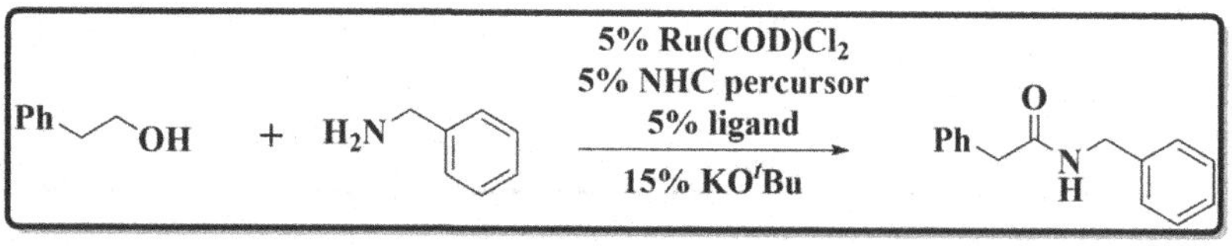

FIGURE 4.19 Direct conversion of primary alcohols into an amide.

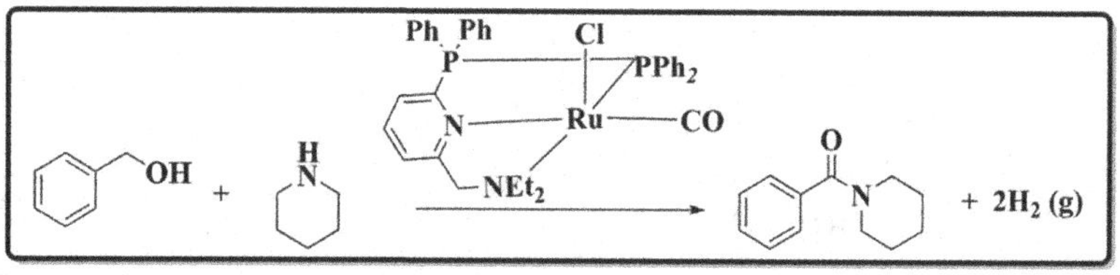

FIGURE 4.20 Direct conversion of primary alcohols into lactam.

Robert H. Crabtree in 2011 developed ruthenium diamine diphosphine complexes as a catalyst for the dehydrogenative synthesis of cyclic amide by coupling benzylic alcohols with cyclic amines.

The same catalyst successfully synthesized lactones from diol and lactams from amino alcohols with a wide range of functionality and good yield.[16]

Designed catalyst complexes can be employed in a dehydrogenative Paal-Knorr synthesis for the synthesis of pyrrole to give 2,5-dimethyl-*N*-alkylpyrroles.

In 2011, David Milstein developed the first ruthenium pincer-based complex controlled by ligand for the synthesis of peptides and pyrazine from b-amino alcohols through alcohol dehydrogenation. Selective cyclic dipeptide can be achieved by self-coupling b-amino alcohols using Ru-PNN complex **1**. In contrast, Ru-PNN complex **2** catalyzed the formation of pyrazine via dehydrogenative coupling of b-amino alcohols.[17]

Out of the three synthesized ligand-controlled catalysts, complex 1 is used for the synthesis of amide, whereas complexes 2 or 3 are used for the synthesis of imine through the dehydrogenative process with the same precursor.

Ruthenium complex 2 or 3 helps in the direct synthesis of piperazine via homocoupling of b-amino alcohol through the dehydrogenation process followed by aromatization.

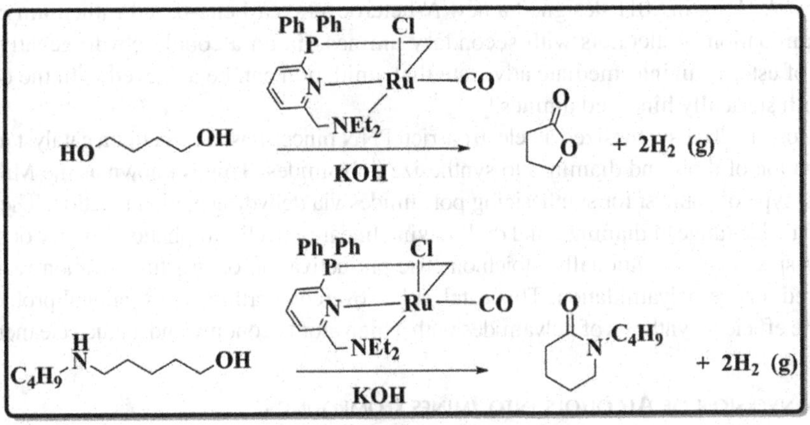

FIGURE 4.21 Direct conversion of alcohols into lactone and lactam.

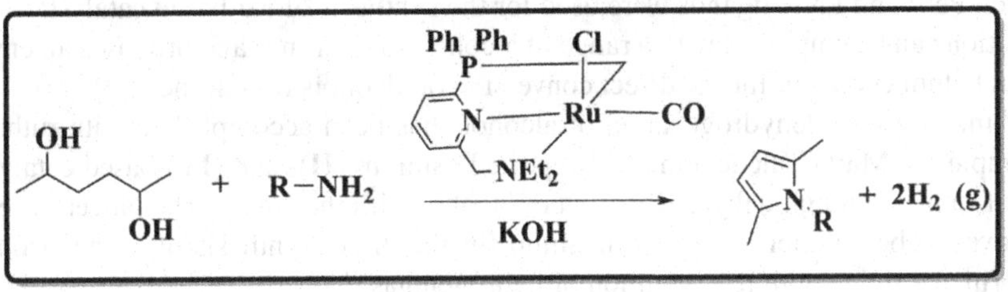

FIGURE 4.22 One-step synthesis of pyrrole through dehydrogenation of diols.

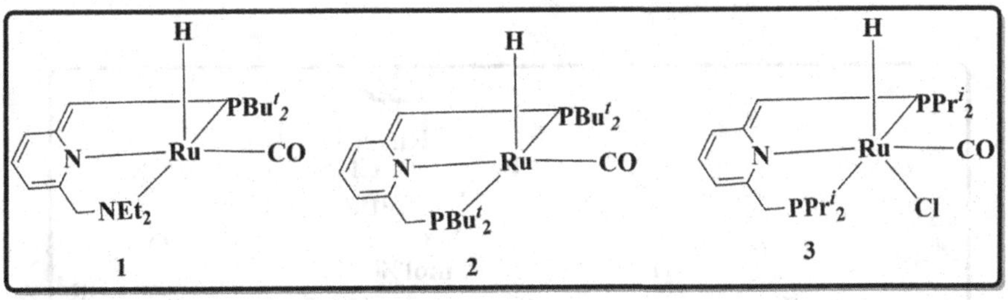

FIGURE 4.23 Ruthenium-based catalyst.

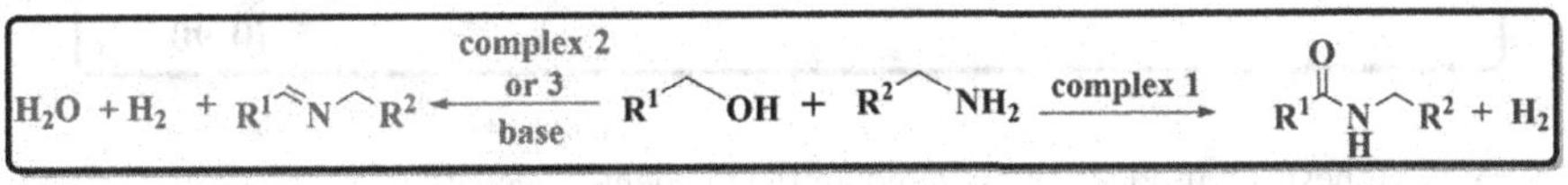

FIGURE 4.24 Ligand-controlled synthesis of imine and amide through dehydrogenation of alcohols.

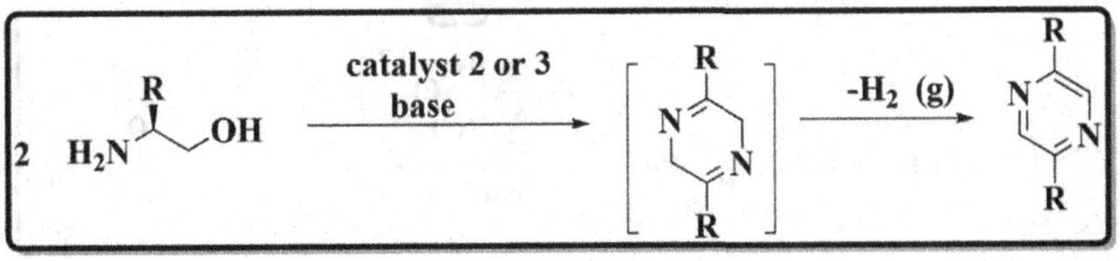

FIGURE 4.25 Direct synthesis of pyrazine via dehydrogenative homocoupling of b-amino alcohols.

Soon Hyeok Hong in 2011 designed a new *N*-heterocyclic carbene-based ruthenium catalyst for the direct amidation of alcohols with secondary amines via an alcohol dehydrogenative process. Formation of ester is an intermediate advocate that amidation can be achieved with the coupling of alcohols with sterically hindered amines.[18]

Zhibin Guan in 2011 synthesized an electron-rich PNN pincer-based ruthenium catalyst used for the dehydrogenation of diols and diamines to synthesize polyamides. This is known as the Milstein catalyst, the first type of catalyst for synthesizing polyamides via dehydrogenation reaction. The catalyst is relevant for a wide range of diamines and diols having linear or cyclic, aliphatic (primary or secondary) or aromatic spacers. Conventionally, stoichiometric pre-activation or in situ activation reagents have been required for the polyamidation. This catalyst has overcome all the conventional problems and is useful for the efficient synthesis of polyamides with a high atom economy and a much cleaner process.[19]

4.5.4 CONVERSION OF ALCOHOLS INTO IMINES COMPOUNDS

David Milstein in 2010 developed a new catalyst for the synthesis of imine through acceptorless dehydrogenation of alcohol with amine having high turnover numbers and without waste products. Electron-rich PNP and PNN ligands were used for the synthesis of Ru-based catalyst.

The reaction can be employed with a range of alcohols and amines and proposes an environmentally pleasant atom economy for the direct conversion of alcohols into imine.[20]

Most of the selective dehydrogenation of alcohols has been accomplished with ruthenium and iridium complexes. Marta Valencia in 2011 reported osmium (**II**)- and (**IV**)-based catalysts for the formation of imines through dehydrogenation of alcohol with the amine. The pincer-based osmium catalyst proves to be an alternative to ruthenium for the direct synthesis of imines from alcohols including primary amine with the evolution of hydrogen gas.[21]

In the same year, Robert Madsen developed a ruthenium *N*-heterocyclic carbene complex as a catalyst for the formation of imines through dehydrogenation of alcohol with the amine. The

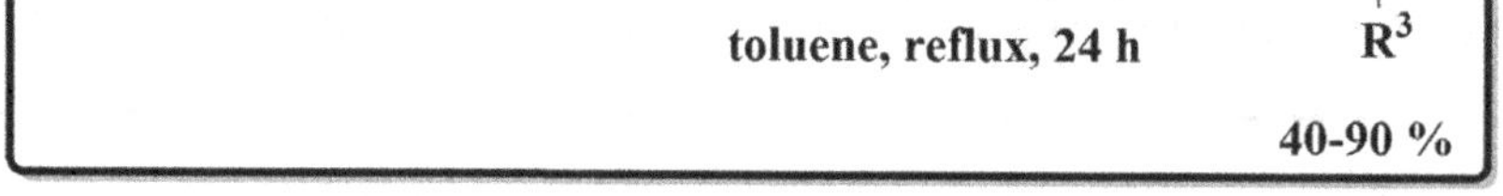

FIGURE 4.26 Synthesis of amide with hindered amine via dehydrogenation.

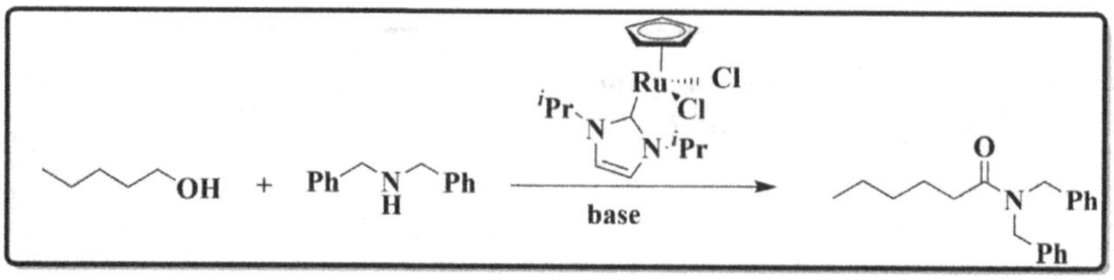

FIGURE 4.27 Synthesis of amide with dehydrogenative coupling of 1° alcohol with 2° amine.

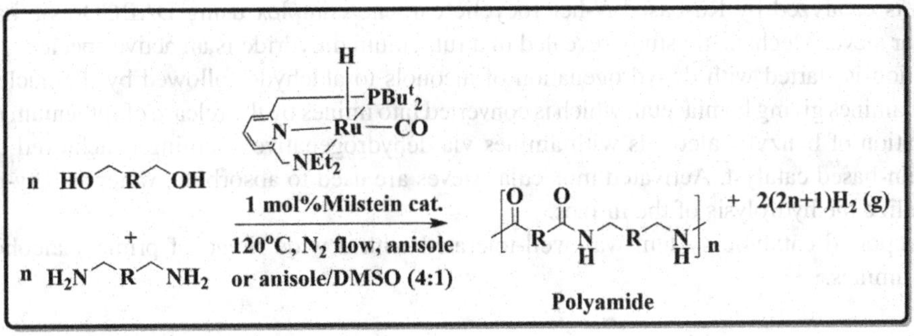

Dehydrogenation Reaction of Aliphatic and Aromatic Alcohols

FIGURE 4.28 Direct synthesis of polyamide.

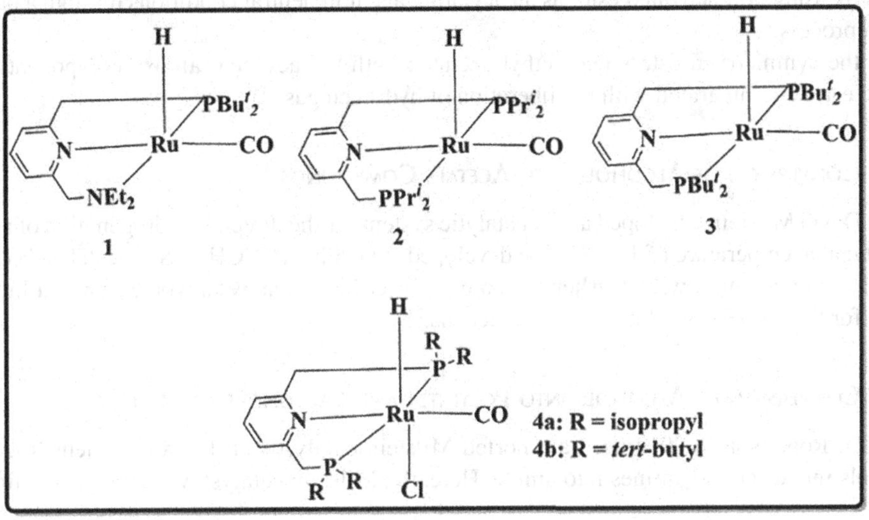

FIGURE 4.29 Various screened Ru-based catalysts.

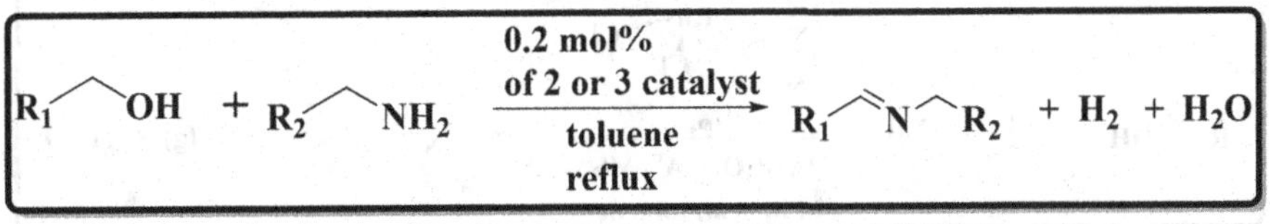

FIGURE 4.30 Synthesis of imine by Ru-based catalyst.

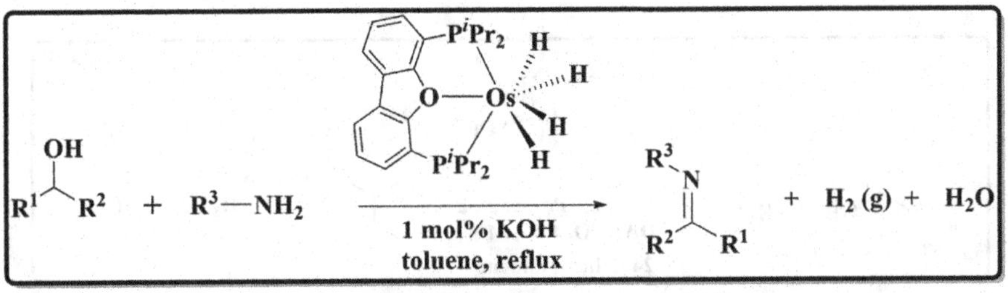

FIGURE 4.31 Synthesis of imine by Os-based catalyst.

reaction is catalyzed by Ru-based *N*-heterocyclic carbene complex using DABCO as a base and molecular sieve. Mechanistic study revealed that ruthenium dihydride is an active species.

Reaction is started with dehydrogenation of alcohols to aldehyde followed by the nucleophilic attack of amines giving hemiacetal, which is converted into imines by the release of ruthenium metal.[22]

Imination of benzylic alcohols with amines via dehydrogenative coupling is achieved through ruthenium-based catalyst. Activated molecular sieves are used to absorb the water as a by-product to neutralize the hydrolysis of the imines.

The reported catalytic system was well-tolerated with a wide range of primary alcohols and primary amines.

4.5.5 CONVERSION OF ALCOHOLS INTO ACYLATED COMPOUNDS

David Milstein in 2010 established a new Ru pincer-based catalyst for the acylation of secondary alcohols by using non-activated ester as an acylating agent in neutral conditions through a dehydrogenation process.

With the symmetrical esters (e.g., ethyl acetate), both the acyl and alkoxy components of the substrate ester are integrated with the liberation of hydrogen gas.[23]

4.5.6 CONVERSION OF ALCOHOLS INTO ACETALS COMPOUNDS

In 2012, David Milstein developed a new catalytic system for the direct dehydrogenation of alcohols into acetal at a temperature of 110 °C. The developed Ru $(PPh_3)_2(NCCH_3)_2(SO_4)$ catalyst is efficient even at low temperature, while earlier ruthenium pincer-based catalysts worked only at high temperature for the conversion of alcohols into acetals.[24]

4.5.7 CONVERSION OF ALCOHOLS INTO POLYESTER AND LACTONES COMPOUNDS

Nicholas J. Robertson in 2012 used a reported Milstein catalyst well-known for dehydrogenation of alcohols into ester and amines into amide. Here, the Milstein catalyst was used for the first time

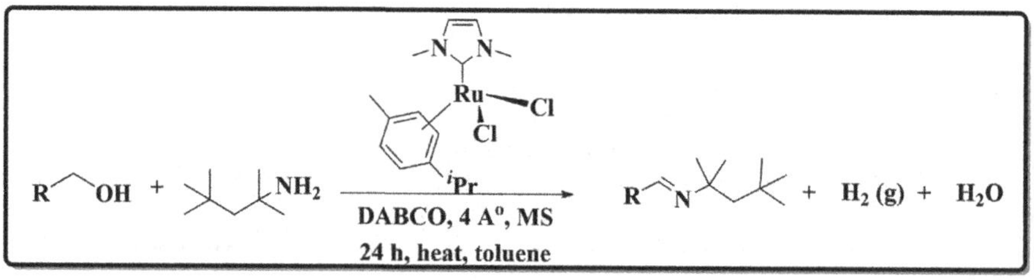

FIGURE 4.32 Imination of alcohols with hindered amine.

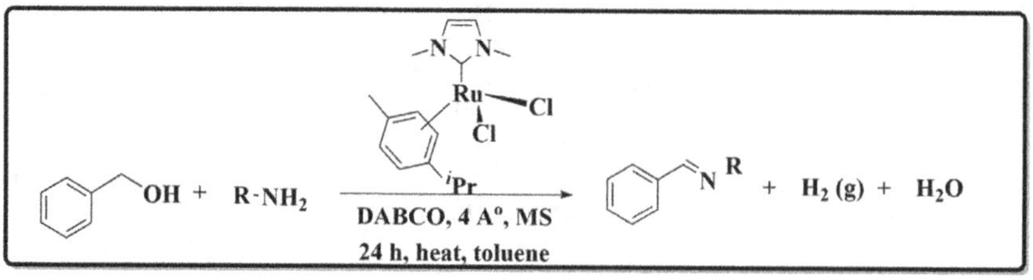

FIGURE 4.33 Imination of benzyl alcohols with amine.

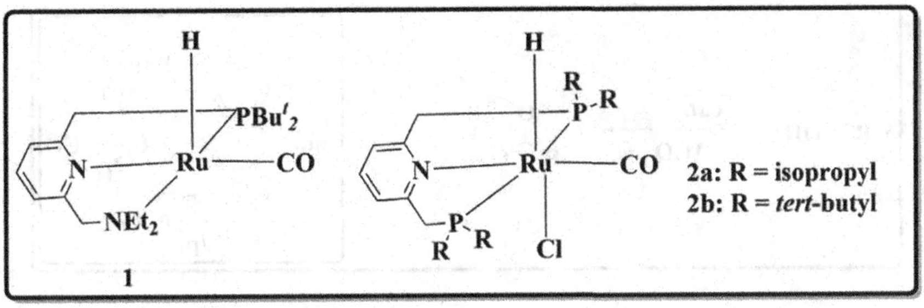

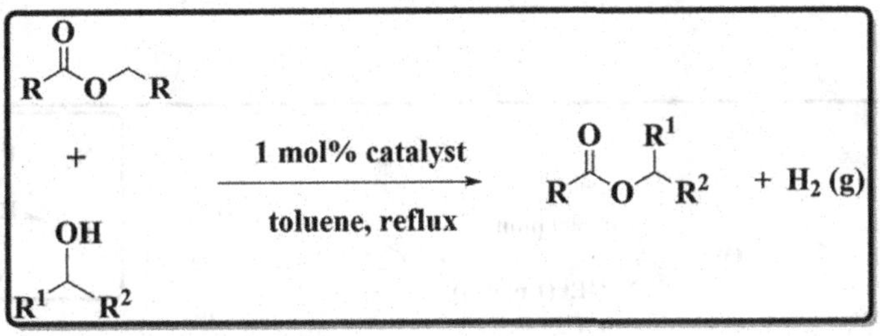

FIGURE 4.34 Acylation of secondary alcohols using ester as acylating agent.

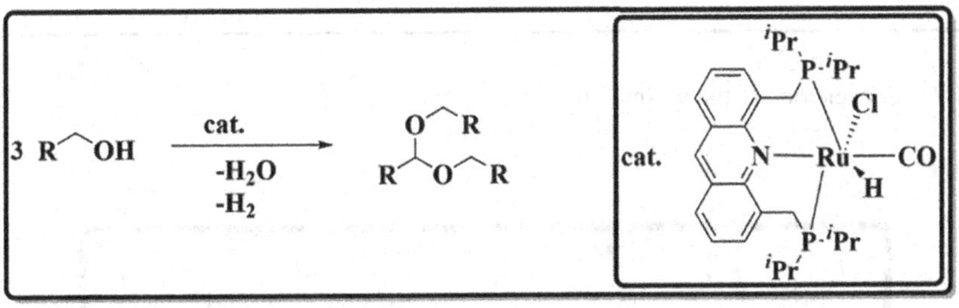

FIGURE 4.35 Acetal formation of alcohols via dehydrogenation of alcohols.

to synthesize high molecular wight polyester under reduced pressure using diol as a precursor. The major advantage of using catalysts under reduced pressure is to effectively remove the by-product, namely, molecular hydrogen. Diols with a smaller number of carbons are converted into lactone, while a higher number of carbons polymerize into polyester. Dehydrogenation of 1,5-pentanediol gives the mixture of polyester and lactone as the boundary between higher and lower diols.[25]

Matthias Beller in 2012 developed a ruthenium PNP pincer-based catalyst for the acceptorless dehydrogenation of ethanol into ethyl acetate. Ethanol plays a dual role of not only reactant but also solvent. Therefore, this is a selective and cost-effective catalyst for directly converting ethanol into ethyl acetate without using additional hydrogen acceptors and solvents.[26]

4.5.8 Direct Synthesis of Pyrrole from Alcohols

David Milstein in 2013 accomplished a protocol for the synthesizing of pyrrole by dehydrogenative cross-coupling between amino alcohols and secondary alcohols using ruthenium pincer-based catalyst. This catalyst is used for successive and selective C-N and C-C bond formations catalyzed by PNN ruthenium complex generated in situ by an air stable complex with base. It is believed that this

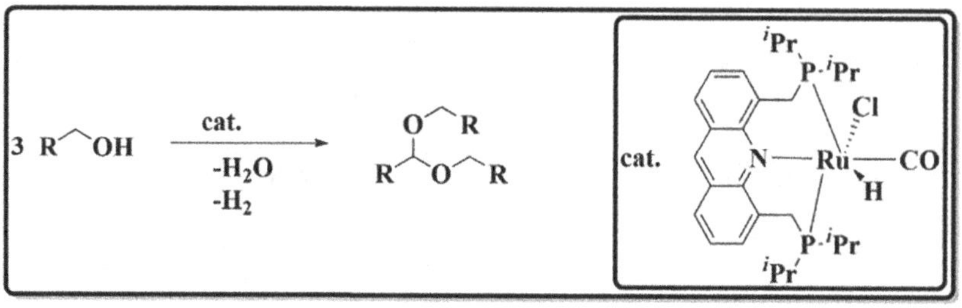

FIGURE 4.36 Synthesis of polyester and lactones under reduced pressure.

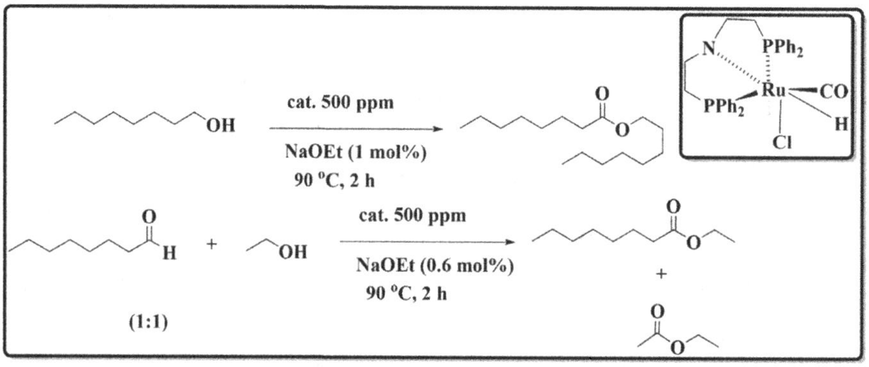

FIGURE 4.37 Conversion of ethanol into ethyl acetate.

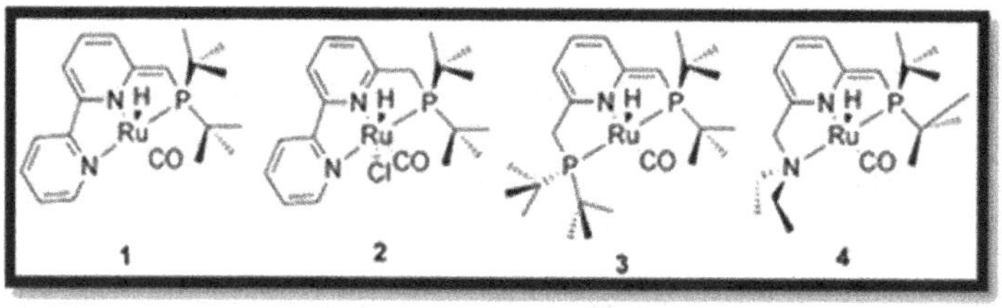

FIGURE 4.38 Screened Ru-based catalyst for the synthesis of pyrrole.

reported protocol has one of the best atom economies, is environment friendly and is an effective procedure for synthesizing pyrrole.[27]

Dehydrogenative coupling of b-amino alcohols with secondary alcohols using potassium *tert.* butoxide as base resulting in 2, 5- disubstituted pyrrole. This is one of the most atom-economical and benign methods for synthesizing pyrrole.

Mechanistic investigation reveals that pyrrole synthesis was achieved by formation of imine followed by dehydrogenation reaction; lastly, a nucleophilic attack of carbanion on electrophilic carbonyl carbon gives a 4-hydroxy 5 membered cyclized product. In the conclusive step, the five-membered cyclized product undergoes dehydration via an aromatization process giving substituted pyrrole with good yield.

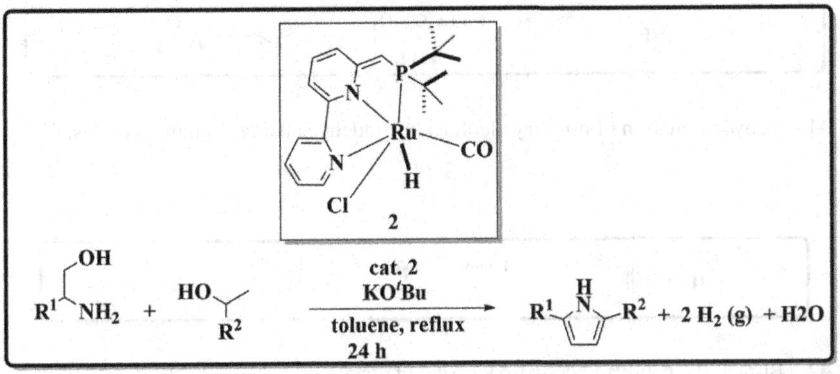

FIGURE 4.39 Direct synthesis of pyrrole through dehydrogenation of b-amino alcohol.

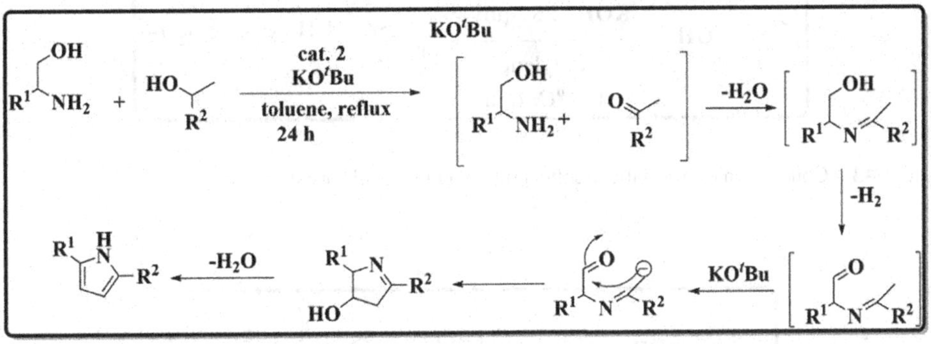

FIGURE 4.40 Mechanism of formation of pyrrole.

4.6 GREEN METHOD FOR THE DEHYDROGENATION OF ALCOHOLS

4.6.1 Dehydrogenation of Alcohols with Nanoparticles

4.6.1.1. Cu Nanocatalyst

Wenjie Shen in 2013 produced green and sustainable development for the dehydrogenation of primary aliphatic alcohols into aldehyde using Cu nanoparticles loaded on $La_2O_2CO_3$ nanorods[28] with styrene as a hydrogen acceptor.

4.6.1.2. Bi-metallic Nanoparticles (Cu-Ni)

In 2014, Jie Xu and his group developed a dehydrogenation of thick range of primary aliphatic alcohols into respective aldehydes by using bi-metallic Cu-Ni/γ-Al_2O_3 catalyst and styrene as a hydrogen acceptor. They also observed that the catalytic activity of monoatomic Cu under the same reaction conditions was lower than the bi-metallic catalyst.[29]

4.6.1.3. Silver as Nanoparticles

In 2017, Robert Madsen reported a new synthetic procedure for dehydrating primary alcohol into carboxylic acid salt without using metal from the platinum group. A wide range of aliphatic and benzylic alcohol groups were treated with 2.5% of Ag_2CO_3 and 2.5–3 equivalent of potassium

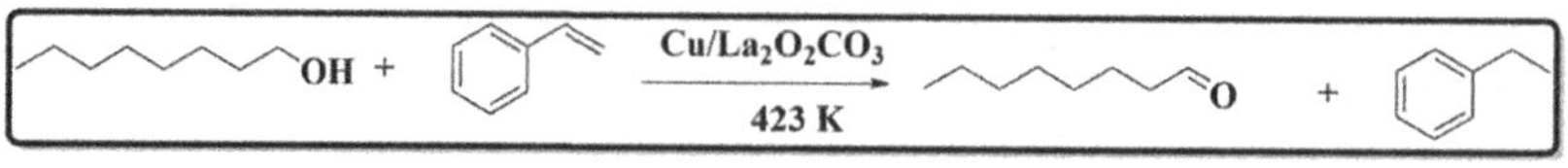

FIGURE 4.41 Dehydrogenation of primary alcohols into aldehyde using Cu nanoparticles.

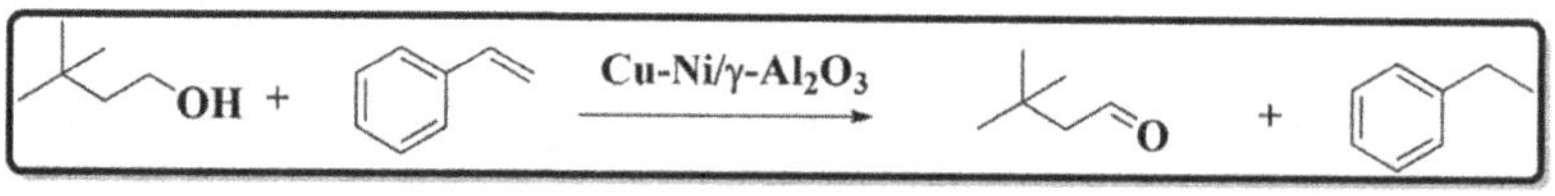

FIGURE 4.42 Bi-metallic catalyzed dehydrogenation of alcohols.

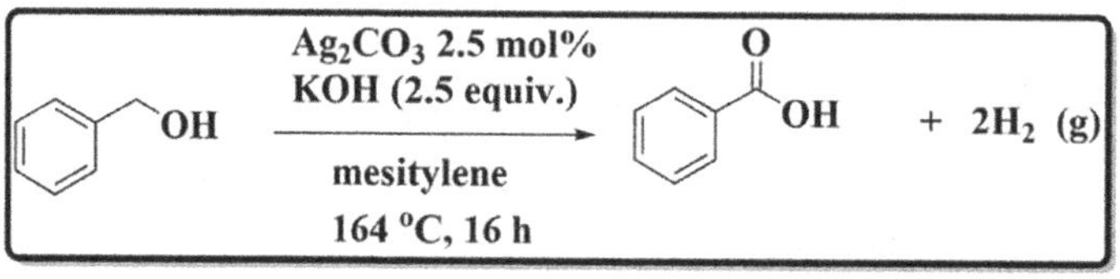

FIGURE 4.43 Conversion of aromatic alcohols into acid using silver carbonate.

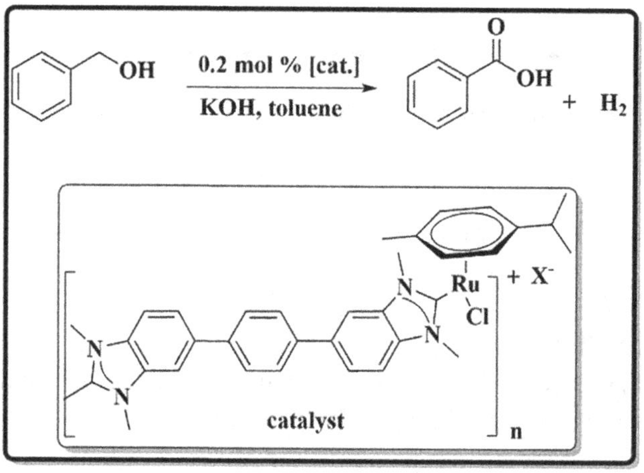

FIGURE 4.44 Conversion of alcohols into acid using solid support.

hydroxide in refluxing mesitylene. It is believed that dehydrogenation, that is, the removal of hydrogen gas, is due to the in situ formation of silver nanoparticles during the reactions.[30]

4.6.1.4. Solid Support

Tao Tu and his team accomplished a protocol for the dehydrogenation of aliphatic and aromatic alcohols into corresponding acids via self-supported NHC-Ru single-site catalysts in 2022. The authors

Dehydrogenation Reaction of Aliphatic and Aromatic Alcohols

claimed that with the low loading catalyst (0.2%), the received TON may be up to 1.8×10^4 without losing capacity up to 20 runs from the milligram to gram scale, with a wide range of alcohols.[31]

4.6.2 Dehydrogenation of Alcohols with Photocatalyst

4.6.2.1. Au Photo-Catalyst

Hiroshi Naka in the same year reported the catalytic dehydrogenation of primary aliphatic alcohols to aldehyde by irradiation of UV–Vis light with gold-loaded titanium dioxide (Au/TiO_2) as photocatalysts.[32]

4.6.2.2. Pt Photo-Catalyst

Chi-Ming Che and his team in 2019 reported novel binuclear platinum (II) diphosphite complexes as photocatalysts for the acceptorless dehydrogenation of aliphatic alcohols and *N*-heterocycles with high yields and a wider substrate scope at room temperature. This novel protocol has mild reaction conditions, scalability power, and moderate to excellent yield.[33]

4.6.2.3. Ni Photo-Catalyst

Motomu Kanai in 2020 developed the first acceptorless dehydrogenation of secondary alcohols into ketones using nickel photo redox catalyst under visible light irradiation at ambient temperature.[34]

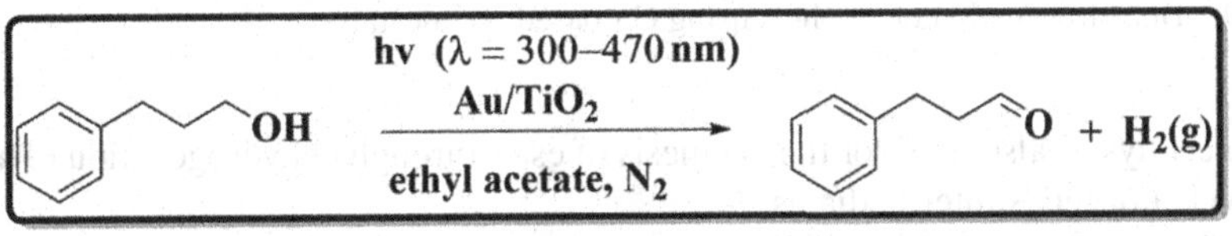

FIGURE 4.45 Dehydrogenation of alcohols using $Au-TiO_2$ photocatalyst.

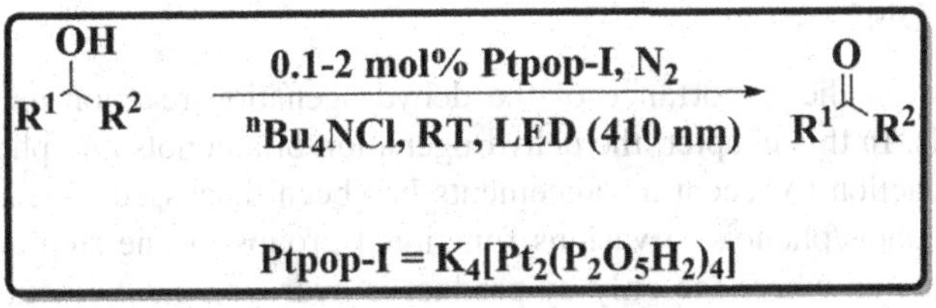

FIGURE 4.46 Dehydrogenation of alcohol using Pt-based photocatalyst.

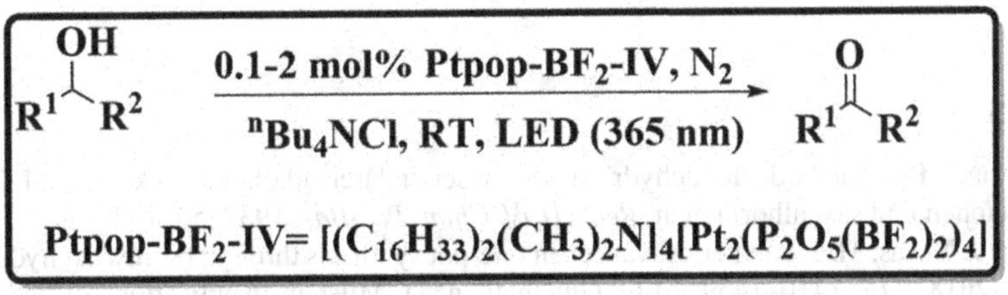

FIGURE 4.47 Dehydrogenation of alcohols using Pt-based photocatalyst.

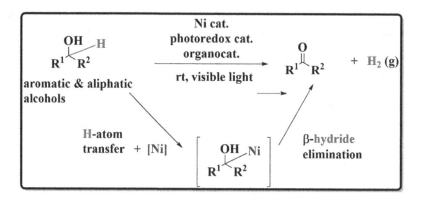

FIGURE 4.48 Dehydrogenation of alcohols using Ni-based photocatalyst.

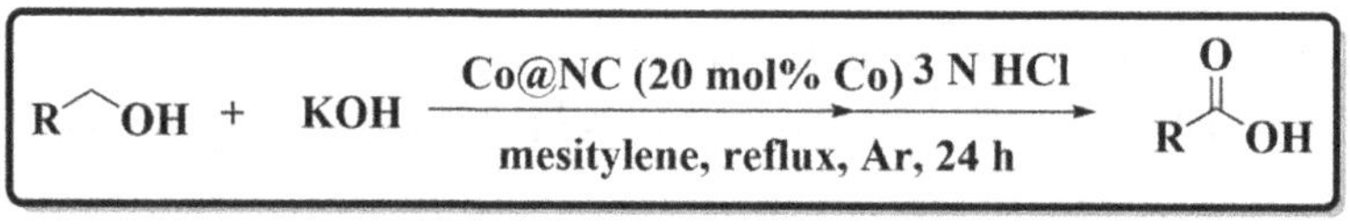

FIGURE 4.49 Dehydrogenation of alcohols using Co-based photocatalyst.

The same catalyst is also used for the synthesis of ester through dehydrogenation of alcohols and aldehydes via hemiacetals intermediates.

In 2022, Cheng Chen et al. developed a cobalt embedded in nitrogen-doped porous carbon (Co@NC) catalyst for the conversion of alcohols into acids. They have used this catalyst at least 15 times without any loss of activity and were easily able to recycle it by external magnetic field.[35]

4.7 CONCLUSIONS

This chapter points out the importance of the dehydrogenation reaction and its application in alcohols and phenols. In this chapter, the dehydrogenation of alcohols and phenols from the very first conventional reaction to recent advancements has been discussed. Specifically, the general conversion from alcohols/phenols to various functional groups in one step and the acceptorless dehydrogenation reaction where the only by-product is hydrogen gas without using stoichiometric and hazardous oxidants have been discussed. This chapter also highlights recent progress in dehydrogenation reactions such as green methods and the use of nano catalyst reactions and visible light-assisted photocatalyst with various metals. Overall, this chapter is intended to provide a complete capsule from the beginning to recent advancements in the dehydrogenation of alcohols and phenols.

REFERENCES

1. R. V. Oppenauer, Eine methode der dehydrierung von sekundären alkoholen zu ketonen. I. Zur herstellung von sterinketonen und sexualhormonen. *Recl. TraV. Chim. Pay-Bas*. **1937**, *56*, 137.
2. a) A. Corma, J. Navas, M. J. Sabater, Advances in one-pot synthesis through borrowing hydrogen catalysis. *Chem. Rev*. **2018**, *118*, 1410–1459; b) C. Gunanathan, D. Milstein, Applications of acceptorless dehydrogenation and related transformations in chemical synthesis. *Science* **2013**, *341*, 1229712; c) S. Bähn, S. Imm, L. Neubert, M. Zhang, H. Neumann, M. Beller, The catalytic amination of alcohols. *Chem. Cat.*

Chem. **2011**, *3*, 1853–1864; d) Y. Obora, Y. Ishii, Iridium-catalyzed reactions involving transfer hydrogenation, addition, n-heterocyclization, and alkylation using alcohols and diols as key substrates. *Synlett* **2011**, *30*, 51.
3. R. Yamaguchi, K. Fujita, M. Zhu, Recent progress of new catalytic synthetic methods for nitrogen heterocycles based on hydrogen transfer reactions. *Heterocycles* **2010**, *81*, 1093–1140; f) A. J. A. Watson, J. M. J. Williams, Recent progress of new catalytic synthetic methods for nitrogen heterocycles based on hydrogen transfer reactions. *Science* **2010**, *329*, 635–636; g) G. E. Dobereiner, R. H. Crabtree, Dehydrogenation as a substrate-activating strategy in homogeneous transition-metal catalysis. *Chem. Rev.* **2010**, *110*, 681–703.
4. G. B. W. L. Ligthart, R. H. Meijer, M. P. J. Donners, J. Meuldijk, J. A. J. M. Vekemans, L. A. Hulshof, Highly sustainable catalytic dehydrogenation of alcohols with evolution of hydrogen gas. *Tetrahedron Lett.* **2003**, *44*, 1507–1509.
5. Jing Zhang, Ekambaram Balaraman, Gregory Leitus, David Milstein, Electron-rich PNP-and PNN-type ruthenium (II) hydrido borohydride pincer complexes. Synthesis, structure, and catalytic dehydrogenation of alcohols and hydrogenation of esters. *Organometallics* **2011**, *30*, 5716–5724.
6. Martin Nielsen, Anja Kammer, Daniela Cozzula, Henrik Junge, Serafino Gladiali, Matthias Beller, Efficient hydrogen production from alcohols under mild reaction conditions. *Angew. Chem. Int. Ed.* **2011**, *50*, 9593–9597.
7. Ken-ichi Fujita, Nobuhide Tanino, Ryohei Yamaguchi, Ligand-promoted dehydrogenation of alcohols catalyzed by Cp* Ir complexes. A new catalytic system for oxidant-free oxidation of alcohols. *Org. Lett.* **2007**, *9*, 109–111.
8. Ken-ichi Fujita, Tetsuya Yoshida, Yoichiro Imori, Ryohei Yamaguchi, Dehydrogenative oxidation of primary and secondary alcohols catalyzed by a Cp* Ir complex having a functional C, N-chelate ligand. *Org. Lett.* **2011**, *13*, 2278–2281.
9. Ryoko Kawahara, Ken-ichi Fujita, Ryohei Yamaguchi, Dehydrogenative oxidation of alcohols in aqueous media using water-soluble and reusable Cp* Ir catalysts bearing a functional bipyridine ligand. *J. Am. Chem. Soc.* **2012**, *134*, 3643–3646.
10. Jing Zhang, Gregory Leitus, Yehoshoa Ben-David, David Milstein, Facile conversion of alcohols into esters and dihydrogen catalyzed by new ruthenium complexes. *J. Am. Chem. Soc.* **2005**, *127*, 10840–10841.
11. Denis Spasyuk, Dmitry G. Gusev, Acceptorless dehydrogenative coupling of ethanol and hydrogenation of esters and imines. *Organometallics* **2012**, *31*, 5239–5242.
12. Dipankar Srimani, Ekambaram Balaraman, Boopathy Gnanaprakasam, Yehosho Ben-David, David Milstein, Ruthenium pincer-catalyzed cross-dehydrogenative coupling of primary alcohols with secondary alcohols under neutral conditions. *Adv. Synth. Catal.* **2012**, *354*, 2403–2406.
13. Chidambaram Gunanathan, Linda J. W. Shimon, David Milstein, Direct conversion of alcohols to acetals and H_2 catalyzed by an acridine-based ruthenium pincer complex. *J. Am. Chem. Soc.* **2009**, *131*, 3146–3147.
14. Ekambaram Balaraman, Eugene Khaskin, Gregory Leitus, David Milstein, Catalytic transformation of alcohols to carboxylic acid salts and H_2 using water as the oxygen atom source. *Nat. Chem.* **2013**, *5*, 122–125.
15. Lars Ulrik Nordstrøm, Henning Vogt, Robert Madsen, Amide synthesis from alcohols and amines by the extrusion of dihydrogen. *J. Am. Chem. Soc.* **2008**, *130*, 17672–17673.
16. Nathan D. Schley, Graham E. Dobereiner, Robert H. Crabtree, Oxidative synthesis of amides and pyrroles via dehydrogenative alcohol oxidation by ruthenium diphosphine diamine complexes. *Organometallics* **2011**, *30*, 4174–4179.
17. Boopathy Gnanaprakasam, Ekambaram Balaraman, Yehoshoa Ben-David, David Milstein, Synthesis of peptides and pyrazines from β-amino alcohols through extrusion of H_2 catalyzed by ruthenium pincer complexes: Ligand-controlled selectivity. *Angew. Chem. Int. Ed.* **2011**, *50*, 12240–12244.
18. Cheng Chen, Yao Zhang, Soon Hyeok Hong, N-heterocyclic carbene based ruthenium-catalyzed direct amide synthesis from alcohols and secondary amines: Involvement of esters. *J. Org. Chem.* **2011**, *76*, 10005–10010.
19. Hanxiang Zeng, Zhibin Guan, Direct synthesis of polyamides via catalytic dehydrogenation of diols and diamines. *J. Am. Chem. Soc.* **2011**, *133*, 1159–1161.
20. Boopathy Gnanaprakasam, Jing Zhang, David Milstein, Direct synthesis of imines from alcohols and amines with liberation of H_2. *Angew. Chem. Int. Ed.* **2010**, 49, 1468–1471.
21. Miguel A. Esteruelas, Nicole Honczek, Montserrat Olivan, Enrique Onate, Marta Valencia, Direct access to pop-type osmium (II) and osmium (IV) complexes: Osmium a promising alternative to ruthenium for the synthesis of imines from alcohols and amines. *Organometallics* **2011**, *30*, 2468–2471.
22. Agnese Maggi, Robert Madsen, Dehydrogenative synthesis of imines from alcohols and amines catalyzed by a ruthenium N-heterocyclic carbene complex. *Organometallics* **2012**, *31*, 451–455.

23. Boopathy Gnanaprakasam, Yehoshoa Ben-David, David Milstein, Ruthenium pincer-catalyzed acylation of alcohols using esters with liberation of hydrogen under neutral conditions. *Adv. Synth. Catal.* **2010**, *352*, 3169–3173.
24. Elizaveta Kossoy, Yael Diskin-Posner, Gregory Leitus, David Milstein, Selective acceptorless conversion of primary alcohols to acetals and dihydrogen catalyzed by the ruthenium (II) complex Ru (PPh$_3$)$_2$ (NCCH$_3$)$_2$(SO$_4$). *Adv. Synth. Catal.* **2012**, *354*, 497–504.
25. David M. Hunsicker, Brian C. Dauphinais, Sean P. Mc Ilrath, Nicholas J. Robertson, Synthesis of high molecular weight polyesters via in vacuo dehydrogenation polymerization of diols. *Macromol. Rapid Commun.* **2012**, *33*, 232–236.
26. Martin Nielsen, Henrik Junge, Anja Kammer, Matthias Beller, Hydrogen generation from the chemical energy carrier methanol. *Angew. Chem. Int. Ed.* **2012**, *51*, 1–4.
27. Dipankar Srimani, Yehoshoa Ben-David, David Milstein, Direct synthesis of pyridines and quinolines by coupling of γ-amino-alcohols with secondary alcohols liberating H 2 catalyzed by ruthenium pincer complexes. *Angew. Chem. Int. Ed.* **2013**, *52*, 1–5.
28. Fei Wang, Ruijuan Shi, Zhi-Quan Liu, Pan-Ju Shang, Xueyong Pang, Shuai Shen Zhaochi Feng, Can Li, Wenjie Shen, Highly efficient dehydrogenation of primary aliphatic alcohols catalyzed by Cu nanoparticles dispersed on rod-shaped La$_2$O$_2$CO$_3$. *ACS Catal.* **2013**, *3*, 890–894.
29. Tianliang Lu, Zhongtian Du, Junxia Liu, Chen Chen, Jie Xu Chinese, Dehydrogenation of primary aliphatic alcohols to aldehydes over Cu-Ni bimetallic catalysts. *J Catal.* **2014**, *35*, 1911–1916.
30. Hajar Golshadi Ghalehshahi, Robert Madsen, Silver-catalyzed dehydrogenative synthesis of carboxylic acids from primary alcohols. *Chem. Eur. J.* **2017**, *23*, 11920–11926.
31. Shenxiang Yin, Qingshu Zheng, Jie Chen, Tao Tu, Acceptorless dehydrogenation of primary alcohols to carboxylic acids by self-supported NHC-Ru single-site catalysts *J. Catal.* **2022**, *408*, 165–172.
32. Masaki Shibata, Ryoko Nagata, Susumu Saito, Hiroshi Naka, Dehydrogenation of primary aliphatic alcohols by Au/TiO$_2$ photocatalysts. *Chem. Select.* **2017**, *46*, 580–582.
33. Jian-Ji Zhong, Wai-Pong To, Yungen Liu, Wei Lua, Chi-Ming Che, Efficient acceptorless photodehydrogenation of alcohols and N-heterocycles with binuclear platinum (II) diphosphite complexes. **Chem. Sci. 2019, 10**, 4883–4889.
34. Hiromu Fuse, Harunobu Mitsunuma, Motomu Kanai, Catalytic acceptorless dehydrogenation of aliphatic alcohols. *J. Am. Chem. Soc.* **2020**, *142*, 4493–4499.
35. Cheng Chen, Zhi-Qin Wang, Yan-Yan Gong, Ji-Chao Wang, Ye Yuan, Hua Cheng, Wei Sang, Somboon Chaemchuen, Francis Verpoort, Cobalt embedded in nitrogen-doped porous carbon as a robust heterogeneous catalyst for the atom-economic alcohol dehydrogenation to carboxylic acids. *Carbon* **2021**, *174*, 284–294.

5 Dehydrogenation Reactions of Hydrocarbons
Alkane, Alkenes, and Aromatic Hydrocarbons

Chandni Pathak and Vijay Bahadur

List of abbreviations

S. No.	Abbreviation	Full Form
1	FCC	Fluid Catalytic Cracking
2	UOP	Pacol Process
3	XANES	X-ray Absorption Near-Edge Structure
4	DFT	Density Functional Theory
5	FBD	Fluidized-Bed Reactor
6	PDH	Propane Dehydrogenation
7	STAR	STeam Active Reforming
8	ODH	Oxidative Dehydrogenation
9	DH	Dehydrogenation
10	ZSM-5	Zeolite Socony Mobil-5
11	GHSV	Gas Hourly Space Velocity
12	TBE	*tert*-butylethylene
13	PCP	Phenylcyclohexylpiperidine
14	COD	Cyclooctadiene
15	DDQ	2,3-dichloro-5,6-dicyano-1,4-benzoquinone
16	NBD	Norbornadiene

5.1 INTRODUCTION

Dehydrogenation (DH) of lower alkanes serves as the best reaction to a wide range of valuable chemicals. Alkanes are easily available feedstocks from petroleum and gas resources. However, the direct conversion of these alkanes to valuable chemicals is challenging as lower alkanes are relatively inert due to a lack of a vacant orbital or lone pair of electrons.[1]

Olefins, particularly propene and ethylene, are amongst the most prominent compounds in the chemical industry as they serve as the feedstock in the production of a wide variety of important chemicals, namely, polyethylene, polypropylene, ethylene glycol, acetaldehyde, acetone, propylene oxide, and important intermediates such as ethylbenzene and propionaldehyde.[2] Over recent years, demand for these building blocks has skyrocketed. The most common methods that are employed for obtaining these light olefines are steam cracking and fluid catalytic cracking (FCC) of naphtha, light diesel, and other oil by-products.[3] The petrochemical industry is encountering innumerable issues including the high-energy demands of these processes, their low selectivity toward the production of particular olefins, declining petroleum reserves, and spiking oil prices. These issues have paved the path for the search/research of further economical feedstock and highly efficient conversion technologies.[3]

DOI: 10.1201/9781003321934-5

Synthesis of the lower molecular weight olefins via DH of lower alkanes is the prominent reaction of recent times because of its inevitable contribution to the fuel and petroleum industries. Additionally, the alkene yields in DH reactions are limited due to the thermodynamics of the reaction. The reaction being endothermic requires increased temperature, but the temperature decreases with the increase in the chain length of alkanes.[4] However, the reaction yields become enhanced upon escalating temperature but hamper the alkene selectivity. The tendencies of dehydrocyclization or aromatization, thermal cracking, and coke formation increase with the chain length of the alkanes.[5]

As discussed earlier, the direct conversion of lower alkanes to the corresponding olefine demands elevated energy. Scientists are exploring the catalyst that, on the one hand, fulfills the lower energy requirement of the DH reaction but, on the other hand, spikes the yield of olefins with higher selectivity. Non-oxidative dehydrogenation (non-ODH), oxidative dehydrogenation (ODH), and the use of pincer complexes have been suggested to lower the energy of the reaction, which we examine in this chapter.

5.2 NON-OXIDATIVE DEHYDROGENATION

Non-ODH reaction of smaller alkanes seems like a straightforward reaction as shown by its stoichiometry (Scheme 5.1),

On the contrary, non-ODH reaction has many thermodynamic obstacles and kinetic hurdles in terms of the vital stages of the reaction. The reaction is a thermodynamically uphill battle, specifically, it is endothermic and requires around 113–134 kJ/mol of energy for the elimination of two hydrogen atoms depending upon the chain length of the parent alkane.[5] Many commercially used patented non-oxidative technologies have been used for alkane DH at the industrial level. Based on the type of catalyst, design of the reactor, alkane composition, and mode of the heat energy input, these methods can be categorized. The outline for this is shown in Table 5.1.[5] These are cyclic processes that involve the repeated DH reaction and the rejuvenation of a catalyst as represented in Table 5.1.

DIFFERENT NON-OXIDATIVE ALKANE DEHYDROGENATION INDUSTRIAL CATALYST

5.2.1 Platinum-Based Catalyst

The first industrial-based alkane DH technology is from 1968 and is known as the Pacol Process (UOP) that employs a platinum-based catalyst for the non-ODH of the alkane chain. Platinum was one of the very first noble metals to be used in commercial petroleum refining due to its C-H bond

$$C_nH_{2n+2} \longrightarrow C_nH_{2n} + H_2$$

SCHEME 5.1 Schematic representation of dehydrogenation of alkanes

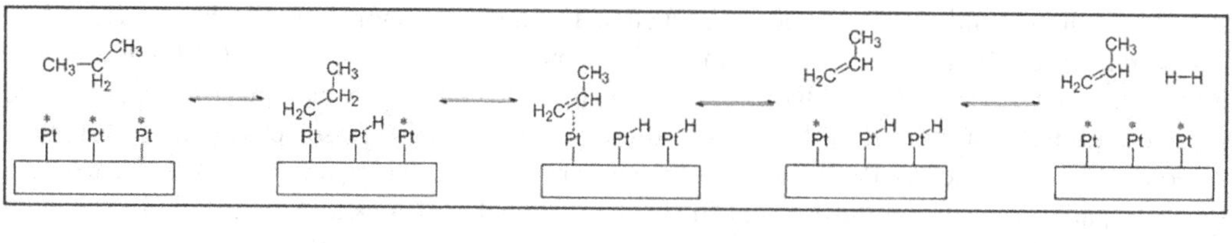

FIGURE 5.1 Dehydrogenation of propane as catalysed by Pt-Sn nanoparticle supported over Al_2O_3.

Dehydrogenation Reactions of Hydrocarbons

activation property.[3] Platinum shows high selectivity toward C-H bond activation and low activity toward C-C bond cleavage activity. Hence, all of these properties make it a better catalyst for the DH reaction of alkane. The platinum catalyst used for DH purposes is typically supported with tin to increase its reactivity. The exact role of tin is unknown, but some XANES (X-ray Absorption Near-Edge Structure) and DFT (Density Functional Theory) data provide for the electronic interaction prevailing between the two metals. A very little quantity of tin is adequate for the electronic interactions between the metals.[6,7]

The preparation of platinum catalysts for commercial use is conducted by infusing the support material with a solution of tin and platinum precursors (Figure 5.1).[3] The support material that is commonly used for the industrial platinum-based DH catalyst is alumina.

5.2.2 Chromium Oxide-Based Catalyst

Frey and Hupke reported chromium oxide as one of the exceptional oxide-based catalysts for the non-oxidative alkane DH reaction. These transition metal oxides can activate C-H bonds through the extraction of the hydrogen atom. Chromium oxide-based DH catalyst that is used in industrial

TABLE 5.1
Commercial Dehydrogenation Processes for Alkanes.[7]

	Catofin	FBD	Linde PDH	UOP oleflex	STAR
Reactor design	Adiabatic fixed-bed reactors	Fluidized-bed reactor	Isothermal fixed bed	Adiabatic moving bed	DH reactor + adiabatic oxyreactor
Catalyst type	Cr_2O_3/Al_2O_3 alkali promoter	Cr_2O_3/Al_2O_3	Cr_2O_3/Al_2O_3, Pt/Sn on Zirconia	$Pt/Sn/Al_2O_3$ with alkaline promote	Pt/Sn on $ZnAl_2O_4$/ $CaAl_2O_4$
Regeneration mode	Air oxidation	Air oxidation	Air oxidation	Air oxidation and reduction in H_2	Air oxidation and reduction in H_2
Operation	Cyclical alternation of reaction and regeneration cycles	Catalyst circulates continuously from the bottom of the reactor to the top of the regenerator	Cyclical	Continuous transport of catalysts between regenerator and reaction zone	Cyclical
Temperature (°C)	590–650	550–600	600	550–620	DH: 550–590; ODH: <600
Pressure (bar)	0.3–0.5	1.1–1.5	>1	2–5	DH: 5–6; ODH: <6
Cycle time (min)	15–30	NA	540 h	NA	NA
Conversion (%)	48–65	40	30	25	40
Selectivity (%)	82–87	89	30	89–91	89
Mode of heat supply	Heat formed in the catalyst regeneration	Fuel added during regeneration	Heating of the reactors	Fired furnaces are placed between one reactor and another	Heating of the DH reactor
Licensor/ developer	SüdChemie/ABB Lummus	Snamprogetti-Yarsinetz	Linde (BASF)	UOP Inc.	Uhde

NA = Not Available

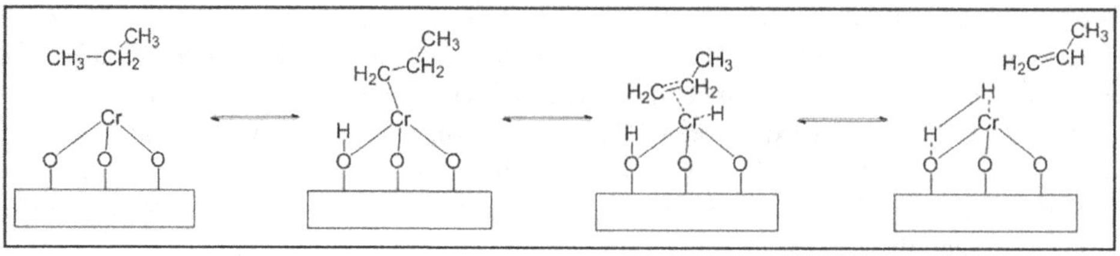

FIGURE 5.2 Dehydrogenation of propane as catalysed by CrO_x supported over Al_2O_3 showing the coordinatively unsaturated Cr^{3+} species catalyse the reaction in the active site on the catalyst.

petroleum refining is prepared by the saturation of an aqueous solution of chromate salt of 5 and 15 wt.% Cr loading on supporting material that have a high surface area[8,9]. Multiple supporting materials have been employed in the process, namely, alumina, silica, and zirconia. The ideal supporting material is one that shows high thermal and chemical stability, and in this regard, zirconia is the preferred supporting material over the other two, while alumina is always preferred over silica.

The catalytic active site of the chromium oxide-based catalyst consists of coordinatively unsaturated Cr^{+3} and Cr^{+2} ions.[10] The activation of the C-H bond of the substrate occurs in these active sites of the catalyst where coordinatively unsaturated Cr^{+3} catalyse the DH reaction (Figure 5.2)[3]. The C-H bond scission is also supported by "surface oxygen transfer"[11,12].

5.2.3 Vanadium Oxide-Based Catalyst

Despite the exceptional catalytic activity of the platinum and chromium oxide-based catalyst in alkane DH, many disadvantages are associated with the use of these catalysts, for example, catalyst poisoning, the use of expensive platinum, and the adverse effect of chromium on the environment.[7] All of these drawbacks associated with the use of these catalysts have led scientists to search for more appropriate and environmentally friendly catalysts. One such catalyst is vanadium oxide. It is known for its activity in many hydrocarbon oxidation reactions and DH reactions. The very first citation of vanadium oxide was of ODH using vanadium-magnesium mixed oxides in 1980.[13,14]

There are multiple factors that decide the type of vanadium utilized in the catalyst, namely, the carrier type, the surface area of the supporting material, the metal loading, and the oxidation state of the vanadium metal ion.[15] Researchers have concluded that vanadium is present as V^{+3} or V^{+4} in the catalytically active species supported over alumina and is responsible for the DH reaction of alkanes.[16-19] At the beginning of the DH reaction, the vanadium is present as V^{+5}; furthermore, the reduction of vanadium catalyst by hydrocarbon leads to the formation of V^{+3} and V^{+4}.[20]

5.2.4 Molybdenum Oxide-Based Catalysts

Another widely used catalyst in alkane DH reaction is molybdenum oxide. The very first citation of a molybdenum oxide-based catalyst supported over alumina (MoO_3/Al_2O_3) was reported in 1946 for the dehydrocyclization of n-heptane.[21] This study led to the wide use of this catalyst in the non-ODH of alkanes.

The chemistry of this catalyst is similar to that of the vanadium catalyst as molybdenum can also present in different forms as monomers, polymers, or crystallites depending upon the type of supporting material used and the conditions in which the catalyst is prepared.[22]

5.2.5 Gallium Oxide-Based Catalyst

The very first DH reaction catalyzed by gallium oxide supported over ZSM-5 was the conversion of "propane to aromatics". Ever since its report in 1980, it has become a catalyst of interest for scientists.[23,24] Gallium oxide (Ga_2O_3) catalysts in DH reactions can be utilized in the form of bulk oxide and with supporting materials.

Chen et al. demonstrated that the criteria for the DH reaction as catalyzed by gallium oxide is the presence of Lewis acid sites, that is, the presence of coordinately unsaturated tetrahedral Ga^{+3} ions, which work as catalytic active sites for the hydrocarbon's DH reaction.[25]

The supporting material that shows prominent results is zeolite, which contains a low concentration of strong acid sites and a high concentration of weak acid sites and makes a stable gallium oxide catalyst for alkane DH.[26]

5.2.6 Carbon-Based Catalyst

Carbon-based materials have been employed as supporting materials in a variety of catalysts due to their high surface area, cost-effectiveness, and environmentally friendly nature. The activated carbon has also shown activity toward the DH of alkanes.[27] McGregor and his co-workers have shown that VO_x/Al_2O_3 catalyst encapsulated in coke deposit is catalytically active towards the DH of butane.[28]

5.3 OXIDATIVE DEHYDROGENATION

The ODH of an alkane is an exothermic reaction. The olefin yield of the ODH reaction as catalyzed by different catalysts depends upon the type of alkane and reaction condition. In non-ODH, the C-H bond is directly activated by the metal, and in ODH, the C-H bond is activated by the oxygen species through hydrogen or proton abstraction.

5.3.1 Groups V and VI Transition Metal Oxides

The transition metals belonging to groups V and VI have been successfully utilized in the ODH of ethane and propane as bulk catalysts or as supporting materials possessing high surface area. Out of all the metals, vanadium has come across as the most active metal in ODH, and particularly, the VMgO system has gained more attention due to its higher propene yields and selectivity. The catalytic activity of the VMgO system towards ODH can be augmented when supported with a redox metal ion. In addition, VMgO, vanadium hydroxycalcium apatite, has also been found to be active toward the ODH of alkanes.[29] Compared to VMgO, the supported vanadium catalyst shows lower activity, and the catalytic activity of the supported vanadium catalyst increases with increased concentration of VO_x.

The kinetics of the ODH reaction of propane as catalyzed by ZrO_2 supported over VOx, MoOx, NdOx, and WOx can be explained by the Mars-van Krevelen model.[30,31] According to this model, the reaction of alkane with the lattice oxygen of the oxo metal species leads to the formation of oxygen vacancies in the catalyst. The oxygen-deficient metal species then re-oxidize to their original state with adsorbed O_2. The adsorbed O_2 molecule can be introduced into the oxo metal species through subsequent reduction processes.[32]

5.3.2 Ni-Based Catalyst Systems

It is well-known that the non-stoichiometric oxides of metal show redox properties that can be utilized for many catalytic applications. In line with this fact, the non-stoichiometric oxide of nickel, NiO_x, having Ni in variable oxidation states, has been examined in the ODH of alkanes. Some of the examples of the ODH of ethane and propane as catalysed by NiO_x are shown in Table 5.2.[5]

Even though the ODH reaction of alkane as catalysed by NiO_x follows the Mars-van Krevelen mechanism,[33] the direct participation of the nickel centre in the C-H bond activation cannot be ruled out.

TABLE 5.2
Comparison of Different Ni-Based Catalysts for the ODH of Alkane.

Catalyst	Preparation method	Substrate	Reaction conditions	Yield (%)
NiO-LiCl/SZ	Precipitation and impregnation	Ethane	$C_3H_8:O_2:N_2 = 10:10:80$ Flow rate = 3,600 ml h^{-1} T = 650 °C	73.5
		Propane	$C_3H_8:O_2:N_2 = 10:10:80$ Flow rate = 3,600 ml h^{-1} T = 600 °C	10.4
CeNi$_{0.5}$OY	Coprecipitation using triethylamine (TEA)	Propane	$C_3H_8:O_2:N_2 = 5:15:80$ Flow rate = 6,000 ml h^{-1} T = 375 °C	5.4
CeNi$_2$O	Oxalate gel coprecipitation method	Propane	$C_3H_8:O_2:N_2 = 0.7:1:12$ GHSV = 5,480 h^{-1} T = 275 °C	10.6
Ce$_{0.015}$Nb$_{0.03}$NiO	Modified sol-gel method using citric acid as a ligand	Propane	$C_3H_8:O_2:He = 1.2:1.0:1.2$ Flow rate = 10,000 ml h^{-1}/g cat T = 250 °C	10.4
NiO–Bi$_2$O$_3$–ZrO$_2$	Coprecipitation	Propane	$C_3H_8:O_2:N_2 = 10:5:5$ GHSV = 15,000 ml h^{-1}/g-cat T = 400 °C	10.4
Mesoporous NiO	Surfactant-assisted route	Propane	$C_3H_8:O_2:He = 1.2:1.0:1.2$ GHSV = 10,000 ml h^{-1}/g-cat T = 450 °C	13.2
Ni–Nb–O (Nb/Ni = 0.11–0.18)	Evaporation method	Ethane	$C_2H_6:O_2:He = 9.1:9.1:81.8$ GHSV = 6,667 ml h^{-1}/g-cat T = 400 °C	46.0

5.3.3 Lithium- and Halide-Containing Catalysts

Lithium- and halide-containing catalysts display ODH activity and alkene selectivity at a higher temperature. The alkene selectivity of the catalyst increases with the increase in temperature. The catalytic reaction follows the Eley-Rideal mechanism. The alkene selectivity is mostly dependent upon the gas-phase radical chemistry since the C-H bond activation takes place at the surface of the catalyst while the transformation of the alkene from the alkyl radical takes place in the gas phase. A few examples of lithium- and halide-catalyzed ODH reactions are given in Table 5.3.[5]

5.4 DEHYDROGENATION OF ALKANES BY PINCER COMPLEXES

Alkane DH reaction has always been a challenge for scientist as the controlled conversion of an alkane to corresponding monoenes with regioselectivity is very difficult. These challenges have made organic chemists come up with catalysts that can overcome these problems. In this regard, the very first transition metal-mediated DH of alkane was discussed by Crabtree and associates in 1979. They reported the conversion of cyclopentane and cyclooctane to corresponding cyclopentadienyl and cyclooctadiene (COD) as catalyzed by iridium complexes in the presence of a hydrogen acceptor, namely, *tert*-butylethylene (TBE), under reflux conditions[34–36] (Scheme 5.2).

As the alkane DH reaction is thermodynamically uphill, it is necessary to couple the reaction with a hydrogen transfer reaction to decrease the energy requirement of the reaction. Ever since Crabtree, the utilization of TBE as a sacrificial hydrogen acceptor has become a trend in these reactions.

TABLE 5.3
Lithium- and Halide-Containing Catalysts for the ODH of Alkane.

Catalyst	Substrate	Reaction conditions	Yield (%)
Li/Ni/CaO	Ethane	$C_2H_6:O_2:N_2 = 15.2:8.6:76.2$ GHSV = 1,000 h^{-1} T = 620 °C	34.8
3.5 wt.% LiCl/SZ	Ethane	$C_2H_6:O_2:N_2 = 10:10:80$ Flow rate = 3,600 ml h^{-1} T = 650 °C	68.1
Mg–Dy–Li–Cl	Butane	$C_4H_{10}:O_2:He = 5-7:10-12:81-85$ WHSV = 0.2 h^{-1} T = 580 °C	50.2
Sn–Mn–LiCl/MgO-YSZ	Ethane	$C_2H_6:O_2:N_2 = 1:1:8$ GHSV = 3,600 ml h^{-1}/g cat T = 662 °C	65.0
Li/Dy/Mg/O/Cl	Ethane	$C_3H_8:O_2 = 1:1$ WHSV 1.8 h^{-1} T = 550 °C	62.0
Li/Dy/Mg/O/Cl	Propane	$C_2H_6:O_2 = 1:1$ WHSV = 0.8 h^{-1} T = 585 °C	14.0

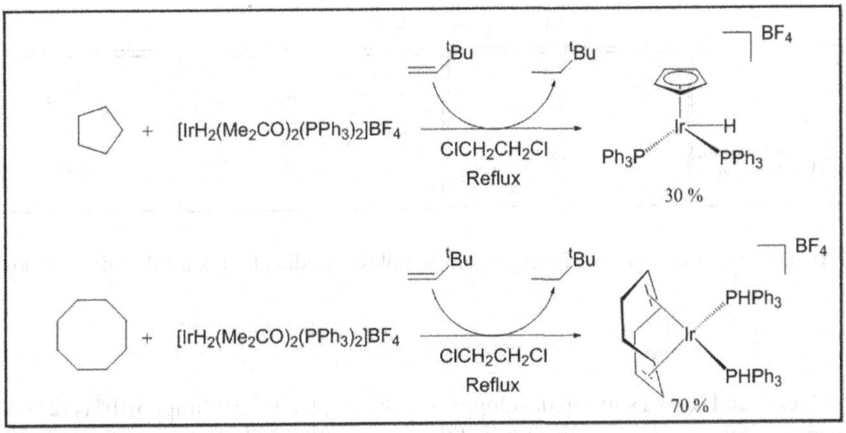

SCHEME 5.2 Conversion of cyclopentane and cyclooctane to corresponding cyclopentadienyl and cyclooctadiene as catalyzed by [IrH$_2$(Me$_2$CO)$_2$(PPh$_3$)$_2$]BF$_4$.

5.4.1 Dehydrogenation of Alkane by Pincer Iridium Complexes

The discovery of Phenylcyclohexylpiperidine (PCP)-based pincer iridium complex, (^{tBu}$_4$PCP)IrH$_2$ (**1**), by Kaska and Jensen as a more efficient catalyst towards alkane transfer DH than its rhodium analogs[37] made many scientists take notice and paved the path for the advancement in the area of pincer-based iridium complex.[38,39] Because of the thermal stability of the complex, Kaska, Jensen, and Goldman were able to successfully catalyze the acceptorless DH of cyclodecane (Scheme 5.3) with the help of iridium complex ((^{tBu}$_4$PCP)IrH$_2$ (**1**)) resulting in 360 TOs in 24 hours at 200 °C.[40] Furthermore, Goldman showed that the isopropyl analog of (**1**) (^{iPr4}PCP)IrH$_4$ (**2**) was found to be more reactive than its parent complex as a catalyst in the acceptorless DH of cyclodecane.

Many analogs of (^{tBu4}PCP)Ir (**1**) have been synthesized in recent years for alkene DH reactions (Figure 5.3)[41,42] by modifying the aryl group,[43–46] substituting the CH$_2$ linker,[43,47] and ligating

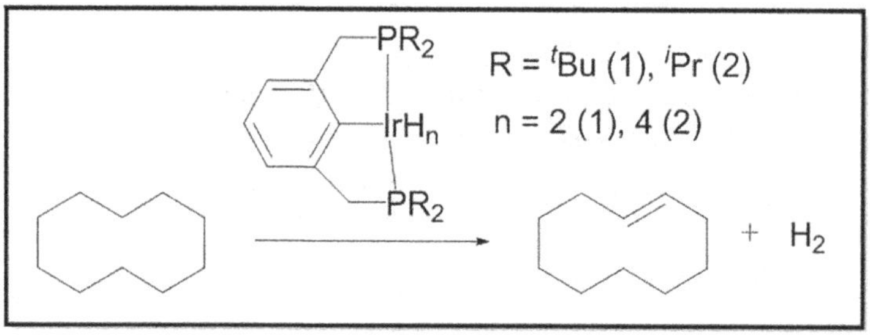

SCHEME 5.3 Dehydrogenation of cyclodecane as catalyzed by $(^{tBu}{}_4PCP)IrH_2$ (1).

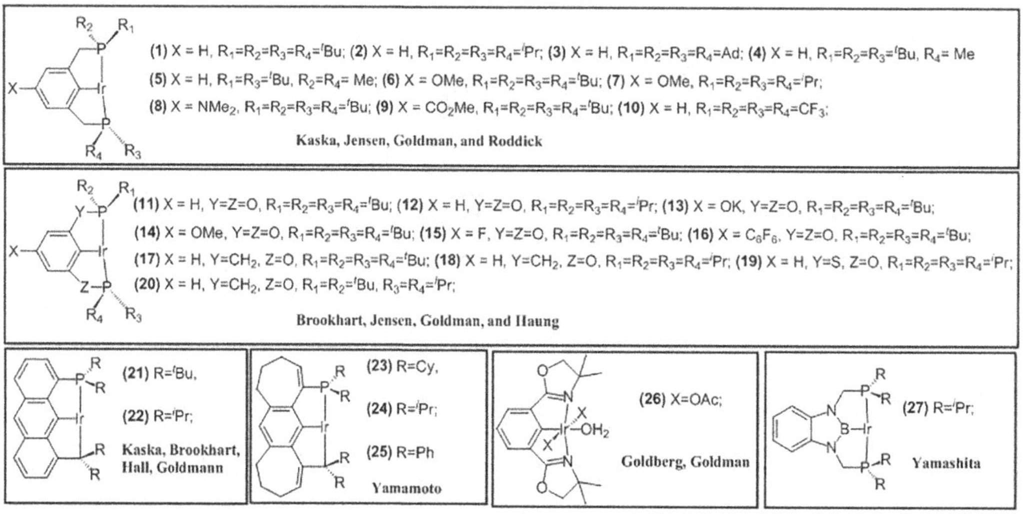

FIGURE 5.3 Different variations of pincer iridium complexes studied for alkane dehydrogenation.

atoms.[48,49] Haenel and Kaska's group developed a new complex $(^{tBu}Anthraphos)IrH_2$ (**21**) that shows higher thermal stability and was active toward the acceptorless DH of alkanes.[50]

Furthermore, Goldman and his group studied the effects of sterically hindered substitutions on the phosphine group and its corresponding effects on the catalytic efficiency of the complex.[48] The results are summarized in Table 5.4.

5.4.2 Pincer-Ruthenium Complexes as Catalysts for Alkane Dehydrogenation

Roddick and his co-workers in 2011 reported (^{CF_3}PCP) Ru(COD)H complex (**28**) as a catalyst for alkane DH. The catalyst showed superior yields of olefin at a higher temperature. The requirement of high temperature was a consequence of the dimerization of the catalyst, which led to its deactivation[52] (Scheme 5.4).

Huang and his group developed a series of pincer-ruthenium complexes and tested them in the transfer DH of alkanes as catalysts.[53] The complex $(^{iPr}{}_4POCOP)Ru(NBD)H$ (**29**, NBD = norbornadiene) showed maximum turnover for the conversion of TBE to TBA at significantly lower concentrations of TBE for the transfer DH of COD at 200 °C.

TABLE 5.4
Alkane Dehydrogenation by Pincer Iridium Complex with Varying Steric Hindrance and at Different TBE Concentrations.[51]

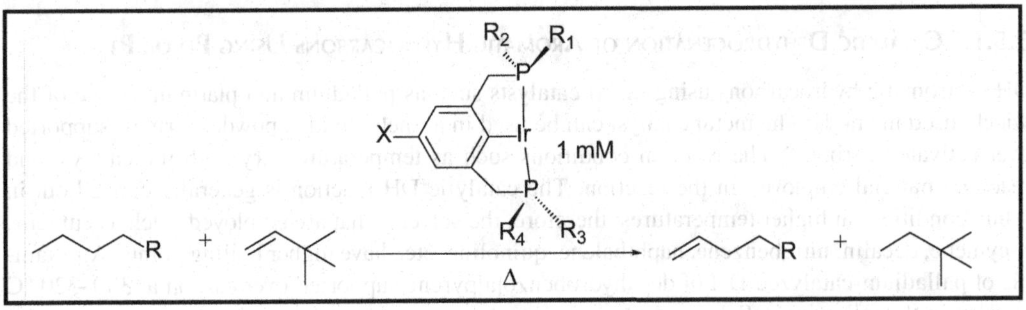

S. No	R	T (°C)	Turn over number after 10 mins (total olefin; 1 mM = 1 TO)			
			Complex 1	Complex 2	Complex 4	Complex 5
1.	n-Bu	157	31	86	126	76
2.	CH_3	200	55	665	110	395

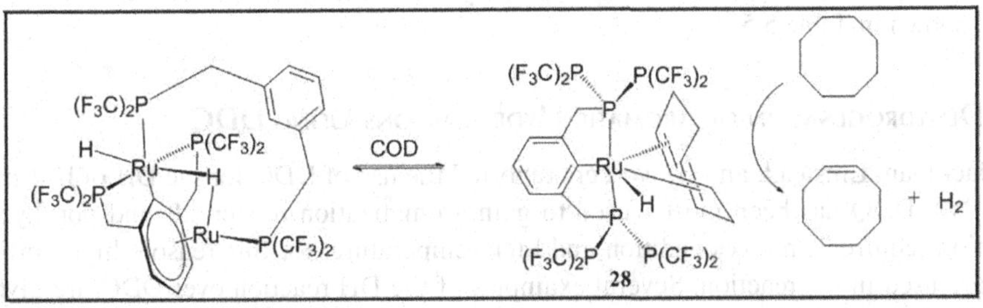

SCHEME 5.4 COD dehydrogenation as catalysed by $(^{CF_3}PCP)Ru(COD)H$ (**28**).

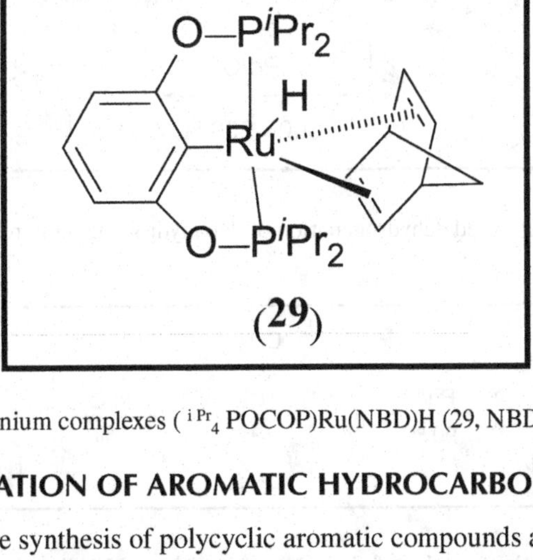

SCHEME 5.5 Pincer-ruthenium complexes ($^{iPr}_4$POCOP)Ru(NBD)H (29, NBD = norbornadiene).

5.5 DEHYDROGENATION OF AROMATIC HYDROCARBONS

Usually, the last step in the synthesis of polycyclic aromatic compounds and their derivatives is the DH reaction.[54] Commonly, the DH reaction is undertaken using some catalysts such as sulphur and selenium or metal catalysts such as palladium and platinum. The use of these catalysts gives good yields but at the expense of selectivity. These methods cannot be employed for the synthesis of sensitive compounds because of the drastic conditions employed in the reaction. Given the drawbacks of using these catalysts, researchers have developed milder methods by using high oxidation potential quinones such as 2,3-dichloro-5,6-dicyano-1,4-benzoquinone (DDQ) and chloranil.

5.5.1 Catalytic Dehydrogenation of Aromatic Hydrocarbons Using Pd or Pt

DH of aromatic hydrocarbons using metal catalysts such as palladium and platinum is one of the widely used methods. The metal catalyst can be used in a finely divided powder form or supported over activated carbon.[55] The reaction conditions such as temperature vary with the catalyst and reaction material employed in the reaction. The catalytic DH reaction is generally carried out in reflux conditions at higher temperatures; therefore, the solvents that are employed, such as cumene, p-cymene, decalin, nitrobenzene, naphthalene, quinoline, etc., have higher boiling points. An example of palladium-catalyzed DH of decahydrobenzo[a]pyrene supported over carbon at 300–320 °C is given in the following[56] (Scheme 5.6).

Substitution of the bulky groups on the ring affects the rate of the reaction and the temperature requirement. They interfere with the adsorption of the catalyst, which results in the decline of the rate of the reaction and an increment in the required temperature. The yields for the DH reaction of cis and trans isomer of 9,10-dimethyl-9,10-dihydroanthracene to 9,10-dimethylanthracene as catalysed by 10% Pd/C were found to be 90% and 20%, respectively, because of the steric hindrance of the trans methyl group with the catalyst, which blocks the association of the catalyst[57] (Scheme 5.7).

A few examples of aromatic DH as catalysed by metal catalysts (Pd and Pt) with comparative yields are shown in Table 5.5.[54]

5.5.2 Dehydrogenation of Aromatic Hydrocarbons Using DDQ

Braude, Jackman, Linstead, and co-workers reported the use of DDQ for the DH of hydroaromatic compounds.[58] DDQ has been mostly used to gain aromatization in steroids and porphyrin rings. The reaction requires a reflux condition and high temperature; for this reason, high boiling point solvents are used in the reaction. Several examples of the DH reaction over DDQ are given in the following[59] (Scheme 5.8).

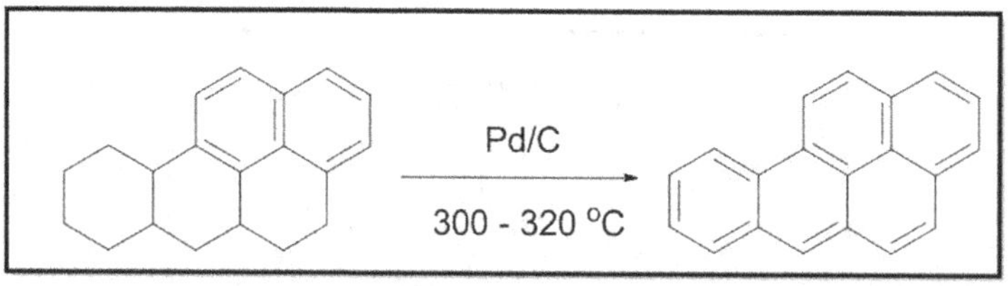

SCHEME 5.6 Palladium-catalysed dehydrogenation of decahydrobenzo[a]pyrene supported over carbon.

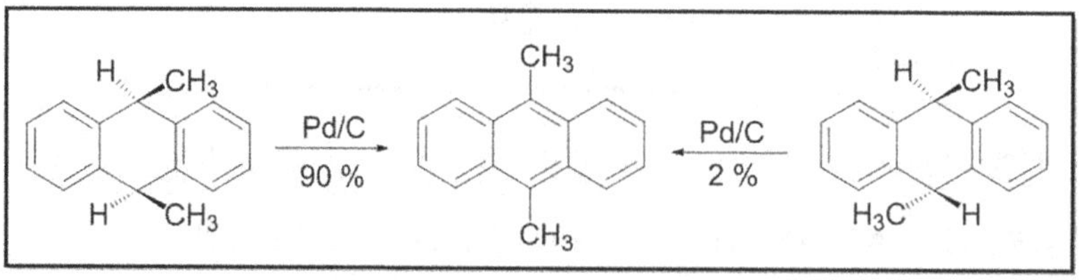

SCHEME 5.7 Dehydrogenation reaction of cis and trans isomer of 9,10-dimethyl-9,10-dihydroanthracene as catalysed by 10% Pd/C.

TABLE 5.5
Catalytic Dehydrogenation of Unsubstituted and Substituted Polycyclic Hydrocarbons.

Substrate	Product	Catalyst	Reaction condition	Yield (%)
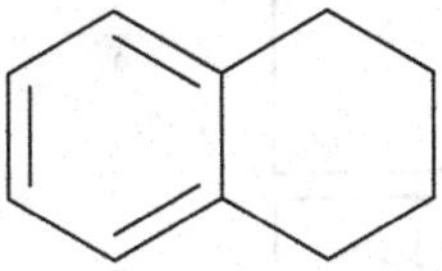	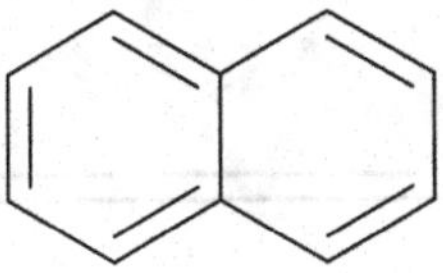	Pd/C	Reflux, 22 h	97
		Pt/C	Reflux, 21 h	78
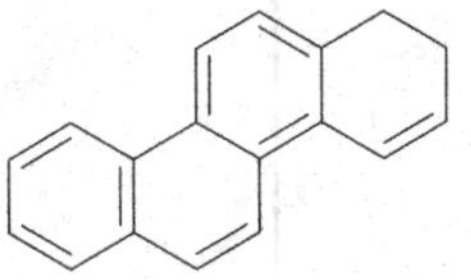	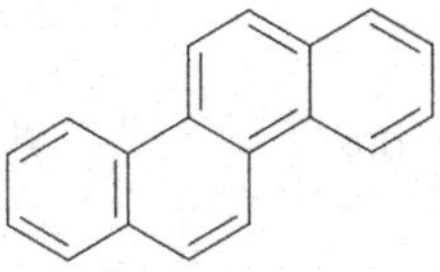	Pd/C	300–320 °C, 1 h	93
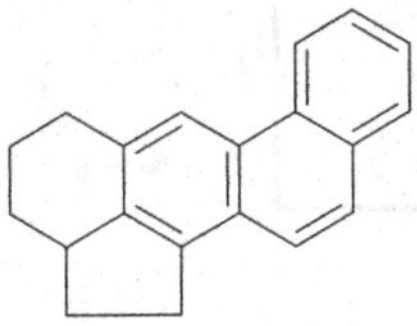	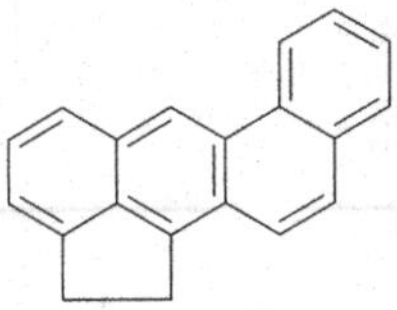	Pd/C	300 °C, 0.5 h	80
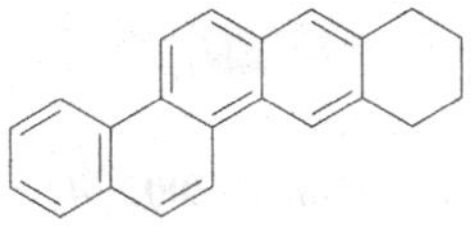	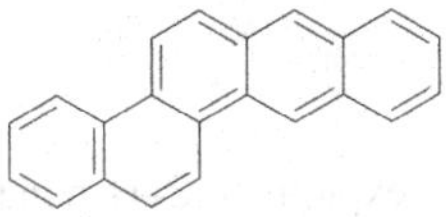	Pd	300 °C, 3 h	
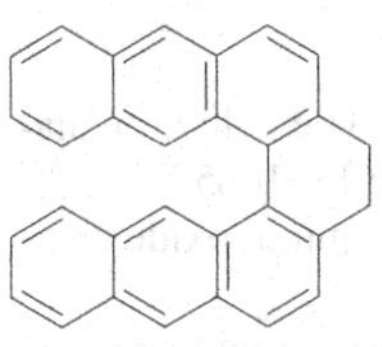	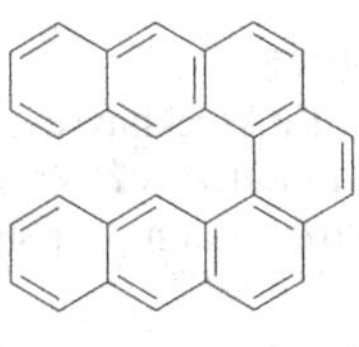	Pd/C	p-Cumene, reflux, 4 h	80
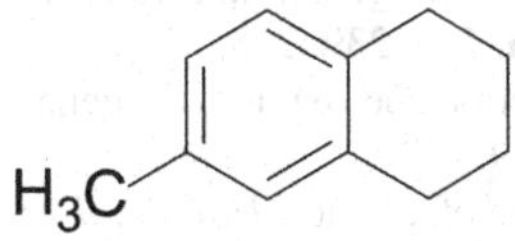	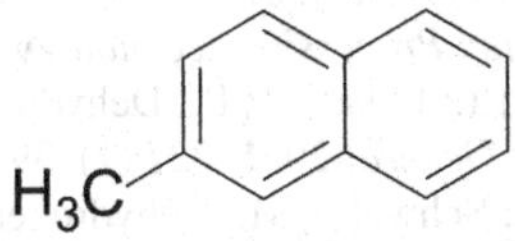	Pd/C	Reflux	80
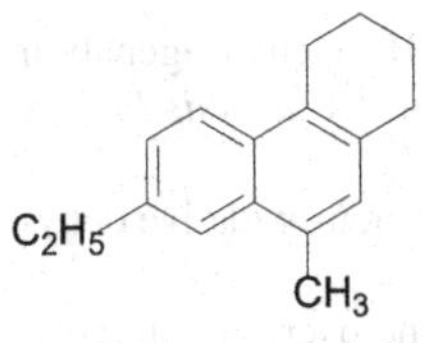	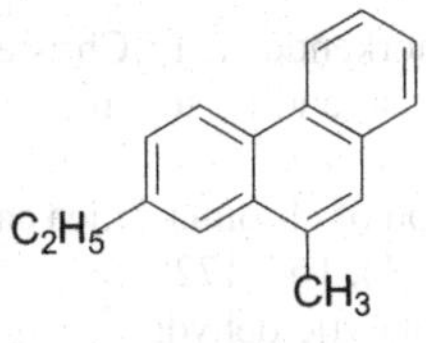	Pd/C	300–320 °C, 1 h	93
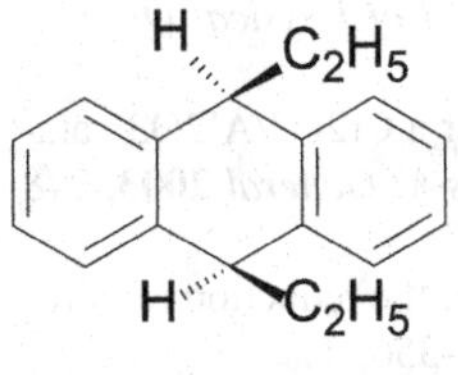	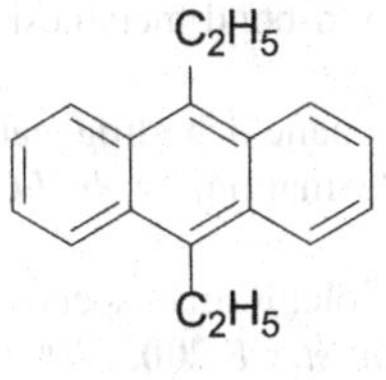	Pd/C	Diglyme, reflux, 12 h	90
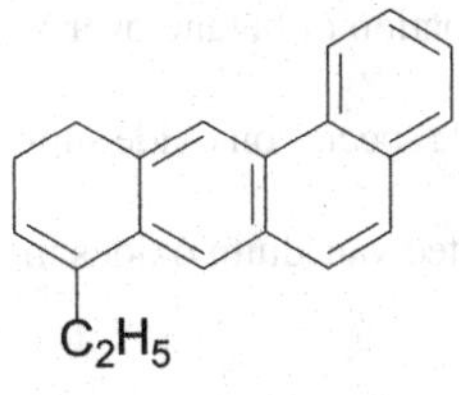	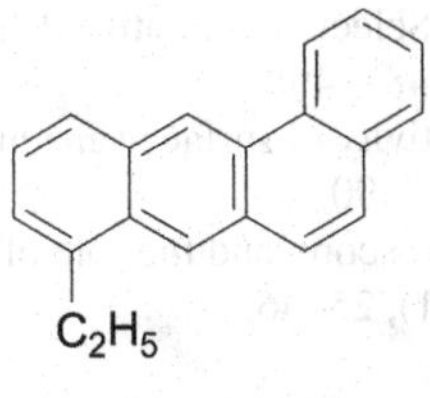	Pd/C	330–340 °C, 1 h	70

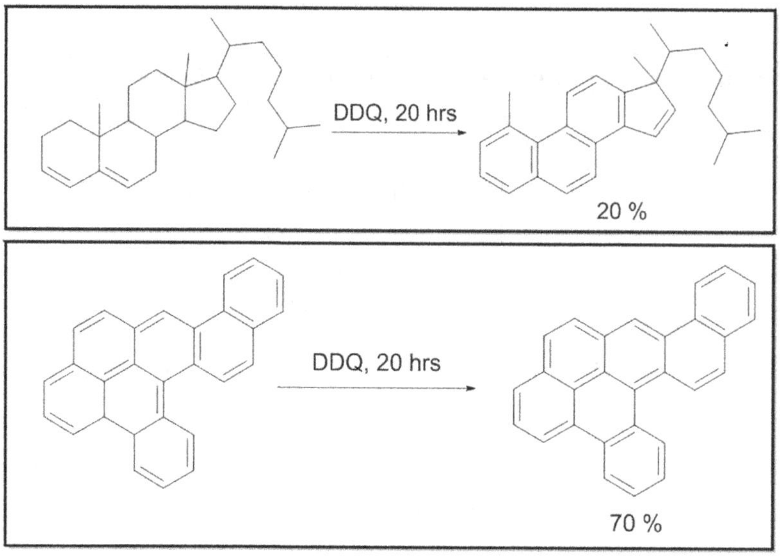

SCHEME 5.8 Dehydrogenation of polycyclic hydroaromatic compounds using DDQ.

REFERENCES

1. Labinger, J. A.; Bercaw, J. E., Understanding and exploiting C–H bond activation, *Nature* **2002**, *417* (6888), 507–514.
2. *Chemical & Engineering News Archive* **2006**, *84* (28), 59–68.
3. Sattler, J. J. H. B.; Ruiz-Martinez, J.; Santillan-Jimenez, E.; Weckhuysen, B. M., Catalytic dehydrogenation of light alkanes on metals and metal oxides, *Chemical Reviews* **2014**, *114* (20), 10613–10653.
4. Weckhuysen, B. M.; Schoonheydt, R. A., Alkane dehydrogenation over supported chromium oxide catalysts, *Catalysis Today* **1999**, *51* (2), 223–232.
5. James, O. O.; Mandal, S.; Alele, N.; Chowdhury, B.; Maity, S., Lower alkanes dehydrogenation: Strategies and reaction routes to corresponding alkenes, *Fuel Processing Technology* **2016**, *149*, 239–255.
6. Bhasin, M. M.; McCain, J. H.; Vora, B. V.; Imai, T.; Pujadó, P. R., Dehydrogenation and oxydehydrogenation of paraffins to olefins, *Applied Catalysis A: General* **2001**, *221* (1), 397–419.
7. Caspary, K. J.; Gehrke, H.; Heinritz-Adrian, M.; Schwefer, M., Dehydrogenation of alkanes, *Handbook of Heterogeneous Catalysis: Online* **2008**, 3206–3229.
8. Rovik, A. K.; Hagen, A.; Schmidt, I.; Dahl, S.; Chorkendorff, I.; Christensen, C. H., Dehydrogenation of light alkanes over rhenium catalysts on conventional and mesoporous MFI supports, *Catalysis Letters* **2006**, *109* (3), 153–156.
9. Martyanov, I.; Sayari, A., Sol–Gel assisted preparation of chromia–silica catalysts for non-oxidative dehydrogenation of propane, *Catalysis Letters* **2008**, *126* (1), 164–172.
10. Lillehaug, S.; Børve, K. J.; Sierka, M.; Sauer, J., Catalytic dehydrogenation of ethane over mononuclear Cr(III) surface sites on silica. Part I. C-H activation by σ-bond metathesis, *Journal of Physical Organic Chemistry*, **2004**, 990–1006.
11. Gascón, J.; Téllez, C.; Herguido, J.; Menéndez, M., Propane dehydrogenation over a Cr2O3/Al2O3 catalyst: Transient kinetic modeling of propene and coke formation, *Applied Catalysis A: General* **2003**, *248* (1), 105–116.
12. Sullivan, V. S.; Jackson, S. D.; Stair, P. C., In situ ultraviolet Raman spectroscopy of the reduction of chromia on alumina catalysts, *The Journal of Physical Chemistry B* **2005**, *109* (1), 352–356.
13. Chaar, M. A.; Patel, D.; Kung, M. C.; Kung, H. H., Selective oxidative dehydrogenation of butane over V Mg O catalysts, *Journal of Catalysis* **1987**, *105* (2), 483–498.
14. Wachs, I. E.; Weckhuysen, B. M., Structure and reactivity of surface vanadium oxide species on oxide supports, *Applied Catalysis A: General* **1997**, *157* (1), 67–90.
15. Weckhuysen, B. M.; Keller, D. E., Chemistry, spectroscopy and the role of supported vanadium oxides in heterogeneous catalysis, *Catalysis Today* **2003**, *78* (1), 25–46.

16. Wu, Z.; Kim, H.-S.; Stair, P. C.; Rugmini, S.; Jackson, S. D., On the structure of vanadium oxide supported on aluminas: UV and visible Raman spectroscopy, UV– visible diffuse reflectance spectroscopy, and temperature, *The Journal of Physical Chemistry B* **2005**, *109* (7), 2793–2800.
17. Wu, Z.; Stair, P. C., UV Raman spectroscopic studies of V/θ-Al_2O_3 catalysts in butane dehydrogenation, *Journal of Catalysis* **2006**, *237* (2), 220–229.
18. Jackson, S. D.; Rugmini, S., Dehydrogenation of n-butane over vanadia catalysts supported on θ-alumina, *Journal of Catalysis* **2007**, *251* (1), 59–68.
19. McGregor, J.; Huang, Z.; Shiko, G.; Gladden, L. F.; Stein, R. S.; Duer, M. J.; Wu, Z.; Stair, P. C.; Rugmini, S.; Jackson, S. D., The role of surface vanadia species in butane dehydrogenation over VOx/Al_2O_3, *Catalysis Today* **2009**, *142* (3), 143–151.
20. Harlin, M. E.; Niemi, V. M.; Krause, A. O. I., Alumina-supported vanadium oxide in the dehydrogenation of butanes, *Journal of Catalysis* **2000**, *195* (1), 67–78.
21. Russell, A. S.; Stokes, J. J., Role of surface area in dehydrocyclization catalysis, *Industrial & Engineering Chemistry* **1946**, *38* (10), 1071–1074.
22. Xie, S.; Chen, K.; Bell, A. T.; Iglesia, E., Structural characterization of molybdenum oxide supported on zirconia, *The Journal of Physical Chemistry B* **2000**, *104* (43), 10059–10068.
23. Gnep, N. S.; Doyemet, J. Y.; Seco, A. M.; Ribeiro, F. R.; Guisnet, M., Conversion of light alkanes to aromatic hydrocarbons: II. Role of gallium species in propane transformation on GaZSM5 catalysts, *Applied Catalysis* **1988**, *43* (1), 155–166.
24. Meitzner, G. D.; Iglesia, E.; Baumgartner, J. E.; Huang, E. S., The chemical state of gallium in working alkane dehydrocyclodimerization catalysts. In situ gallium K-edge X-ray absorption spectroscopy, *Journal of Catalysis* **1993**, *140* (1), 209–225.
25. Chen, M.; Xu, J.; Su, F.-Z.; Liu, Y.-M.; Cao, Y.; He, H.-Y.; Fan, K.-N., Dehydrogenation of propane over spinel-type gallia–alumina solid solution catalysts, *Journal of Catalysis* **2008**, *256* (2), 293–300.
26. Xu, B.; Li, T.; Zheng, B.; Hua, W.; Yue, Y.; Gao, Z., Enhanced stability of HZSM-5 supported Ga_2O_3 catalyst in propane dehydrogenation by dealumination, *Catalysis Letters* **2007**, *119* (3), 283–288.
27. Shimada, H.; Akazawa, T.; Ikenaga, N.-O.; Suzuki, T., Dehydrogenation of isobutane to isobutene with iron-loaded activated carbon catalyst, *Applied Catalysis A: General* **1998**, *168* (2), 243–250.
28. McGregor, J.; Huang, Z.; Parrott, E. P. J.; Zeitler, J. A.; Nguyen, K. L.; Rawson, J. M.; Carley, A.; Hansen, T. W.; Tessonnier, J.-P.; Su, D. S.; Teschner, D.; Vass, E. M.; Knop-Gericke, A.; Schlögl, R.; Gladden, L. F., Active coke: Carbonaceous materials as catalysts for alkane dehydrogenation, *Journal of Catalysis* **2010**, *269* (2), 329–339.
29. Sugiyama, S.; Osaka, T.; Hirata, Y.; Sotowa, K.-I., Enhancement of the activity for oxidative dehydrogenation of propane on calcium hydroxyapatite substituted with vanadate, *Applied Catalysis A: General* **2006**, *312*, 52–58.
30. Grabowski, R., Kinetics of oxidative dehydrogenation of C2-C3 alkanes on oxide catalysts, *Catalysis Reviews* **2006**, *48* (2), 199–268.
31. Shee, D.; Rao, T. V. M.; Deo, G., Kinetic parameter estimation for supported vanadium oxide catalysts for propane ODH reaction: Effect of loading and support, *Catalysis Today* **2006**, *118* (3), 288–297.
32. Gellings, P. J.; Bouwmeester, H. J. M., Solid state aspects of oxidation catalysis, *Catalysis Today* **2000**, *58* (1), 1–53.
33. Heracleous, E.; Lemonidou, A. A., Ni–Nb–O mixed oxides as highly active and selective catalysts for ethene production via ethane oxidative dehydrogenation. Part II: Mechanistic aspects and kinetic modeling, *Journal of Catalysis* **2006**, *237* (1), 175–189.
34. Crabtree, R. H.; Mihelcic, J. M.; Quirk, J. M., Iridium complexes in alkane dehydrogenation, *Journal of the American Chemical Society* **1979**, *101* (26), 7738–7740.
35. Crabtree, R. H.; Demou, P. C.; Eden, D.; Mihelcic, J. M.; Parnell, C. A.; Quirk, J. M.; Morris, G. E., Dihydrido olefin and solvento complexes of iridium and the mechanisms of olefin hydrogenation and alkane dehydrogenation, *Journal of the American Chemical Society* **1982**, *104* (25), 6994–7001.
36. Crabtree, R. H.; Mellea, M. F.; Mihelcic, J. M.; Quirk, J. M., Alkane dehydrogenation by iridium complexes, *Journal of the American Chemical Society* **1982**, *104* (1), 107–113.
37. Budavari, S., *The Merck index: An encyclopedia of chemicals, drugs, and biologicals*. Merck: Whitehouse Station, NJ, 1996.
38. Ghatak, T.; Sarkar, M.; Dinda, S.; Dutta, I.; Rahaman, S. M. W.; Bera, J. K., Olefin oxygenation by water on an iridium center, *Journal of the American Chemical Society* **2015**, *137* (19), 6168–6171.
39. Rahaman, S. M. W.; Dinda, S.; Ghatak, T.; Bera, J. K., Carbon monoxide induced double cyclometalation at the iridium center, *Organometallics* **2012**, *31* (15), 5533–5540.

40. Xu, W.-W.; Rosini, G. P.; Krogh-Jespersen, K.; Goldman, A. S.; Gupta, M.; Jensen, C. M.; Kaska, W. C., Thermochemical alkane dehydrogenation catalyzed in solution without the use of a hydrogen acceptor, *Chemical Communications* **1997**, *23*, 2273–2274.
41. Choi, J.; Goldman, A. S., Ir-Catalyzed Functionalization of C–H Bonds. In *Iridium catalysis*, Andersson, P. G., Ed. Springer Berlin Heidelberg: Berlin, Heidelberg, 2011; pp. 139–167.
42. Choi, J.; MacArthur, A. H. R.; Brookhart, M.; Goldman, A. S., Dehydrogenation and related reactions catalyzed by iridium pincer complexes, *Chemical Reviews* **2011**, *111* (3), 1761–1779.
43. Zhu, K.; Achord, P. D.; Zhang, X.; Krogh-Jespersen, K.; Goldman, A. S., Highly effective pincer-ligated iridium catalysts for alkane dehydrogenation. DFT calculations of relevant thermodynamic, kinetic, and spectroscopic properties, *Journal of the American Chemical Society* **2004**, *126* (40), 13044–13053.
44. Huang, Z.; Brookhart, M.; Goldman, A. S.; Kundu, S.; Ray, A.; Scott, S. L.; Vicente, B. C., Highly active and recyclable heterogeneous iridium pincer catalysts for transfer dehydrogenation of alkanes, *Advanced Synthesis & Catalysis* **2009**, *351* (1–2), 188–206.
45. Kuklin, S. A.; Sheloumov, A. M.; Dolgushin, F. M.; Ezernitskaya, M. G.; Peregudov, A. S.; Petrovskii, P. V.; Koridze, A. A., Highly active iridium catalysts for alkane dehydrogenation. Synthesis and properties of iridium bis (phosphine) pincer complexes based on ferrocene and ruthenocene, *Organometallics* **2006**, *25* (22), 5466–5476.
46. Bézier, D.; Brookhart, M., Applications of PC(sp3)P iridium complexes in transfer dehydrogenation of alkanes, *ACS Catalysis* **2014**, *4* (10), 3411–3420.
47. Göttker-Schnetmann, I.; Brookhart, M., Mechanistic studies of the transfer dehydrogenation of cyclooctane catalyzed by iridium bis(phosphinite) p-XPCP pincer complexes, *Journal of the American Chemical Society* **2004**, *126* (30), 9330–9338.
48. Kundu, S.; Choliy, Y.; Zhuo, G.; Ahuja, R.; Emge, T. J.; Warmuth, R.; Brookhart, M.; Krogh-Jespersen, K.; Goldman, A. S., Rational design and synthesis of highly active pincer-iridium catalysts for alkane dehydrogenation, *Organometallics* **2009**, *28* (18), 5432–5444.
49. Jia, X.; Zhang, L.; Qin, C.; Leng, X.; Huang, Z., Iridium complexes of new NCP pincer ligands: catalytic alkane dehydrogenation and alkene isomerization, *Chemical Communications* **2014**, *50* (75), 11056–11059.
50. Haenel, M. W.; Oevers, S.; Angermund, K.; Kaska, W. C.; Fan, H.-J.; Hall, M. B., *Angewandte Chemie International Edition* **2001**, *40* (19), 3596–3600.
51. Kumar, A.; Bhatti, T. M.; Goldman, A. S., Thermally stable homogeneous catalysts for alkane dehydrogenation, *Chemical Reviews* **2017**, *117* (19), 12357–12384.
52. Gruver, B. C.; Adams, J. J.; Warner, S. J.; Arulsamy, N.; Roddick, D. M., Acceptor pincer chemistry of ruthenium: Catalytic alkane dehydrogenation by (CF3PCP)Ru(cod)(H), *Organometallics* **2011**, *30* (19), 5133–5140.
53. Zhang, Y.; Fang, H.; Yao, W.; Leng, X.; Huang, Z., Synthesis of pincer hydrido ruthenium olefin complexes for catalytic alkane dehydrogenation, *Organometallics* **2016**, *35* (2), 181–188.
54. Fu, P. P.; Harvey, R. G., Dehydrogenation of polycyclic hydroaromatic compounds, *Chemical Reviews* **1978**, *78* (4), 317–361.
55. Linstead, R. P.; Thomas, S. L. S., Dehydrogenation. Part II. The elimination and migration of methyl groups from quaternary carbon atoms during catalytic dehydrogenation, *Journal of the Chemical Society (Resumed)* **1940**, 1127–1134.
56. Phillips, D. D.; Chatterjee, D. N., Polynuclear aromatic hydrocarbons. IX.1 The synthesis of 3,4-benzpyrene and 7-methyl-3,4-benzpyrene, *Journal of the American Chemical Society* **1958**, *80* (16), 4360–4364.
57. Harvey, R. G.; Arzadon, L.; Grant, J.; Urberg, K., Metal-ammonia reduction. IV. Single-stage reduction of polycyclic aromatic hydrocarbons, *Journal of the American Chemical Society* **1969**, *91* (16), 4535–4541.
58. Braude, E. A.; Brook, A. G.; Linstead, R. P., Hydrogen transfer. Part IV. The use of quinones of high potential as dehydrogenation reagents, *Journal of the Chemical Society (Resumed)* **1954**, 3569–3574.
59. Walker, D.; Hiebert, J. D., 2, 3-Dichloro-5, 6-dicyanobenzoquinone and its reactions, *Chemical Reviews* **1967**, *67* (2), 153–195.

6 Dehydrogenation Reactions of Aliphatic and Aromatic Amines

Nandini Mukherjee and Sauvik Chatterjee

6.1 INTRODUCTION

The introduction of an olefin group into a molecule is a lucrative approach for scientists due to the wide scope of functionalization of the double bond. Formation of a double bond particularly in an N-containing molecule is interesting because of its importance as a platform chemical for biologically significant molecules [1]. Dehydrogenation is a highly significant way of introducing a double bond into a saturated molecule [2]. In this context, dehydrogenation of amines, both aliphatic and aromatic, needs to be considered. Due to the formation of hydrogen as a by-product of the dehydrogenation reaction, the saturated N-heterocycles are also considered important liquid organic hydrogen carriers (LOHCs), which has very high importance for blue hydrogen fuel production that has two major advantages over conventional LOHCs, such as $NH_3:BH_3$, $NaBH_4$, and metal hydrides [3]. Compared to ammonia borane and sodium borohydride, the N-heterocyclic LOHCs are naturally abundant and economically efficient. In addition, the dehydrogenation process of N-heterocyclic is endothermic, which makes the reaction more controllable and eliminates the possibility of generating undesired by-products [4, 5].

Nitriles are also a class of compounds derived from the double dehydrogenation of amines. Nitriles are very important from a pharmaceutical standpoint [6] and are intermediates for the synthesis of a wide variety of organic compounds [7]. Nitriles synthesis from amides, aldehydes, aldoximes, etc. involves higher waste materials and additional energy and costs. It has been observed that oxidative dehydrogenation stands out as a clean and energy-efficient process for the synthesis of nitriles [8].

6.2 MECHANISTIC CONSIDERATION

To improve the efficiency of any chemical process, the first task is to understand the mechanism of the reactions, the underlying conditions, and the deterministic parameters. Investigation of both kinetic and thermodynamic aspects leads to the tuning of reaction parameters.

The dehydrogenation process, as it indicates, involves the production of hydrogen molecules as the end product. However, the variety of reactions that requires removal of two H atoms leading to the formation of a double bond to the organic molecules does not form H_2 but involves other hydride acceptors.

The basic process of the reaction involving a radical initiator is as follows (Figure 6.1).

In both of the proposed paths, formation of a radical takes place. Although a hydrogen atom transfer (HAT) occurs in the presence of $^tBuO•$, a radical formed out of tertiarybutyl hydrogen peroxide (TBHP) followed by a single electron transfer (SET) lead to the formation of a radical ion (compound **B** in Figure 6.1). The same radical ion forms in the presence of TsN, but the steps occur in reverse order. In the next step, a sequestering of a H^+ produces enamine [9].

In a typical metal-mediated catalysis, the N of the aliphatic amines gets attached to the metal centres followed by a transfer of an H atom to the metal centre, forming an imine bond. Bernskoetter and Brookhart isolated the formation of amine metal complex adduct with Ir as the metal centre

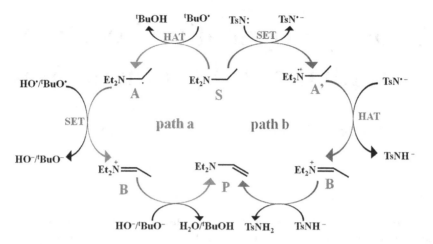

FIGURE 6.1 Mechanism of metal-free dehydrogenation of amine in a radical pathway.

Source: Adapted from ref [9]

[10]. They proposed an Ir (I) amine to have swift equilibrium with an Ir (III) amine hydride complex with a very high preference for the former complex (Figure 6.2). In the presence of an H acceptor, in this case, *tert*-butylethylene (TBE), the equilibrium shifts irreversibly towards a β-hydride elimination that leads to the formation of an imine bond. The same sequence takes place again to form a nitrile adduct. The nitrile adduct is more susceptible of undergoing a substitution reaction by the aliphatic amines present in the medium. This indicates that the presence of some residual saturated amines (2,2-dimethylbutane (TBA) in the given example) is necessary for the isolation of dehydrogenated products.

The presence of any hydrogen acceptor acts as a driving force to move the reaction forward. However, it has been observed that the change in TBE concentration does not change any rate of the reaction. The zero-order dependence on TBE and certain other isotope labelling experiments suggests that the β-hydride elimination or N-H oxidative addition is likely to be the rate-determining step [10].

The Choi group has conducted extensive work on transition metal-mediated electrocatalysis for the transformation of organic molecules where the metal catalyst is present in the form of a solid electrode. They demonstrated Ni(III) catalyst for potential dependent indirect oxidation that achieves dehydrogenation through HAT. More recently, they reported that the Ni(IV) centre of the catalyst undergoes dehydrogenation through hydride transfer [11]. The catalysis occurs once the reactant is adsorbed on the catalyst. They investigated the stepwise mechanism of electrochemical dehydrogenation and based on their studies, proposed the following scheme (Figure 6.3) where they introduced a proton transfer followed by a hydride transfer:

6.3 DEHYDROGENATION REACTIONS OF ALIPHATIC AMINES

6.3.1 Ru-Catalyzed Dehydrogenation

Panda et al. developed a series of Ru-complex catalysts for different transformations of 2-picolyamines (PA). Amongst them, [Ru^{II}(Cl)(H)(CO)(PPh$_3$)$_3$] undergoes oxidative dehydrogenation with C$_α$-substituted 2-picolyamine (PA) in the presence of phenyl-(pyridin-2-yl)methanamine and tetrahydro-1,1'-bis-(isoquinoline) to generate the corresponding imine with high yield (78% and 80%, respectively) (Figure 6.4). The reaction condition involves strong alkaline media to facilitate α-H abstraction. In the absence of alkaline media, the reaction does not proceed to generate

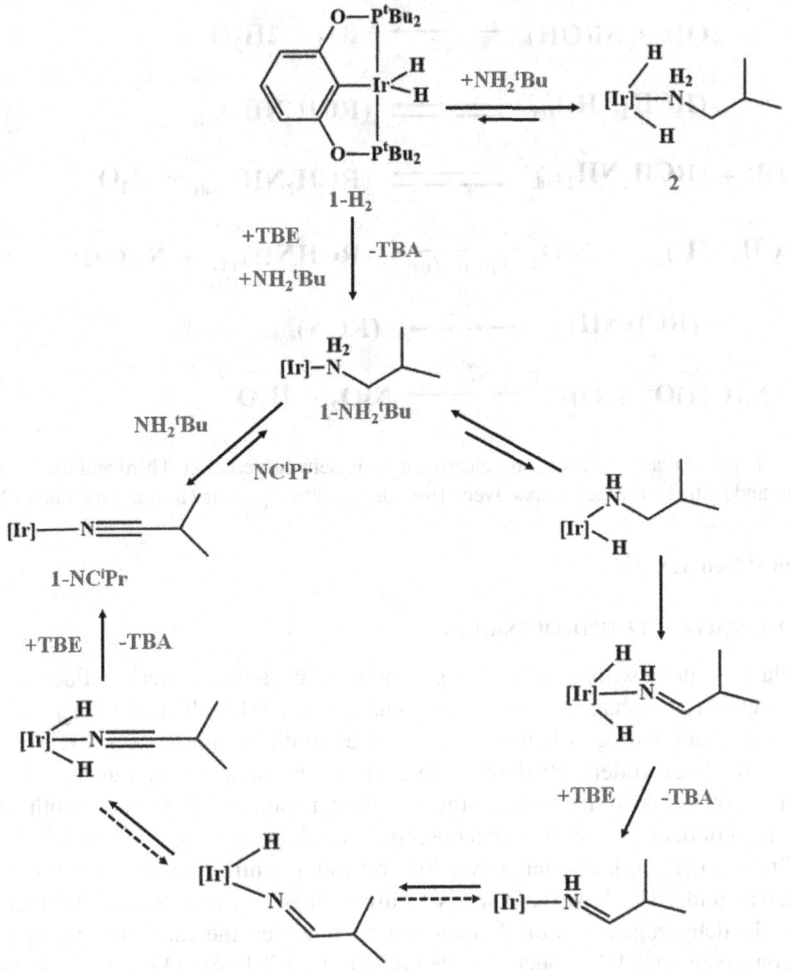

Dehydrogenation Reactions of Aliphatic and Aromatic Amines

FIGURE 6.2 General reaction mechanism of dehydrogenation of amine using a transition metal-based catalyst.

Source: Adapted from ref [10]

the desired dehydrogenation product, which indicates the process to be base-induced oxidative dehydrogenation.

Ruthenium has also been used for the dehydrogenation of amine borane adduct. Friedrich et al. reported Ru-PNP amido pincer complex to give cyclic dimer upon dehydrogenation unlike the formation of olefin (Figure 6.5) [14].

Ru-benzene dichloride dimer has been reported to be used for the double dehydrogenation of amine to nitrile and the highly selective dehydrogenation of secondary amine to imine with more than 90% yield [15]. These sets of reactions have been co-catalyzed by hexamethylenetetramine as a hydride donor to the ruthenium catalyst.

An interesting work by Ray et al. demonstrates the dependence of the outcome on the ligand of the Ru-complex catalyst under TEMPO/air conditions. Although the bis (imino)acenaphthene, that is, BIAN complex, [Ru(II)(R-BIAN)(PPh$_3$)$_2$(CO)(H)]ClO$_4$, leads to the formation of nitrile from primary amine, phenanthroline (phen) complex, [Ru(II)(phen)(PPh$_3$)$_2$(CO)(H)]ClO$_4$, gives secondary imine under the same conditions (Figure 6.6) [16].

$$2\text{OH}^- + \text{Ni(OH)}_2 \xrightarrow{-2e^-} \text{NiO}_2 + 2\text{H}_2\text{O} \qquad (1)$$

$$(\text{RCH}_2\text{NH}_2)_{\text{sol}} \rightleftharpoons (\text{RCH}_2\text{NH}_2)_{\text{ads}} \qquad (2)$$

$$\text{OH}^- + (\text{RCH}_2\text{NH}_2)_{\text{ads}} \rightleftharpoons (\text{RCH}_2\text{NH}^-)_{\text{ads}} + \text{H}_2\text{O} \qquad (3)$$

$$(\text{RCH}_2\text{NH}^-)_{\text{ads}} + \text{NiO}_2 \xrightarrow{\text{hydride transfer}} (\text{RCHNH})_{\text{ads}} + \text{Ni(OH)O}^- \qquad (4)$$

$$(\text{RCHNH})_{\text{ads}} \xrightarrow{-2e^-} (\text{RCN})_{\text{sol}} \qquad (5)$$

$$\text{Ni(OH)O}^- + \text{OH}^- \xrightarrow{-2e^-} \text{NiO}_2 + \text{H}_2\text{O} \qquad (6)$$

FIGURE 6.3 Stepwise reaction scheme for electrocatalytic dehydrogenation. Third and fourth steps indicate proton transfer and hydride transfer, respectively, from the adsorbed primary amine substrate to Ni^{4+} catalytic site.

Source: Adapted from ref. [12]

6.3.2 Mo-Catalyzed Dehydrogenation

Mo-Halide clusters, along with several other groups 5 and 6 transition metal halides, were reported by Kamiguchi et al. for aliphatic amine dehydrogenation [17]. $(H_3O)_2[(Mo_6Cl_8)Cl_6] \cdot 6H_2O$ crystals, activated by and under hydrogen flow, were introduced to diethlyamine at 673 K to produce 72% selectivity towards N-ethylidene ethylamine. In contrast, the same reaction under helium flow and activation shows 60% selectivity towards the dehydrogenation product. Under both conditions, a considerable level of dealkylation was also observed, which leads to the production of ethyl amine. With $[(Mo_6Br_8)(OH)_4(H_2O)_2]$, a better conversion and almost similar selectivity was obtained under H_2 flow, whereas under He flow, the conversion drops. It is pertinent to mention that Mo metals also catalyze the dehydrogenation of aliphatic amine; however, the catalytic activity is poor with around 4% conversion and 45% selectivity. $(H_3O)_2[(Mo_6Cl_8)Cl_6] \cdot 6H_2O$ gives a 21% conversion of pyrrolidine to pyrroline with 81% selectivity under H_2 flow but only an 8% conversion of piperidine to 2,3,4,5-tetrahydropyridine.

6.3.3 Ni-Catalyzed Dehydrogenation

Ni is one of the most common dehydrogenating agents and has been used for a long time. Inorganic Ni-based catalysts were used earlier for the dehydrogenation reactions of smaller aliphatic amines. Yamazaki and Yamazaki, back in 1990, reported $NiSO_4$ in the presence of $S_2O_8^{-2}$ as an oxidant that showed more than 90% yield for the double dehydrogenation of primary amines to yield nitriles [18].

Ni in the presence of Ir gives the dehydrogenation of a N-heterocyclic-saturated compound under photochemical conditions following a β-hydride elimination pathway [19].

6.3.4 Ir-Based Catalyst for Dehydrogenation

Ir-PCP pincer ligand complex is a popular catalyst for the transfer dehydrogenation of secondary amines. Reactions carried out at a 473-K temperature and in sealed tubes have shown good yield in the presence of alkenes. Although it is predictive that the reaction occurs through the transfer dehydrogenation mechanism, the exact reaction pathway has not yet been established (Figure 6.7) [20].

A Cp*Ir complex containing a pyridyl-triazolylidene ligand in 1,2-dichlorobenzene (1,2-DCB) media was reported by Albrecht and coworkers for the dehydrogenation of a wide range of

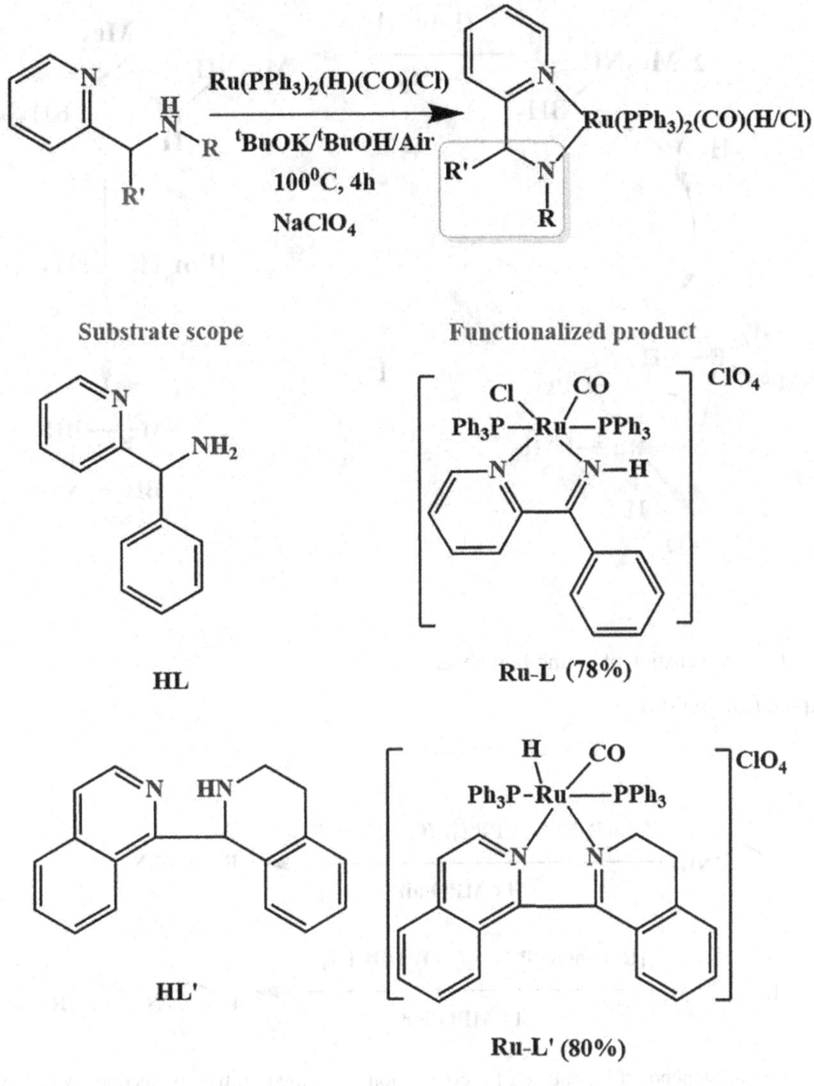

FIGURE 6.4 Dehydrogenation of amine leading to complexation.

Source: Adapted from ref [13]

benzylamine, which gives a mixture of secondary imine, secondary amine, and tertiary amines (Figure 6.8). Out of the two investigated complexes having pyridyl or phenyl pendant attachment, the former shows higher selectivity towards secondary imine [21].

6.4 DEHYDROGENATION REACTIONS OF AROMATIC AMINES

A ruthenium-catalyzed transfer dehydrogenation has been reported in the presence of 2,6-dimethoxybenzoquinone under mild conditions for the transformation of aromatic amine into an imine. The reaction was realized under toluene reflux conditions.

The presence of +R group as R or R' facilitates the reaction by reducing the time and improving the yield. The presence of MnO_2 is essential for the redox cycle of 2,6-dimethoxybenzoquinone (Figure 6.9), whose absence inhibits the reactions with yields below 10% even after 24 h [22].

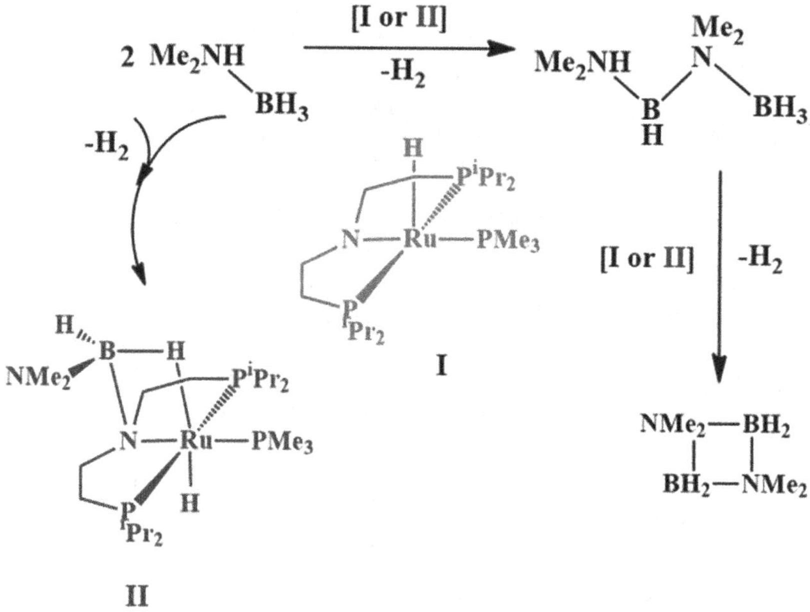

FIGURE 6.5 Dehydrogenation of amine-borane adduct.
Source: Adapted from ref [14]

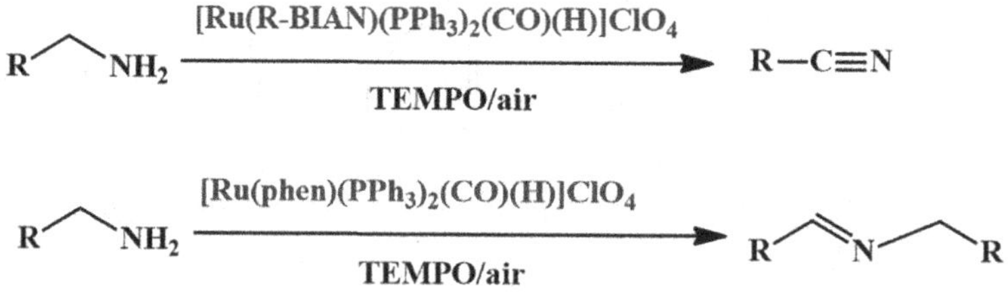

FIGURE 6.6 Ligand-dependent Ru catalyst for conversion of amine to nitrile or secondary imine.
Source: Adapted from ref [16]

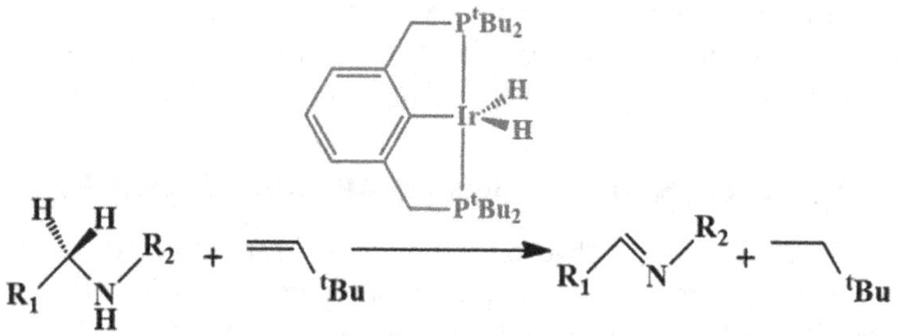

FIGURE 6.7 Dehydrogenation of secondary amine with an Ir-catalyst.
Source: Adapted from ref [20]

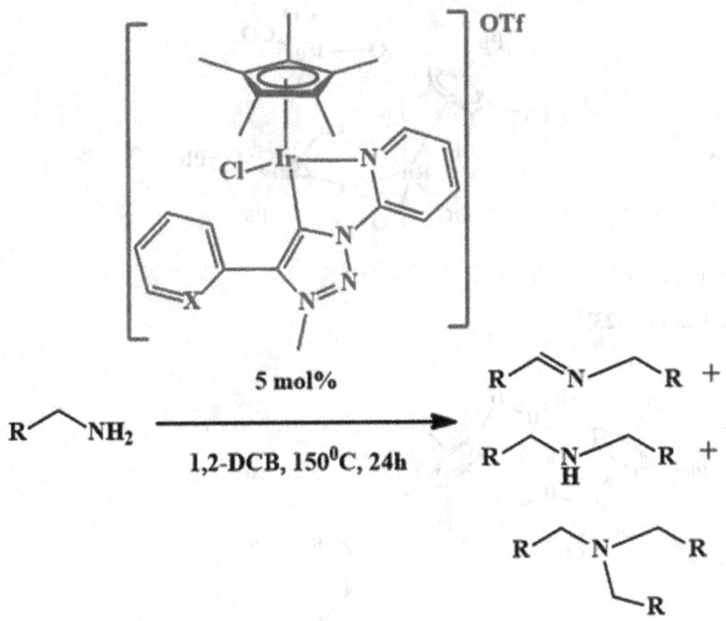

Dehydrogenation Reactions of Aliphatic and Aromatic Amines

FIGURE 6.8 Cp*Ir complex containing pyridyl-triazolylidene for the oxidative conversion of primary amine.

Source: Adapted from ref [21]

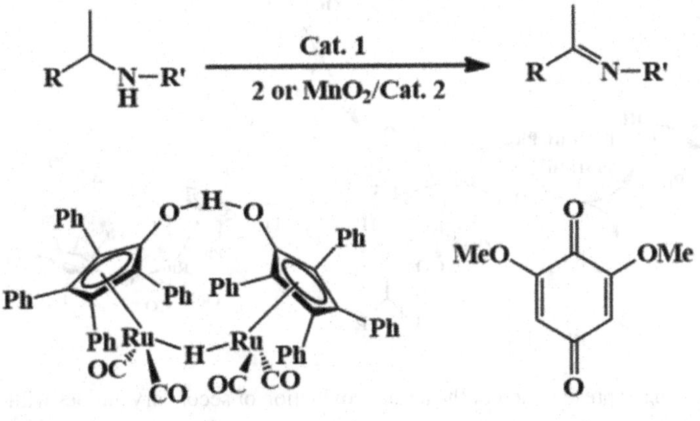

FIGURE 6.9 Conversion of secondary amine to imine.

Source: Adapted from ref [22]

A much simpler catalyst was reported by Samec et al. to give a much faster and efficient yield of ketamine and aldamines (Figure 6.10) [23].

Here, in place of MnO_2, a Co-salen type of co-catalyst was used for the activation of 2,6-dimethoxybenzoquinone leading to faster kinetics. A schematic mechanism proposed by the authors shows the reaction to follow the aerobic oxidation of secondary amines with aromatic amines (Figure 6.11).

The acceptorless dehydrogenation of secondary amines to imines using $RuH_2(CO)(PPh_3)_3$ and Shvo's catalyst systems was reported by Hong and co-workers [24]. Ru(II) hybrid complex with

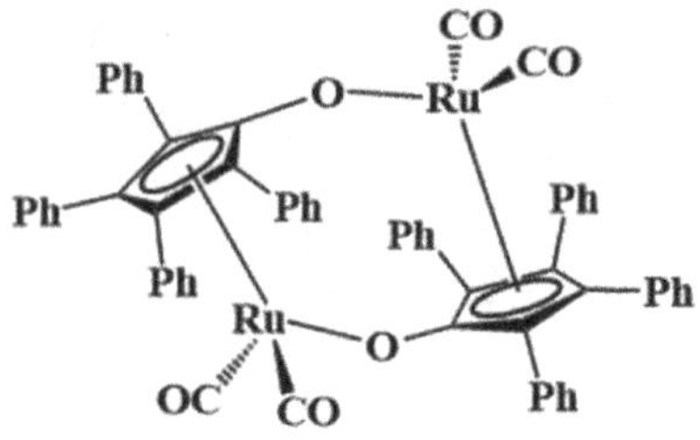

FIGURE 6.10 Structure of the Ru-catalyst.

Source: Adapted from ref [23]

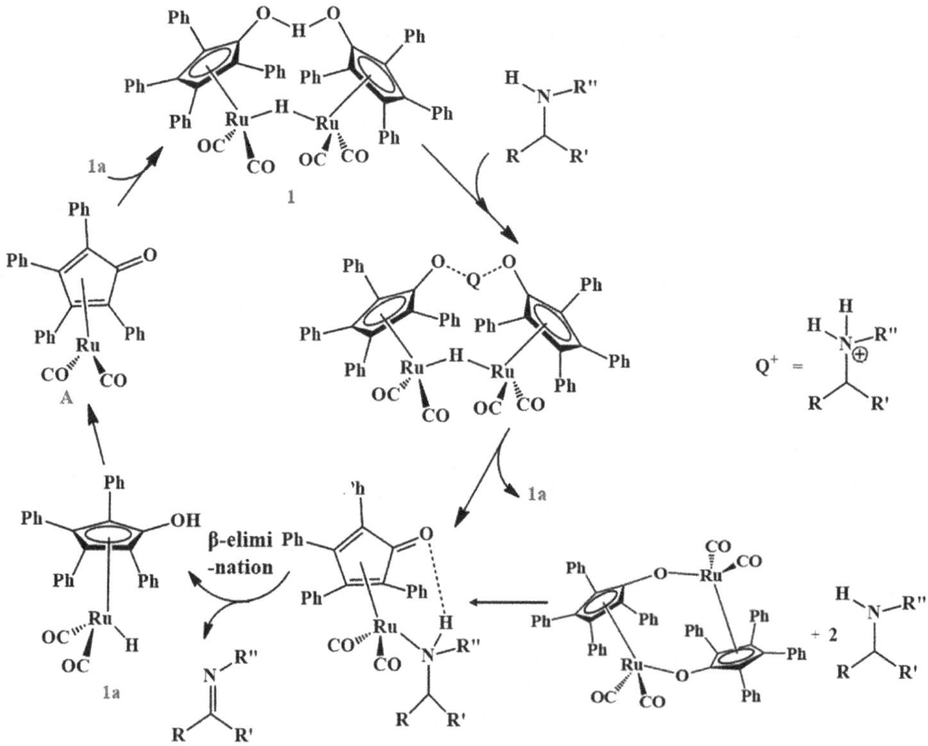

FIGURE 6.11 Schematic presentation of the aerobic oxidation of secondary amines with aromatic amines.

Source: Adapted from ref [23]

pyrazolyl-(2-indol-1-yl)-pyridine ligand has been shown to be an efficient catalyst for the acceptorless dehydrogenation of N-heterocycles with a broad substrate scope (Figure 6.12) [25]. The mechanism of the transformation is similar to that of aliphatic amine, which involves oxidative hydride transformation through the metal centre, and the double dehydrogenation results in the aromatization of the final product.

The mechanism proposed by the authors is shown in Figure 6.13:

N-benzylhexamethylenetetramine, an easily available commercial compound along with $RuCl_3 \cdot nH_2O$, was reported very recently by Kannan et al. to give more than 70% yield to the dehydrogenation reaction in the acceptorless pathway for both primary and secondary amines. The primary amines produce corresponding nitriles, whereas the secondary amines give imines as the final product along with gaseous hydrogen under refluxing conditions with toluene as the solvent. The schematic

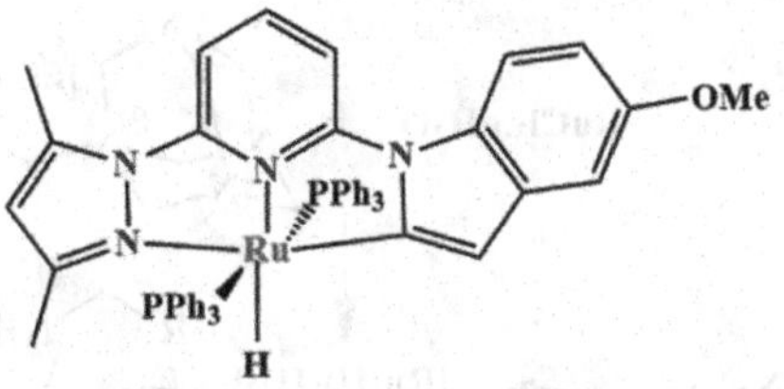

Dehydrogenation Reactions of Aliphatic and Aromatic Amines 109

FIGURE 6.12 Structure of the Ru-catalyst.

Source: **Adapted from ref [25]**

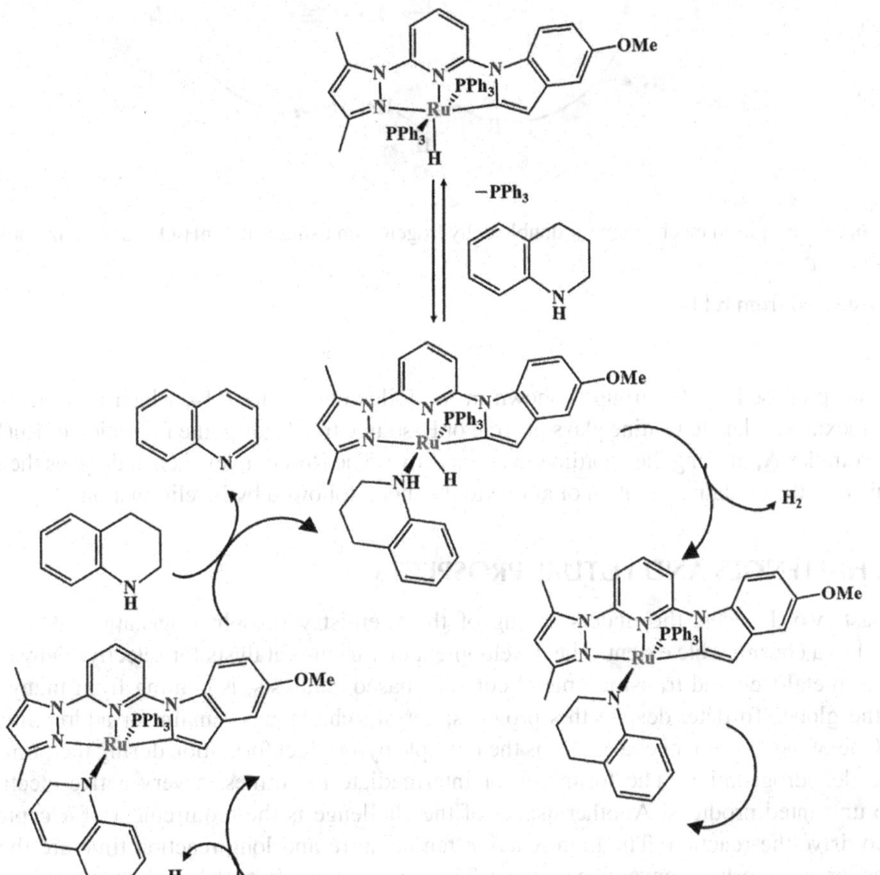

FIGURE 6.13 Proposed reaction pathway for N-heterocyclic aromatic amines using an Ru-complex catalyst.

Source: **Adapted from ref [25]**

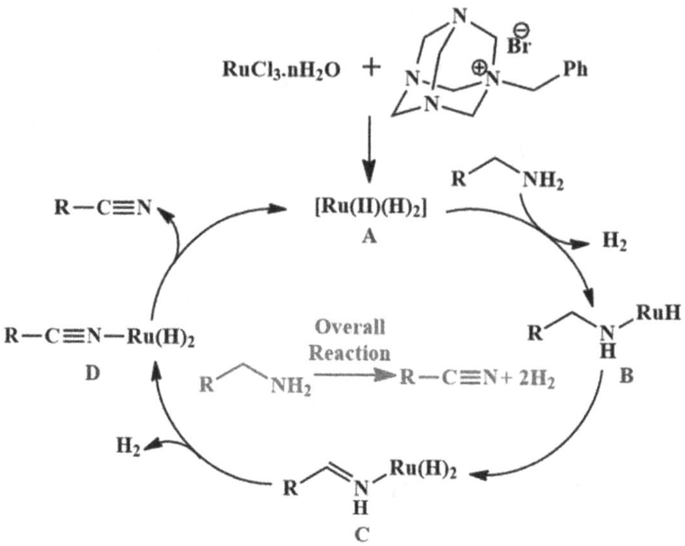

FIGURE 6.14 Proposed mechanism for double dehydrogenation using $RuCl_3.nH_2O$ and N-benzylhexamethylenetetramine.

Source: Adapted from ref [26]

mechanism proposed by the group is shown in the following Figure 6.14, which indicates that the N-benzylhexamethylenetetramine plays the role of H-source that leads to the formation of $Ru(II)(H)_2$, namely, complex **A**, making the coordination sites active. The Ru-complex then undergoes the general mechanism of the oxidative addition of amine to the metal followed by H_2 elimination [26].

6.5 CHALLENGES AND FUTURE PROSPECTS

In the last two decades, the understanding of the chemistry of dehydrogenation of amine has increased to a considerable extent. The development of various catalysts for selective dehydrogenation, both metal-free and transition metal complex-based catalysts, is coming from many groups across the globe. To date, despite this progress, certain challenges remain for addressing in the future. One of the prominent challenges is the multiple by-product formation during metal-mediated catalytic dehydrogenation. The formation of intermediate aldimines, a very active electrophile, leads to unwanted products. Another aspect of the challenge is the requirement of acceptor molecules to drive the reaction. The high reaction temperature and long reaction time are the major concerns for acceptorless conversion systems. The design of catalysts that will promote dehydrogenation in a tandem manner along with β-hydride elimination needs to be one of the focus areas of future research. This will also, in turn, facilitate green hydrogen production. The most conventional challenges for any catalytic reaction, specifically, the issue of the recyclability of the catalyst and the usage of the solvent, are also major points of concern. More efforts towards making heterogeneous catalytic systems will perhaps open new avenues for the better reusability of catalysts and less solvent-intensive processes.

REFERENCES

[1] N. Kerru, L. Gummidi, S. Maddila, K. K. Gangu, S. B. Jonnalagadda, A Review on Recent Advances in Nitrogen-Containing Molecules and Their Biological Applications. *Molecules* 25 (2020) 1909. (Kerru et al. 2020)

[2] H. A. Wittcoff, B. G. Reuben, J. S. Plotkin, *Industrial Organic Chemicals*, Second Edition, Wiley Online Library, 2004. (Wittcoff et al. 2004)

[3] S. A. Stepanenko, D. M. Shivtsov, A. P. Koskin, I. P. Koskin, R. G. Kukushkin, P. M. Yeletsky, V. A. Yakovlev, N-Heterocyclic Molecules as Potential Liquid Organic Hydrogen Carriers: Reaction Routes and Dehydrogenation Efficacy. *Catalysts* 12 (2022) 1260. (Shivtsov et al. 2022)

[4] P. G. Campbell, L. N. Zakharov, D. J. Grant, D. A. Dixon, S. -Y. Liu, Hydrogen Storage by Boron–Nitrogen Heterocycles: A Simple Route for Spent Fuel Regeneration. *J. Am. Chem. Soc.* 132 (2010) 3289–3291. (Campbell et al. 2010)

[5] R. H. Crabtree, Nitrogen-Containing Liquid Organic Hydrogen Carriers: Progress and Prospects. *ACS Sustain. Chem. Eng.* 5 (2017) 4491–4498. (Crabtree et al. 2017)

[6] F. F. Fleming, L. Yao, P. C. Ravikumar, L. Funk, B. C. Shook, Nitrile-Containing Pharmaceuticals: Efficacious Roles of the Nitrile Pharmacophore. *J. Med. Chem.* 53 (2010) 7902–7917. (Fleming et al. 2010)

[7] N. K. Bhattacharyya, S. Jha, S. Jha, T. Y. Bhutia, G. Adhikary, Dehydration of Amides to Nitriles: A Review. *Int. J. Chem. Appl.* 4 (2012) 295–304. (Bhattacharyya et al. 2012)

[8] R. Patil, M. K. Gupta, Methods of Nitriles Synthesis from Amines through Oxidative Dehydrogenation. *Adv. Synth. Catal.* 362 (2020) 3987. https://doi.org/10.1002/adsc.202000635 (Patil et al. 2020)

[9] A. Rouzi, R. Hudabaierdi, A. Wusiman, Synthesis of N-Sulfonylformamidines by Tert-butyl Hydroperoxide–Promoted, Metal-Free, Direct Oxidative Dehydrogenation of Aliphatic Amines. *Tetrhedron* 74 (2018) 2475. https://doi.org/10.1016/j.tet.2018.03.074 (Rouzi et al. 2018)

[10] W. H. Bernskoetter, M. Brookhart, Kinetics and Mechanism of Iridium-Catalyzed Dehydrogenation of Primary Amines to Nitriles. *Organometallics* 27 (2008) 2036–2045. (Brenskoetter et al. 2008)

[11] M. T. Bender, R. E. Warburton, S. Hammes-Schiffer, K.-S. Choi, Understanding Hydrogen Atom and Hydride Transfer Processes During Electrochemical Alcohol and Aldehyde Oxidation. *ACS Catal.* 11 (2021) 15110–15124. https://doi.org/10.1021/acscatal.1c04163 (Bender et al. 2021)

[12] M. T. Bender, K. S. Choi, Electrochemical Dehydrogenation Pathways of Amines to Nitriles on NiOOH. *JACS Au.* 2 (2022) 1169–1180. (Bender et al. 2022)

[13] S. Panda, A. S. Hazari, M. Gogia, G. K. Lahiri, Diverse Functionalization of Ruthenium-Chelated 2-Picolylamines: Oxygenation, Dehydrogenation, Cyclization, and N-Dealkylation. *Inorg. Chem.* 59 (2020) 1355–1363. https://doi.org/10.1021/acs.inorgchem.9b0306 (Panda et al. 2020)

[14] A. Friedrich, M. Drees, S. Schneider, Ruthenium-Catalyzed Dimethylamineborane Dehydrogenation: Stepwise Metal-Centered Dehydrocyclization. *Chem. Eur. J.* 15 (2009) 10339–10342. https://doi.org/10.1002/chem.200901372 (Friedrich et al. 2009)

[15] M. Kannan, S. Muthaiah, Extending the Chemistry of Hexamethylenetetramine in Ruthenium Catalyzed Amine Oxidation. *Organometallics* 38 (2019) 3560–3567. (Kannan et al. 2019)

[16] R. Ray, S. Chandra, V. Yadav, P. Mondal, D. Maiti, G. K. Lahiri, Ligand Controlled Switchable Selectivity in Ruthenium Catalyzed Aerobic Oxidation of Primary Amines. *Chem. Commun.* 53 (2017) 4006–4009. (Ray et al. 2017)

[17] S. Kamiguchi, A. Nakamura, A. Suzuki, M. Kodomari, M. Nomura, Y. Iwasawa, T. Chihara, Catalytic Dehydrogenation of Aliphatic Amines to Nitriles, Imines, or Vinylamines and Dealkylation of Tertiary Aliphatic Amines Over Halide Cluster Catalysts of Group 5 and 6 Transition Metals. *J. Catal.* 230 (2005) 204–213. (Kamiguchi et al.2005)

[18] Y. Shigekazu, Y. Yasuyuki, Nickel-Catalyzed Dehydrogenation of Amines to Nitriles. *Bull. Chem. Soc. Jpn.* 63 (1990) 301–303. (Shigekazu et al. 1990)

[19] Ritu, S. Das, Y.-M. Tian, T. Karl, N. Jain, B. König, Photocatalyzed Dehydrogenation of Aliphatic N-Heterocycles Releasing Dihydrogen. *ACS Catal.* 12 (2022) 10326–10332. (Ritu et al. 2022)

[20] D. L. J. Broere, Transition Metal-Catalyzed Dehydrogenation of Amines. *Phys. Sci. Rev.* (2018) 20170029. (Broere et al. 2018)

[21] M. Valencia, A. Pereira, H. Müller-Bunz, T. R. Belderraín, P. J. Pérez, M. Albrecht. Triazolylidene-Iridium Complexes with a Pendant Pyridyl Group for Cooperative Metal–Ligand Induced Catalytic Dehydrogenation of Amines. *Chem. Eur. J.* 23 (2017) 8901–8911. (Valencia et al. 2017)

[22] A. H. Éll, J. S. M. Samec, C. Brasse, J.-E. Bäckvall, Dehydrogenation of Aromatic Amines to Imines Via Ruthenium-Catalyzed Hydrogen Transfer. *Chem. Commun.* (2002) 1144–1145. (Éll et al. 2002)

[23] J. S. M. Samec, A. H. Éll, J.-E. Bäckvall, Efficient Ruthenium-Catalyzed Aerobic Oxidation of Amines by Using a Biomimetic Coupled Catalytic System. *Chem. Eur. J.* 11 (2005) 2327–2334. (Samac et al. 2005)

[24] S. Muthaiah, S. H. Hong, Acceptorless and Base-Free Dehydrogenation of Alcohols and Amines Using Ruthenium-Hydride Complexes. *Adv. Synth. Catal.* 354 (2012) 3045–3053. (Muthaiah et al. 2012)

[25] Q. Wang, H. Chai, Z. Yu, Acceptorless Dehydrogenation of N-Heterocycles and Secondary Alcohols by Ru(II)-NNC Complexes Bearing a Pyrazoyl-indolyl-pyridine Ligand. *Organometallics* 37 (2018) 584–591. (Wang et al. 2018)

[26] M. Kannan, P. Barteja, P. Devi, S. Muthaiah, Acceptorless Dehydrogenation of Amines and Alcohols Using Simple Ruthenium Chloride. *J. Catal.* 386 (2020) 1–11. (Kannan et al. 2020)

7 Dehydrogenation Reactions of Aliphatic and Aromatic Carboxylic Acids and Their Derivatives

Megha Balha

7.1 INTRODUCTION

Dehydrogenation is the process of removal of hydrogen from an organic compound to form a new chemical compound. That is, saturated compounds are converted into unsaturated ones. The dehydrogenation of the carboxylic acid is shown in Figure 7.1. The dehydrogenation of carboxylic acids furnishes α,β-unsaturated carboxylic acids.

In organic chemistry, α,β-unsaturated carbonyl moieties are important building blocks of primary importance for synthesis because of their broad-ranging reactivities [1]. Conjugated structures are present in various natural products [2] and chemicals of medicinal importance [2, 3]. Selected examples of α,β-unsaturated carbonyl motifs available in natural products are demonstrated in Scheme 7.1. As a result, the advancement of structured methods for their synthesis is one of the key tasks.

The most likely method for forming unsaturated compounds with high atom efficiency is through the dehydrogenation of adjacent two sp^3 C-H bonds [4]. To accomplish this reaction, the major problem is over-dehydrogenation. After the dehydrogenation, the obtained alkene contains more reactive allyl sp^3C-H bonds. Despite these challenges, various carbonyl dehydrogenation reactions are achieved by employing transition metals (Cu or Pd) or stoichiometric reagents based on S, Se, Br, or I. Direct dehydrogenation of carbonyl compounds is quite favorable for furnishing α,β-unsaturated carbonyl compounds such as ketones [5, 6], aldehydes [5], esters [7], and amides [8]. One of the most useful dehydrogenation methods in synthesis is derived by exploring *syn*-eliminations of sulfoxide and selenoxide intermediates [9–11]. Other organic reagents that have been developed to prepare enones are hypervalent iodine, *N*-oxoammonium salts, and *N-tert*-butyl phenylsulfinimidoyl chloride [12–14]. Recently, the desaturation of carbonyl compounds was also achieved by the electrochemical method [15]. A widely studied catalytic pathway for the dehydrogenation of ketones is through the generation of Pd(II) enolates followed by β-hydride elimination, furnishing unsaturated compounds [16–20]. Dehydrogenation reactions of carboxylic acids are also quite important methods in chemical industries. Despite its importance, the dehydrogenation of α,β-unsaturated

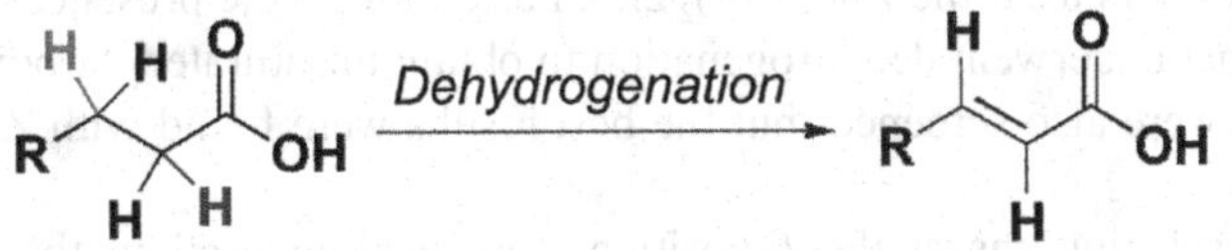

FIGURE 7.1 Schematic representation of the dehydrogenation of carboxylic acids.

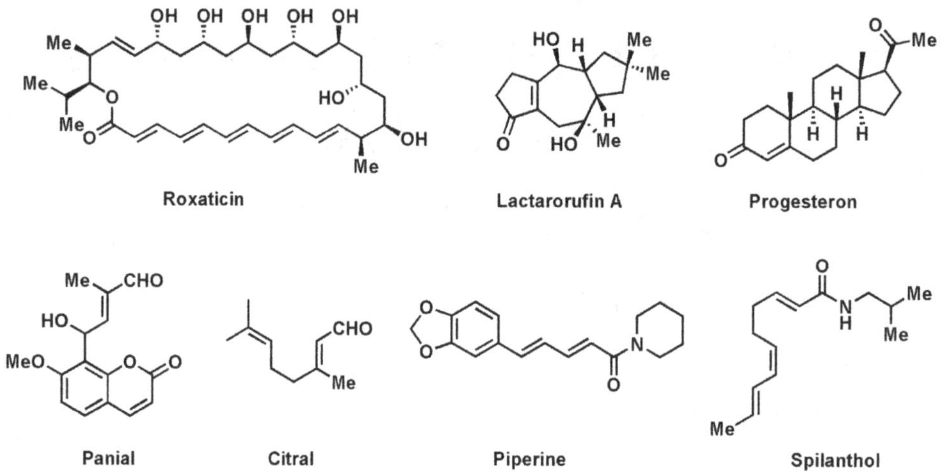

SCHEME 7.1 Natural products having α,β-unsaturated carbonyl moieties.

carboxylic acids is limited compared to the dehydrogenation of aldehydes and ketones. A few approaches have been developed for the dehydrogenation of α,β-unsaturated carboxylic acids [21].

In this chapter, we discuss the dehydrogenation reactions of aromatic and aliphatic carboxylic acids for the formation of α,β-unsaturated carboxylic acids.

7.2. DEHYDROGENATION REACTIONS OF ALIPHATIC AND AROMATIC CARBOXYLIC ACIDS

In 1973, Ronchi and group reported the dehydrogenation of α-anions of carboxylic acids by 2,3-Dichloro-4,5-dicyanobenzoquinone (DDQ) for efficient synthesis of α,β-unsaturated acid derivatives (Scheme 7.2) [22]. α-Anions of carboxylic acids were synthesized from lithium diisopropylamide and sodium carboxylates. The authors observed a lower yield on substituting DDQ with 10-phenylisoalloxazine. Butanoic, hexanoic, octanoic, and dodecanoic acids are well-endured in the reaction condition, and the corresponding α,β-unsaturated acids were isolated with moderate yields (up to 30%). The lower yield was observed for the desired products because a considerable amount of starting material remained unreacted and, thus, recovered.

In 2014, Jr and group reported the selective catalytic oxidative dehydrogenation of carboxylic acids for the synthesis of unsaturated acids at C2 and C3 carbon (Scheme 7.3) [23]. The transformation was catalyzed by Au nanoparticles supported on TiO_2. The authors discussed the higher activity and selectivity of Au/TiO_2 towards the C-H bonds at the C2 and C3 carbons of aliphatic linear organic acids than Pd/Pt catalysts. The authors confirmed the synthesis of crotonate and acrylate from butyric acid and propionic acid, respectively, through infrared (IR) and density functional theory (DFT).

In 2017, Newhouse and group reported the α,β-dehydrogenation of carboxylic acids via enediolates using allyl-palladium catalysis. The reaction afforded the major products with excellent yields (up to 90%) and diastereoselectivities (> 20:1) (Scheme 7.4) [24]. To avoid the classical decarboxylation pathway, the authors used the Zn(TMP)$_2$·2LiCl as a base in the presence of $ZnCl_2$ (excess) to generate dianions that underwent dehydrogenation to obtain unsaturated carboxylic acids. Various bases and additives were also screened, but the best results were found with Zn(TMP)$_2$·2LiCl and $ZnCl_2$ (Table 7.1).

Substrates with substitutions at the β-position were well-endured in the reaction condition. Terminal F-, Cl-, Br-, and MOMO-substituted desired products were isolated with 76, 74, 94, and 90% yield, respectively. Other functionalities such as internal epoxide and N-methyl indole were

Dehydrogenation Reactions of Carboxylic Acids

SCHEME 7.2 Dehydrogenation of carboxylic acids using lithium diisopropylamide and DDQ by Ronchi et al.

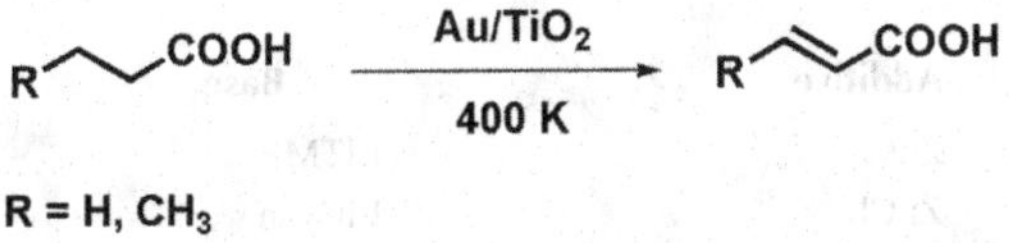

SCHEME 7.3 Oxidative-dehydrogenation of carboxylic acids by Jr et al.

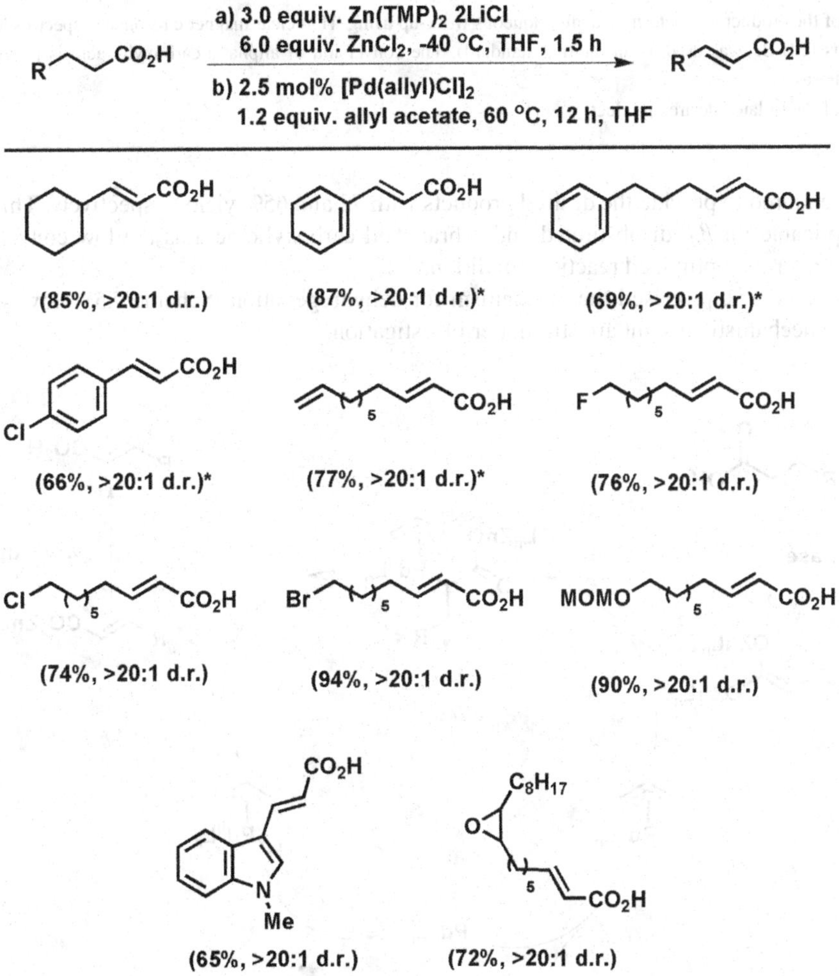

SCHEME 7.4 Dehydrogenation of carboxylic acids.

*2.3 equiv. Zn(TMP)$_2$·2LiCl, ZnCl$_2$ not added, and after 3 h, the reaction was quenched.

TABLE 7.1
Optimization of Carboxylic Acid Dehydrogenation

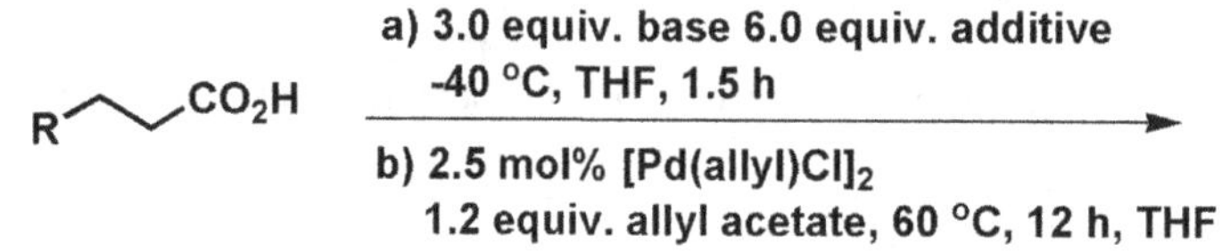

R-CH2-CO2H → R-CH=CH-CO2H

a) 3.0 equiv. base 6.0 equiv. additive, −40 °C, THF, 1.5 h
b) 2.5 mol% [Pd(allyl)Cl]$_2$, 1.2 equiv. allyl acetate, 60 °C, 12 h, THF

Entry	Additive	Base	Yield [%][a]
1	ZnCl$_2$	LiTMP	40 (48)
2	ZnCl$_2$	LiCyan	<5 (<5)
3	-	Zn(TMP)$_2$	52 (55)
4	-	Zn(TMP)$_2$·2LiCl	51 (55)
5	ZnCl$_2$	Zn(TMP)$_2$·2LiCl	85[b] (99)
6	LiCl	Zn(TMP)$_2$·2LiCl	19 (68)

[a] Yield of the product was determined after aqueous work-up using ^{1}H nuclear magnetic resonance spectroscopy (NMR) analysis (CH$_2$Br$_2$ was used as an internal standard). The conversion of aliphatic carboxylic acid is provided within parentheses.
[b] Yield of the isolated desired product.

also well-endured to provide the desired products with 72 and 65% yield, respectively. This method is not applicable for β,β-disubstituted and α-branched carboxylic acid as the low conversion was observed under the optimized reaction condition.

The authors have proposed the mechanism for dehydrogenation (Scheme 7.5). However, a few aspects of mechanistic insight are still under investigation.

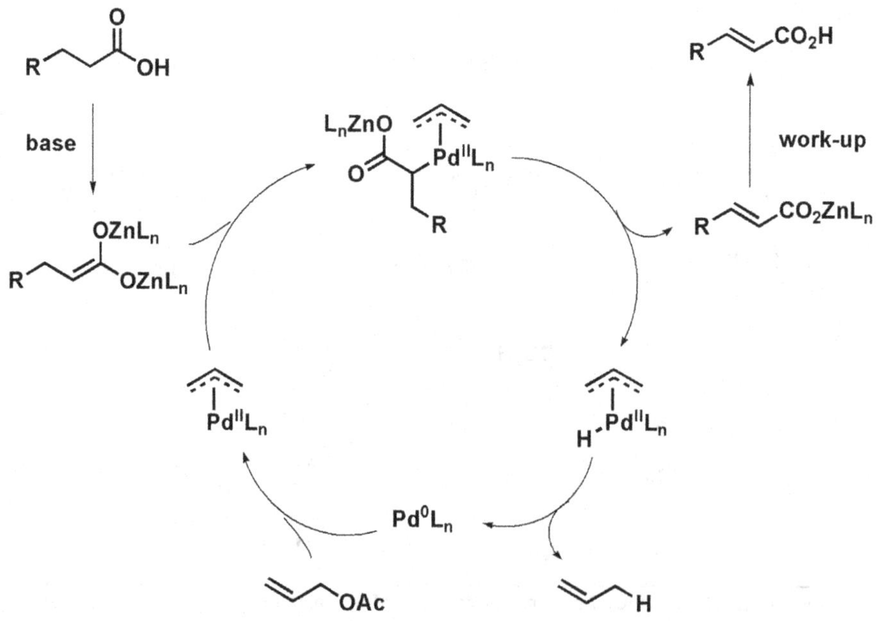

SCHEME 7.5 Proposed mechanism for dehydrogenation.

Dehydrogenation Reactions of Carboxylic Acids

In 2021, Yu et al. reported the dehydrogenation reactions of carboxylic acids using the bidentate pyridine-pyridone ligand (Scheme 7.6) [25]. The reaction proceeded through β-methylene C-H activation of carboxylic acids, which results in the formation of α,β-unsaturated products. Based on a literature survey and theoretical studies, the authors found that 2-pyridones are a suitable class of ligands for promoting Pd-catalyzed C-H activation through ligand activation. Thus, the authors prepared various bidentate pyridine-pyridone ligands and employed them in the reaction. 6-(Quinolin-2-yl)pyridin-2(1H)-one ligand was found to be the best. Various solvents and cosolvents were also screened, but 1,4-dioxane with tert-amyl alcohol was the best combination for this methodology (Table 7.2). Cosolvent lowered the Pd loading from 10% to 4%.

This method was applied to simple linear and branched aliphatic acids, and corresponding major products were isolated with high yields predominantly with *E*-isomer. Ring systems as substituents

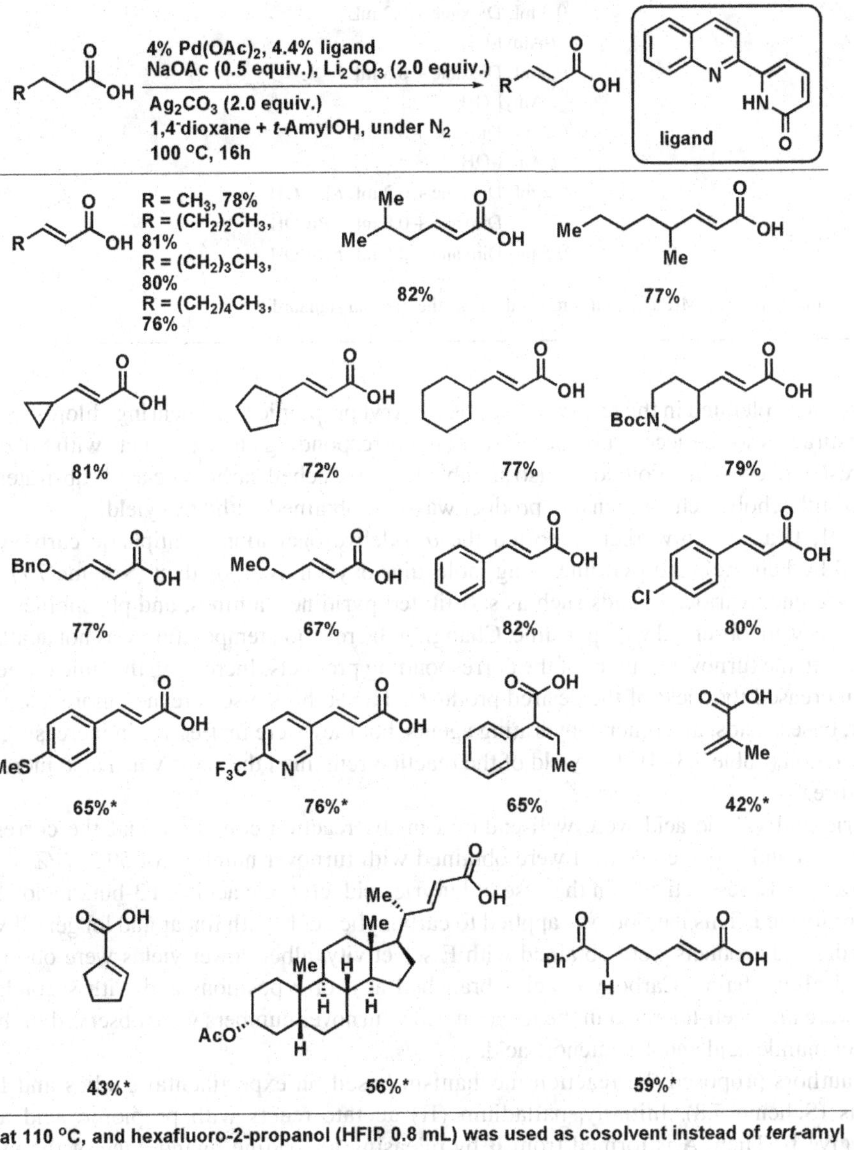

*at 110 °C, and hexafluoro-2-propanol (HFIP 0.8 mL) was used as cosolvent instead of *tert*-amyl alcohol.

SCHEME 7.6 Ligand-controlled dehydrogenation reaction of carboxylic acids.

TABLE 7.2
Solvent Effect on the Dehydrogenation Reactions of Carboxylic Acids

Entry	Solvent	Yield [%]
1	1.0 mL Dioxane	49
2	0.2 mL Dioxane + 0.2 mL *t*-Amyl-OH	47
3	0.5 mL Dioxane + 0.5 mL *t*-Amyl-OH	61
4	0.2 mL Dioxane + 0.8 mL *t*-Amyl-OH	70
5	0.2 mL Dioxane + 0.2 mL *t*-Bu-OH	49
6	0.2 mL Dioxane + 0.8 mL *t*-Bu-OH	53
7	0.5 mL Dioxane + 0.5 mL *t*-Bu-OH	50

†Yields determined by ^{1}H NMR using dibromomethane as the internal standard.

were also well-tolerated in the reaction conditions. Aryl propionic acids bearing chloro- and methylthio-substituents were successfully desaturated to corresponding enoic products with 80% and 65% yields, respectively. This protocol is also suitable for α-branched carboxylic acid substrates. Acetyl-protected lithocholic dehydrogenated product was also obtained with 56% yield.

In 2021, Ura and co-workers reported the α,β-dehydrogenation of aliphatic carboxylic acids catalyzed by Pd(OAc)$_2$ and pyridine using molecular oxygen as an oxidant (Scheme 7.7) [26]. The authors screened various ligands such as substituted pyridines, amines, and phosphines; however, lower yields were observed with pyridine. Changing the reaction temperature was not at all effective in increasing the turnover number of the corresponding products. Increasing the time of the reaction to 72 h increased the yield of the desired products. The authors also screened more additives such as water, bases, acids, and other dehydrating agents, but they were ineffective in increasing the yield of the reaction (Table 7.3–4). The yield of the reaction remained the same when also increasing the O$_2$ pressure.

Butyric and valeric acid were well-endured in the reaction condition, and the corresponding crotonic acid and 2-pentenoic acid were obtained with turnover numbers of 59.2 (*E/Z* = 12.1) and 28.2 (*E/Z* = 18:1), respectively. In the case of butyric acid, crotonic acid and 3-butanoyloxybutanoic acid were formed. This method was applied to carboxylic acids with linear and longer alkyl chains, and the desired products were obtained with E selectivity, albeit lower yields were observed with elongated alkyl chains. Carboxylic acids branched at β or γ positions and with secondary alkyl chains, were also well-tolerated in the reaction. Low turnover numbers were observed in the case of dihydrocinnamic acid and 4-pentenoic acid.

The authors proposed the reaction mechanism based on experimental studies and DFT calculations (Scheme 7.8). Initially, palladium (II) acetate reacts with propionic acid and pyridine to give **B**. Then, **A** is formed from **B** by releasing a pyridine ligand. One of the propionate ligands coordinates to Pd (**C**) through β-C(sp^3)-H bonds. Then, the β-C(sp^3)-H bond is activated through the concerted metalation deprotonation mechanism to form five-membered complex **D**.

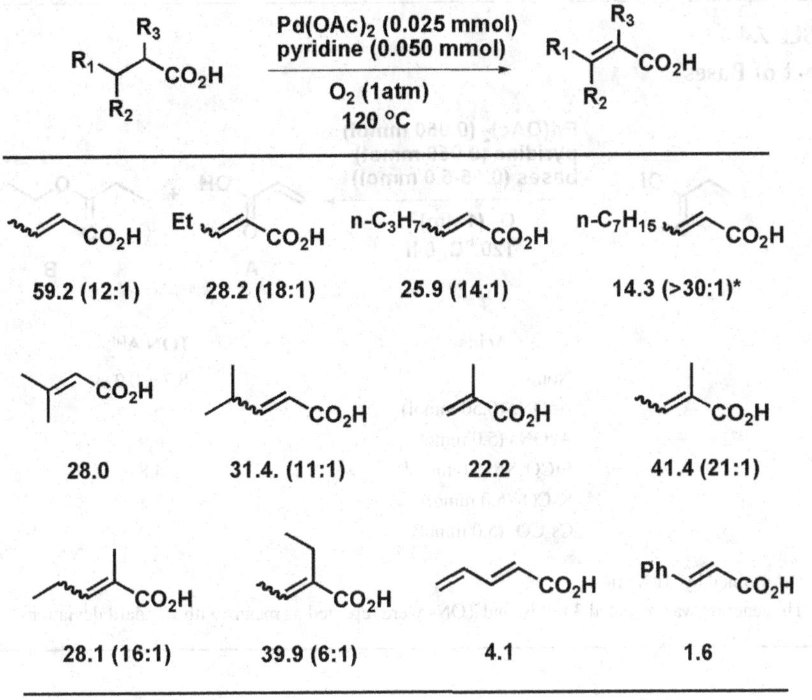

Dehydrogenation Reactions of Carboxylic Acids

$$\underset{R_2}{\overset{R_3}{R_1}}\!\!\!\!\!\!\!\!\!\text{CO}_2\text{H} \xrightarrow[\substack{\text{O}_2\,(1\,\text{atm}) \\ 120\,°\text{C}}]{\substack{\text{Pd(OAc)}_2\,(0.025\,\text{mmol}) \\ \text{pyridine}\,(0.050\,\text{mmol})}} \underset{R_2}{\overset{R_3}{R_1}}\!\!\!\!\!\!\!\!=\!\!\!\!\text{CO}_2\text{H}$$

⌢=CO₂H	Et-⌢=CO₂H	n-C₃H₇-⌢=CO₂H	n-C₇H₁₅-⌢=CO₂H
59.2 (12:1)	28.2 (18:1)	25.9 (14:1)	14.3 (>30:1)*
Me₂C=CH-CO₂H	iPr-CH=CH-CO₂H	(CH₃)₂C=CH-CO₂H	Me-C(=CH₂)-CO₂H (methacrylic)
28.0	31.4 (11:1)	22.2	41.4 (21:1)
(E)-MeCH=C(Me)CO₂H	Et-C(=CH-)CO₂H	CH₂=CH-CH=CH-CO₂H	Ph-CH=CH-CO₂H
28.1 (16:1)	39.9 (6:1)	4.1	1.6

The values in the parentheses are E/Z ratios. * Pd(OAc)₂ (0.013 mmol) and pyridine (0.025 mmol).

SCHEME 7.7 Pd-catalyzed dehydrogenation reaction.

TABLE 7.3
Effect of Acids

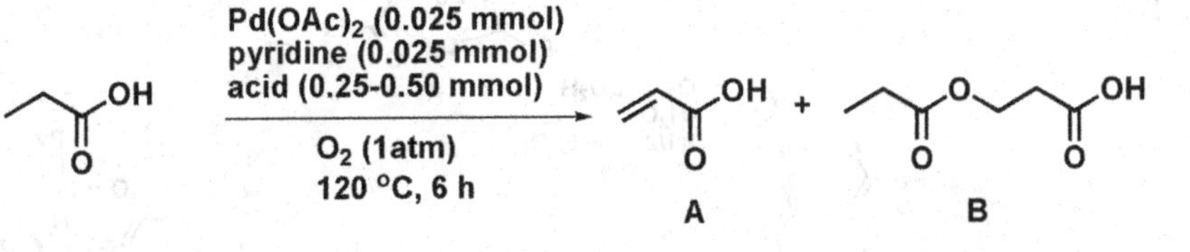

$$\text{EtCH}_2\text{CO}_2\text{H} \xrightarrow[\substack{\text{O}_2\,(1\,\text{atm}) \\ 120\,°\text{C},\,6\,\text{h}}]{\substack{\text{Pd(OAc)}_2\,(0.025\,\text{mmol}) \\ \text{pyridine}\,(0.025\,\text{mmol}) \\ \text{acid}\,(0.25\text{--}0.50\,\text{mmol})}} \text{CH}_2=\text{CHCO}_2\text{H}\;(\textbf{A}) + \text{EtC(O)OCH}_2\text{CH}_2\text{CO}_2\text{H}\;(\textbf{B})$$

Entry	Acids	Turn over number (TON) A[a]	TON B[a]
1	None	17.3 ± 1.2	4.4 ± 0.6
2	CF₃CO₂H (0.50 mmol)	6.8	1.8
3	CCl₃CO₂H (0.50 mmol)	0	0.3
4	B(OⁿBu)₃ (0.25 mmol)	12.7	3.3
5	Al(OEt)₃ (0.25 mmol)	12.2	2.7
6	Ti(OⁱPr)₄ (0.25 mmol)	8.7	2.2

[a] Determined by ¹H NMR.
[b] The reaction was repeated 5 times, and TONs were reported as means with standard deviations.

TABLE 7.4
Effect of Bases

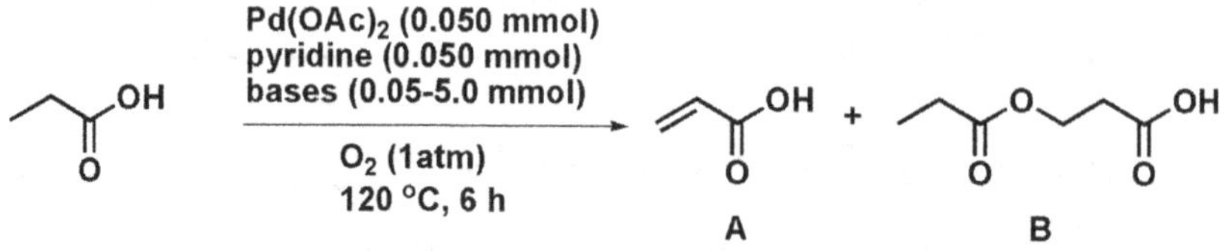

Entry	Acids	TON A[a]	TON B[a]
1	None	8.7 ± 0.2	2.3 ± 0.4
2	AcONa (0.50 mmol)	6.4	3.8
3	AcONa (5.0 mmol)	4.9	6.8
4	EtCO$_2$Na (5.0 mmol)	4.8	7.0
5	K$_2$CO$_3$ (5.0 mmol)	3.2	5.0
6	Cs$_2$CO$_3$ (5.0 mmol)	3.2	5.7

[a] Determined by ^{1}H NMR.
[b] The reaction was repeated 3 times, and TONs were reported as means with standard deviations.

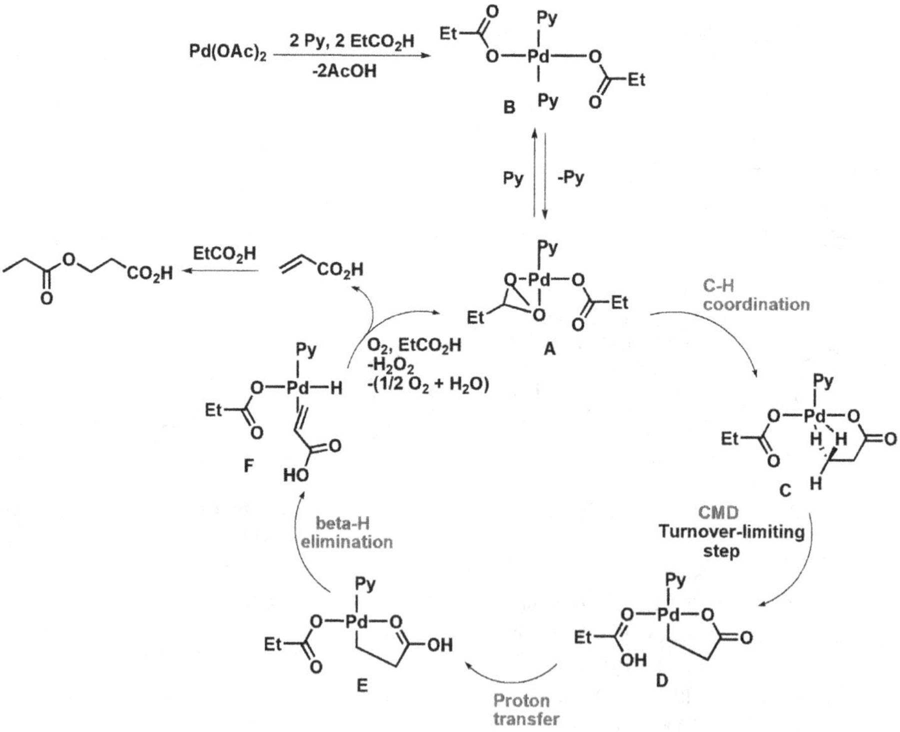

SCHEME 7.8 Proposed Mechanism.

Complex **D** undergoes proton transfer to give **E**, which undergoes β-hydride elimination to furnish **F**. **F** furnishes acrylic acid. **A** is regenerated from **F** by reductive elimination and reoxidation [27]. Furthermore, the synthesized acrylic acid reacts slowly with propionic acid to form β-propionyloxylated propionic acid.

7.3 CONCLUSION

For the synthesis of pharmaceutically important conjugated carbonyl moieties, α,β-dehydrogenation is the most versatile and sturdy method. Various novel methodologies have been developed in recent years for the dehydrogenation of these carbonyl moieties, which we discussed. In this chapter, we have shown the dehydrogenation of carboxylic acids using DDQ, Au nanoparticles supported on TiO_2, allyl-palladium catalysis, bidentate pyridine-pyridone ligand, and $Pd(OAc)_2$ and pyridine ligand for the synthesis of unsaturated acids. Thus, this chapter shows the potential of unsaturated acids to develop new drugs and biologically active compounds. We expect that greener methodologies will be developed in the near future for the synthesis of these valuable conjugated acids.

REFERENCES

[1] For reviews, see: a) S. R. Harutyunyan, T. den Hartog, K. Geurts, A. J. Minnaard, B. L. Feringa, *Chem. Rev.* **2008**, *108*, 2824–2852; b) N. A. Keiko, N. V. Vchislo, *Asian J. Org. Chem.* **2016**, *5*, 1169–1197; c) V. Marcos, J. Alemán, *Chem. Soc. Rev.* **2016**, *45*, 6812–6832; d) G. Desimoni, G. Faita, P. Quadrelli, *Chem. Rev.* **2018**, *118*, 2080–2248; e) O. K. Drosik, J. W. Suwinski, *Curr. Org. Chem.* **2018**, *22*, 345–361; f) K. Zheng, X. Liu, X. Feng, *Chem. Rev.* **2018**, *118*, 7586–7656; g) R. A. Farrar-Tobar, A. Dell'Acqua, S. Tin, J. G. de Vries, *Green Chem.* **2020**, *22*, 3323–3357; h) C. Brenninger, J. D. Jolliffe, T. Bach, *Angew. Chem. Int. Ed.* **2018**, *57*, 14338–14349; *Angew. Chem.* **2018**, *130*, 14536–14547; i) M. Kotora, R. Betík, *Catalytic Asymmetric Conjugate Reactions* (Ed. A. Córdova), Wiley-VCH, Weinheim, **2010**, 71–144; j) S. Zhang, W. Wang, *Catalytic Asymmetric Conjugate Reactions* (Ed. A. Córdova), Wiley-VCH, Weinheim, **2010**, 295–319; k) A. Lattanzi, *Catalytic Asymmetric Conjugate Reactions* (Ed. A. Córdova), Wiley-VCH, Weinheim, **2010**, 351–391; l) Y. Yamamoto, T. Nishikata, N. Miyaura, *Pure Appl. Chem.* **2008**, *80*, 807–817; m) T. J. Sommer, *Synthesis* **2004**, 161–201.

[2] a) S. Amslinger, *ChemMedChem* **2010**, *5*, 351–356; b) L. Arshad, I. Jantan, S. N. A. Bukhari, M. A. Haque, *Front. Pharmacol.* **2017**, *8*, 22; c) N. M. Xavier, A. P. Rauter, *Carbohydr. Res.* **2008**, *343*, 1523–1539; d) C. Zhuang, W. Zhang, C. Sheng, W. Zhang, C. Xing, Z. Miao, *Chem. Rev.* **2017**, *117*, 7762–7810.

[3] a) M. Hossain, U. Das, J. R. Dimmock, *Eur. J. Med. Chem.* **2019**, *183*, 111687; b) M. Gehringer, S. A. Laufer, *J. Med. Chem.* **2019**, *62*, 5673–5724; c) P. A. Jackson, J. C. Widen, D. A. Harki, K. M. Brummond, *J. Med. Chem.* **2017**, *60*, 839–885.

[4] a) A. Kumar, T. M. Bhatti, A. S. Goldman, *Chem. Rev.* **2017**, *117*, 12357–12384; b) V. Arun, S. De Sarkar, *Curr. Org. Chem.* 2019, 23, 1005–1018; c) J. M. Venegas, W. P. McDermott, L. Hermans, *Acc. Chem. Res.* **2018**, *51*, 2556–2564. d) S. Hati, U. Holzgrabe, S. Sen, *Belstein J. Org. Chem.* **2017**, *13*, 1670–1692; e) J. J. H. B. Sattler, J. Ruiz-Martinez, E. Santillan-Jimenez, B. M. Weckhuysen, *Chem. Rev.* **2014**, *114*, 10613–10653; f) C. Gunanathan, D. Milstein, *Science* **2013**, *341*, 249–260; g) J. Choi, A. H. R. MacArthur, M. Brookhart, A. S. Goldman, *Chem. Rev.* **2011**, *111*, 1761–1779.

[5] a) A. Toshimitsu, H. Owada, S. Uemura, M. Okano, *Tetrahedron Lett.* **1982**, *23*, 2105–2108; b) T. Mukaiyama, M. Ohshima, T. Nakatsuka, *Chem. Lett.* **1983**, *12*, 1207–1210; c) Y. Shvo, A. H. I. Arisha, *J. Org. Chem.* **1998**, *63*, 5640–5642; d) K. C. Nicolaou, T. Montagnon, P. S. Baran, *Angew. Chem. Int. Ed.* **2002**, *41*, 1386–1389; *Angew. Chem.* **2002**, *114*, 1444–1447; e) K. C. Nicolaou, T. Montagnon, P. S. Baran, *Angew. Chem. Int. Ed.* **2002**, *41*, 993–996; *Angew. Chem.* **2002**, *114*, 1035–1038; f) J. Liu, J. Zhu, H. Jiang, W. Wang, J. Li, *Chem. Asian J.* **2009**, *4*, 1712–1716; g) J. Zhu, J. Liu, R. Ma, H. Xie, J. Li, H. Jiang, W. Wang, *Adv. Synth. Catal.* **2009**, *351*, 1229–1232; h) T. Diao, T. J. Wadzinski, S. S. Stahl, *Chem. Sci.* **2012**, *3*, 887–891; i) W. Gao, Z. He, Y. Qian, J. Zhao, Y. Huang, *Chem. Sci.* **2012**, *3*, 883–886; j) M.-M. Wang, X.-S. Ning, J.-P. Qu, Y.-B. Kang, *ACS Catal.* **2017**, *7*, 4000–4003; k) G.-F. Pan, X.-Q. Zhu, R.-L. Guo, Y.-R. Gao, Y.-Q. Wang, *Adv. Synth. Catal.* **2018**, *360*, 4774–4783.

[6] a) J. Carretto, M. Simalty, *Tetrahedron Lett.* **1973**, *14*, 3445–3448; b) D. H. R. Barton, D. J. Lester, S. V. Ley, *J. Chem. Soc. Chem. Commun.* **1978**, 130–131; c) D. H. R. Barton, J. W. Morzycki, W. B. Motherwell, S. V. Ley, *J. Chem. Soc. Chem. Commun.* **1981**, 1044–1045; d) S.-I. Murahashi, T. Tsumiyama, Y. Mitsue, *Chem. Lett.* **1984**, *13*, 1419–1422; e) S.-I. Murahashi, Y. Mitsue, T. Tsumiyama, *Bull. Chem. Soc. Jpn.* **1987**, *60*, 3285–3290; f) A. Martin, M.-P. Jouannetaud, J.-C. Jacquesy, *Tetrahedron Lett.* **1996**, *37*, 7731–7734; g) T. Diao, S. S. Stahl, *J. Am. Chem. Soc.* **2011**, *133*, 14566–14569; h) Y. Izawa, D. Pun, S. S. Stahl, *Science* **2011**, *333*, 209–213; i) M. Schittmayer, A. Glieder, M. K. Uhl, A. Winkler, S. Zach, J. H. Schrittwieser, W. Kroutil, P. Macheroux, K. Gruber, S. Kambourakis, J. D. Rozzell, M. Winkler, *Adv. Synth. Catal.* **2011**, *353*, 268–274; j) T. A. Hamlin, C. B. Kelly, N. E. Leadbeater, *Eur. J. Org. Chem.* **2013**, 3658–3661; k) Y. Chen, D. Huang, Y. Zhao, T. R. Newhouse, *Angew. Chem. Int. Ed.* **2017**, *56*, 8258–8262; *Angew. Chem.* **2017**, *129*, 8370–8374; l) M. Chen, G. Dong, *Angew. Chem. Int. Ed.* **2021**, *60*, 7956–7961; *Angew. Chem.* **2021**, *133*, 8035–8040.

[7] a) J.-I. Matsuo, Y. Aizawa, *Tetrahedron Lett.* **2005**, *46*, 407–410; b) Y. Chen, J. P. Romaire, T. R. Newhouse, *J. Am. Chem. Soc.* **2015**, *137*, 5875–5878; c) S. M. Szewczyk, Y. Zhao, H. A. Sakai, P. Dube, T. R. Newhouse, *Tetrahedron* **2018**, *74*, 3293–3300.

[8] Y. Chen, A. Turlik, T. R. Newhouse, *J. Am. Chem. Soc.* **2016**, *138*, 1166–1169.

[9] K. B. Sharpless, R. F. Lauer, A. Y. Teranishi, *J. Am. Chem. Soc.* **1973**, *95*, 6137–6139.

[10] H. J. Reich, L. L. Reich, J. M. Renga, *J. Am. Chem. Soc.* **1973**, *95*, 5813–5815.

[11] B. M. Trost, T. N. Salzmann, K. Hiroi, *J. Am. Chem. Soc.* **1976**, *98*, 4887–4902.

[12] T. Mukaiyama, J.-I. Matsuo, H. Kitagawa, *Chem. Lett.* **2000**, *29*, 1250–1251.

[13] K. C. Nicolaou, Y. L. Zhong, P. S. Baran, *J. Am. Chem. Soc.* **2000**, *122*, 7596–7597.

[14] M. Hayashi, M. Shibuya, Y. Iwabuchi, *Org. Lett.* **2012**, *14*, 154–157.

[15] S. Gnaim, Y. Takahira, H. R. Wilke, Z. Yao, J. Li, D. Delbrayelle, P.-G. Echeverria, J. C. Vantourout, P. S. Baran, *Nat. Chem.* **2021**, *13*, 367–372.

[16] R. J. Theissen, *J. Org. Chem.* **1971**, *36*, 752–757.

[17] Y. Ito, T. Hirao, T. Saegusa, *J. Org. Chem.* **1978**, *43*, 1011–1013.

[18] J.-Q. Yu, H.-C. Wu, E. J. Corey, *Org. Lett.* **2005**, *7*, 1415–1417.

[19] T. Diao, S. S. Stahl, *J. Am. Chem. Soc.* **2011**, *133*, 14566–14569.

[20] M. Chen, G. Dong, *J. Am. Chem. Soc.* **2017**, *139*, 7757–7760.

[21] S. Gnaim, J. C. Vantourout, F. Serpier, P.-G. Echeverria, P. S. Baran, *ACS Catal.* **2021**, *11*, 883–892.

[22] G. Cainelli, G. Cardillo, A. U. Ronchi, *J. Chem. Soc. Chem. Commun.* **1973**, *3*, 94–95.

[23] M. McEntee, W. Tang, M. Neurock, J. T. Yates, Jr, *J. Am. Chem. Soc.* **2014**, *136*, 5116–5120.

[24] Y. Zhao, Y. Chen, T. R. Newhouse, *Angew. Chem. Int. Ed.* **2017**, *129*, 13302–13305.

[25] Z. Wang, L. Hu, N. Chekshin, Z. Zhuang, S. Qian, J. X. Qiao, J.-Q. Yu, *Science* **2021**, *374*, 1281–1285.

[26] A. Shibatani, Y. Kataoka, Y. Ura, *Asian J. Org. Chem.* **2021**, *10*, 3285–3289.

[27] a) M. M. Konnick, S. S. Stahl, *J. Am. Chem. Soc.* **2008**, *130*, 5753–5762; b) B. V. Popp, S. S. Stahl, *Chem. Eur. J.* **2009**, *15*, 2915–2922.

8 Dehydrogenation Reactions of Heterocyclic Compounds and Their Derivatives

Prakash Chandra and Syed Shahabuddin

8.1 INTRODUCTION

Heterocyclic compounds are an important component of the diversity of pharmaceuticals, fine chemicals, materials chemistry and natural products [1–5]. As a consequence of this diversity, chemical routes and materials have been exploited for the economical and facile synthesis of these heterocyclic compounds. In the recent past, multifarious pioneering synthetic protocols were developed to synthesize a variety of heterocyclic compounds [6]. However, these synthetic protocols are limited by the usage of high-cost materials, the liberation of hazardous chemicals into the environment, a mundane multistep synthesis process and harsh functional group protection/deprotection processes [7, 8]. Application of these methodologies to synthesize complicated organic molecules with sensitive functional groups further complicates the efficiency and yield for the desired heterocyclic compounds [9, 10].

These shortcomings have resulted in the development of more economical, environmentally benign and cost-efficient methods for the synthesis of a variety of heterocyclic compounds. Acceptorless dehydrogenation (AD) involves application of economical and easily available starting materials such as alcohols, carbonyl compounds (aldehydes/ketones), amines, etc. and has been explored for the synthesis of a diversity of heterocyclic compounds [11, 12]. In most cases, alcohols are important precursors to synthesize heterocyclic compounds via AD reactions. Moreover, there are manifold advantages of using the AD of alcohols for the synthesis of heterocyclic compounds, for example, (a) the generated hydrogen and water molecules are the only by-products of the reaction, and (b) hydrogen as the important by-product can be used for future clean energy storage molecules [13]. Transition metal (TM)-based catalysts have been studied for the AD reaction, and a one-pot atom-economical and benign technique has emerged to synthesize valuable heterocyclic compounds and hydrogen generation [14]. Recently, a TM-catalyzed AD reaction was recently used as an efficient technique for the C-X (X = O, S, N) bond-forming reaction, and multiple papers are published every year based on this topic [15, 16]. A diversity of homogeneous and heterogeneous TM catalysts have been investigated for the synthesis of a variety of heterocyclic compounds. However, to make the process more atom-economical and eco-friendly, heterogeneous TM catalysts have been examined to synthesize this variety of substituted heterocyclic compounds, namely, indoles, pyrazine, pyrrole, quinoline, benzaimidazole, pyrimidine, benzofuran, pyrrolidine, piperidine, morpholine, benzodiazepine, quinoxaline, quinazolinones, 2-alkylaminoquinolines, etc. [17].

8.2 TRANSITION METAL-CATALYZED SYNTHESIS OF HETEROCYCLIC COMPOUNDS

Heterogeneous TM catalysts have been investigated to synthesize a diversity of heterocyclic compounds. This section provides brief insight into the synthesis of many heterocyclic compounds

DOI: 10.1201/9781003321934-8

such as oxygen-containing heterocyclic compounds (lactones, benzofurans and chromones) and nitrogen-containing heterocyclic compounds (indoles, benzimidazoles, quinazolinones and pyrroles) [17].

8.2.1 SYNTHESIS OF LACTONES BY HETEROGENEOUS TM CATALYSTS

Heterogeneous TM catalysts have been used to synthesize a diversity of substituted lactones. Lactones are important building blocks, intermediates and solvents used for organic transformations. The mechanism for the heterogeneous TM-catalyzed lactones synthesis involves Brønsted acid-mediated esterification via the dehydrogenation of α-hydroxyl-substituted carboxylic acid. Alternative techniques for the synthesis of lactones involve reactions of epoxides/alcohols with carbon monoxide or reactions of allenols/butadiene with carbon monoxide. Ruthenium-based homogeneous catalysts have been developed to synthesize lactones via the lactonization of 1, 4-butanediol. Copper catalyst supported over hydrotalcite effectively promotes the lactonization of diols through the AD mechanism. Catalytic material efficiently promotes the dehydrogenation of a variety of liner, cyclic and aromatic diols. 0.01 mmol $Pt/SnO2$ was used as the heterogeneous and recyclable catalyst for the dehydrogenative lactonization of 1,2-diols at 180 °C under inert atmosphere. Many liner, benzylic and branched aliphatic alcohols were transformed into cyclic lactones with good yields (60–90% yield). Both the reactants (1,6-hexandiol is produced by the hydrogenation of bioderived 5-hydroxymethyl-2-furfural or furfural) and products are important and fulfil the criteria for the biorefinery concept. The aforementioned catalyst effectively promoted the dehydrogenative lactonization of 1,6-hexanediol to form valorized ε-caprolactone (the transformation of ε-caprolactone into ε-caprolactam) (see Figure 8.1) [18, 19]. $Pt/TiO2$ was used as the photocatalyst for the dehydrogenation of diols to lactones by Yoshida and co-workers. In the catalytic system, rutile demonstrates superior catalytic performance compared to anatase titania [20].

8.2.2 BENZOFURANS AND CHROMONES FROM ORTHO-SUBSTITUTED PHENOLS

Heterogeneous TM catalysts have been explored for oxygen-containing heterocycles such as benzofurans and chromones via the AD of ortho-substituted phenols. The oxygen-containing heterocycles are valuable ingredients of a diversity of pharmaceutical compounds. Benzofuran is the key component of a variety of natural products and pharmaceutical compounds. A facile route to synthesize benzofurans involves intramolecular C-O bond formation via coupling the phenolic −OH group with the vinyl carbon ortho-substituted alkenyl moiety. 2,3-dichloro-5,6-dicyano-p-benzoquinone (DDQ)-based sacrificial hydrogen acceptors were previously investigated for the synthesis of the substituted benzofurans, which frequently results in the generation of undesired by-products [21]. Recently, the AD technique was used to synthesize in a green and economical way many substituted benzofurans by the intramolecular C-O bond formation of o-alkylated phenols without the application of an oxidant or sacrificial agents. A diversity of electron-deficient and electron-rich aromatic rings of o-alkylated phenols undergo intermolecular coupling to form a furan ring. Pd-based homogeneous catalysts have been studied for the dehydrogenative synthesis of benzofurans [22]. Pd nanoparticles dispersed over carbonaceous support (Pd/C) efficiently catalyzed dehydrogenative benzofuran synthesis via a C-O bond-forming reaction without using sacrificial oxidants or dehydrogenating reagents [23]. Moreover, the catalytic system promotes electron-rich and electron-poor phenols efficiently to undergo dehydrogenative coupling to form a variety of corresponding furans (Figure 8.2) [24].

Moreover, Pd/C catalyst has been investigated for the synthesis of many chromones through the AD of ortho-acyl phenols via AD reaction without involving any oxidants or hydrogen acceptors. The AD proceeds with hydrogen elimination from the ortho-acyl side chain and substitution with phenolic hydroxyl oxygen to form a six-membered heterocyclic flavonoid ring via dehydrogenation (Figure 8.3) [25]. Interestingly, substituents present in the α-position slightly affect the fate of the

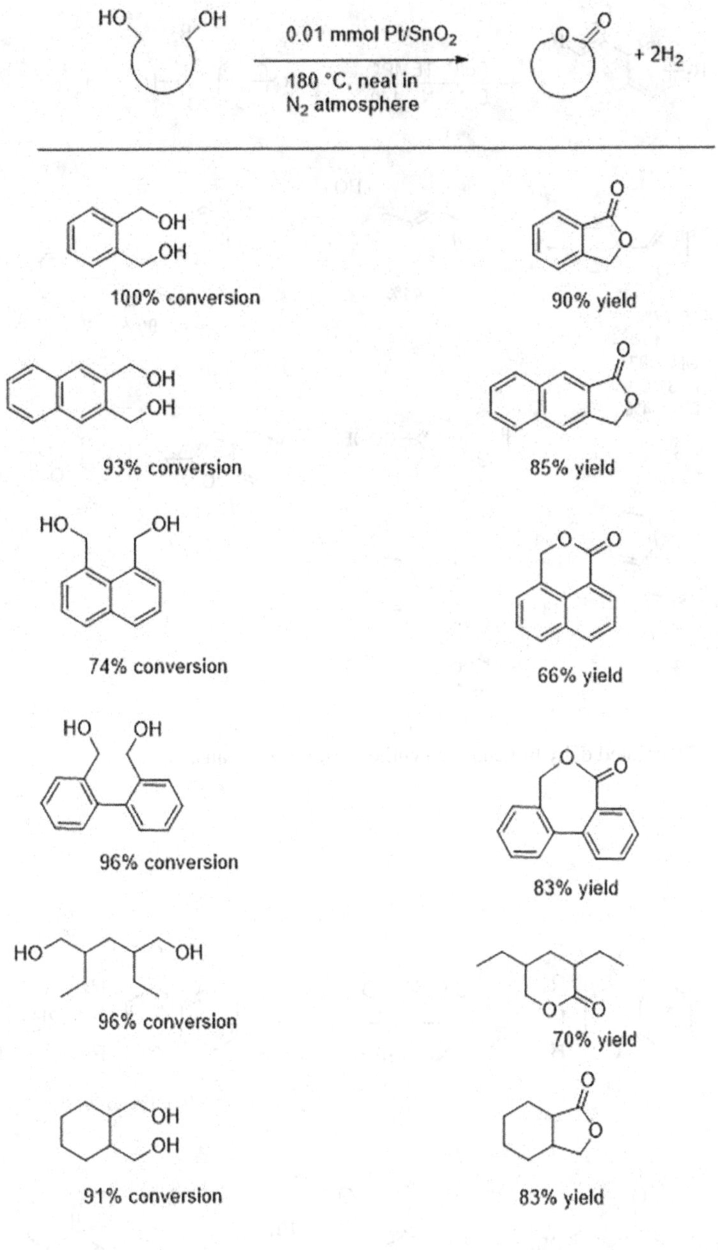

FIGURE 8.1 Pt/SnO2 effectively promoted the acceptorless dehydrogenative lactonization of diols.

catalytic reaction. The presence of the β-aryl and benzyl groups assist in the generation of chromones with high yields, whereas the absence of such substituents is responsible for lower yields (50–59% yield).

8.2.3 Nitrogen-Containing Heterocycles by Heterogeneous TM Catalysts

Heterogeneous TM catalysts have been examined as a green, atom-efficient and economical route to synthesize nitrogen-containing heterocycles, which are important components of the diversity of organic material such as indoles, benzimidazoles, quinazolinones and pyrroles.

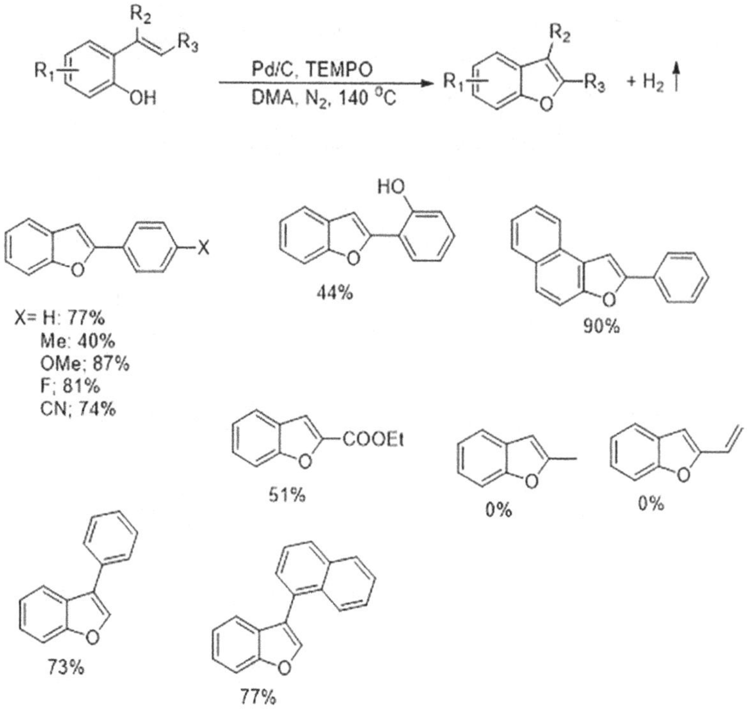

FIGURE 8.2 Pd/C mediated dehydrogenative synthesis of benzofurans.

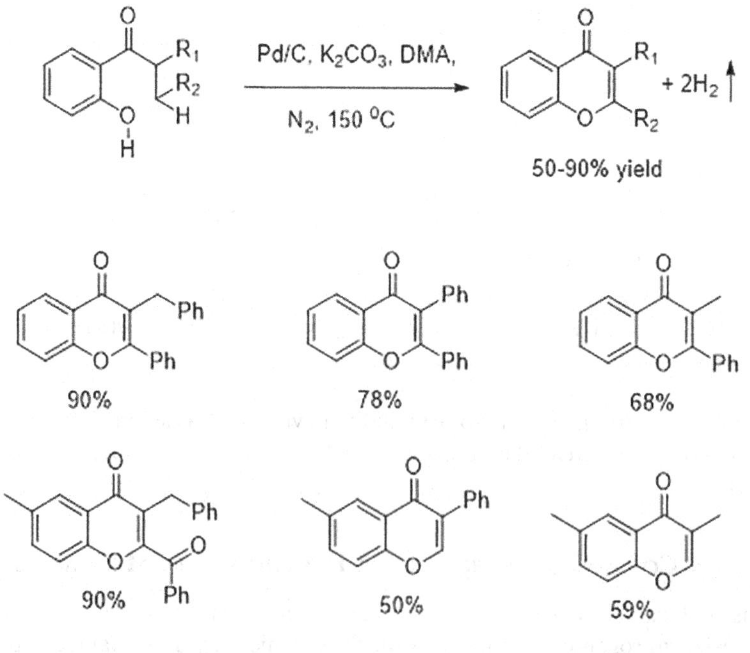

FIGURE 8.3 Pd/C-catalyzed dehydrogenative conversion of ortho-acyl phenols to chromones.

Nitrogen-containing heterocycles are essential ingredients occurring in a variety of biological systems and processes.

8.2.4 Indoles, Benzimidazoles, Quinazolinones and Pyrroles

Indoles are important organic compounds that occur in a wide variety of natural products, fine chemicals, pharmaceuticals, agrochemicals and functional materials. Dehydrogenative/oxidative cyclization 2-(2-aminophenyl) ethanol is one of the most promising ways to synthesize many substituted indoles, which are particularly significant. Ru loaded over metal oxide catalysts such as CeO2 and ZrO2 effectively promoted intramolecular dehydrogenation of substituted 2-(2-aminophenyl) ethanol to form substituted indoles. Other supports such as Al2O3, TiO2 and MgO were less effective in promoting this reaction. Amongst multifarious catalysts, Ru/CeO2 prepared by calcination at 200 °C outperformed other catalysts with > 99% yield for substituted indoles at 140 °C. Interestingly, the catalyst was recycled with reproducible results. The most plausible precursor active sites responsible for the interaction among $Ru^{IV} = O$, CeO2 and ZrO2 assisted the dehydrogenative cyclization of substituted 2-(2-aminophenyl) ethanol to corresponding substituted indoles (Figure 8.4) [26]. Uniformly dispersed Pt nanoparticles over solid acid supports such as Nb2O5 or H-beta for Pt/Nb2O5 or Pt/H-beta were used to synthesize 2-(2-aminophenyl) ethanol with 76% and 90% yield, respectively. However, Pt/Nb2O5 presents superior catalytic performance compared to the Pt/H-beta for the recycle studies.

Multifarious five-membered heterocyclic ring systems, namely, benzothiazoles, benzimidazoles and benzoxazoles, are useful components and building blocks occurring in a diversity of pharmaceuticals, natural products and agrochemicals. The conventional technique for the synthesis of 2-substituted benzazoles involves the condensation of aldehydes such as 2-aminophenols and 2-aminothiophenols or 1,2-phenelynediamines followed by oxidation with strong oxidizing agents. AD provides an interesting alternative to synthesize heterocyclic compounds via the dehydrogenation of alcohols; it is also a greener option for the synthesis of heterocyclic compounds. Ir/TiO2 was used as the catalyst for the dehydrogenative synthesis of benzimidazole via the condensation of 1,2-phenelynediamines and a variety of primary alcohols. Many substituted 1,2-phenylenediamine and primary alcohols were converted into the corresponding substituted benzimidazoles with moderate to good yield. Additionally, the catalyst was recycled multiple times without any significant dip in catalysis. Pt/TiO2 was used as the photocatalytic dehydrogenative condensation

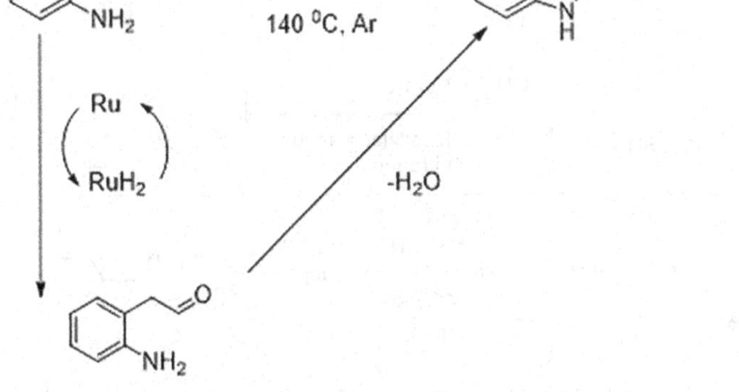

FIGURE 8.4 Synthesis of indole from 2-(2-aminophenyl)ethanol using a Ru/CeO2 catalyst.

under photoirradiation (λ > 300 nm). The method affords the synthesis of a diversity of substituted benzimidazole syntheses from a variety of 1,2-phenelynediamines and many primary and benzylic alcohols (Figure 8.5) [27]. 2-Substituted benzothiazoles are synthesized with high yields using Pt/Al2O3 under reflux conditions with mesitylene as a solvent at 140 °C via the dehydrogenative coupling of 2-aminothiophenols with 1-octanol (Figure 8.6) [28].

Quinazolinone is a principal component that occurs in many drugs, pharmaceuticals and biological compounds. Quinazolinone is prepared by the condensation of o-aminobenzamides and aldehydes via the formation of aminal intermediates and the application of hazardous oxidants. Homogeneous iridium complex [Cp*IrCl2]2 promotes the dehydrogenative condensation of primary alcohols and o-aminobenzamides to form quinazolinones under oxidant-free conditions. Heterogeneous TM catalysts have also been investigated for the economical and sustainable

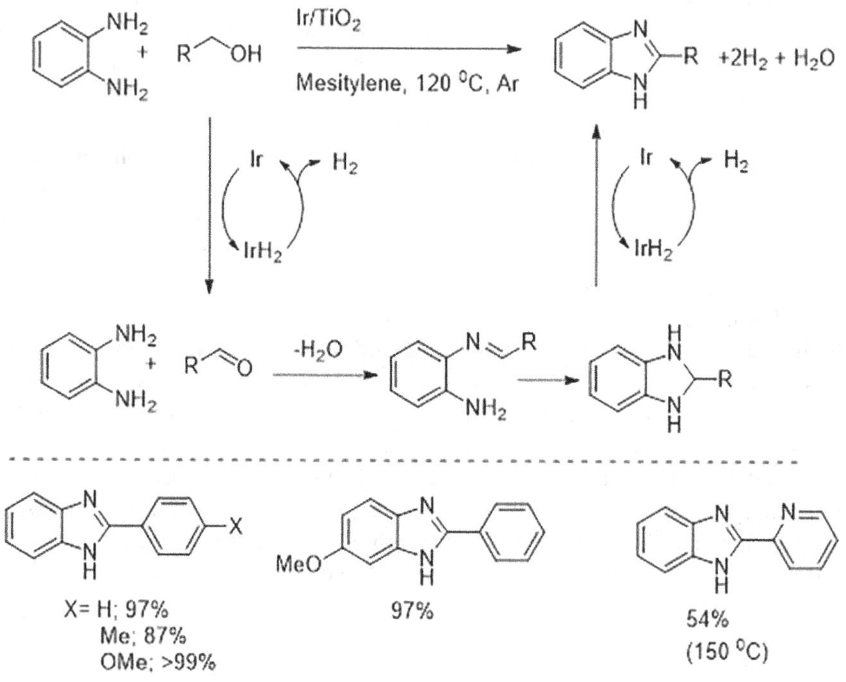

FIGURE 8.5 Synthesis of 2-substituted benzimidazoles from 1,2-phenylenediamine and primary alcohols using an Ir/TiO2 catalyst.

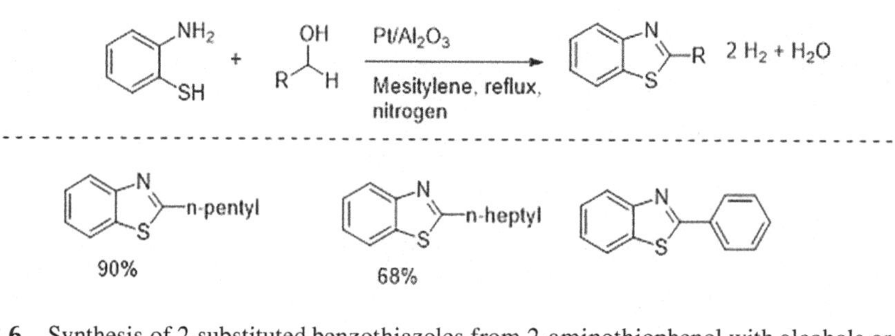

FIGURE 8.6 Synthesis of 2-substituted benzothiazoles from 2-aminothiophenol with alcohols or aldehydes using a Pt/Al2O3 catalyst.

synthesis of quinazolinone. A variety of Pt nanoparticles loaded over metal oxide (Al2O3, MgO, CeO2, ZrO2, TiO2, SnO2, SiO2, Y2O3, Nb2O5, SiO2-Al2O3, H-ZSM-5 and H-beta zeolites) and carbonaceous supports were used as the heterogeneous and recyclable catalyst to synthesize dehydrogenative quinazolinones with good yields. Amongst various catalytic systems, H-beta outperformed the rest. Substituted benzyl alcohols (both electron-donating and electron-withdrawing), cyclohexyl methanol and linear aliphatic alcohols were efficiently transformed into the corresponding substituted quinazolinones with good yields. The turnover number (TON) of heterogeneous catalysts was 20 times higher than the corresponding homogeneous catalysts. Water and hydrogen are the by-products of the catalytic reactions.

2-quinazolines are important pharmaceutical ingredients occurring in many biologically active compounds. The AD technique has been exploited for the synthesis of 2-quinazolines via hydrogen auto-transfer reaction. Heterogeneous TM catalysts have been synthesized through the dehydrogenative reaction of diamines and alcohols. Pt/CeO2 efficiently catalyzed 2-quinazoline synthesis by the reaction of primary alcohols and 2-aminomethyl-phenylamine. Interestingly, aliphatic primary alcohols are more easily converted into the corresponding quinazolines than benzyl alcohols (Figure 8.7) [28, 29]. Mechanistic studies demonstrate that the AD of alcohols transforms into aldehydes

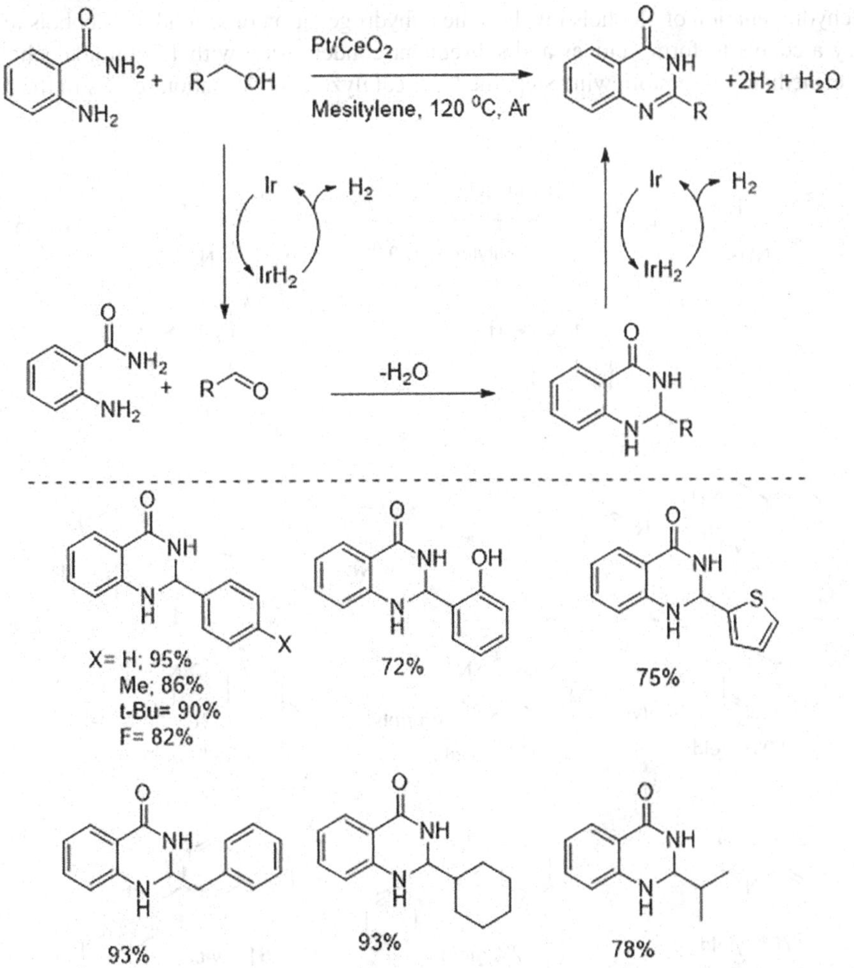

FIGURE 8.7 Synthesis of quinazolinones from o-aminobenzamide and primary alcohols using a Pt/H-beta catalyst.

followed by dehydrogenative condensation in the presence of 2-aminomethyl-phenylamine to form 2-quinazolines derivatives. Experimental analysis performed using XPS spectroscopy clearly demonstrates that an increase in the electron density O1s sites of metal oxide support decreases the binding energy and basicity of the materials. O1s basic sites with low binding energies are predominant active sites for the dehydrogenative 2-quinazolines synthesis.

Pyrrole rings are an important component that occur in a diversity of pharmaceuticals and fine chemicals. Homogeneous Ir and Ru complexes in the presence of bases such as KO-t-Bu have been studied to synthesize a variety of substituted pyrroles with moderate to good yields. However, these homogeneous catalysts suffer from poor recyclability and longevity of the catalytic materials. Ir@SiCN-based heterogeneous catalyst was used for the AD synthesis of pyrroles from 1,2-aminoalcohols and secondary alcohols in the presence of KO-t-Bu as the base. Pt nanoparticles supported over the carbonaceous support in the presence of KO-t-Bu have been investigated for the synthesis of many pyrroles. The catalytic materials prepared by the facile impregnation method present a high TON and were used for multiple cycles with reproducible results. The catalytic materials were active for different types of alcohols, namely, primary, secondary, cyclic and acyclic alcohols. Moreover, the catalytic materials were also active for a variety of amino-alcohols such as 2-amino-3-methyl-1-butanol, R-(−)-2-aminomehyl-1-butanol and (S)-(+)-2-amino-2-phenylethanol in corroboration with many secondary alcohols with moderate to good yields. A plausible mechanism for the dehydrogenation of alcohols involves the dehydrogenation of secondary alcohols to produce secondary alcohols to form ketones and subsequent condensation with 1,2-aminoalcohol to form imine intermediate. In the following step, the base-catalyzed condensation results in the formation

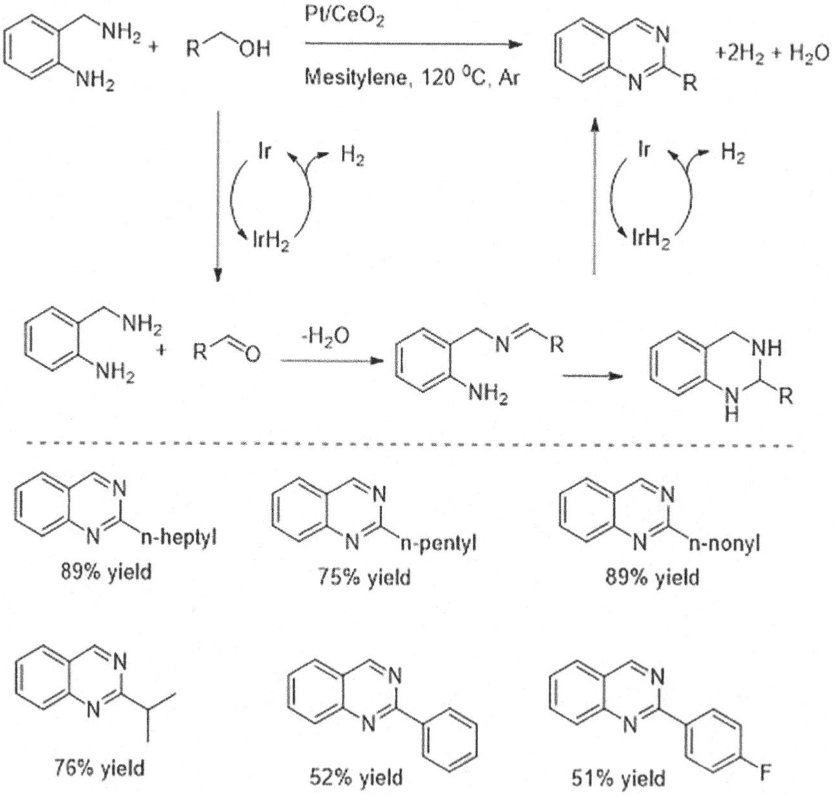

FIGURE 8.8 Synthesis of 2-substituted quinazolines from 2-aminobenzylamine and primary alcohols using a Pt/CeO2 catalyst.

Dehydrogenation Reactions of Heterocyclic Compounds

of 2,5-disubstituted pyrrole. In the reaction, KO-t-Bu promotes the dehydrogenation and deprotonation of the −OH group (Figure 8.9) [30].

8.3 SUMMARY AND OUTLOOK

N-heterocyclic compounds are noteworthy organic compounds because of their incredible pertinence to numerous arenas that encompass industry and pharmacological medicine. In recent decades, the AD approach has emerged as the necessary technique to manufacture uncomplicated and complicated organic compounds from readily available and low-cost organic compounds. The present chapter summarizes the application of TM-based heterogeneous catalysts to synthesize heterocyclic compounds through an AD reaction. A wide array of heterogeneous TM catalysts have been explored for the synthesis of these heterocyclic compounds via the coupling of simple organic molecules such as alcohols with nitro/nitrile/amine-derivatives involving a tandem and one-step process and avoiding the tedious purification process. Moreover, the outcome of the catalytic process has been ameliorated by modulating the catalyst synthesis procedure and understanding the reaction mechanism.

Despite incredible progress in the field, some challenges still need to be addressed. One of the important constraints is the development of a competent enantioselective heterogeneous catalyst to synthesize the methods for the highly in-demand enantiomeric heterocyclic compound relevant

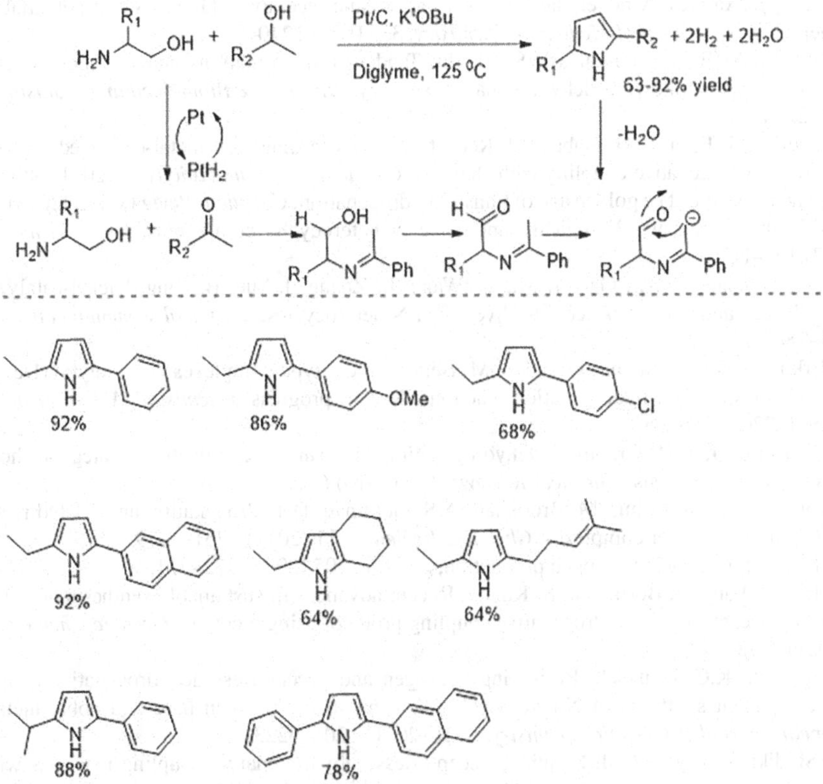

FIGURE 8.9 Synthesis of 2,5-disubstituted pyrroles from 1,2-aminoalcohols and secondary alcohols using a Pt/C catalyst.

to the pharmaceutical industry. Compared to the other contemporaneous routes available for the synthesis of enantiomeric heterocyclic compounds. Their synthesis via the application of heterogeneous TM catalysts has been less explored to date. Normally, acceptorless dehydrogenation coupling (ADC) reactions are performed at high temperature and involve stoichiometric reagents. However, asymmetric reactions desire milder reaction conditions to obtain products possessing high chirality. Thus, a cautious chiral catalyst fabrication technique is highly desirable to synthesize highly enantioselective and pure heterocycle. Moreover, noble metal-based homogeneous and heterogeneous catalysts have been widely investigated for the synthesis of these heterocyclic compounds. However, investigations of earth-abundant heterogeneous catalysts belonging to the 3d-TM series that can give competent results to those of noble metal catalysts are highly desirable.

REFERENCES

[1] A. Gomtsyan, Heterocycles in drugs and drug discovery, *Chemistry of Heterocyclic Compounds*, 48 (2012) 7–10.
[2] J.G. Cannon, *Comprehensive Heterocyclic Chemistry II*, edited by A.R. Katritzky, C.W. Rees, and E.F.V. Scriven, Pergamon Press, Elsevier Science, Ltd., Tarrytown, NY. 1996; 12 Volumes: Vol. 1A, xxvii+ 505 pp; Vol. 1B, 860 pp; Vol. 2, xiii+ 1102 pp; Vol. 3, xiii+ 932 pp; Vol. 4, xxiii+ 1006 pp; Vol. 5, xxiii+ 794 pp; Vol. 6, xxiii+ 1307 pp; Vol. 7, xxiii+ 1044 pp; Vol. 8, xv+ 1326 pp; Vol. 9, xvii+ 1146 pp; Vol. 10, xix+ 729 pp; Vol. 11, xi+ 596 pp. 19.5× 28 cm. ISBN 0-08-042724-3; 0-08-042725-1; 0-08-042726-X; 0-08-042727-8; 0-08-042728-6; 0-08-042729-4; 0-08-042730-8; 0-08-042731-6; 0-08-042732-4; 0-08-042965-3; 0-08-042987-4. $6345.00 (set), ACS Publications, 1997.
[3] R.A. Sheldon, I. Arends, U. Hanefeld, *Green Chemistry and Catalysis*, John Wiley & Sons, 2007.
[4] J.P. Michael, Quinoline, quinazoline and acridone alkaloids, *Natural Product Reports*, 25 (2008) 166–187.
[5] R. Sarges, H.R. Howard, R.G. Browne, L.A. Lebel, P.A. Seymour, B.K. Koe, 4-Amino [1, 2, 4] triazolo [4, 3-a] quinoxalines. A novel class of potent adenosine receptor antagonists and potential rapid-onset antidepressants, *Journal of Medicinal Chemistry*, 33 (1990) 2240–2254.
[6] B. Paul, M. Maji, K. Chakrabarti, S. Kundu, Tandem transformations and multicomponent reactions utilizing alcohols following dehydrogenation strategy, *Organic & Biomolecular Chemistry*, 18 (2020) 2193–2214.
[7] D. Banerjee, S. Bera, L.M. Kabadwal, Recent advances in transition metal-catalyzed (1, n) annulation using (de)-hydrogenative coupling with alcohols, *Chemical Communications* 57 (2021) 9807–9819.
[8] D. Wang, D. Astruc, The golden age of transfer hydrogenation, *Chemical Reviews*, 115 (2015) 6621–6686.
[9] A. Al-Mulla, A review: Biological importance of heterocyclic compounds, *Der Pharma Chemica*, 9 (2017) 141–147.
[10] P. Yang, C. Zhang, W.-C. Gao, Y. Ma, X. Wang, L. Zhang, J. Yue, B. Tang, Nickel-catalyzed borrowing hydrogen annulations: Access to diversified N-heterocycles, *Chemical Communications*, 55 (2019) 7844–7847.
[11] S. Werkmeister, J. Neumann, K. Junge, M. Beller, Pincer-type complexes for catalytic (De) hydrogenation and transfer (De) hydrogenation reactions: Recent progress, *Chemistry: A European Journal*, 21 (2015) 12226–12250.
[12] G.E. Dobereiner, R.H. Crabtree, Dehydrogenation as a substrate-activating strategy in homogeneous transition-metal catalysis, *Chemical Reviews*, 110 (2010) 681–703.
[13] J. Choi, A.H.R. MacArthur, M. Brookhart, A.S. Goldman, Dehydrogenation and related reactions catalyzed by iridium pincer complexes, *Chemical Reviews*, 111 (2011) 1761–1779.
[14] J.A. Turner, Sustainable hydrogen production, *Science*, 305 (2004) 972–974.
[15] M. Maji, D. Panja, I. Borthakur, S. Kundu, Recent advances in sustainable synthesis of N-heterocycles following acceptorless dehydrogenative coupling protocol using alcohols, *Organic Chemistry Frontiers*, 8 (2021) 2673–2709.
[16] N. Hofmann, K.C. Hultzsch, Borrowing hydrogen and acceptorless dehydrogenative coupling in the multicomponent synthesis of N-heterocycles: A comparison between base and noble metal catalysis, *European Journal of Organic Chemistry*, 2021 (2021) 6206–6223.
[17] S.H. Siddiki, T. Toyao, K.-I. Shimizu, Acceptorless dehydrogenative coupling reactions with alcohols over heterogeneous catalysts, *Green Chemistry*, 20 (2018) 2933–2952.

[18] A. Touchy, K.-I.J.O. Shimizu, Acceptorless dehydrogenative lactonization of diols by Pt-loaded SnO2 catalysts, *RSC Adv.* 5 (2015) 29072–29075. (d) K.-I. Fujita, W. Ito, R. Yamaguchi, Dehydrogenative lactonization of diols in aqueous media catalyzed by a water-soluble iridium complex bearing a functional bipyridine ligand, *ChemCatChem*, 6 (2014) 109–112. (e) K.-N.T. Tseng, J.W. Kampf, N.K. Szymczak, 32 (2013) 2046–2049.
[19] A.S. Touchy, K.-I. Shimizu, Acceptorless dehydrogenative lactonization of diols by Pt-loaded SnO2 catalysts, *RSC Advances*, 5 (2015) 29072–29075.
[20] E. Wada, A. Tyagi, A. Yamamoto, H. Yoshida, Dehydrogenative lactonization of diols with a platinum-loaded titanium oxide photocatalyst, *Photochemical & Photobiological Sciences*, 16 (2017) 1744–1748.
[21] G. Cardillo, R. Cricchio, L. Merlini, Reaction of ortho alkenyl-and alkylphenols with 2, 3-dichloro-5, 6-dicyanobenzoquinone (DDQ): Syntheses of 2, 2-dialkylchromenes, *Tetrahedron*, 27 (1971) 1875–1883.
[22] B. Xiao, T.-J. Gong, Z.-J. Liu, J.-H. Liu, D.-F. Luo, J. Xu, L. Liu, Synthesis of dibenzofurans via palladium-catalyzed phenol-directed C–H activation/C–O cyclization, *Journal of the American Chemical Society*, 133 (2011) 9250–9253.
[23] X. Zhao, J. Zhou, S. Lin, X. Jin, R. Liu, C–H functionalization via remote hydride elimination: Palladium catalyzed dehydrogenation of ortho-acyl phenols to flavonoids, *Organic Letters*, 19 (2017) 976–979.
[24] D. Yang, Y. Zhu, N. Yang, Q. Jiang, R.J.A.S. Liu, One-step synthesis of substituted benzofurans from ortho-alkenylphenols via palladium-catalyzed C-H functionalization, *Catalysis*, 358 (2016) 1731–1735.
[25] X. Zhao, J. Zhou, S. Lin, X. Jin, R.J.O.L. Liu, C–H functionalization via remote hydride elimination: Palladium catalyzed dehydrogenation of ortho-acyl phenols to flavonoids, *Organic Letters*, 19 (2017) 976–979.
[26] S. Shimura, H. Miura, K. Wada, S. Hosokawa, S. Yamazoe, M.J.C.S. Inoue, Ceria-supported ruthenium catalysts for the synthesis of indole via dehydrogenative N-heterocyclization, *Catalysis Science & Technology*, 1 (2011) 1340–1346.
[27] K. Tateyama, K. Wada, H. Miura, S. Hosokawa, R. Abe, M.J.C.S. Inoue, Dehydrogenative synthesis of benzimidazoles under mild conditions with supported iridium catalysts, *Catalysis Science & Technology*, 6 (2016) 1677–1684.
[28] C. Chaudhari, S.H. Siddiki, K.-I.J.T.L. Shimizu, Acceptorless dehydrogenative synthesis of benzothiazoles and benzimidazoles from alcohols or aldehydes by heterogeneous Pt catalysts under neutral conditions, *Tetrahedron Letters*, 56 (2015) 4885–4888.
[29] S.H. Siddiki, K. Kon, A.S. Touchy, K.-I.J.C.S. Shimizu, Direct synthesis of quinazolinones by acceptorless dehydrogenative coupling of o-aminobenzamide and alcohols by heterogeneous Pt catalysts, *Catalysis Science & Technology*, 4 (2014) 1716–1719.
[30] S. Siddiki, A.S. Touchy, C. Chaudhari, K. Kon, T. Toyao, K.-I.J.O.C.F. Shimizu, Synthesis of 2, 5-disubstituted pyrroles via dehydrogenative condensation of secondary alcohols and 1, 2-amino alcohols by supported platinum catalysts, *Organic Chemistry Frontiers*, 3 (2016) 846–851.

9 Recent Advances in Dehydrogenative Technique for Hydrogen Energy Storage and Utilization

Prakash Chandra and Syed Shahabuddin

9.1 INTRODUCTION

Fossil fuels are the primary non-renewable resources of energy and are limited and depleting fast at the current rate of consumption.[1-4] Moreover, these fossil fuel-based fuel sources are primarily responsible for global climatic change. If this scenario continues, then the end of the fossil fuel-based energy era may end sooner.[5] Governing bodies globally are busy implementing new policies for the reduction dependency on non-fossil fuel-based resources and for fostering renewable resources to fulfil the energy demands of humanity.[6] It is estimated that renewable energy resources such as wind, tidal, hydropower and solar energy will supply 18% of total energy by 2035.

Despite recent developments in renewable energy resources, fossil fuel-based energy resources are still in demand because renewable energy resources (wind and solar energies) are unstable and fluctuate. Therefore, energy storage systems are necessary to balance the fluctuations. There are well-known energy storage systems including kinetic (flywheel),[7] magnetic (superconducting magnets),[8] thermochemical (molten salts),[9] electrochemical (fuel cells, batteries and supercapacitors)[10,11] and chemical (hydrogen, metal hydrides, metal organic frameworks, etc.).[12-14] However, most of these energy storage systems present either poor life cycles or shelf life or mediocre energy storage densities.

Hydrogen is currently regarded as the clean and renewable energy resource for the fuel cells for stationary/mobile and greenhouse-free portable power applications.[15,16] Hydrogen generation technologies are intensively investigated, and diverse sources for hydrogen generation such as reforming fossil fuels, biological routes and decomposition of water are very mature.[17-21] However, all of these hydrogen generation techniques are high energy processes and require expensive catalysts; therefore, these hydrogen storage techniques are imperiling widespread application of hydrogen as a future fuel.[22] Other major shortcomings hampering widespread application of hydrogen energy storage are the low energy density of hydrogen molecules (0.08988 g/L at 1 atm) and unavailability of suitable hydrogen carriers.[23,24] The low specific density of hydrogen makes the hydrogen storage process a formidable task. Other obstacles and threats associated with traditional hydrogen storage are transportation challenges, boil-off challenges and high hydrogen losses and high cost.[25] To counter these shortcomings, the scientific community is exploring more economical and robust methods for hydrogen storage.

The characteristic feature for hydrogen storage materials is the surplus supply of hydrogen energy at a bargain basement price (40 g/L and 5.5 wt.% hydrogen, as recommended by the US department of energy). Chemical hydrogen storage in the form of inorganic hydrides are potential future candidates for energy storage.[26-27] The shortcomings can be overcome through the hydrolysis of metals or their hydrides, which has recently emerged as the economical, facile operation conditions and high hydrogen

Dehydrogenative Technique for Hydrogen Energy Storage

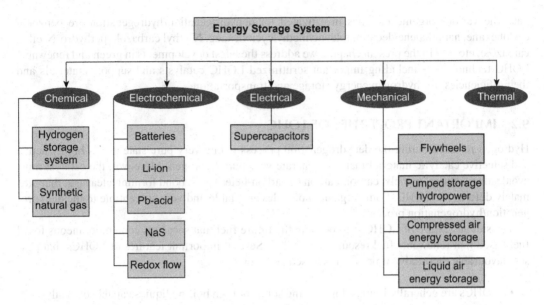

FIGURE 9.1 Multifarious energy storage systems.

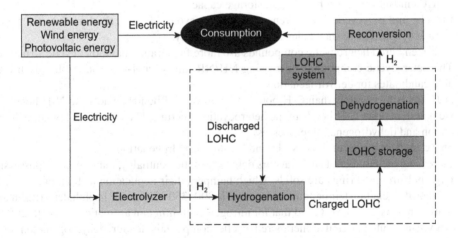

FIGURE 9.2 Multifarious renewable energy storage methodologies based on LOHC technology.

yielding alternatives for the hydrolysis of metal or their hydrides under simple operating conditions. The serious limitations can be overcome by using potent hydrogen storage materials such as inorganic hydrides, metal organic frameworks (MOFs), liquid organic hydrogen carriers (LOHCs), etc.[28–30]

To counter the shortcomings, LOHCs have recently emerged as interesting materials for the economical and facile transportation of hydrogen.[31–33] In the present chapter, we discuss the significance of LOHCs as important hydrogen storage materials of the future.[34–36] An LOHC system is composed of one pair of organic components, that is, hydrogen lean (LOHC−) and hydrogen rich (LOHC+) components. LOHCs involve a hydrogenation-dehydrogenation cycle where hydrogen is stored *via* the catalytic hydrogenation of (LOHC−) to (LOHC+) in the hydrogenation cycle and (LOHC+) to (LOHC−) in the dehydrogenation cycle. LOHC+ compounds are stable and store hydrogen for very time without getting discharged. This tendency of LOHC molecules facilitates prolonged hydrogen storage and transportation of hydrogen to remote areas at minimum risk. To date, the various organic moieties investigated for dehydrogenation/hydrogenation are benzene/cyclohexane, naphthalene/decalin, toluene/methylcyclohexane, N-ethyl carbazole/perhydro-N-ethyl carbazole, etc.[37–45] In the present chapter, we address the latest developments in green and renewable LOHC technologies, including important scrutinized LOHC catalysts and support materials and their efficiencies, for hydrogen energy storage and transportation.

9.2 IMPORTANT PROPERTIES OF LOHC

Hydrogen released during the dehydrogenation process is in a very pure state and is highly active and selective catalytic material that can generate very pure hydrogen. Moreover, the catalyst must avoid side reactions, namely, carbon-carbon or carbon-heteroatom bond forming cleavage, that ultimately deform the LOHC. Some organic molecules are stable and very susceptible to the dehydrogenation/hydrogenation process.

The salient feature of LOHC is to become the future fuel that succeeds contemporaneous fossil fuel-based non-renewable fuel resources.[31,33,35,46–49] Several important features of LOHCs that possess favourable thermodynamics are discussed as follows:

- LOHCs are generally liquids, low-boiling solids or high boiling liquids capable of catalyst-promoted reversible hydrogenation/dehydrogenation at high temperature.
- Moreover, during the catalytic reversible dehydrogenation/hydrogenation process, the structure of the LOHC remains intact.
- LOHCs must possess high hydrogen storage capacity.
- Bicyclic and polycyclic heteroarenes are highly favourable structural motives for energy storage because these molecules have low aromaticity.
- Five-membered heterocyclic compounds are more favourable than other cyclic structures.
- The heterocyclic compounds possessing NH/NR moiety at 1-position are highly favourable candidates for dehydrogenation.
- N-H bonds are weaker than C-H bonds; therefore, C-H bonds adjacent to N-H bonds are weaker than other C-H bonds in the heterocyclic structure. These features favour hydrogenation and dehydrogenation processes.
- Steric hindrance around the N-H bond favours dehydrogenation.
- Increasing the number of rings favours dehydrogenation enthalpy, but molecules possessing large polyaromatic rings are solids, which hampers their application as fuel. For example, the dehydrogenation temperature of benzene is 300–350 °C, that for naphthalene (fused aromatic rings) is 250–300 °C and that for nitrogen containing an aromatic ring is 50–200 °C.
- Alkyl-substituted and five-membered cyclic compounds favour dehydrogenation more than six-membered rings.
- LOHCs must possess a low heat of hydrogen desorption (42–54 kJ/mol-H_2) and a low dehydrogenation temperature.
- LOHCs must be compatible with present-day infrastructure.
- LOHCs must be economical and have a low production cost and facile technical access.
- LOHCs must have low toxicity and eco-toxicity during the transportation process.

In the following section, we discuss various LOHC systems and the catalytic systems that promote efficient and selective dehydrogenation/hydrogenation systems.

9.3 MONO- AND POLYAROMATIC SYSTEMS FOR LOHC APPLICATIONS

9.3.1 BENZENE-CYCLOHEXANE SYSTEM

LOHC is formed by the hydrogenation of benzene and the dehydrogenation of cyclohexane. Cyclohexane has a hydrogen storage capacity of ~7.19 wt.% and is liquid under ambient conditions. However, a low boiling point and high volatility make cyclohexane more volatile. Benzene, the dehydrogenation product of cyclohexene, is highly toxic. Despite these shortcomings, a benzene/cyclohexane system has often been examined as an LOHC. Different heterogeneous transition metal catalysts have been explored for LOHC systems. Ni-Pt-based bimetallic catalyst supported over activated carbon catalysts (ACCs) effectively promotes LOHC dehydrogenation with 99.7% selectivity for hydrogen. The reaction was performed using a spray-pulsed reactor at 300 °C. 10 wt.% Ag/ACC in corroboration with a noble metal catalyst (1 wt.% Rh, Pt, Pd) improves the hydrogen generation performance.[50–52]

Xia and co-workers promoted cyclohexane dehydrogenation Ni-Cu/SiO$_2$ (17.3 mol% Ni and 3.6 mol.% Cu). The reaction was performed using a plug flow reactor at 250 °C with 95% cyclohexane conversion, 99.4% selectivity for benzene and a hydrogen production rate (HPR) of 54 mmol/h/g$_{cat}$. The catalytic performance was improved using Ni-Cu/SBA-15 (4.9 wt.% Ni, 3.5 wt.% Cu) with 99.7% cyclohexene and 99% benzene selectivity of 61 mmol/h/g$_{cat}$. Cyclohexane dehydrogenation was performed using 2 wt.% Pt/activated carbon (AC) and 3 wt.% Pt/Al$_2$O$_3$ using a batch type reactor at 300 °C with HPRs of 910 mmol/h/g$_{cat}$ and 1,800 mmol/h/g$_{cat}$, respectively.[52–55]

9.3.2 Toluene-Methylcyclohexane System

A toluene-methylcyclohexane system has been investigated for dehydrogenative hydrogen generation. Nickel nanoparticles dispersed over Al$_2$O$_3$ or Al$_2$O$_3$-TiO$_2$ were studied for hydrogen generation. Experimental investigations demonstrated that Ni/Al$_2$O$_3$-TiO$_2$ exhibited (99%) superior methylcyclohexane conversion compared to Ni/Al$_2$O$_3$ (16.5%). Diverse inorganic supports such as La$_2$O$_3$, TiO$_2$, ZrO$_2$, CeO$_2$, Fe$_2$O$_3$, Al$_2$O$_3$, MnO$_2$ and perovskites (La$_{0.3}$Y$_{0.3}$NiO$_3$) have been examined for methylcyclohexane dehydrogenation. Pt/La$_{0.3}$Y$_{0.3}$NiO$_3$ exhibited complete methylcyclohexane conversion and an HPR of 45 mmol/g$_{met}$/min using a spray pulse reactor at 350 °C. Zhang et al. studied 1 wt.% platinum nanoparticles dispersed over material generated by pyrolysis from tyre char that was used as the catalyst, which exhibited > 95% methylcyclohexane conversion and 100% toluene selectivity using a fixed-bed reactor at 300 °C and an HPR of 342 mmol/g$_{met}$/min. Highly dispersed Pt nanoparticles uniformly dispersed over the MoO$_2$ phase (Pt/Mo(x)-SiO$_2$ (x = 8 wt.% and Pt = 5 wt.%), which was used as the efficient catalyst for the > 80% selectivity for toluene. The reaction was performed using a down flow-fixed bed reactor at 400 °C and 2.2 MPa. Pt/Y$_2$O$_3$ that demonstrated a 98% yield because Y$_2$O$_3$ itself is good catalytic material for the dehydrogenation of methylcyclohexane. Gora and team researched a Pd membrane reactor for methylcyclohexane dehydration at 225 °C using a 1 wt.% Pt/Al$_2$O$_3$ catalyst. The 1 wt.% Pt/Al$_2$O$_3$ catalyst in corroboration with the membrane reactor exhibited more than 70% methylcyclohexane conversion. 1 wt.% Pt uniformly dispersed over stacked-cone carbon nanotubes (1 wt.% Pt SC-CNT and 0.25 wt.% Pt SC-CNT) gave more than 90% methylcyclohexane conversion with 100% selectivity of toluene. Moreover, these results are similar to those of 1 wt.% Pt/Al$_2$O$_3$ catalysts. These results clearly demonstrate that catalytic performance can be manipulated by varying the mixture of supports and textual properties of the catalytic material. Bimetallic catalytic materials outperform monometallic catalysts by modifying the stochiometric ratio of the bimetallic system. Amongst various catalytic systems examined, carbon-based materials exhibit superior catalytic performance compared to other inorganic supports.[56–64]

9.3.3 Decalin-Naphthalene System

Decalin molecules exhibit a 7.3 wt.% hydrogen storage capacity. But the high melting point of hydrogen lean naphthalene complicates the application of LOHC to hydrogen storage. Moreover, naphthalene is solid under ambient conditions, and the dehydrogenation of decalin is an irreversible process. Therefore, decalin can only be used once, and a new batch of decalin is required for each cycle. These shortcomings hamper the widespread application of decalin as a LOHC fuel. Despite

TABLE 9.1
Comparison of Catalytic Performance of Various Catalysts.

Entry	Catalyst	Reactor	Reaction condition	Results	Ref.
			Cyclohexane		
1	10 wt% Pt/CFF-1500S	SPR	330 °C	P = 0.51 mol/gPt/min, FH2 = 55 mmol/min	65
2	3 g/m2 Pt/Al2O3	SPR	350 °C	P = 3.8 mol/gPt/min, FH2 = 89 mmol/min	65
3	20 wt% Ni/ACC	SPR	300 °C	P = 8.5 mmol/gcat/min	66
4	10 wt% Ag/ACC + 1 wt% Pt	SPR	300 °C	P = 14.2 mmol/gmet/mi	67
5	2 wt% Pt/AC	BR	300 °C	P = 910 mmol/gmet/min	68
			Methylcyclohexane		
1	Ni/Al2O3	-	127 °C	XMCH = 16.5%	69
2	1 wt% Pt/La0.7Y0.3NiO3	SPR	350 °C	P = 45.76 mmol/gmet/min	70
3	Pt/V2O5	SPR	350 °C	-	70
4	5 wt% Pt/8 wt% Mo-SiO2	FBR	400 °C	-	71
5	1 wt% Pt/Al	MR	225 °C	dod = 70%	72
6	1 wt% Pt/SC-CNT	FBR	315 °C	dod = 90%	73

these shortcomings, sincere efforts have been examined for the development of efficient methodologies for LOHC hydrogen storage. A diversity of catalytic materials have been investigated, and fused ring systems have also been studied for decalin/naphthalene to be used as an LOHC. Suttisawat and group examined 1 wt.% Pt decorated on AC for electrical- and microwave-assisted decalin conversion to naphthalene. Experimental studies demonstrate that when the contact time during the reaction was 260 min, there was low decalin conversion during both electrical- (10% decalin conversion) and microwave-assisted (13% decalin conversion) conversion. However, when the contact time during the reaction was low (30 min), an 85% decalin conversion was observed. The decrease in the catalytic performance when the reaction was performed with a high contact time was due to the sintering Pt at high temperature. The problem of sintering was overcome by using Sn as a promoter in the catalyst. The addition of Sn promoted a 74% decalin conversion and 90% naphthalene selectivity. In the catalytic system, Sn assists the electronic modification of Pt by donating electrons to the holes present in the 5d band of the Pt. Because of the Pt-electronic modification, C-C bond cleavage does not occur over the catalyst surface. This process assists in the reduction of carbon deposition over the catalyst surface and enhances adsorption/desorption of reactants and products.

3 wt.% Pt/C catalysts prepared by various synthesis procedures such as ion-exchange, precipitation, impregnation and polyol techniques were also investigated as efficient and smart catalysts for decalin dehydrogenation. Amongst all of them, Pt/C prepared by the ion-exchange method presents 19.6% Pt dispersion and an HPR of 45.4 mmol/g_{Pt}/min and exhibits the highest catalytic performance and improved hydrogen production. Higher Pt-dispersion promotes hydrogen productivity. Sebastian et al. explored the effect of Pt dispersed over different carbon supports including active carbon, carbon nanofibers, carbon xerogel and carbon black for decalin dehydrogenation. Amongst all of them, Pt/dispersed over mesoporous carbon atoms facilitates the initial catalytic performance (220 mmol/g_{Pt}/min) for decalin dehydrogenation. The superior catalytic performance of Pt supported over AC was due to the high surface area of the mesoporous carbon (930 m^2/g). Pt nanoparticles dispersed over carbon black with wide pores exhibit lower catalytic performance, prolonged operation times and low catalytic performance (100–140 mmol/gPt/min). All experiments were performed in a batch reactor at 260 °C.[74–78]

9.3.4 Perhydrodibenzyltoluene–Dibenzyltoluene

Dibenzyltoluene is readily available (thousands of tons have been produced annually since 1960), has been investigated as a heat transfer oil and finds multifarious applications in distillation columns, polymerization vessels, heat exchangers, etc. Sasol supplies an isomeric mixture of dibenzyltoluenes under the trade name Marlotherm SH. Dibenzyltoluene exhibits excellent hydrogen storage capacities (6.2 wt.%), and perhydrodibenzyltoluene exhibits high hydrogen storage capacity (57 kg-H_2/m^3) and high power density (2.06 kWh/kg). Moreover, perhydrodibenzyltoluene shows higher volumetric hydrogen storage density than methylcyclohexane (perhydrodibenzyltoluene = 0.91 g/mL; methylcyclohexane = 0.77 g/mL). Other advantages of dibenzyltoluene are that it is low melting (−32 °C) with a high boiling point (390 °C), non-flammable, non-toxic and is regarded as a non-hazardous chemical. Additionally, low melting temperatures facilitate the facile separation of hydrogen via LOHC condensation. Moreover, dibenzyltoluene/perhydrodibenzyltoluene promotes diesel-like properties and outstanding non-toxicity, and the hydrocarbon-like structure facilitates facile handling under available infrastructure. Hydrogen can be released using a Pt-based catalyst at > 250 °C. Furthermore, dibenzyltoluene finds applications in stationary power sources and possesses a reaction enthalpy of 65.4 kJ/mol hydrogen. A mixture of fuel systems with dibenzyltoluene has also been studied for LOHC application.

Mixing dibenzyltoluene with hydrogen and carbon dioxide was screened with Pd/Al_2O_3 and Rh/Al_2O_3 catalysts by Jorschick and co-workers. An experimental investigation demonstrates that the degree of hydrogenation reached 0.8 at 210 °C using Rh/Al_2O_3 and 270 °C using Pd/Al_2O_3. Moreover, a CO_2/CH_4 ratio <1 was obtained using Pd at 120 °C to 270 °C, and Rh gives similar results in the temperature range of 120 °C to 150 °C. Other transition metal catalysts such as Pt and Ru were not selective towards hydrogenation dibenzyltoluene in the presence of carbon dioxide. Moreover, Pt gets poisoned in the presence of trace in-situ CO generated during carbon dioxide hydrogenation. However, methane was produced as a major product when Ru acted as a catalyst. 1 wt.% Pt/C promotes 96% perhydrodibenzyltoluene dehydrogenation at 270 °C, whereas 0.5 wt.% Pt/Al_2O_3 promotes 40% perhydrodibenzyltoluene dehydrogenation under identical conditions. Higher weight loadings of Pt (5 wt.%) demonstrate inferior catalytic performance. Support also plays a significant role in catalytic performance with catalytic activity in the order C > Al_2O_3 > SiO_2.

Jorschick and co-workers also explored the perhydrodibenzyltoluene dehydrogenation/hydrogenation application of a 0.3 wt.% Pt/Al_2O_3 catalyst using a two-in-one reactor. The advantages of the two-phase reactor are that it assisted in saving catalyst and equipment costs for energy storage application in a remote area. The process can be modulated from dehydrogenation to hydrogenation by changing the pressure from 1-bar to 30-bar pressure using the same catalyst. The hydrogenation/dehydrogenation reaction was performed for four cycles, and the experimental studies clearly demonstrate a dip in catalysis after the first cycle. However, the catalytic performance was constant after

TABLE 9.2
List of Catalyst Used in a Two-in-One Reactor.

Entry	Catalyst	Reactor	Reaction condition	Results	Ref.
		Perhydrodibenzyltoluene			
1	0.5 wt% Pt/Al2O3	BR	270 °C	dod = 75%	88
2	1 wt% Pt/C	BR	270 °C	dod = 96%	88
3	1 wt% Pt/C	BR	270 °C	dod = 71%	88
4	0.3 wt% Pt/Al2O3	PSR	291 °C	P = 1.2 gH2/gPt/min	89

the first cycle. Moreover, the generation of heavy products was increased after each cycle from 0.9 to 2.2%, whereas the yield for the lighter products remained constant. An increase in the generation of heavy products resulted in a decrease in product yield because the heavier products blocked the active sites of the catalyst.[79–87]

9.4 HETEROCYCLIC COMPOUNDS

9.4.1 Carbazole Derivatives

N-ethyl carbazole has been one of the most widely investigated heterocyclic LOHCs, and its hydrogenation leads to the formation of dodecahydro-N-ethylcarbazole (perhydrocarbazole). This LOHC was first suggested by the company Air Products and Chemicals. Perhydrocarbazole (50 kJ/mol) possesses low dehydrogenation enthalpy and high hydrogen storage capacity (5.8 wt.%). However, two major shortcomings of N-ethyl carbazole are its high melting point and high lability of the N-alkyl bond at temperatures higher than 270 °C. As a consequence of these shortcomings N-ethyl carbazole degrades faster during the dehydrogenation process. The dehydrogenation efficiency of N-ethyl carbazole was improved by an application packed back reactor possessing a microchannel reactor. The advantages of using a microchannel reactor include improved internal and external mass transfer and heat transfer without any significant pressure drop in the reactor. Moreover, the productivity of the microchannel reactor with 1.5 $gH_2/(g_{Pt}$ min) was higher than the packed-bed reactor with 0.2 $gH_2/(g_{Pt}$ min), and 90% perhydrocarbazole conversion occurred in both cases. A metallic structure prepared by selective electron beam melting (SEBM)-based structured reactors was coated with Pt/Al_2O_3 via a spray-coat and dip-coating technique. In the catalyst coated reactor system, the metal core promoted excellent thermal conductivity and transport from the walls of the reactor to the catalyst sites. Moreover, the catalytic activity was performed using a single reactor and 10-fold parallel reactors. Experimental examinations performed for N-ethyl carbazole dehydrogenation using a single reactor demonstrated hydrogen generation of 1.27 $g_{H2}/(gPt$ min) (feed flow 4 mL/min at 260 °C), whereas the 10-fold parallel reactor exhibited a hydrogen flow of 9.8 N L_{H2}/min (feed flow of 30 mL/min at 250 °C). The power densities generated were 4.32 kW/h and 3.84 kW/L by the single and 10-fold reactors, respectively. Perhydrocarbazole dehydrogenation was performed using a downflow tubular reactor at 260 °C. The catalytic material exhibited an HPR of 10.9 g_{H2}/gPt/min, and the power generated was 21.7 kW power/gPt (temperature below 235 °C). Yang et al. prepared a diversity of an alumina supported (5 wt.%) Pt, Pd, Rh or Ru catalyst for perhydrocarbazole dehydrogenation at 180 °C using a batch reactor system. Experiential study of the catalytic performance in the order of Pd > Pd > Rh > Ru was under identical conditions. According to another report by Smith et al., 5 wt.% Pd/SiO_2 exhibited 100% perhydrocarbazole dehydrogenation and 60% hydrogen yields at 170 °C. Overall, Pd was the best catalyst for this reaction.[31,35,90–94]

TABLE 9.3
Catalysts and Its Reaction Condition for Dodecahydro-N-Ethylcarbazole.

Entry	Catalyst	Reactor	Reaction condition	Results	Ref.
			Dodecahydro-N-ethylcarbazole		
1	0.82 mg Pt/Al2O3	TR	F_{feed} = 0.8 mL/min, 260 °C	YH2 = 20.9%, P = 7.3 gH2/gPt/min,	95
2	5 wt% Pd/SiO2	BR	F_{feed} = 0.4 mL/min, 170 °C, 17 h	YH2 = 35%, XNEC = 100%, YH2 = 60%	96

9.4.2 Pyridines and Quinolines

Pyridine and quinoline are composed of one or more heterocyclic rings containing one nitrogen atom. The hydrogenation and dehydrogenation of pyridines and quinolines have been investigated for the synthesis of pharmaceuticals. Hydrodenitrification of pyridine and quinoline uses nickel- and cobalt-based catalysts during the petroleum refining process.

The hydrogenation product of pyridine leads to the formation of piperidine, which is volatile and flammable and is therefore not favourable for LOHC applications. However, two-membered quinoline and three-membered ring systems such as 4,7-phenanthroline and its derivatives are high-boiling alternatives for LOHC applications below 250 °C. Moreover, quinolines are degradable, colourless and hygroscopic liquids. A diversity of catalytic materials including Ni, Au, Pd, Pt, Rh, Cu and Co are supported over inorganic supports, namely, silica, alumina, titania, coal, $(Ca_{10}(PO_4)_6(OH)_2)$ and poly(4-vinylpyridine). However, the complete hydrogenation of quinoline resulting in the formation of intermediates 1,2,3,4-tetrahydroquinoline and 5,6,7,8- tetrahydroquinoline exhibits decline in catalytic performance by binding over the catalytic surface.

Oh and team studied 2-(N-methylbenzyl) pyridine as a potential candidate for LOHC with 6.15 wt.% hydrogen storage capacity. Pd/C and Ru/Al_2O_3 have been used as the catalyst for this reaction. Moreover, multiple hydrogenation/dehydrogenations were possible without any significant loss in structure of 2-(N-methylbenzyl) pyridine. Conversion of quinoline to tetrahydroquinoline is promoted at 200 °C and 10–70-bar pressure. But the conversion of tetrahydroquinoline to dihydroquinoline took more time even at elevated temperatures (175–260 °C) and pressures (110–210 bar). Moreover, catalyst deactivation was promoted using Bronsted acid-based solvents or a Lewis base because of the competitive adsorption at the active sites. Quinoline hydrogenation was promoted by enhancing the basicity of the support material (MgO< CaO< SrO). The Co_3O_4-Co/NGr@α-Al_2O_3 promoted the dehydrogenation of quinoline to tetrahydroquinoline even at 20 °C. Ruthenium nanoparticles (1–2 nm) supported over poly(4-vinylpyridine) (PVPy) promoted selective quinoline hydrogenation to 1,2,3,4-tetrahydroquinoline at 100–120 °C and 30–40-bar pressure. Polar solvents such as triethylamine and acetic acid assisted in promoting catalytic performance. Rh/Al_2O_3 efficiently catalysed the total hydrogenation of quinolines to form completely hydrogenated dihydroquinoline in the presence of hexafluoroisopropanol as a solvent. Rh-pillared layered catalyst (Rh-PLC) (having Rh = 2 nm) assisted 41% of the hydrogenation of quinoline to dihydroquinoline under 20-bar pressure using IPA as a solvent for 2.5 h. Rh/AlO(OH) catalysed quinoline hydrogenated to dihydroquinoline with 99.3% yield at 125 °C and quinoline/Rh/AlO(OH) in the ratio 120/1. Aprotic solvents were more efficient in promoting DHQ hydrogenation than aprotic solvents due to hydrogen bonding with the π-bond of the aromatic solvent. However, recycle studies clearly demonstrate an increase in particle size to 5.2 nm and a decrease in DHQ selectivity to 71%.[97–103]

9.4.3 Pyrroles and Indoles

Five-membered heterocycles such as pyrroles and fused indoles (benzene ring fused with nitrogen-containing heterocycles) have been examined as interesting LOHC. The hydrodenitrogenation of these five-membered rings find application in petroleum refinery industries. Moreover, pyrrole and its derivatives are too highly volatile to be considered for LOHC applications. However, the indole/octahydroindole pair have a 6.4 wt.% hydrogen storage capacity and are solid at room temperature. Moreover, indoline is an intermediate formed during the hydrogenation of indole because of the occupancy of the carbocyclic ring at the coordination sites of the catalyst. Several substituted pyrroles such as butyl pyrrole (hydrogen storage capacity = 3.14 wt.%), N-ethyl indole and N-methyl indole have been explored as LOHC compounds. N-methyl pyrrole and N-ethyl indole are liquid at room temperature with a hydrogen storage capacity of

5.81 wt.% and 5.23 wt.%, respectively. 5 wt.% Ru/Al$_2$O$_3$ catalyst promotes N-ethyl indole hydrogenation at 160–190 °C at a 90-bar pressure using hexane as a solvent. The rate of N-ethyl indole hydrogenation increased with an increase in temperature. Dehydrogenation of octahydroindole proceeds at 180 °C in 9 h. The activation energy for octahydroindole dehydrogenation was 117.7 kJ/mol. Ru- and Rh-based catalysts effectively promote indoline hydrogenation in the presence of protic solvents at 60 °C, 30 bar and a turnover frequency (TOF) of 100. Pd/C, Ru/C, Rh/Al2O3 and Pd/hydroxyapatite have been studied as catalysts to promote the dehydrogenation of indoline to indole.[104–108]

9.5 INTEGRATION OF LOHC PROCESS

LOHC is primarily an endothermic process; therefore, external heat is required to perform dehydrogenation. Several techniques have been integrated along the LOHC process for hydrogen and electricity generation. The integration of LOHC with a proton exchange membrane fuel cell (PEMFC) has been investigated for the dehydrogenation reaction. During the operation, electrical energy is converted into thermal energy, leading to a 24% reduction in the efficiency of the process (PEMFC electrical output vs heating value of LOHC hydrogen). Alternative techniques have been developed to enhance the efficiency of the process. Hydrogen burners have been integrated to provide the desired thermal energy for dehydrogenation. Hydrogen generated during LOHC provides the thermal energy for the dehydrogenation process. The remaining hydrogen is supplied to PEMFC to generate electricity. Process efficiency is enhanced to 34% when a hydrogen burner is used. The integration of LOHC with solid oxide fuel cell (SOFC) provides a much better and energy-efficient process for hydrogen generation. The process involves high-temperature exothermic SOFC integrated with the endothermic LOHC dehydrogenation process. SOFC exhaust has a temperature of > 600 °C. The high temperature of the exhaust can be reduced to 300 °C by supplying cool air, which creates an optimum temperature for LOHC dehydrogenation and efficient heat integration. Moreover, during the LOHC dehydrogenation process, the hydrogen produced is utilized to run SOFC to generate electricity. The SOFC+ LOHC-based symbiotic system produces electricity with 45% efficiency. Moreover, the experimental examination clearly demonstrated that the symbiotic SOFC+LOHC system produced energy for more than a decade without deterioration of the system.[31,109,110]

9.6 REACTOR FOR LOHC

A diversity of reactors have been investigated to improve LOHC performance for hydrogen generation including fixed-bed reactors, CSTR batch reactors, tubular reactors, pressure swing reactors, spray pulsed reactors and 3D structured monolith reactors (SEBMs). LOHC dehydrogenation reactors have to address the high volume of hydrogen generated. 1 mL of perhydrodibenzyltoluene

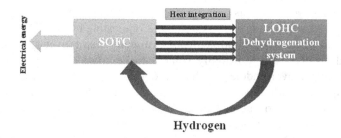

FIGURE 9.3 SOFC+LOHC integrated system for electricity production.

Dehydrogenative Technique for Hydrogen Energy Storage

dehydrogenation generates more than 650 mL of hydrogen in the gas generation device. The high volume of hydrogen generated can easily displace perhydrodibenzyltoluene, leading to poor heat and mass transfer and residence time over the catalyst surface. These phenomena cause a decline in catalyst performance.

A horizontal tubular reactor provides an interesting platform for performing LOHC reactions, where the hydrogen gas generated during the dehydrogenation process moves to the empty gas volumes, whereas the liquid flows to horizontal fixed bed catalyst pellets. Hydrogen evolution creates turbulence in the liquid and therefore improves heat transfer from the hot wall to the liquid. Multitubular or plate reactors are more efficient for large systems. Recently, a radial flow reactor was examined for dibenzyl-toluene using a Pt/Al$_2$O$_3$ catalyst and Pt-Ag membrane reactor. The reactor readily directs the radial flow outwards using microstructure. This type of reactor counters the voluminous release of hydrogen by discharging the hydrogen via multiple reactors.[50–52,64,77,78,92,111]

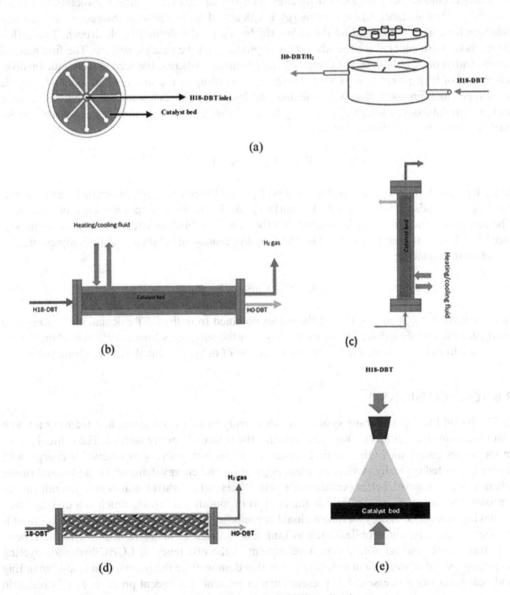

FIGURE 9.4 Different types of reactors investigated for LOHC application.

9.7 THEORETICAL AND COMPUTATIONAL APPROACH

Computational investigations performed using density functional theory (DFT) assist in understanding both the catalyst and the catalytic reaction. The computational and algorithm technique in corroboration with experimental explorations help in understanding the mechanism for catalyst-promoted LOHC reactions. DFT studies are useful in understanding the diverse aspects of catalysis, the effect on different surfaces for high-throughput reactions, important guidelines for the fabrication of catalytic materials and the identification and characterization of active sites present on the catalyst surface. These computational studies provide guidelines to design new reactions with optimized yield. Moreover, these computational techniques are not very expensive. DFT studies also provide useful insight into the interaction and the adsorption characteristics of the reactants, intermediates and products over the catalyst surface.

DFT calculations have been performed for octahydroindole dehydrogenation. The mechanism for octahydroindole dehydrogenation initiates with the adsorption of substrate (octahydroindole) over the catalyst surface. Adsorption energy is calculated for the process. Moreover, similar calculations have been performed for the indole (hydrogen-lean molecule) and hydrogen. Two different methods were adopted for the adsorption of products on the catalyst surface. The first method involves adsorption of both products on the catalyst surface, whereas the second technique involves adsorption of both products separately. Moreover, according to the first method, octahydroindole dehydrogenation products, that is, indole and four hydrogen molecules, are adsorbed on the same surface simultaneously, whereas, according to second method, both products are adsorbed separately on the same or different surfaces.

$$E_{ADS} = E_{S+A} - E_S - E_A$$

In the formula, E_{ADS} is total adsorption energy; E_{S+A} is adsorption energy of the surface and adsorbate; E_S is the adsorption energy of the surface; and E_A is the adsorption energy of adsorbate. The adsorption energy assists in determining the adsorbate binding strength of the catalytic surface.[112–117] The total energy calculated for the dehydrogenation of octahydroindole is represented by the following equation:

$$ER = [E_{H0} + nE_{H2} - E_{H8}]/n$$

In this equation, $E_{H0} + nE_{H2} - E_{H8}$ are the values obtained from the DFT calculations. EH0 is total energy for the surface indole; EH2 is total energy for the surface hydrogen; EH8 is total energy for the octahydroindole system; and n is the total number of dehydrogenated hydrogen molecules.

9.8 CONCLUSIONS

LOHC-based hydrogen storage systems relish sharply growing attention, and teaming up with manifold industrial set-ups makes these systems the future of energy storage. The critical points in the advancement for LOHC as fuel are the (a) better accessibility of renewable energy with hotspots situated typically in sites with less population and energy demand; (b) additional future advancements of novel, better performing electrolyzers; (c) the shift towards greener, minimum-emission technologies, particularly in the context of mobility division, which is a driving force of making renewable energy an international commodity in business; (d) incorporation of mobility sectors, namely, automobiles, such as tank trucks, tank farms or tank ships, in the renewable framework; and (e) noteworthy development in the efficiency of LOHC hydrogen cycling technologies and advancement in technologies that demonstrate the potential to heat-demanding and heat-releasing processes. The present chapter presents the recent progress in this scenario to make the reader mindful about the increasing potential for applications shaped by neoteric results.

Moreover, the present work also provides insight into offering partially investigated technologies. For instance, the performance of the catalytic material for both the dehydrogenation and hydrogenation steps can be further ameliorated, and by-products can be removed from the equation. The chemical and textural aspects of the catalytic material, including its acid-base property, wettability and porosity, can be further modulated to better optimize the yield. These features of catalytic material can be modified by adopting a novel catalyst synthesis procedure (such as varying the catalyst activation and calcination temperature and adopting in-exchange membranes). These steps can make the process more fruitful for future research applications. Additional possible improvements include a reduction in the dehydrogenation temperature and seeking novel LOHC molecules with lower reaction enthalpy. Computational modelling and kinetic studies to obtain a deep understanding of LOHC systems need to be sincerely addressed. From the reactor perspective, the major shortcomings that require attention are the formation of a large volume of gas and proper transfer of the heat generated during dehydrogenation at the catalytic sites that are filled with gases. To overcome these issues, catalytic-coated heat transfer plates that compose a well-organized catalytic system to optimize the hydrodynamics are highly recommended. Finally, process optimization also needs to be addressed that involves the implication of an intelligent coupling design between the exothermic hydrogen consumption (fuel cell or engine) and the endothermic hydrogenation process (LOHC hydrogenation or compression) to improve the performance of the overall process. The LOHC carrier material must possess excellent recyclability and by-product formation must be avoided to increase hydrogen storage economy. Accordingly, LOHC technology can emerge as a new hope for companies associated with mining and supplying hydrogen logistics. OHC technology can also be utilized for rural electrification.

REFERENCES

1 *Climate Change 2013—The Physical Science Basis: Working Group I Contribution to the Fifth Assessment Report of the Intergovernmental Panel on Climate Change*, Intergovernmental Panel on Climate, C., Ed., Cambridge University Press: Cambridge, **2014**. https://doi.org/10.1017/CBO9781107415324.004.
2 Turunen, H. *An Engineering Approach*, dissertation, Acta Universitatis Ouluensis, **2011**.
3 Agency, I. E. *Renewables Information 2015*, Springer, **2015**.
4 Ball, M.; Seydel, P.; Wietschel, M.; Stiller, C. Hydrogen-infrastructure build-up in Europe. *The Hydrogen Economy: Opportunities and Challenges*, Michael Ball and Martin Wietschel, Eds., Cambridge University Press, 385–453, **2009**.
5 Bianchini, C.; Meli, A.; Vizza, F. Hydrogenation of arenes and heteroatoms. *Handbook of Homogeneous Hydrogenation* Prof. Dr. Johannes G. de Vries, Prof. Dr. Cornelis J. Elsevier, Eds., (Vol. 2), Wiley-VCH, Weinheim, Germany, **2007**.
6 Preuster, P.; Papp, C.; Wasserscheid, P. Liquid organic hydrogen carriers (LOHCs): Toward a hydrogen-free hydrogen economy. *Accounts of Chemical Research* **2017**, *50* (1), 74.
7 Yang, Z.; Zhang, J.; Kintner-Meyer, M. C.; Lu, X.; Choi, D.; Lemmon, J. P.; Liu, J. Electrochemical energy storage for green grid. *Chemical Reviews* **2011**, *111* (5), 3577.
8 Louie, H.; Strunz, K. Superconducting magnetic energy storage (SMES) for energy cache control in modular distributed hydrogen-electric energy systems. *IEEE Transactions on Applied Superconductivity* **2007**, *17* (2), 2361.
9 Abedin, A. H.; Rosen, M. A. A critical review of thermochemical energy storage systems. *The Open Renewable Energy Journal* **2011**, *4* (1).
10 Alotto, P.; Guarnieri, M.; Moro, F. Redox flow batteries for the storage of renewable energy: A review. *Renewable and Sustainable Energy Reviews* **2014**, *29*, 325.
11 Kear, G.; Shah, A. A.; Walsh, F. C. Development of the all-vanadium redox flow battery for energy storage: A review of technological, financial and policy aspects. *International Journal of Energy Research* **2012**, *36* (11), 1105.

12 Frenette, G.; Forthoffer, D. Economic & commercial viability of hydrogen fuel cell vehicles from an automotive manufacturer perspective. *International Journal of Hydrogen Energy* **2009**, *34* (9), 3578.

13 Biniwale, R. B.; Rayalu, S.; Devotta, S.; Ichikawa, M. Chemical hydrides: A solution to high capacity hydrogen storage and supply. *International Journal of Hydrogen Energy* **2008**, *33* (1), 360.

14 Chang, P.-L.; Hsu, C.-W.; Lin, C.-Y. Assessment of hydrogen fuel cell applications using fuzzy multiple-criteria decision making method. *Applied Energy* **2012**, *100*, 93.

15 Bockris, J. O. M. The hydrogen economy: Its history. *International Journal of Hydrogen Energy* **2013**, *38* (6), 2579.

16 Jones, L. W. Perspectives on the evolution into a hydrogen economy. *Miami University. First World Hydrogen Energy Conference Proceedings* (Vol. 3), IOP Science, **1976**.

17 Singh, S.; Jain, S.; Venkateswaran, P.; Tiwari, A. K.; Nouni, M. R.; Pandey, J. K.; Goel, S. Hydrogen: A sustainable fuel for future of the transport sector. *Renewable and Sustainable Energy Reviews* **2015**, *51*, 623.

18 Wang, Z.; Naterer, G. F. Integrated fossil fuel and solar thermal systems for hydrogen production and CO2 mitigation. *International Journal of Hydrogen Energy* **2014**, *39* (26), 14227.

19 Hu, B.-B.; Zhu, M.-J. Enhanced hydrogen production and biological saccharification from spent mushroom compost by Clostridium thermocellum 27405 supplemented with recombinant β-glucosidases. *International Journal of Hydrogen Energy* **2017**, *42* (12), 7866.

20 Thungklin, P.; Sittijunda, S.; Reungsang, A. Sequential fermentation of hydrogen and methane from steam-exploded sugarcane bagasse hydrolysate. *International Journal of Hydrogen Energy* **2018**, *43* (21), 9924.

21 Zhang, Z.; Gao, C.; Li, Y.; Han, W.; Fu, W.; He, Y.; Xie, E. Enhanced charge separation and transfer through Fe2O3/ITO nanowire arrays wrapped with reduced graphene oxide for water-splitting. *Nano Energy* **2016**, *30*, 892.

22 Felderhoff, M.; Weidenthaler, C.; von Helmolt, R.; Eberle, U. Hydrogen storage: The remaining scientific and technological challenges. *Physical Chemistry Chemical Physics* **2007**, *9* (21), 2643.

23 Frischauf, N.; Acosta-Iborra, B.; Harskamp, F.; Moretto, P.; Malkow, T.; Honselaar, M.; Steen, M.; Hovland, S.; Hufenbach, B.; Schautz, M. The hydrogen value chain: Applying the automotive role model of the hydrogen economy in the aerospace sector to increase performance and reduce costs. *Acta Astronautica* **2013**, *88*, 8.

24 Tzimas, E.; Filiou, C.; Peteves, S.; Veyret, J. Hydrogen storage: State-of-the-art and future perspective. *EU Commission*, JRC Petten, EUR 20995EN, **2003**.

25 Mazloomi, K.; Gomes, C. Hydrogen as an energy carrier: Prospects and challenges. *Renewable and Sustainable Energy Reviews* **2012**, *16* (5), 3024.

26 von Helmolt, R.; Eberle, U. Fuel cell vehicles: Status 2007. *Journal of Power Sources* **2007**, *165* (2), 833.

27 Wang, Y.; Wang, Y. Recent advances in additive-enhanced magnesium hydride for hydrogen storage. *Progress in Natural Science: Materials International* **2017**, *27* (1), 41.

28 Teichmann, D.; Arlt, W.; Wasserscheid, P. Liquid organic hydrogen carriers as an efficient vector for the transport and storage of renewable energy. *International Journal of Hydrogen Energy* **2012**, *37* (23), 18118.

29 Gianotti, E.; Taillades-Jacquin, M. L.; Rozière, J.; Jones, D. J. High-purity hydrogen generation via dehydrogenation of organic carriers: A review on the catalytic process. *ACS Catalysis* **2018**, *8* (5), 4660.

30 Teichmann, D.; Arlt, W.; Wasserscheid, P.; Freymann, R. A future energy supply based on Liquid Organic Hydrogen Carriers (LOHC). *Energy & Environmental Science* **2011**, *4* (8), 2767.

31 Modisha, P. M.; Ouma, C. N.; Garidzirai, R.; Wasserscheid, P.; Bessarabov, D. The prospect of hydrogen storage using liquid organic hydrogen carriers. *Energy & Fuels* **2019**, *33* (4), 2778.

32 Sotoodeh, F.; Huber, B. J. M.; Smith, K. J. The effect of the N atom on the dehydrogenation of heterocycles used for hydrogen storage. *Applied Catalysis A: General* **2012**, *419–420*, 67.

33 Dean, D.; Davis, B.; Jessop, P. G. The effect of temperature, catalyst and sterics on the rate of N-heterocycle dehydrogenation for hydrogen storage. *New Journal of Chemistry* **2011**, *35* (2), 417.

34 Crabtree, R. H. Hydrogen storage in liquid organic heterocycles. *Energy & Environmental Science* **2008**, *1* (1), 134.

35 Von Wild, J.; Friedrich, T.; Cooper, A.; Toseland, B.; Muraro, G.; TeGrotenhuis, W.; Wang, Y.; Humble, P.; Karim, A. *18th World Hydrogen Energy Conference*, Essen, Germany, 2010.

36 Moores, A.; Poyatos, M.; Luo, Y.; Crabtree, R. H. Catalysed low temperature H2 release from nitrogen heterocycles. *New Journal of Chemistry* **2006**, *30* (11), 1675.

37 Sotoodeh, F.; Smith, K. J. Kinetics of hydrogen uptake and release from heteroaromatic compounds for hydrogen storage. *Industrial & Engineering Chemistry Research* **2010**, *49* (3), 1018.

38 Morawa Eblagon, K.; Tam, K.; Yu, K. K.; Zhao, S.-L.; Gong, X.-Q.; He, H.; Ye, L.; Wang, L.-C.; Ramirez-Cuesta, A. J.; Tsang, S. C. Study of catalytic sites on ruthenium for hydrogenation of N-ethylcarbazole: Implications of hydrogen storage via reversible catalytic hydrogenation. *The Journal of Physical Chemistry C* **2010**, *114* (21), 9720.
39 Jiang, N.; Rao, K. R.; Jin, M.-J.; Park, S.-E. Effect of hydrogen spillover in decalin dehydrogenation over supported Pt catalysts. *Applied Catalysis A: General* **2012**, *425*, 62.
40 Hodoshima, S.; Arai, H.; Saito, Y. Liquid-film-type catalytic decalin dehydrogeno-aromatization for long-term storage and long-distance transportation of hydrogen. *International Journal of Hydrogen Energy* **2003**, *28* (2), 197.
41 Oda, K.; Akamatsu, K.; Sugawara, T.; Kikuchi, R.; Segawa, A.; Nakao, S.-I. Dehydrogenation of methylcyclohexane to produce high-purity hydrogen using membrane reactors with amorphous silica membranes. *Industrial & Engineering Chemistry Research* **2010**, *49* (22), 11287.
42 Ferreira-Aparicio, P.; Rodriguez-Ramos, I.; Guerrero-Ruiz, A. On the performance of porous Vycor membranes for conversion enhancement in the dehydrogenation of methylcyclohexane to toluene. *Journal of Catalysis* **2002**, *212* (2), 182.
43 Ali, J. K.; Rippin, D. W. Comparing mono-and bimetallic noble-metal catalysts in a catalytic membrane reactor for methylcyclohexane dehydrogenation. *Industrial & Engineering Chemistry Research* **1995**, *34* (3), 722.
44 Koutsonikolas, D.; Kaldis, S.; Zaspalis, V.; Sakellaropoulos, G. Potential application of a microporous silica membrane reactor for cyclohexane dehydrogenation. *International Journal of Hydrogen Energy* **2012**, *37* (21), 16302.
45 Kou, Z.; Zhi, Z.; Xu, G.; An, Y.; He, C. Investigation of the performance and deactivation behavior of Raney-Ni catalyst in continuous dehydrogenation of cyclohexane under multiphase reaction conditions. *Applied Catalysis A: General* **2013**, *467*, 196.
46 Bourane, A.; Elanany, M.; Pham, T. V.; Katikaneni, S. P. An overview of organic liquid phase hydrogen carriers. *International Journal of Hydrogen Energy* **2016**, *41* (48), 23075.
47 Aakko-Saksa, P. T.; Cook, C.; Kiviaho, J.; Repo, T. Liquid organic hydrogen carriers for transportation and storing of renewable energy–Review and discussion. *Journal of Power Sources* **2018**, *396*, 803.
48 Markiewicz, M.; Zhang, Y.; Bösmann, A.; Brückner, N.; Thöming, J.; Wasserscheid, P.; Stolte, S. Environmental and health impact assessment of Liquid Organic Hydrogen Carrier (LOHC) systems–challenges and preliminary results. *Energy & Environmental Science* **2015**, *8* (3), 1035.
49 Cooper, A. C.; Fowler, D. E.; Scott, A. R.; Abdourazak, A. H.; Cheng, H.; Wilhelm, F. C.; Toseland, B. A.; Campbell, K. M.; Pez, G. P. Hydrogen storage and delivery by reversible hydrogenation of liquid-phase hydrogen carriers. *Papers of the American Chemical Society* **2005**, *50*, 271.
50 Kariya, N.; Fukuoka, A.; Ichikawa, M. Efficient evolution of hydrogen from liquid cycloalkanes over Pt-containing catalysts supported on active carbons under "wet–dry multiphase conditions". *Applied Catalysis A: General* **2002**, *233* (1–2), 91.
51 Biniwale, R. B.; Kariya, N.; Ichikawa, M. Dehydrogenation of cyclohexane over Ni based catalysts supported on activated carbon using spray-pulsed reactor and enhancement in activity by addition of a small amount of Pt. *Catalysis Letters* **2005**, *105* (1), 83.
52 Pande, J. V.; Shukla, A.; Biniwale, R. B. Catalytic dehydrogenation of cyclohexane over Ag-M/ACC catalysts for hydrogen supply. *International Journal of Hydrogen Energy* **2012**, *37* (8), 6756.
53 Xia, Z.; Lu, H.; Liu, H.; Zhang, Z.; Chen, Y. Cyclohexane dehydrogenation over Ni-Cu/SiO2 catalyst: Effect of copper addition. *Catalysis Communications* **2017**, *90*, 39.
54 Xia, Z.; Liu, H.; Lu, H.; Zhang, Z.; Chen, Y. Study on catalytic properties and carbon deposition of Ni-Cu/SBA-15 for cyclohexane dehydrogenation. *Applied Surface Science* **2017**, *422*, 905.
55 Kariya, N.; Fukuoka, A.; Utagawa, T.; Sakuramoto, M.; Goto, Y.; Ichikawa, M. Efficient hydrogen production using cyclohexane and decalin by pulse-spray mode reactor with Pt catalysts. *Applied Catalysis A: General* **2003**, *247* (2), 247.
56 Okada, Y.; Sasaki, E.; Watanabe, E.; Hyodo, S.; Nishijima, H. Development of dehydrogenation catalyst for hydrogen generation in organic chemical hydride method. *International Journal of Hydrogen Energy* **2006**, *31* (10), 1348.
57 Okada, Y.; Shimura, M. Development of large-scale H_2 storage and transportation technology with Liquid Organic Hydrogen Carrier (LOHC). *Proceedings of the 21st Joint GCC-Japan Environment Symposium*, Doha, Qatar, 5–6, **2013**.
58 Zhu, G.; Yang, B.; Wang, S. Nanocrystallites-forming hierarchical porous Ni/Al_2O_3–TiO_2 catalyst for dehydrogenation of organic chemical hydrides. *International Journal of Hydrogen Energy* **2011**, *36* (21), 13603.

59. Shukla, A. A.; Gosavi, P. V.; Pande, J. V.; Kumar, V. P.; Chary, K. V.; Biniwale, R. B. Efficient hydrogen supply through catalytic dehydrogenation of methylcyclohexane over Pt/metal oxide catalysts. *International Journal of Hydrogen Energy* **2010**, *35* (9), 4020.
60. Shukla, A.; Pande, J. V.; Biniwale, R. B. Dehydrogenation of methylcyclohexane over Pt/V2O5 and Pt/Y2O3 for hydrogen delivery applications. *International Journal of Hydrogen Energy* **2012**, *37* (4), 3350.
61. Zhang, C.; Liang, X.; Liu, S. Hydrogen production by catalytic dehydrogenation of methylcyclohexane over Pt catalysts supported on pyrolytic waste tire char. *International Journal of Hydrogen Energy* **2011**, *36* (15), 8902.
62. Boufaden, N.; Akkari, R.; Pawelec, B.; Fierro, J.; Zina, M. S.; Ghorbel, A. Dehydrogenation of methylcyclohexane to toluene over partially reduced silica-supported Pt-Mo catalysts. *Journal of Molecular Catalysis A: Chemical* **2016**, *420*, 96.
63. Gora, A.; Tanaka, D. A. P.; Mizukami, F.; Suzuki, T. M. Lower temperature dehydrogenation of methylcyclohexane by membrane-assisted equilibrium shift. *Chemistry Letters* **2006**, *35* (12), 1372.
64. Wang, Y.; Shah, N.; Huffman, G. P. Pure hydrogen production by partial dehydrogenation of cyclohexane and methylcyclohexane over nanotube-supported Pt and Pd catalysts. *Energy & Fuels* **2004**, *18* (5), 1429.
65. Kariya, N.; Fukuoka, A.; Ichikawa, M. J. A. C. A. G. Efficient evolution of hydrogen from liquid cycloalkanes over Pt-containing catalysts supported on active carbons under "wet–dry multiphase conditions". *Applied Catalysis A: General* **2002**, *233* (1–2), 91.
66. Biniwale, R. B.; Kariya, N.; Ichikawa, M. J. C. L. Dehydrogenation of cyclohexane over Ni based catalysts supported on activated carbon using spray-pulsed reactor and enhancement in activity by addition of a small amount of Pt. *Catalysis Letters* **2005**, *105* (1), 83.
67. Pande, J. V.; Shukla, A.; Biniwale, R. B. J. I. J. O. H. E. Catalytic dehydrogenation of cyclohexane over Ag-M/ACC catalysts for hydrogen supply. *International Journal of Hydrogen Energy* **2012**, *37* (8), 6756.
68. Kariya, N.; Fukuoka, A.; Utagawa, T.; Sakuramoto, M.; Goto, Y.; Ichikawa, M. J. A. C. A. G. Efficient hydrogen production using cyclohexane and decalin by pulse-spray mode reactor with Pt catalysts. *Applied Catalysis A: General* **2003**, *247* (2), 247.
69. Zhu, G.; Yang, B.; Wang, S. J. I. J. O. H. E. Nanocrystallites-forming hierarchical porous Ni/Al2O3–TiO2 catalyst for dehydrogenation of organic chemical hydrides. *International Journal of Hydrogen Energy* **2011**, *36* (21), 13603.
70. Shukla, A.; Pande, J. V.; Biniwale, R. B. J. I. J. O. H. E. Dehydrogenation of methylcyclohexane over Pt/V2O5 and Pt/Y2O3 for hydrogen delivery applications. *International Journal of Hydrogen Energy* **2012**, *37* (4), 3350.
71. Boufaden, N.; Akkari, R.; Pawelec, B.; Fierro, J.; Zina, M. S.; Ghorbel, A. J. J. O. M. C. A. C. Dehydrogenation of methylcyclohexane to toluene over partially reduced silica-supported Pt-Mo catalysts. *Journal of Molecular Catalysis A: Chemical* **2016**, *420*, 96.
72. Gora, A.; Tanaka, D. A. P.; Mizukami, F.; Suzuki, T. M. J. C. L. Lower temperature dehydrogenation of methylcyclohexane by membrane-assisted equilibrium shift. *Chemistry Letters* **2006**, *35* (12), 1372.
73. Wang, Y.; Shah, N.; Huffman, G. P. J. E. Pure hydrogen production by partial dehydrogenation of cyclohexane and methylcyclohexane over nanotube-supported Pt and Pd catalysts. *Energy Fuels* **2004**, *18* (5), 1429.
74. Hodoshima, S.; Takaiwa, S.; Shono, A.; Satoh, K.; Saito, Y. Hydrogen storage by decalin/naphthalene pair and hydrogen supply to fuel cells by use of superheated liquid-film-type catalysis. *Applied Catalysis A: General* **2005**, *283* (1–2), 235.
75. Cacciola, G.; Giordano, N.; Restuccia, G. Cyclohexane as a liquid phase carrier in hydrogen storage and transport. *International Journal of Hydrogen Energy* **1984**, *9* (5), 411.
76. Suttisawat, Y.; Sakai, H.; Abe, M.; Rangsunvigit, P.; Horikoshi, S. Microwave effect in the dehydrogenation of tetralin and decalin with a fixed-bed reactor. *International Journal of Hydrogen Energy* **2012**, *37* (4), 3242.
77. Lee, G.; Jeong, Y.; Kim, B.-G.; Han, J. S.; Jeong, H.; Na, H. B.; Jung, J. C. Hydrogen production by catalytic decalin dehydrogenation over carbon-supported platinum catalyst: Effect of catalyst preparation method. *Catalysis Communications* **2015**, *67*, 40.
78. Sebastián, D.; Alegre, C.; Calvillo, L.; Pérez, M.; Moliner, R.; Lázaro, M. J. Carbon supports for the catalytic dehydrogenation of liquid organic hydrides as hydrogen storage and delivery system. *International Journal of Hydrogen Energy* **2014**, *39* (8), 4109.
79. Modisha, P. M. *Evaluation of Perhydrodibenzyltoluene Dehydrogenation Parameters and Durability Using Noble Metal Catalysts for Hydrogen Production*, doctoral dissertation, North-West University, South Africa, **2021**.
80. Leinweber, A.; Müller, K. Hydrogenation of the liquid organic hydrogen carrier compound monobenzyl toluene: Reaction pathway and kinetic effects. *Energy Technology* **2018**, *6* (3), 513.

81 Heller, A.; Rausch, M. H.; Schulz, P. S.; Wasserscheid, P.; Fröba, A. P. Binary diffusion coefficients of the liquid organic hydrogen carrier system dibenzyltoluene/perhydrodibenzyltoluene. *Journal of Chemical & Engineering Data* **2016**, *61* (1), 504.

82 Amende, M.; Kaftan, A.; Bachmann, P.; Brehmer, R.; Preuster, P.; Koch, M.; Wasserscheid, P.; Libuda, J. Regeneration of LOHC dehydrogenation catalysts: In-situ IR spectroscopy on single crystals, model catalysts, and real catalysts from UHV to near ambient pressure. *Applied Surface Science* **2016**, *360*, 671.

83 Emel'yanenko, V. N.; Varfolomeev, M. A.; Verevkin, S. P.; Stark, K.; Müller, K.; Müller, M.; Bösmann, A.; Wasserscheid, P.; Arlt, W. Hydrogen storage: Thermochemical studies of N-alkylcarbazoles and their derivatives as a potential liquid organic hydrogen carriers. *The Journal of Physical Chemistry C* **2015**, *119* (47), 26381.

84 Dürr, S.; Müller, M.; Jorschick, H.; Helmin, M.; Bösmann, A.; Palkovits, R.; Wasserscheid, P. Carbon dioxide-free hydrogen production with integrated hydrogen separation and storage. *ChemSusChem* **2017**, *10* (1), 42.

85 Jorschick, H.; Bösmann, A.; Preuster, P.; Wasserscheid, P. Charging a liquid organic hydrogen carrier system with H2/CO2 gas mixtures. *ChemCatChem* **2018**, *10* (19), 4329.

86 Brückner, N.; Obesser, K.; Bösmann, A.; Teichmann, D.; Arlt, W.; Dungs, J.; Wasserscheid, P. Evaluation of industrially applied heat-transfer fluids as liquid organic hydrogen carrier systems. *ChemSusChem* **2014**, *7* (1), 229.

87 Fikrt, A.; Brehmer, R.; Milella, V.-O.; Müller, K.; Bösmann, A.; Preuster, P.; Alt, N.; Schlücker, E.; Wasserscheid, P.; Arlt, W. Dynamic power supply by hydrogen bound to a liquid organic hydrogen carrier. *Applied Energy* **2017**, *194*, 1.

88 Brückner, N.; Obesser, K.; Bösmann, A.; Teichmann, D.; Arlt, W.; Dungs, J.; Wasserscheid, P. J. C. Evaluation of industrially applied heat-transfer fluids as liquid organic hydrogen carrier systems. *ChemSusChem* **2014**, *7* (1), 229.

89 Jorschick, H.; Preuster, P.; Dürr, S.; Seidel, A.; Müller, K.; Bösmann, A.; Wasserscheid, P. J. E.; Science, E. Hydrogen storage using a hot pressure swing reactor. *Energy & Environmental Science* **2017**, *10* (7), 1652.

90 Müller, K.; Stark, K.; Emel'yanenko, V. N.; Varfolomeev, M. A.; Zaitsau, D. H.; Shoifet, E.; Schick, C.; Verevkin, S. P.; Arlt, W. Liquid organic hydrogen carriers: Thermophysical and thermochemical studies of benzyl-and dibenzyl-toluene derivatives. *Industrial & Engineering Chemistry Research* **2015**, *54* (32), 7967.

91 Gleichweit, C.; Amende, M.; Schernich, S.; Zhao, W.; Lorenz, M. P.; Höfert, O.; Brückner, N.; Wasserscheid, P.; Libuda, J.; Steinrück, H. P. Dehydrogenation of dodecahydro-N-ethylcarbazole on Pt (111). *ChemSusChem* **2013**, *6* (6), 974.

92 Peters, W.; Eypasch, M.; Frank, T.; Schwerdtfeger, J.; Körner, C.; Bösmann, A.; Wasserscheid, P. Efficient hydrogen release from perhydro-N-ethylcarbazole using catalyst-coated metallic structures produced by selective electron beam melting. *Energy & Environmental Science* **2015**, *8* (2), 641.

93 Yang, M.; Dong, Y.; Fei, S.; Ke, H.; Cheng, H. A comparative study of catalytic dehydrogenation of perhydro-N-ethylcarbazole over noble metal catalysts. *International Journal of Hydrogen Energy* **2014**, *39* (33), 18976.

94 Sotoodeh, F.; Smith, K. J. An overview of the kinetics and catalysis of hydrogen storage on organic liquids. *The Canadian Journal of Chemical Engineering* **2013**, *91* (9), 1477.

95 Peters, W.; Seidel, A.; Herzog, S.; Bösmann, A.; Schwieger, W.; Wasserscheid, P. J. E. Macrokinetic effects in perhydro-N-ethylcarbazole dehydrogenation and H 2 productivity optimization by using eggshell catalysts. *Energy & Environmental Science* **2015**, *8* (10), 3013.

96 Sotoodeh, F.; Smith, K. J. An overview of the kinetics and catalysis of hydrogen storage on organic liquids. *The Canadian Journal of Chemical Engineering* **2013**, *91* (9), 1477.

97 Gutiérrez, O. Y.; Hrabar, A.; Hein, J.; Yu, Y.; Han, J.; Lercher, J. A. Ring opening of 1,2,3,4-tetrahydroquinoline and decahydroquinoline on MoS2/γ-Al2O3 and Ni–MoS2/γ-Al2O3. *Journal of Catalysis* **2012**, *295*, 155.

98 Zhao, J.; Chen, H.; Xu, J.; Shen, J. Effect of surface acidic and basic properties of the supported nickel catalysts on the hydrogenation of pyridine to piperidine. *The Journal of Physical Chemistry C* **2013**, *117* (20), 10573.

99 Oh, J.; Jeong, K.; Kim, T. W.; Kwon, H.; Han, J. W.; Park, J. H.; Suh, Y.-W. 2-(N-Methylbenzyl)pyridine: A potential liquid organic hydrogen carrier with fast H2 release and stable activity in consecutive cycles. *ChemSusChem* **2018**, *11* (4), 661.

100 Campanati, M.; Vaccari, A.; Piccolo, O. Mild hydrogenation of quinoline: 1. Role of reaction parameters. *Journal of Molecular Catalysis A: Chemical* **2002**, *179* (1), 287.

101 Campanati, M.; Casagrande, M.; Fagiolino, I.; Lenarda, M.; Storaro, L.; Battagliarin, M.; Vaccari, A. Mild hydrogenation of quinoline: 2. A novel Rh-containing pillared layered clay catalyst. *Journal of Molecular Catalysis A: Chemical* **2002**, *184* (1), 267.

102 Chen, F.; Surkus, A.-E.; He, L.; Pohl, M.-M.; Radnik, J.; Topf, C.; Junge, K.; Beller, M. Selective catalytic hydrogenation of heteroarenes with N-graphene-modified cobalt nanoparticles (Co3O4–Co/NGr@α-Al2O3). *Journal of the American Chemical Society* **2015**, *137* (36), 11718.

103 Fan, G.-Y.; Wu, J. Mild hydrogenation of quinoline to decahydroquinoline over rhodium nanoparticles entrapped in aluminum oxy-hydroxide. *Catalysis Communications* **2013**, *31*, 81.

104 Hara, T.; Mori, K.; Mizugaki, T.; Ebitani, K.; Kaneda, K. Highly efficient dehydrogenation of indolines to indoles using hydroxyapatite-bound Pd catalyst. *Tetrahedron Letters* **2003**, *44* (33), 6207.

105 Dong, Y.; Yang, M.; Yang, Z.; Ke, H.; Cheng, H. Catalytic hydrogenation and dehydrogenation of N-ethylindole as a new heteroaromatic liquid organic hydrogen carrier. *International Journal of Hydrogen Energy* **2015**, *40* (34), 10918.

106 Brayton, D. F.; Jensen, C. M. Dehydrogenation of pyrrolidine based liquid organic hydrogen carriers by an iridium pincer catalyst, an isothermal kinetic study. *International Journal of Hydrogen Energy* **2015**, *40* (46), 16266.

107 Cui, Y.; Kwok, S.; Bucholtz, A.; Davis, B.; Whitney, R. A.; Jessop, P. G. The effect of substitution on the utility of piperidines and octahydroindoles for reversible hydrogen storage. *New Journal of Chemistry* **2008**, *32* (6), 1027.

108 Araujo, C. M.; Simone, D. L.; Konezny, S. J.; Shim, A.; Crabtree, R. H.; Soloveichik, G. L.; Batista, V. S. Fuel selection for a regenerative organic fuel cell/flow battery: Thermodynamic considerations. *Energy & Environmental Science* **2012**, *5* (11), 9534.

109 Teichmann, D.; Stark, K.; Müller, K.; Zöttl, G.; Wasserscheid, P.; Arlt, W. Energy storage in residential and commercial buildings via Liquid Organic Hydrogen Carriers (LOHC). *Energy & Environmental Science* **2012**, *5* (10), 9044.

110 Preuster, P.; Fang, Q.; Peters, R.; Deja, R.; Blum, L.; Stolten, D.; Wasserscheid, P. Solid oxide fuel cell operating on liquid organic hydrogen carrier-based hydrogen–making full use of heat integration potentials. *International Journal of Hydrogen Energy* **2018**, *43* (3), 1758.

111 Jorschick, H.; Preuster, P.; Dürr, S.; Seidel, A.; Müller, K.; Bösmann, A.; Wasserscheid, P. Hydrogen storage using a hot pressure swing reactor. *Energy & Environmental Science* **2017**, *10* (7), 1652.

112 Saeys, M.; Reyniers, M.-F.; Neurock, M.; Marin, G. B. Density functional theory analysis of benzene (de) hydrogenation on Pt (111): Addition and removal of the first two H-atoms. *The Journal of Physical Chemistry B* **2003**, *107* (16), 3844.

113 Delbecq, F.; Vigne-Maeder, F.; Becker, C.; Breitbach, J.; Wandelt, K. New insights in adsorption and dehydrogenation of cyclohexene on Pt (111) and ordered Pt–Sn surface alloys: Experiment and theory. *The Journal of Physical Chemistry C* **2008**, *112* (2), 555.

114 Tsuda, M.; Diño, W. A.; Watanabe, S.; Nakanishi, H.; Kasai, H. Cyclohexane dehydrogenation catalyst design based on spin polarization effects. *Journal of Physics: Condensed Matter* **2004**, *16* (48), S5721.

115 Ouma, C. N.; Modisha, P. M.; Bessarabov, D. Catalytic dehydrogenation of the liquid organic hydrogen carrier octahydroindole on Pt (1 1 1) surface: Ab initio insights from density functional theory calculations. *Applied Surface Science* **2019**, *471*, 1034.

116 Humbert, M. P.; Chen, J. G. Correlating hydrogenation activity with binding energies of hydrogen and cyclohexene on M/Pt (111)(M= Fe, Co, Ni, Cu) bimetallic surfaces. *Journal of Catalysis* **2008**, *257* (2), 297.

117 Tsuda, M.; Diño, W. A.; Nakanishi, H.; Watanabe, S.; Kasai, H. First principles interpretation of cyclohexane dehydrogenation process using Pt. *Japanese Journal of Applied Physics* **2005**, *44* (1R), 402.

10 Dehydrogenation Reactions and Inspirations from Nature for the Synthesis of Building Blocks Leading to Valued Pharmaceutical Compounds

Pravin R. Bhansali, Vijayendran K. K. Praneeth, and Ronald E. Viola

10.1 INTRODUCTION

In addition to the widely observed aromatic hydrocarbons, heterocycles are an important class of organic compounds for life sciences. Heterocyclic compounds are found in numerous natural products including vitamins, hormones, and antibiotics.[1] Interestingly, heterocycles are also frequently present as scaffolds in prescribed pharmaceutical drugs available for treating various diseases and disorders. Due to the diverse therapeutical properties of heterocyclic compounds, these frameworks have long served as attractive synthetic targets for pharmaceutical chemists. Elaborate synthetic strategies involving multicomponent reactions are used to prepare various heterocyclic compounds, with these Figure (Scheme)s designed to maximize yield and efficiency of the protocol. Subsequent aromatization of these heterocycles is used to drive these reaction Figure (Scheme)s to produce the desired heteroaromatic derivatives with this energetically favorable transformation frequently achieved through dehydrogenation reactions. Additionally, dehydrogenation reactions can be utilized to carry out various functional group transformations (including the conversion of alcohols to aldehydes or ketones).[2]

These dehydrogenation reactions can be classified into three general types[2]: 1) oxidative dehydrogenations (ODs), 2) hydrogen transfer reactions; and 3) acceptorless dehydrogenations. These different classes of dehydrogenation reactions, using alcohol **1** as the model substrate, are depicted in **Figure 10.1**. The first type, OD reaction, involves the use of a stoichiometric oxidant such as O_2/H_2O_2/o-iodoxybenzoic acid/2,3-dichloro-5,6-dicyano-1,4-benzoquinone (DDQ) or many others to convert the organic substrate into the corresponding dehydrogenated product **2** (**Figure 10.1A**). In this reaction, toxic by-products are frequently generated, which can be a major concern. The second type of dehydrogenation is hydrogen transfer reaction (HT) that involves the use of metal catalysts to effectively perform the dehydrogenation reaction with the assistance of a stoichiometric amount of sacrificial hydrogen acceptors, usually an alkene (**Figure 10.1B**) used to yield the dehydrogenated product **2**. Similarly, in these cases, waste by-products are generated. However, in the third type, called acceptorless dehydrogenation reaction (AD) (**Figure 10.1C**), sacrificial hydrogen acceptors are not needed. In these cases, molecular hydrogen is successfully removed from the substrate by using novel metal catalysts and the liberated H_2, which is a very valuable clean source of energy that can be stored and subsequently utilized. Importantly, the acceptorless dehydrogenation strategy is an atom-economical method without requiring stoichiometric reagents (oxidant or

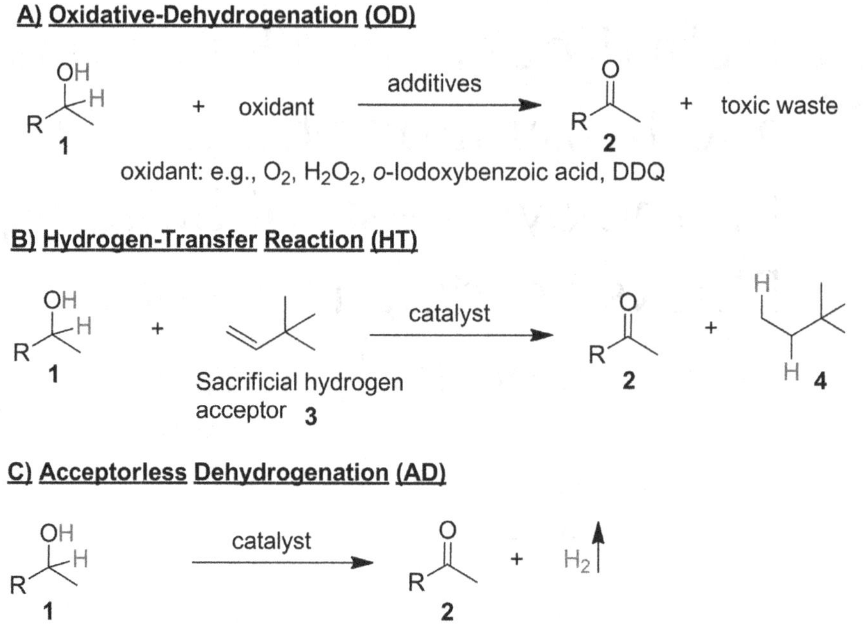

A) Oxidative-Dehydrogenation (OD)

R–CH(OH)(H) **1** + oxidant —(additives)→ R–C(=O) **2** + toxic waste

oxidant: e.g., O_2, H_2O_2, o-Iodoxybenzoic acid, DDQ

B) Hydrogen-Transfer Reaction (HT)

R–CH(OH)(H) **1** + sacrificial hydrogen acceptor **3** —(catalyst)→ R–C(=O) **2** + **4**

C) Acceptorless Dehydrogenation (AD)

R–CH(OH)(H) **1** —(catalyst)→ R–C(=O) **2** + $H_2\uparrow$

FIGURE 10.1 Classification of dehydrogenation reactions.

sacrificial hydrogen acceptors) that at the same time produces valuable H_2 gas from the substrate along with the desired dehydrogenated product.

Importantly, the construction of complex organic molecules through carbon-carbon (C-C) and carbon-nitrogen (C-N) bond formation can be initiated in conjunction with the dehydrogenation reactions of organic substrates followed by inter- or intramolecular condensation reactions with either amines or alkenes. This intermolecular dehydrogenative coupling reaction is a very useful general strategy to construct various bioactive molecules with applications throughout medicinal and pharmaceutical chemistry. For example, Milstein *et al.* reported the synthesis of complex amides **6** by combining a range of primary alcohols **5** and either alkyl or aryl amines by employing the dearomatized Ru-pincer catalyst **[Ru(H)(PNN)(CO)]** through an intermolecular dehydrogenative coupling reaction with the resulting liberation of H_2 (**Figure 10.2A**).[3] A dehydrogenative coupling reaction can also be used to produce useful intermolecular C-C bond formations. Brookhart and co-workers reported C-C bond formation through tandem catalysis by using alkanes **7** and **8**.[4] Initially, the iridium catalyst **[Ir(C$_2$H$_4$)(PNN)]** converts an alkane into the respective alkene through a dehydrogenation reaction with the liberation of H_2. In the next step, the alkene formed reacts with another olefin substrate through an olefin metathesis reaction using Mo-based Schrock's catalyst to yield the corresponding coupled alkenes **9** and **10** as the products (**Figure 10.2B**).

10.2 DEHYDROGENATION REACTIONS FOUND IN NATURE

The ethanol 11 present in alcoholic beverages is a toxic chemical in higher doses in humans, which needs to be eliminated. A special class of human enzymes that can process the alcohols to aldehydes or ketones is alcohol dehydrogenase (ADH) and is found in the liver. Liver-ADH is a zinc-based enzyme that in conjunction with the oxidized form of the cofactor nicotinamide adenine dinucleotide (NAD$^+$), interconverts ethanol into aldehyde 12. In this process, the eliminated hydrogens from ethanol are transferred to NAD$^+$ to form nicotinamide adenine dinucleotide hydrogen (reduced) (NADH) (Figure 10.3). However, the ADH present in yeast catalyzes the reverse reaction, that is, the conversion of an aldehyde or ketone to alcohol in a process called fermentation. Yeast

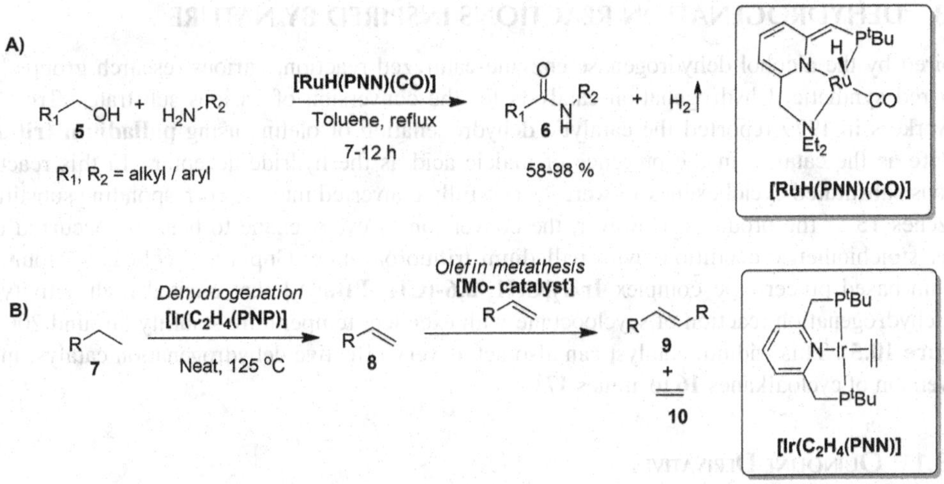

Dehydrogenation Reactions Leading to Pharmaceutical Compounds

FIGURE 10.2 Intermolecular dehydrogenative coupling reactions through the formation of intermolecular C-N bond formation (A) and C-C bond formation (B).

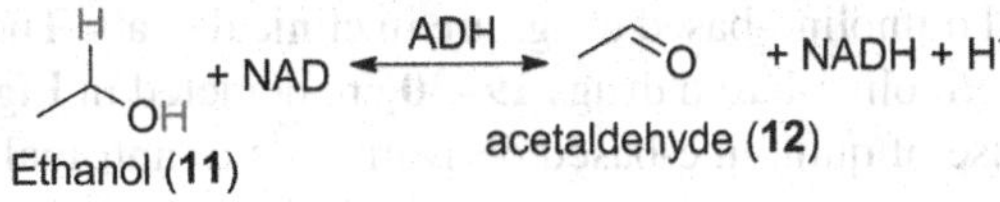

FIGURE 10.3 ADH catalyzed interconversion of ethanol and acetaldehyde.

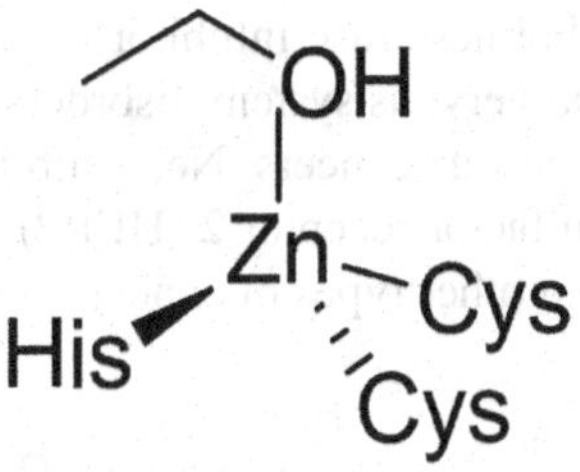

FIGURE 10.4 Active site structure of liver alcohol dehydrogenase (**13**).

alcohol dehydrogenase utilizes NADH for the reaction, and in the process, NAD^+ is formed. In this reverse reaction, various chiral alcohols can be produced, which is of great synthetic importance. The pharmaceutical industry utilizes chiral alcohols as intermediates for manufacturing various drugs including calcium, potassium channel blockers, and anti-viral and anti-arrhythmic drugs.[5]

Liver alcohol dehydrogenase is the most extensively studied of this enzyme family. X-ray crystal structure studies have revealed that the active site consists of zinc in a tetrahedral geometry coordinated with two cysteine and one histidine amino acid residues with a water molecule occupying the fourth ligand position (**Figure 10.4**). The catalysis occurs when the substrate alcohol replaces the water molecule in the presence of NAD^+. The key steps involved in the catalysis are as follows: (i) binding of NAD^+ to the active site, (ii) displacement of H_2O molecule by an alcohol, (iii) formation of zinc alkoxide intermediate, (iv) hydride transfer to form an aldehyde and to generate NADH, and (v) displacement of aldehyde by water.

10.3 DEHYDROGENATION REACTIONS INSPIRED BY NATURE

Inspired by the alcohol dehydrogenase enzyme-catalyzed reaction, various research groups have prepared synthetic dehydrogenation catalysts for the conversion of various substrates. Trost and co-workers in 1979 reported the catalytic dehydrogenation of olefins using **palladium trifluoro acetate** as the catalyst in the presence of maleic acid as the hydride acceptor.[6] In this reaction, various substituted cyclohexenes 14 were successfully converted into the corresponding substituted benzenes 15 as the products. However, the conversion of cyclohexane to benzene occurred only under stoichiometric conditions with palladium trifluoroacetate. Gupta and colleagues found an iridium-based pincer-type complex $IrH_2\{C_6H_3\text{-}2,6\text{-}(CH_2\text{-}PBu^t_2)_2\}$ that elicited high activity for the dehydrogenation reaction of scyclooctane with excellent temperature stability (around 200 °C) (**Figure 10.5**). This iridium catalyst can also act as very effective dehydrogenation catalyst in the conversion of cycloalkanes 16 to arenes 17.[7]

10.3.1 QUINOLINE DERIVATIVES

Quinoline and its derivatives are found in many natural and bioactive compounds and are considered a valuable scaffold in many drug molecules. Many marketed pharmaceutical drug molecules that function as anti-malarial, anti-fungal, anti-viral, anti-mycobacterial, antibacterial, anti-epileptic, anti-inflammatory, and many more agents contain a quinoline scaffold in their structures. Moreover, many additional quinoline-based drugs are in clinical trials. The structures of a selection of some of the important quinoline-based drugs 19–30 are depicted in **Figure 10.6**.[8]

The most widespread use of quinoline-based derivatives is in anti-malarial drugs. Chloroquine was the first quinoline containing the discovered drug and has been the front-line treatment for malaria infections. Other derivatives such as amodiaquine, mefloquine, and hydroxychloroquine were subsequently discovered and used because of the development of resistance against chloroquine. Additionally, quinoline derivatives have been used in the treatment of various other diseases. For example, tacrine is used as a cholinesterase inhibitor in the treatment of Alzheimer's disease and is also used against other central nervous system disorders. Quinoline drugs are also available on the market for the treatment of various cancers. Neratinib is approved for use in breast cancer therapy in human epidermal growth factor receptor 2 (HER2)-positive cases.[9] Currently, neratinib is undergoing clinical trials for use in other types of cancer.[10]

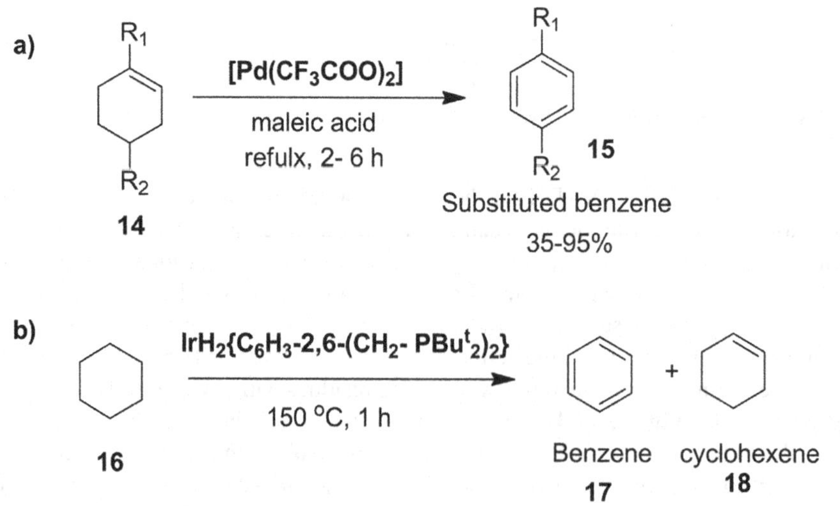

FIGURE 10.5 Catalytic dehydrogenation reaction to form arenes from cycloalkenes and cycloalkanes.

Dehydrogenation Reactions Leading to Pharmaceutical Compounds

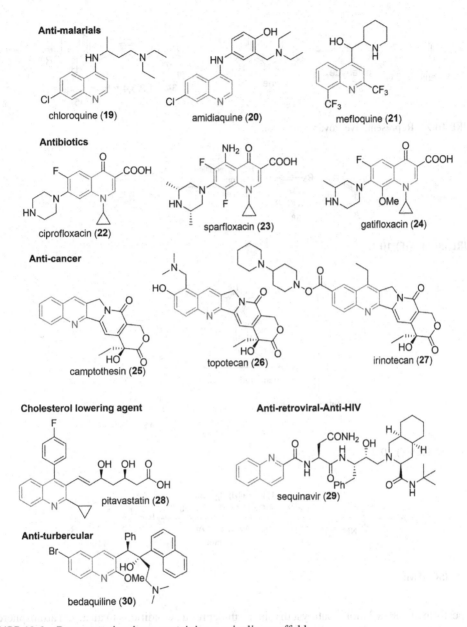

FIGURE 10.6 Representative drugs containing a quinoline scaffold.

Due to the importance of quinoline derivatives in various drugs, pharmaceutical chemists are actively involved in the synthesis and testing of a wide range of quinoline moieties. Dehydrogenation chemistry is one of the key synthetic methodologies used to easily and effectively achieve quinoline moiety.

The Povarov reaction (**Figure 10.7**) followed by oxidation could be utilized for quinoline derivative synthesis. A modified Povarov reaction-aerobic OD was exploited by Guchhait et al. for the synthesis of an array of quinoline derivatives **39** (**Figure (Scheme) 10.1**).[11]

In 2009, Yamaguchi et al. reported the catalytic dehydrogenation of nitrogen-based heterocycles using a novel iridium-based homogeneous catalyst **[Ir(2-OH-Py)Cp*Cl]** that has a 2-hydroxypyridine ligand system (**Figure (Scheme) 10.2**). This catalyst effectively transforms various tetrahydroquinolines **40** into the corresponding quinolines **41** with moderate to good yields. The reactions

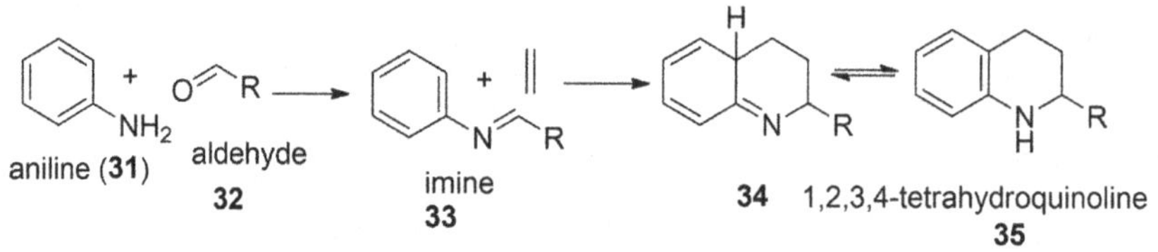

FIGURE 10.7 Representative Povarov reaction.

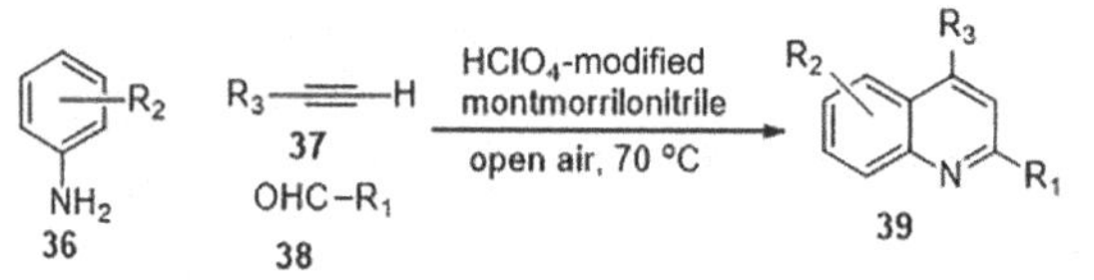

FIGURE (SCHEME) 10.1.

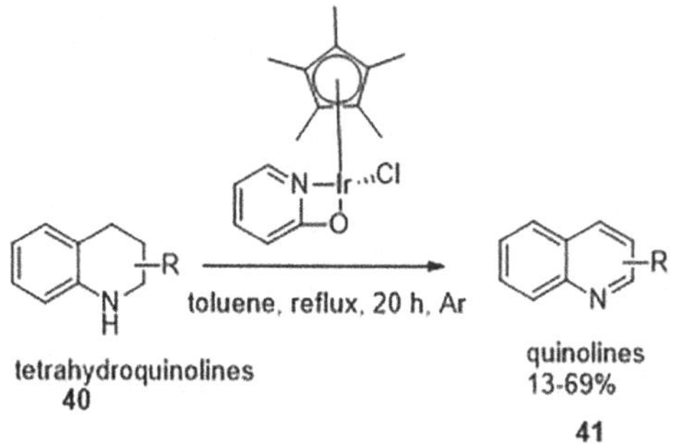

FIGURE (SCHEME) 10.2.

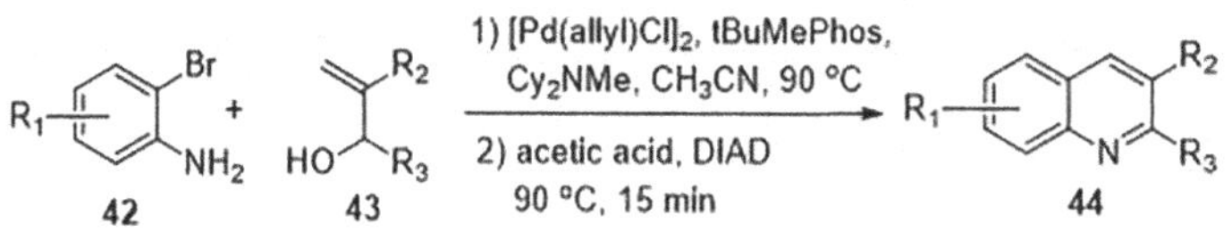

FIGURE (SCHEME) 10.3.

were performed with a 2 mol% catalyst in toluene under reflux conditions in an argon atmosphere for 20 h. The excellent activity of the catalyst is attributed to the effect of 2-hydroxypyridine ligand.[12]

Quinolines are also present in various bioactive compounds and in numerous natural products. Stone *et al.* utilized a modified Larock synthesis for the assembly of a range of quinoline derivatives **44** prompted through a Heck C-C coupling protocol of 2-bromoanilines **42** with allylic alcohols **43** followed by a dehydrogenation protocol of diisopropyl azodicarboxylate (DIAD) (**Figure (Scheme) 10.3**).[13]

Chakraborty *et al.* employed an iron-based homogeneous catalyst **[FeH(PNP)(CO)]** for the conversion of tetrahydroquinolines to quinolines (**Figure (Scheme) 10.4**).[14] The utilization of an iron-containing system is favored compared to the previously described noble metal system due to the high earth abundancy and inexpensive nature of iron. Similar to the [Ir(2-OH-Py)Cp*Cl] catalyst, the iron catalyst also effectively converts various tetrahydroquinolines **45** into the corresponding quinolines **46** with moderate to good yields. These dehydrogenation reactions were performed with a 3–5 mol% catalyst in xylene under reflux conditions in an argon atmosphere for 30 h.

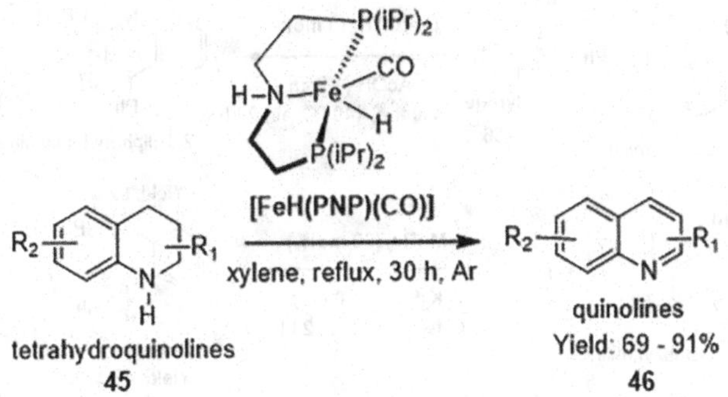

FIGURE (SCHEME) 10.4.

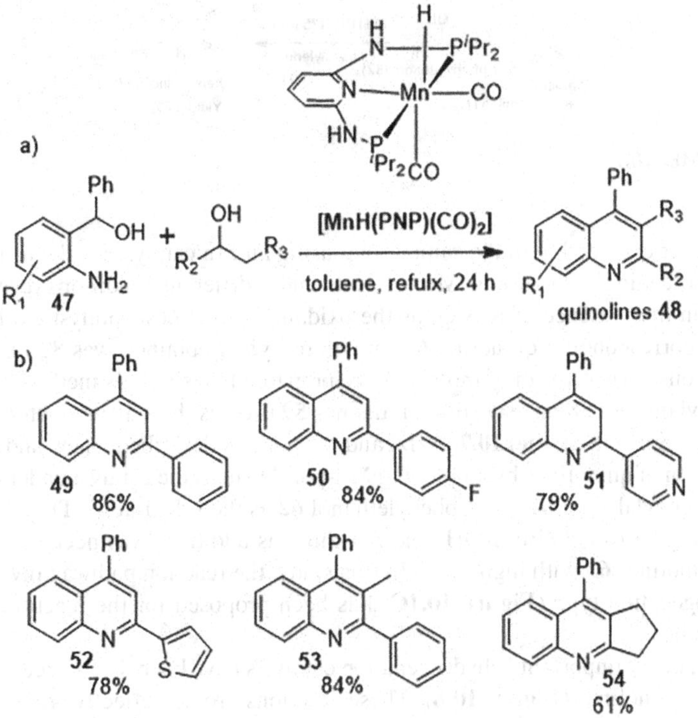

FIGURE (SCHEME) 10.5.

A manganese-based homogeneous catalyst has also been utilized in dehydrogenation reactions for preparing quinoline derivatives. As shown in **Figure (Scheme) 10.5**, Mastalir et al.[15] employed a well-defined Mn-pincer complex (5 mol%) for this reaction using 2-aminobenzyl alcohols **47** and secondary alcohols as the substrates in the presence of a toluene base at 140 °C for 24 h. The 2-aminobenzyl alcohols and secondary alcohols were successfully combined to form various derivatives of quinolines **49–54** containing a wide-range of functional groups with good yield.

Yang and group in 2018 reported a simple iron salt, $Fe(OTf)_3$, that can catalyze the dehydrogenation reaction to produce quinolines. Treatment of benzylaniline **55** with styrene **56** in the presence of $Fe(Otf)_3$ and acetic acid under air or O_2 produced the 2,4-diphenylquinoline **57** with 82% yield (**Figure (Scheme) 10.6**). The reaction proceeds through a dehydrogenative [4+2] cycloaddition reaction to yield the quinoline product. By employing a variety of N-alkylanilines and substituted alkenes, various substituted quinolines were synthesized with good to excellent yield.[16] This

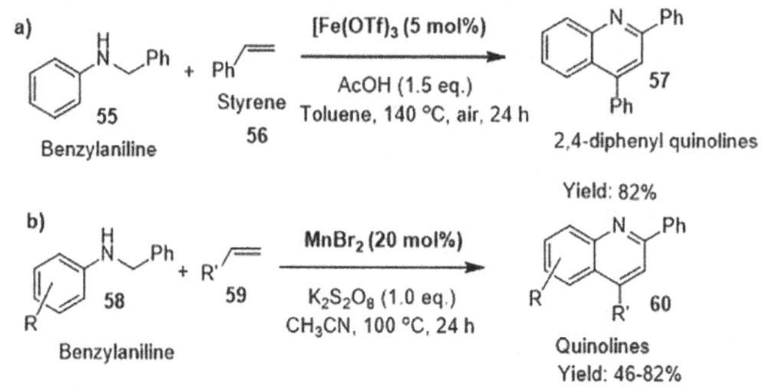

FIGURE (SCHEME) 10.6.

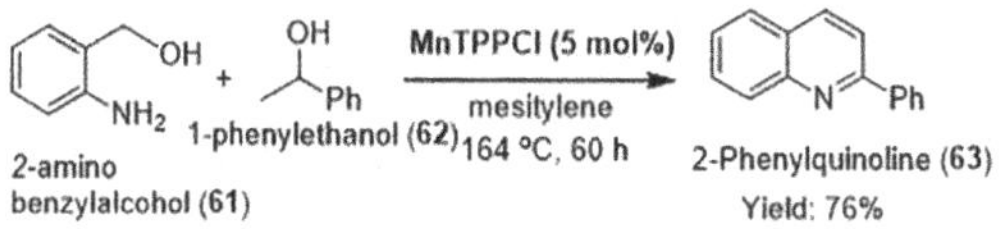

FIGURE (SCHEME) 10.7.

approach was later extended with the same group using an Mn catalyst for the formation of quinoline. Reaction screening with various Mn catalysts under different conditions revealed that MnBr$_2$ as the catalyst in the presence of K$_2$S$_2$O$_8$ as the oxidant was the best catalyst-oxidant combination to produce the corresponding quinoline **60** (maximum yield obtained was 82%). Under the same conditions, various derivatives of quinolines were prepared by using this method, with a number of different N-alkylanilines **58** and substituted-alkenes **59** used as the starting materials.[17]

As shown in **Figure (Scheme) 10.7**, Azizi and co-workers described a new and efficient method for the preparation of quinolines by employing Mn-porphyrin as the catalyst under basic conditions, using 2-aminobenzylalcohol **61** and 1-phenylethanol **62** as the substrates.[18] They found that a combination of strong bases (KO^{t}Bu, KOH, and pyridine) as additives was needed for the successful synthesis of quinolines **63** with high yield. In this study, the reaction pathway involving the acceptorless dehydrogenation type (**Figure 10.1C**) has been proposed for the reaction through several mechanistic studies.

Accordingly, many important dehydrogenation catalysts have been discovered and optimized for the synthesis of quinolines (**Figure 10.8**). These reactions provide effective routes for the conversion of various tetrahydroquinolines to produce an array of quinolines.

10.3.2 Pyrrole Derivatives

Pyrrole is also considered to be a very important scaffold as it is present in many biological compounds with diverse biological and therapeutical activities. Many drugs containing a pyrrole substructure possess therapeutic utilities including anti-cancer, antibiotic, anti-fungal, β-adrenergic antagonist, anti-psychotic, and anti-malarial compounds.[19] Some examples of pyrrole-based pharmaceutical drugs are shown in **Figure 10.9**. Lipitor **64**, a very successful cholesterol-lowering drug, is built from a pyrrole unit.[20] Sunitinib **67** is an anti-cancer drug used to treat renal cancer and gastrointestinal tumors.[21] Tolmetin **65** is an anti-inflammatory drug used to treat acute rheumatoid arthritis and osteoarthritis, and it also possesses analgesic properties. Similarly, ketorolac **66** is used as an analgesic and anti-inflammatory drug.[22]

Due to the obvious importance of pyrrole derivatives in the pharmaceutical and chemical industries, various synthetic protocols have been devised for the synthesis of pyrrole frameworks.

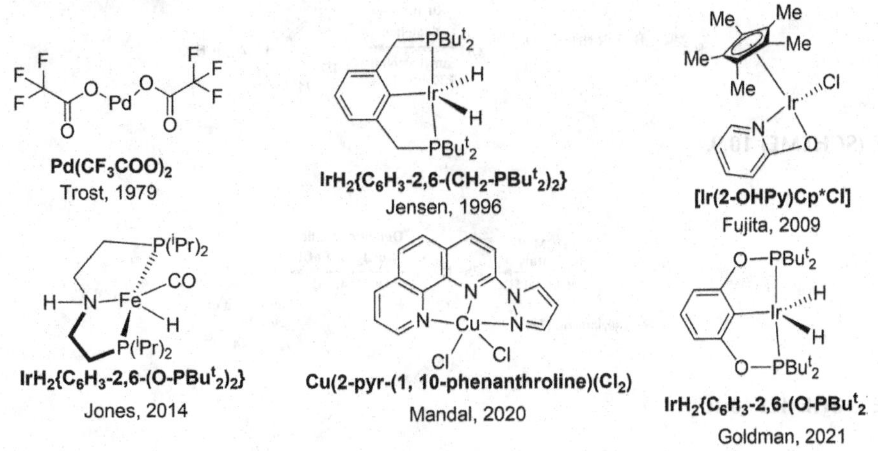

FIGURE 10.8 Metal complexes used as homogeneous catalysts for the dehydrogenation of organic substrates.

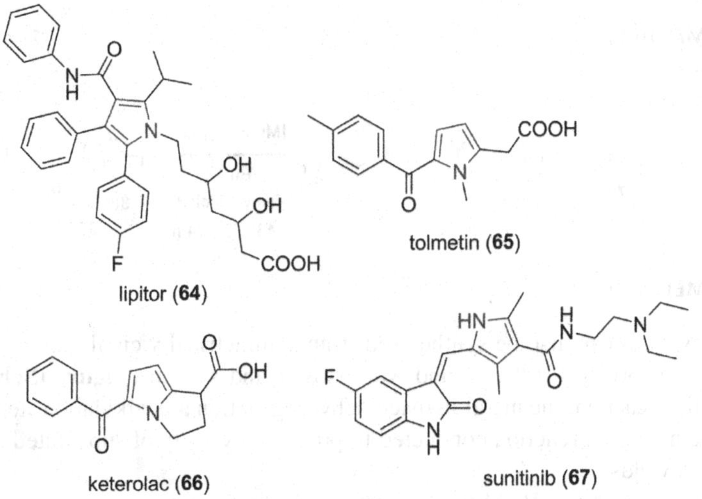

FIGURE 10.9 Some of the important pyrrole-based pharmaceutical drugs.

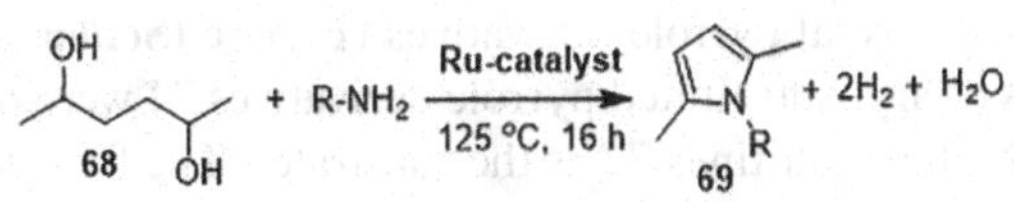

FIGURE (SCHEME) 10.8.

Paal-Knorr, Clauson-Kass, and ring closing metathesis (RCM) reactions have been the approaches generally used to carry out the synthesis of pyrrole derivatives.

In 2011, Schley and group reported a dehydrogenative Paal-Knorr-type synthesis of various N-alkyl pyrroles using a ruthenium catalyst.[23] In this study, dehydrogenative oxidation of 2,6-hexanediols **68** in the presence of alkyl amines was carried out to produce various pyrrole derivatives **69** (**Figure (Scheme) 10.8 and Table 10.1**) but only with a maximum yield of 48%.

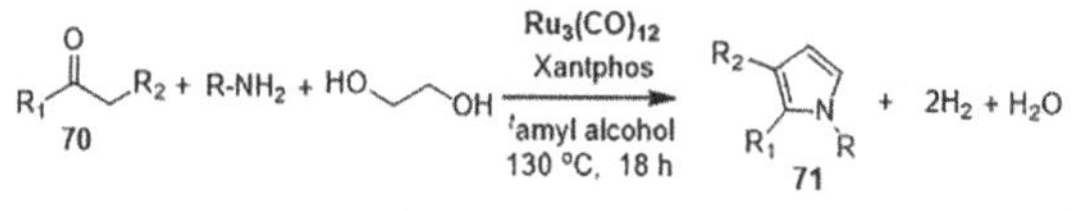

FIGURE (SCHEME) 10.9.

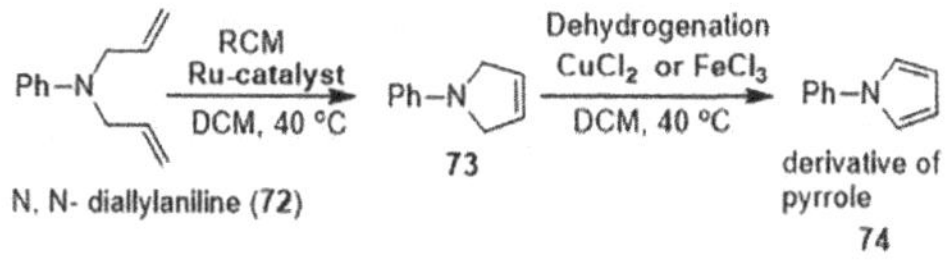

FIGURE (SCHEME) 10.10.

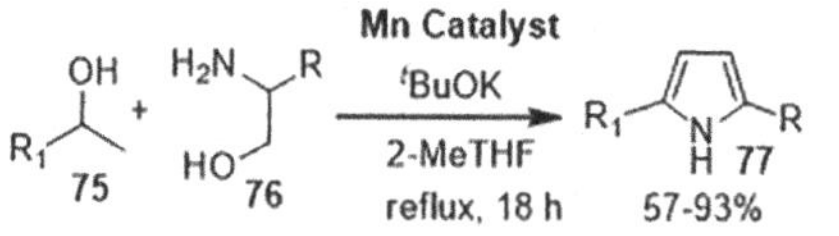

FIGURE (SCHEME) 10.11.

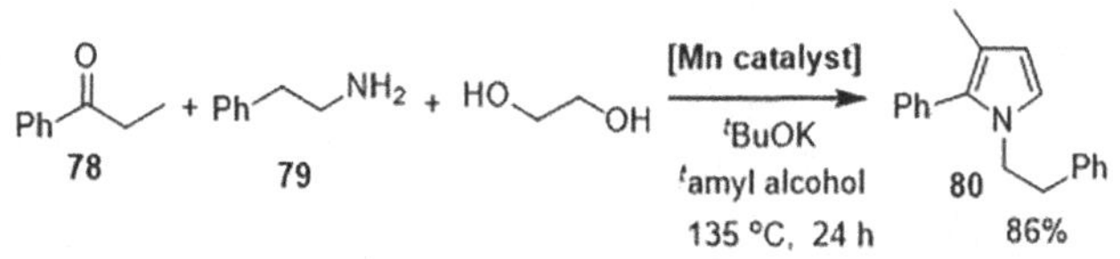

FIGURE (SCHEME) 10.12.

Substituted pyrroles can also be synthesized from symmetrical vicinal diols by using a tandem catalysis method involving $Ru_3(CO)_{12}$ and Xantphos ligand system (**Figure (Scheme) 10.9, and Table 10.1**). In this reaction, the metal-assisted dehydrogenation and alkylation steps were followed according to the hydration reaction conducted to produce a variety of substituted pyrroles **71** with moderate to good yields.[24]

For each of these methods, RCM is a powerful strategy for easy access to the assembly of pyrrole derivatives with the dehydrogenation step, an indispensable part of the process. Linear diene substrates can be converted into pyrrole derivatives through RCM and the subsequent dehydrogenation reaction by using metal catalysts. In this context, Chen et al.[25] used Cu or Fe catalysts for dehydrogenation to produce several pyrrole derivatives (**Figure (Scheme) 10.10 and Table 10.1**). In these reactions, a series of aryl-substituted pyrrole derivatives **74** were obtained with good yields using aryl-substituted N, N-diallylanilines **72** as the substrates. The first step involves an RCM step using a ruthenium carbene catalyst. This was then followed by the dehydrogenation step with either $CuCl_2$ or $FeCl_3$ as the catalyst to deliver pyrrole **74**.

As shown in **Figure (Scheme) 10.11 and Table 10.1,** Kallmeier et al. in 2017 reported the reaction of secondary alcohols and amino alcohols using a Mn pincer catalyst to produce a series of substituted pyrroles **77** with yields up to 93% under mild conditions.[26] However, the related Fe and Co complexes of the same ligand framework did not show activity under the identical reaction conditions.

In a related work, Borghs and co-workers utilized acceptorless dehydrogenation coupled with a hydrogen auto-transfer catalysis method for the construction of pyrrole derivatives. These researchers used a novel manganese pincer catalyst made of a PNP[Ph] ligand system. The catalyst

Dehydrogenation Reactions Leading to Pharmaceutical Compounds

TABLE 10.1
Metal Complexes Used as Homogeneous Catalysts for the Dehydrogenation of Organic Substrates

	Reaction	Catalyst	Ref.
A	**68** (diol) + R-NH$_2$ $\xrightarrow{\text{Ru-catalyst, 125 °C, 16 h}}$ **69** (pyrrole) + 2H$_2$ + H$_2$O	Fe-Ru complex with H$_2$N, P, Cl, pyridine ligands	23
B	**70** (R$_1$COCH$_2$R$_2$) + R-NH$_2$ + HOCH$_2$CH$_2$OH $\xrightarrow{\text{Ru}_3\text{(CO)}_{12}, \text{Xantphos}, \text{tamyl alcohol}, 130 °C, 18 h}$ **71** + 2H$_2$ + H$_2$O	Ru$_3$(CO)$_{12}$ + Xantphos (ligand)	24
C	N,N-diallylaniline (**72**) $\xrightarrow{\text{RCM Ru-catalyst, DCM, 40 °C}}$ **73** $\xrightarrow{\text{Dehydrogenation CuCl}_2 \text{ or FeCl}_3, \text{DCM, 40 °C}}$ **74** derivative of pyrrole	CuCl$_2$ or FeCl$_3$	25
D	**75** (R$_1$-CH(OH)-) + **76** (H$_2$N-R, HO-) $\xrightarrow{\text{Mn Catalyst, }^t\text{BuOK, 2-MeTHF, reflux, 18 h}}$ **77** (57–93%)	Mn pincer complex with triazine, (i-Pr)$_2$HP, CO ligands	26
E	**78** (PhCOCH$_2$CH$_3$) + **79** (PhCH$_2$CH$_2$NH$_2$) + HOCH$_2$CH$_2$OH $\xrightarrow{\text{[Mn catalyst], }^t\text{BuOK, tamyl alcohol, 135 °C, 24 h}}$ **80** (86%)	Mn complex with PHPh$_2$, CO, pyridine ligands	27

is air-stable and thus enables inert atmosphere-free reaction conditions. By treating propiophenone (**78**), 2-phenylethylamine (**79**), and ethylene glycol in the presence of 20 mol% of the Mn-catalyst, a substituted pyrrole **80** (**Figure (Scheme) 10.12 and Table 10.1**) was obtained with 86% yield.[27]

10.3.3 β-Carboline Derivatives

The aromatic β-carboline moiety is also a very important structural scaffold that is found in various natural products and pharmaceutical drugs. These β-carboline-containing compounds show activity against various infections such as viral infections, HIV, and malaria.[28] Because of the significance of the β-carboline unit, several synthetic strategies have been employed to prepare carboline derivatives (**Figure (Scheme) 10.13**). As on example, Gaikwad et al. employed a metal-free OD method to prepare a range of β-carboline derivatives **82** with excellent yields using I$_2$ and H$_2$O$_2$ in DMSO.[29] Mechanistic studies revealed that both of the oxidants (I$_2$ and H$_2$O$_2$) and DMSO are indispensable for the reaction to proceed. These researchers displayed the utility of this method in the preparation of several bioactive alkaloids, namely, kumujian-C, eudistomin, norharmane, and harmane.

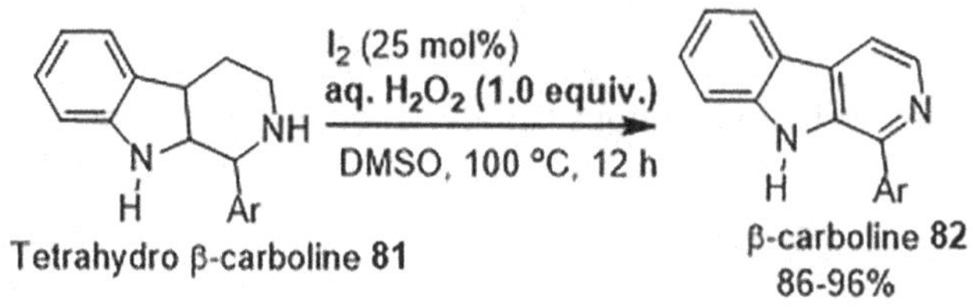

FIGURE (SCHEME) 10.13.

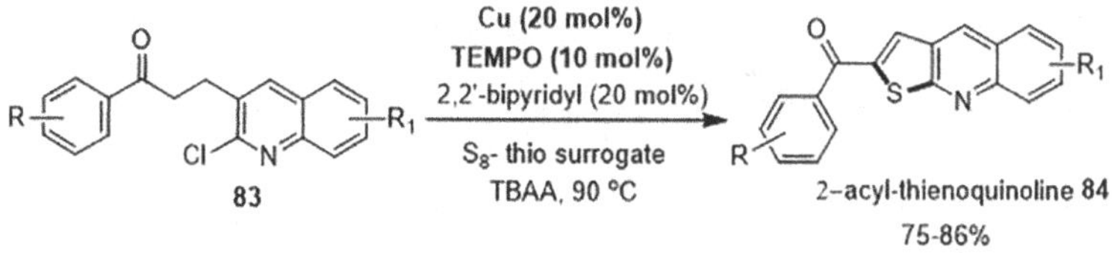

FIGURE (SCHEME) 10.14.

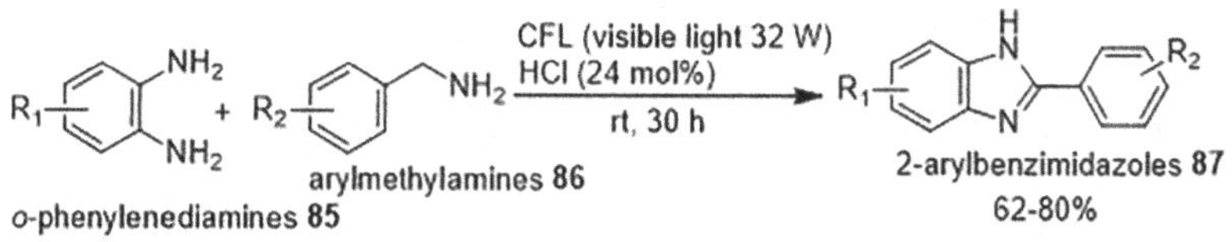

FIGURE (SCHEME) 10.15.

10.3.4 THIENOQUINOLINES DERIVATIVES

Sulfur-containing heterocycles fused to quinolines are called thienoquinolines. These moieties are present in a variety of bioactive natural products and pharmaceutical drugs including a tyrosine kinase EGFR inhibitor and a protein kinase inhibitor.[30] However, the possible synthetic routes for the preparation of thienoquinolines has not been explored to a great extent. In 2019, Teja and co-workers developed an inexpensive Cu-TEMPO-mediated OD catalysis to achieve the construction of various thienoquinolines **84** (**Figure (Scheme) 10.14**). These researchers used 20 mol% of the Cu catalyst, 10 mol% of TEMPO (oxidant), 20 mol% of 2,2-bipyridine (ligand) for the dehydrogenation reaction, followed by C-H functionalization using elemental sulphur (as a thio surrogate) for the synthesis of 2-acyl-thienoquinolines.[31]

10.3.5 BENZIMIDAZOLES DERIVATIVES

The benzimidazole moiety is one of the key heterocycles found in bioactive molecules with diverse activities, including anti-microbial, anti-diabetic, anti-convulsant, anti-parasitic and anti-cancer.[32] Sofi and group developed a general and efficient visible-light-promoted dehydrogenation-deaminative cyclocondensation strategy for the transformation of *o*-phenylenediamines (**85**) and arylmethyl amines **86** into the corresponding 2-arylbenzimidazoles **87** with high yields.[33] The reaction has a wide substrate scope and tolerates various functional groups, thereby allowing many derivatives of benzimidazoles to be synthesized through this methodology (**Figure (Scheme) 10.15**). In addition, in this method, greener protocols were followed by avoiding the use of any metals (transition metal photocatalyst) and using air as an oxidant at room temperature.

10.3.6 GALANTAMINE DERIVATIVES

Galantamine is a specific and reversible acetylcholinesterase inhibitor with neuroprotective properties. It is also a key drug used to slow Alzheimer's disease with recent worldwide clinical trials showing positive results in the application of this drug against Alzheimer's.[34] Therefore, effective synthetic methods need to be developed for the more efficient production of the drug. In this aspect, Zhang et al. in 2020 illustrated a method for the synthesis of galantamine intermediate using a palladium-catalyzed dehydrogenation reaction through late-stage C-H activation.[35] A subsequent reaction Figure (Scheme) with L-selectride and then treatment with LiAlH$_4$ afforded galantamine **90** as the product with 67% yield (**Figure (Scheme) 10.16**).

10.3.7 PYRAZOLONE AND PYRAZOLE DERIVATIVES

Pyrazole and Pyrazolone moieties are quite common structural features present in numerous drug molecules.[36] As shown in **Figure (Scheme) 10.17**, Zhu et al. used a palladium-catalyzed OD reaction in the presence of AMS (co-catalyst) to readily prepare various derivatives of pyrazoles **92, 95** and pyrazolones.[37] In this method, various substrates with alkyl, alkoxy, halogen atoms, an ester group, and electron-withdrawing CF$_3$ were tolerated, and each afforded the dehydrogenation products with good yield. As a bonus, this method also operates under mild reaction conditions.

10.4 METABOLIC OXIDATIVE DEHYDROGENATION REACTIONS

Almost anything entering the human body is considered potential foreign material. Biological transformations are frequently required to detoxify and remove any non-native substances from the

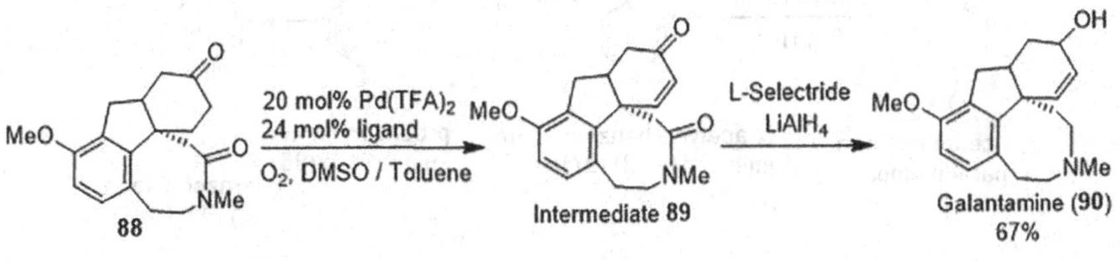

FIGURE (SCHEME) 10.16.

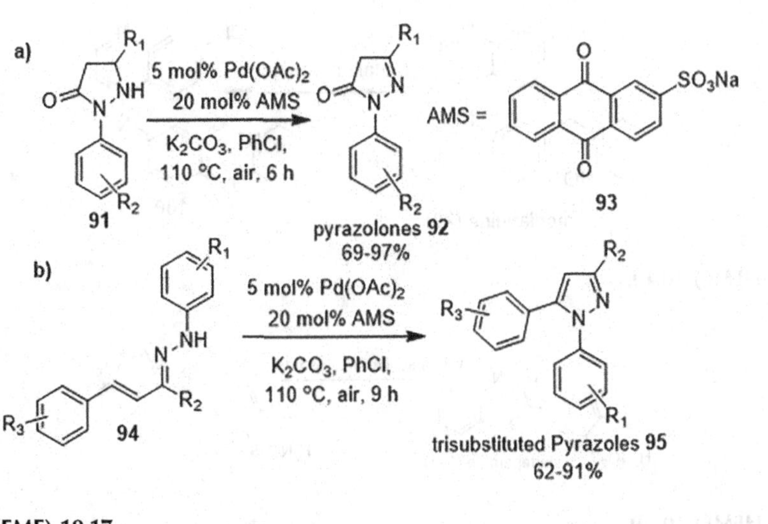

FIGURE (SCHEME) 10.17.

body. These metabolic events generally involve the conversion of water insoluble materials to water soluble compounds. To determine the metabolic fate of administered pharmaceutical drugs in the human body, many model systems are devised to determine the *in vitro* effect of these detoxification systems on the drug metabolic fate of drug molecules. Bernadou *et al.* used metalloporphyrin coupled with KHSO$_5$ (as a water-soluble oxygen carrier) as a model for the peroxidase or cytochrome P-450 detoxification systems present in our body. Paracetamol/acetaminophen (**96**) was subjected to metalloporphyrin along with KHSO$_5$. Upon oxidation, paracetamol was transformed initially into 1) N-acetyl-p-benzoquinone-imine (NAPQI) (**97**) (dehydrogenation reaction product), then into 2) p-benzoquinone-imine (PQI) (**98**), and ultimately, into 3) p-benzoquinone (BQ) (**99**) (**Figure (Scheme) 10.18**). The proportion of these metabolites that are formed is affected by the reaction time, pH, and the metal system present in the porphyrin ring. Paracetamol is an extensively utilized medicine to reduce elevated body temperatures (anti-pyretic) and to a limited extent, to lower inflammation (anti-inflammatory) and alleviate pain (analgesic).[38]

In a similar metabolic study by Johansson and group,[39] researchers modelled the cytochrome P450 system with (1) electrochemical (EC) oxidation, (2) an electrochemically assisted Fenton reaction (EC-Fenton), and (3) synthetic metalloporphine-catalyzed oxidations. Amodiaquine (**20**), a 4-amino quinoline derivative, is an anti-malarial drug that consists of substituted 4-aminophenol moiety and that of acetaminophen (**Figure (Scheme) 10.19**). It was found that once this compound metabolized, it undergoes a series of dehydrogenation reactions to generate a substituted quinone imine **100**.

Indapamide (**101**), a diuretic anti-hypertensive agent, has an indoline heterocycle present in its structure, and this 2,3-dihydroindole moiety is present in many investigational drugs. It has been

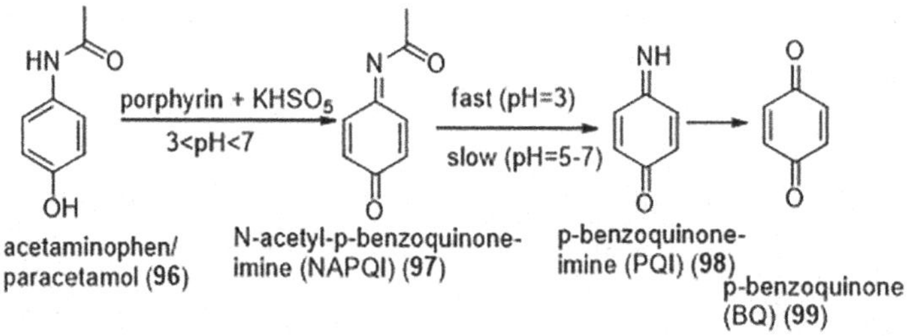

FIGURE (SCHEME) 10.18.

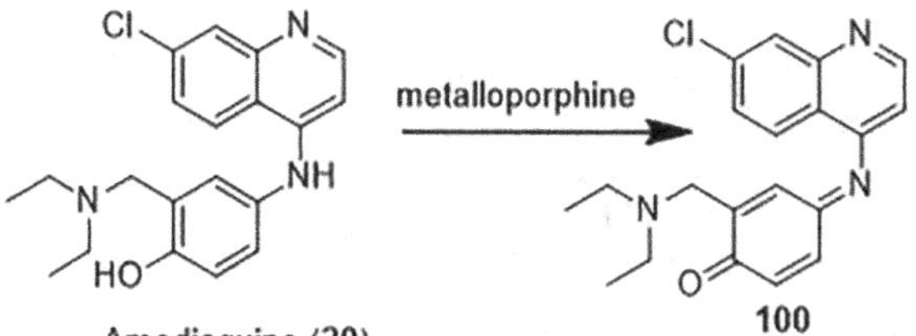

FIGURE (SCHEME) 10.19.

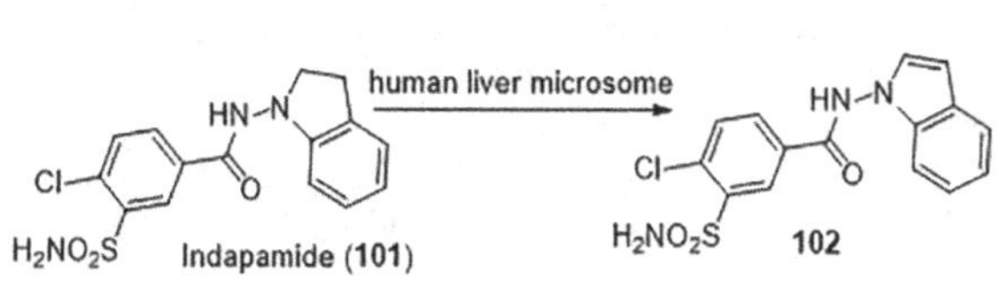

FIGURE (SCHEME) 10.20.

shown that indapamide undergoes an OD metabolic fate in the human body.[40] It has been observed that when indoline is treated with isolated human liver microsomes, the microsomal cytochrome P450 system carries out OD at the 2 and 3 positions of indoline to yield the indole derivative **102** as the product (**Figure (Scheme) 10.20**).[41];

10.5 MISCELLANEOUS DEHYDROGENATION REACTIONS

Gibberellic acid is a plant hormone that promotes the growth of plant cells. This hormone is also found in fungal species. Interestingly, oxidation/dehydrogenation with the assistance of an iron-based catalytic system produced the lactone ring, a cyclized ester **104** as shown in **Figure (Scheme) 10.21**. This study of "Biologically inspired oxidation catalysis" was carried out by Cue *et al.* to learn from a natural catalyst and the catalysis processes.[42]

Heterocyclic cores are essential components of many pharmaceutically active drugs, and have therefore attracted many chemists' attention to design routes for the synthesis of various heterocycles. Once produced, these new heterocyclic-containing compounds are screened to determine if they are associated with some therapeutic benefits. For example, Balamurugan *et al.* synthesized a range of 2-amino-5-arylthieno-[2,3-b]-thiophenes derivatives **109** by using the Gewald reaction–dehydrogenation domino reactions (**Figure (Scheme) 10.22**) and then tested them for anti-tubercular activity. Fascinatingly, these synthesized compounds elicited significant activity in the form of minimum inhibitory concentrations (MICs) in the range of 1.1–97.6 µM against multi-drug-resistant TB (MDR-TB).[43] The general synthetic route for this Gewald reaction is depicted in **Figure 10.10**.

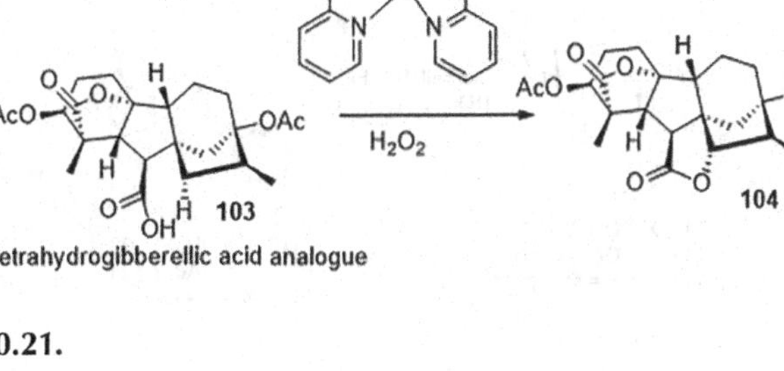

FIGURE (SCHEME) 10.21.

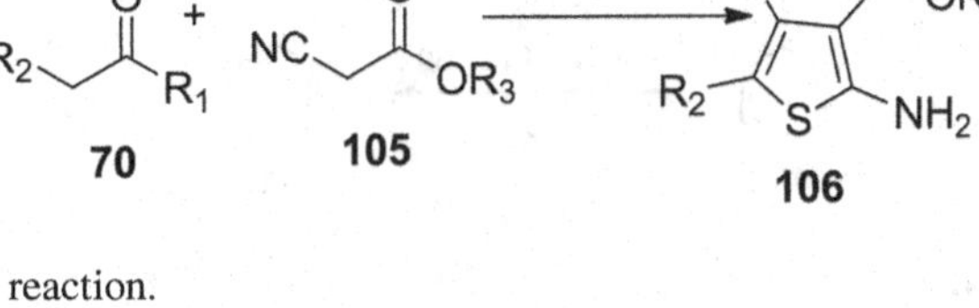

FIGURE 10.10 Gewald reaction.

FIGURE (SCHEME) 10.22.

Anti-cancer compounds have been identified against a wide range of planned targets, and some of these drugs have been synthesized by a Figure (Scheme) that involves a dehydrogenation reaction. Steroidal drugs generally function by interfering with the activity of natural substrates such as oestrogen, progesterone, and testosterone, which have a similar steroidal nucleus. The anti-cancer compound exemestane, which acts by inhibiting the enzyme aromatase, was synthesized by Marcos-Escribano et al. in a reaction involving dehydrogenation by using chloranil, bis(trimethylsilyl) trifluoroacetamide (BSTFA), and triflic acid (TfOH) in refluxing conditions in toluene solvent[44] (**Figure (Scheme) 10.23**).

Sulphonamides are versatile molecules that are found in various diuretics, anti-bacterials, HIV protease inhibitors, anti-convulsants, and anti-diabetic compounds. C-N bond formation is an important step in sulphonamide synthetic Figure (Scheme)s and generally involves the generation of toxic by-products. As depicted in **Figure (Scheme) 10.24**, Shi et al. used a dehydrogenation-condensation-hydrogenation sequence in one pot that avoids these toxic issues by coupling alcohols **116** with sulphonamides to produce an array of substituted sulphonamides **117**.[45] The spirooxindole heterocyclic nucleus is present in many natural products, and some of them have been shown to have numerous therapeutic activities.[46] Compounds including alstonisine, horsfiline, and elacomine are used in indigenous medicine; mitraphylline and spirotryprostatins A and B possess anti-tumor activity, and rhynchophylline is associated with anti-pyretic activity. Assembly of numerous spirooxiindole derivatives was carried out by Bhaskar et al.[47] followed by a determination of these synthetic compounds for their anti-microbial properties against 8 different bacterial and 3 different fungal species. These compounds were assembled through the cycloaddition reaction of azomethine ylide and dipolarophile 1,4-naphthoquinone followed by the products of this cycloaddition reaction undergoing dehydrogenation to provide spirooxiindole derivatives **121**. As one example, azomethine ylide is synthesized from substituted isatin **118** and sarcosine or L-proline[47] (**Figure (Scheme) 10.25**).

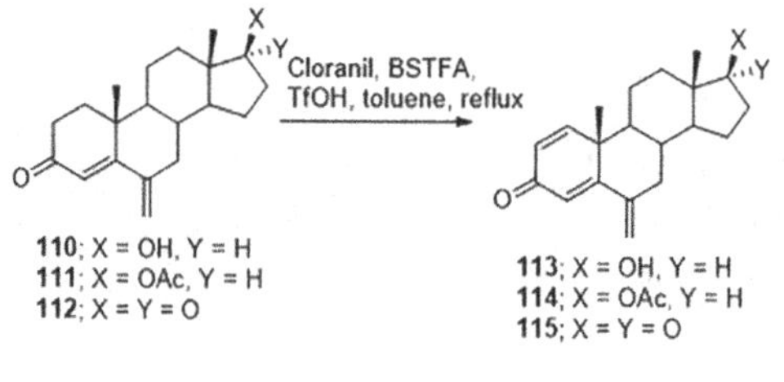

FIGURE (SCHEME) 10.23.

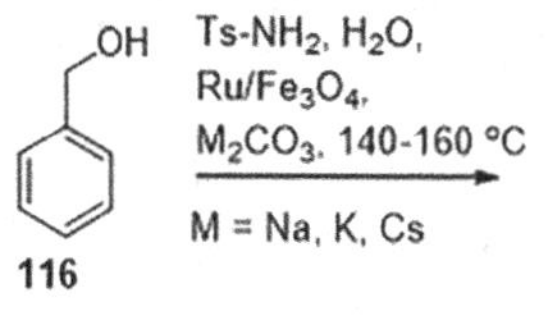

FIGURE (SCHEME) 10.24.

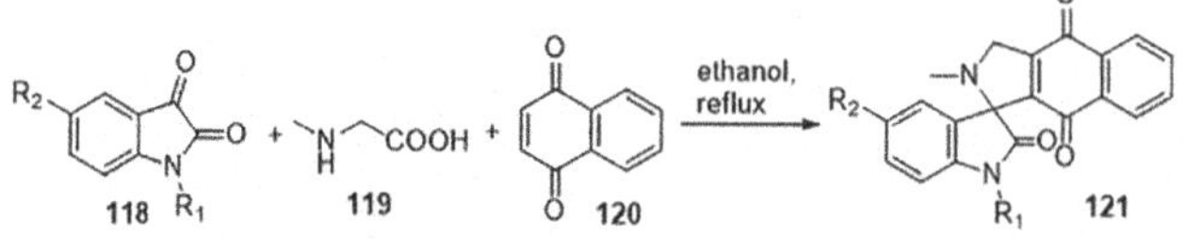

FIGURE (SCHEME) 10.25.

Various pyridazinone-containing drugs such as zardaverine, emorfazone, and irdabisant are available on the market and are used for their anti-platelet, anti-inflammatory, and GPCR H3R antagonistic properties, respectively. Different pyridazinones derivatives **123** were synthesized by the dehydrogenation of 6-substituted dihydropyridazinones **122** by using $Cu(OAc)_2$, Na_2CO_3 as a base, the additive pyridine, toluene as the solvent, and oxygen at 100 °C, producing the final compounds in a range of 65–93% overall yield[48] (**Figure (Scheme) 10.26**).

2-aminobenzothiazole acts as a building block for various bioactive compounds. For example, riluzole (**125**) is useful in the treatment of the symptoms of amyotropic lateral sclerosis and other motor neuron diseases. R-(-)-lubeluzole **124** provides neuroprotection especially in hypoxic conditions, while pramipexole **126**, with a partly saturated benzene ring, is useful in the treatment of Parkinson's disease[49] (**Figure 10.11**). Zhao and group have devised a very efficient metal-free protocol for assembly of the 2-aminobenzothiazole (**129**) moiety, with most targeted compounds produced with a 62–95% yield. To synthesize these target compounds, cyclohexanones **127** were reacted with thioureas **128** in the presence of I_2 and O_2 using para-toluenesulphonic acid (PTSA) via a dehydrogenation route[50] (**Figure (Scheme) 10.27**).

Some drugs including raloxifene, which is useful in osteoporosis treatment, zileuton, an anti-asthmatic drug, sertaconazole, which is effective as an anti-fungal agent, and mobam, as a lice controller contain a benzo[b]thiophene heterocyclic moiety in their structure. Many newly designed experimental drugs also possess this benzo[b]thiophene heterocycle pharmacophoric unit.[51] Chemists have subsequently devised various metal-free protocols for its assembly. Zhang et al. assembled these heterocycle benzo[b]thiophenes **131** by reacting substituted 2-halostyrene **130** with K_2S, which was heated either in the presence or absence of additives (**Figure (Scheme) 10.28**).[52]

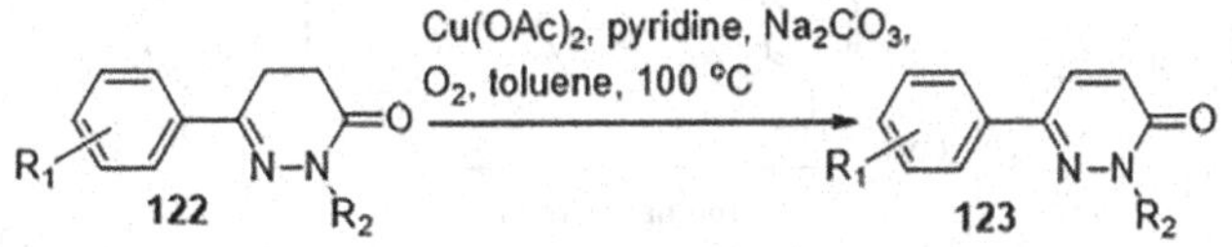

FIGURE (SCHEME) 10.26.

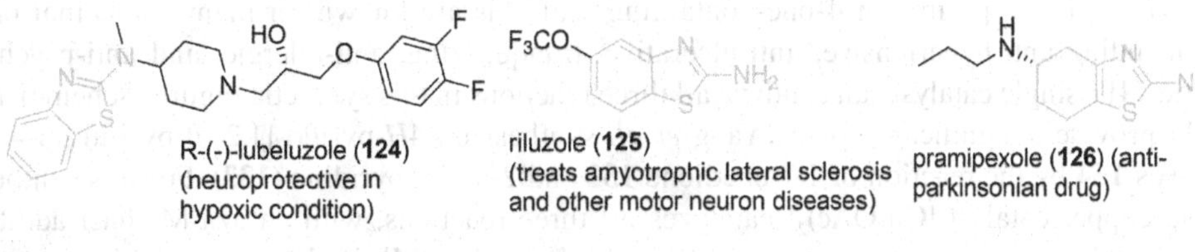

FIGURE 10.11 2-aminobenzothiazole-containing medication.

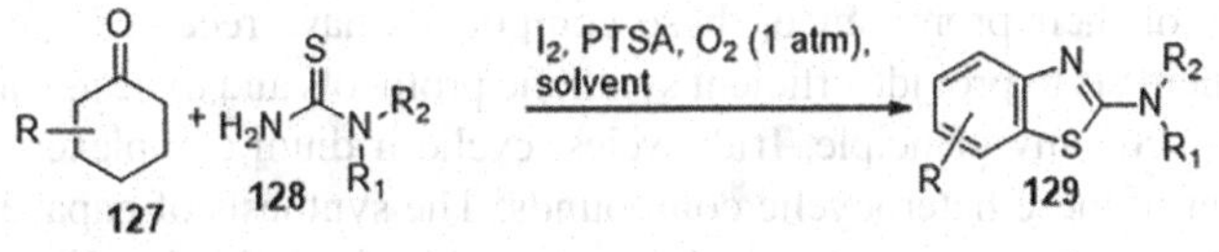

FIGURE (SCHEME) 10.27.

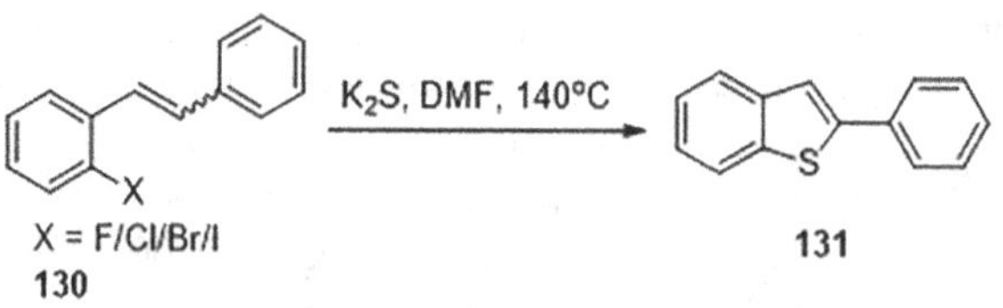

FIGURE (SCHEME) 10.28.

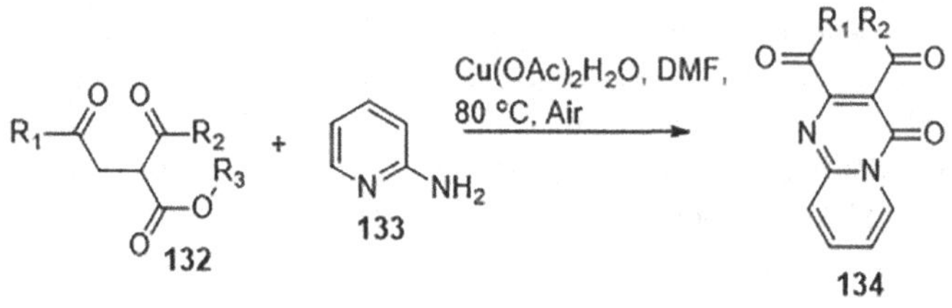

FIGURE (SCHEME) 10.29.

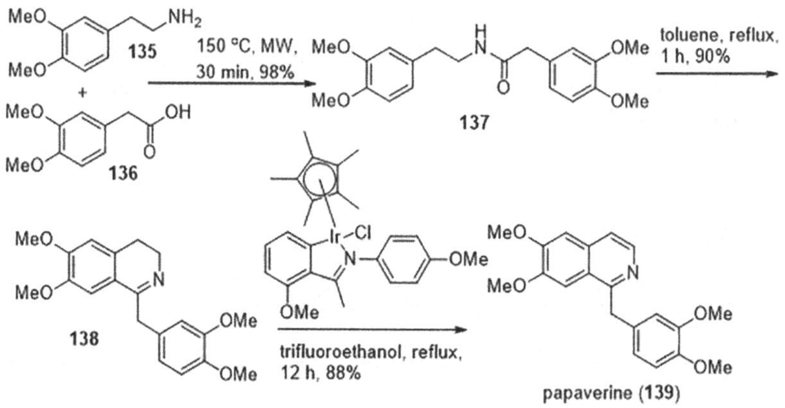

FIGURE (SCHEME) 10.30.

4H-pyrido-[1,2-a]-pyrimidin-4-one-containing scaffolds are known for many medicinal properties including anti-hypertensive, anti-ulcerative, tranquilizing, anti-allergic, and anti-psychotic properties. If a single catalyst can catalyze all three reactions in this synthetic Figure (Scheme), then it would provide a significant bonus. Yang et al. synthesized 4H-pyrido-[1,2-a]-pyrimidin-4-one derivatives **134** by the reaction of 1,4-enedione **132** and 2-aminopyridine (**133**). For this synthesis, a simple copper catalyst [Cu(OAc)$_2$] catalyzes all three reactions, with an aza-Michael addition, aerobic dehydrogenation and an intramolecular amidation in DMF in the presence of air at 80 °C (**Figure (Scheme) 10.29**).[53]

Nitrogenous heterocycles have a ubiquitous presence in drugs that are used to treat a variety of diseases and disorders such as anti-cancer, anti-malarial, anti-biotic, anti-viral, and numerous other properties. Because of their prominence, these compounds have received special attention from synthetic organic chemists to provide efficient synthetic protocols and to design an array of catalysts that serve the atom economy principle. Iridacycles, cyclic iridium complexes, have been utilized to compile a number of these heterocyclic compounds. The synthesis of papaverine (**139**), an antispasmodic alkaloid present in opium poppy, is illustrated in the following **Figure (Scheme) 10.30** that utilizes iridacycles as a catalyst for the dehydrogenation reaction.[54]

Quinazolinone alkaloids are very common in plants, and this scaffold has been reported to occur in more than 100 natural products. An alkaloid isolated from Peganum *harmala* known as pegamine was found to weaken cytotoxicity, which could be utilized as a lead for further drug design and development. Mathaqualone was introduced as a sedative and hypnotic drug but was subsequently withdrawn from the market because of its addictive properties. Halofuginone is an alkaloid from *Hydrangea febrifuga* that has received orphan drug status by the US FDA for the treatment of scleroderma. Prazosin is an α1 blocker that is used as an anti-hypertensive drug. Fenquizone is another anti-hypertensive drug, but it is in the sulphonamide diuretic category. Gefitinib is an anti-cancer drug used for non-small cell lung cancer (NSCLC) and especially utilized for the treatment of malignancies. Idelalisib is an anti-cancer agent that is used for the treatment of blood cancer. Albaconazole is being developed as an anti-fungal drug, while several quinazolinones have shown promise to be developed as anti-bacterial agents. Considering the diverse biological properties associated with quinazolinones, these compounds have attracted the interest of chemists for the design of new synthetic routes. Cheng *et al.* assembled quinazolinones **141** from various anthranilamides **140** and aldehydes in a one-pot reaction using i) PTSA in THF, rt, 10 min and ii) phenyliodine diacetate [PhI(OAc)$_2$, PIDA], rt, 1 h[55] (**Figure (Scheme) 10.31**).

Biobased compounds can also play a role in the synthesis of new materials. Butanol can be utilized to synthesize butadiene, a Figure (Scheme) involving dehydration followed by a dehydrogenation reaction. Butadiene acts as a precursor in the synthesis of various reagents including chloroprene (involved in neoprene polymer synthesis), 1,5-cyclooctadiene, 1,4-butanediol, tetrahydrofuran, sulfolane, and an array of polymers including 1,4-cis-polybutadiene, styrene-butadiene polymer, and co-polymer of acrylonitrile and butadiene.[56]

The 2-aminopyrimidine scaffold is widely represented in anti-cancer drugs such as pazopanib, imatinib, ceritinib, osimertinib, and nilotinib; in HIV drugs rilpivirine amd etravirine; in anxiolytic buspirone; in anti-bacterial timethoprim and sulphadiazine; in anti-protozoal pyrimethamine; in erectile dysfunction drug avanafil; in metastatic melanoma molecule debrafenib; in hypercholesterolemic medication rosuvastatin; and in diarrhea-associated carcinoid syndrome, drug telotristat ethyl.[57] Than *et al.* described a synthetic protocol for producing an array of 2-aminopyrimidine derivatives **143** featuring dehydrogenation reactions using Pd(OAc)$_2$, Cu(I)-thiophene-2-carboxylate (CuTC), and LiHMDS at 100 °C in toluene solvent under argon conditions (**Figure (Scheme) 10.32**).[58]

FIGURE (SCHEME) 10.31.

FIGURE (SCHEME) 10.32.

10.6 CONCLUSIONS AND PERSPECTIVES

This chapter summarized the prominent role of dehydrogenation reactions (catalytic/stoichiometric process) in the synthesis of diverse heterocyclic scaffolds, pharmaceutical drugs, and drug intermediates. Although noble metal catalysts have been used primarily for these catalytic dehydrogenation reactions, various base metal catalysts have recently been discovered and are being successfully employed for improved reaction conditions. Compared to the traditional dehydrogenation reaction, which produces toxic waste, the acceptorless dehydrogenation strategy can successfully eradicate the generation of these waste products in the reaction. Importantly, acceptorless dehydrogenation can be combined with hydrogen transfer reactions or condensation reactions with other substrates. In this way, desired complex drug molecules can be prepared with excellent reaction efficiency in an atom-economical setting. Accordingly, this chapter examined the development of dehydrogenation reactions and the evolution of dehydrogenation catalysts for the synthesis of important pharmaceutical drugs and bioactive compounds. Further development of these dehydrogenation reactions is still needed to successfully scale up to produce the several hundreds of kilogram quantities of the target drug molecules needed and to do this with good yield. This review on the role of the dehydrogenation reaction provides valuable background to form a strong basis supporting additional research aimed at the discovery of novel catalysts and new reaction conditions for the preparation of various value-added chemicals, natural products, and pharmaceutical drug molecules.

REFERENCES

1. (a) Arora, P.; Arora, V.; Lamba, H. S.; Wadhwa, D., Importance of heterocyclic chemistry: A review. *Int. J. Pharm. Sci. Res.* **2012**, *3* (9), 2947–2954; (b) Kaushik, N. K.; Kaushik, N.; Attri, P.; Kumar, N.; Kim, C. H.; Verma, A. K.; Choi, E. H., Biomedical importance of indoles. *Molecules* **2013**, *18* (6), 6620–6662.
2. Gunanathan, C.; Milstein, D., Applications of acceptorless dehydrogenation and related transformations in chemical synthesis. *Science* **2013**, *341* (6143), 1229712.
3. Gunanathan, C.; Ben-David, Y.; Milstein, D., Direct synthesis of amides from alcohols and amines with liberation of H_2. *Science* **2007**, *317* (5839), 790–792.
4. Goldman, A. S.; Roy, A. H.; Huang, Z.; Ahuja, R.; Schinski, W.; Brookhart, M., Catalytic alkane metathesis by tandem alkane dehydrogenation-olefin metathesis. *Science* **2006**, *312* (5771), 257–261.
5. de Miranda, A. S.; Milagre, C. D. F.; Hollmann, F., Alcohol dehydrogenases as catalysts in organic synthesis. *Front. Catal.* **2022**, 2.
6. Trost, B. M.; Metzner, P. J., Reaction of olefins wficith palladium trifluoroacetate. *J. Am. Chem. Soc.* **1980**, *102* (10), 3572–3577.
7. Gupta, M.; Hagen, C.; Kaska, W. C.; Cramer, R. E.; Jensen, C. M., Catalytic dehydrogenation of cycloalkanes to arenes by a dihydrido iridium P–C–P pincer complex. *J. Am. Chem. Soc.* **1997**, *119* (4), 840–841.
8. Matada, B. S.; Pattanashettar, R.; Yernale, N. G., A comprehensive review on the biological interest of quinoline and its derivatives. *Bioorg. Med. Chem.* **2021**, *32*, 115973.
9. Chan, A., Neratinib in HER-2-positive breast cancer: Results to date and clinical usefulness. *Ther. Adv. Med. Oncol.* **2016**, *8* (5), 339–350.
10. Chilà, G.; Guarini, V.; Galizia, D.; Geuna, E.; Montemurro, F., The clinical efficacy and safety of neratinib in combination with capecitabine for the treatment of adult patients with advanced or metastatic HER2-positive breast cancer. *Drug. Des. Devel. Ther.* **2021**, *15*, 2711–2720.
11. Guchhait, S. K.; Jadeja, K.; Madaan, C., A new process of multicomponent Povarov reaction–aerobic dehydrogenation: Synthesis of polysubstituted quinolines. *Tetrahedron Lett.* **2009**, *50* (49), 6861–6865.
12. Yamaguchi, R.; Ikeda, C.; Takahashi, Y.; Fujita, K., Homogeneous catalytic system for reversible dehydrogenation-hydrogenation reactions of nitrogen heterocycles with reversible interconversion of catalytic species. *J. Am. Chem. Soc.* **2009**, *131* (24), 8410–8412.
13. Stone, M. T., An improved Larock synthesis of quinolines via a Heck reaction of 2-bromoanilines and allylic alcohols. *Org. Lett.* **2011**, *13* (9), 2326–2329.
14. Chakraborty, S.; Brennessel, W. W.; Jones, W. D., A molecular iron catalyst for the acceptorless dehydrogenation and hydrogenation of N-heterocycles. *J. Am. Chem. Soc.* **2014**, *136* (24), 8564–8567.
15. Mastalir, M.; Glatz, M.; Pittenauer, E.; Allmaier, G.; Kirchner, K., Sustainable synthesis of quinolines and pyrimidines catalyzed by manganese PNP pincer complexes. *J. Am. Chem. Soc.* **2016**, *138* (48), 15543–15546.

16. Yang, J.; Meng, X.; Lu, K.; Lu, Z.; Huang, M.; Wang, C.; Sun, F., Acid-promoted iron-catalysed dehydrogenative [4 + 2] cycloaddition for the synthesis of quinolines under air. *RSC Adv.* **2018**, *8* (55), 31603–31607.
17. Wang, C.; Yang, J.; Meng, X.; Sun, Y.; Man, X.; Li, J.; Sun, F., Manganese(ii)-catalysed dehydrogenative annulation involving C-C bond formation: Highly regioselective synthesis of quinolines. *Dalton Trans.* **2019**, *48* (14), 4474–4478.
18. Azizi, K.; Akrami, S.; Madsen, R., Manganese(III) porphyrin-catalyzed dehydrogenation of alcohols to form imines, tertiary amines and quinolines. *Chemistry* **2019**, *25* (25), 6439–6446.
19. Dai, Y.; Cai, X.; Bi, X.; Liu, C.; Yue, N.; Zhu, Y.; Zhou, J.; Fu, M.; Huang, W.; Qian, H., Synthesis and anti-cancer evaluation of folic acid-peptide- paclitaxel conjugates for addressing drug resistance. *Eur. J. Med. Chem.* **2019**, *171*, 104–115.
20. Thompson, R. B., Foundations for blockbuster drugs in federally sponsored research. *FASEB J.* **2001**, *15* (10), 1671–1676.
21. Papaetis, G. S.; Syrigos, K. N., Sunitinib: A multitargeted receptor tyrosine kinase inhibitor in the era of molecular cancer therapies. *BioDrugs* **2009**, *23* (6), 377–389.
22. Martindale, W.; Parfitt, K. *Martindale: The Complete Drug Reference.* London: Pharmaceutical Press, **1999**.
23. Schley, N. D.; Dobereiner, G. E.; Crabtree, R. H., Oxidative synthesis of amides and pyrroles via dehydrogenative alcohol oxidation by ruthenium diphosphine diamine complexes. *Organometallics* **2011**, *30* (15), 4174–4179.
24. Zhang, M.; Neumann, H.; Beller, M., Selective ruthenium-catalyzed three-component synthesis of pyrroles. *Angew. Chem. Int. Ed. Engl.* **2013**, *52* (2), 597–601.
25. Chen, W.; Wang, J., Synthesis of pyrrole derivatives from diallylamines by one-pot tandem ring-closing metathesis and metal-catalyzed oxidative dehydrogenation. *Organometallics* **2013**, *32* (6), 1958–1963.
26. Kallmeier, F.; Dudziec, B.; Irrgang, T.; Kempe, R., Manganese-catalyzed sustainable synthesis of pyrroles from alcohols and amino alcohols. *Angew. Chem. Int. Ed. Engl.* **2017**, *56* (25), 7261–7265.
27. Borghs, J. C.; Azofra, L. M.; Biberger, T.; Linnenberg, O.; Cavallo, L.; Rueping, M.; El-Sepelgy, O., Manganese-catalyzed multicomponent synthesis of pyrroles through acceptorless dehydrogenation hydrogen autotransfer catalysis: Experiment and computation. *ChemSusChem* **2019**, *12* (13), 3083–3088.
28. (a) Yu, X.; Lin, W.; Li, J.; Yang, M., Synthesis and biological evaluation of novel beta-carboline derivatives as Tat-TAR interaction inhibitors. *Bioorg. Med. Chem. Lett.* **2004**, *14* (12), 3127–3130; (b) Moradi, M. T.; Karimi, A.; Rafieian-Kopaei, M.; Fotouhi, F., In vitro antiviral effects of Peganum harmala seed extract and its total alkaloids against Influenza virus. *Microb. Pathog.* **2017**, *110*, 42–49.
29. Gaikwad, S.; Kamble, D.; Lokhande, P., Iodine-catalyzed chemoselective dehydrogenation and aromatization of tetrahydro-β-carbolines: A short synthesis of Kumujian-C, Eudistomin-U, Norharmane, Harmane Harmalan and Isoeudistomine-M. *Tetrahedron Lett.* **2018**, *59* (25), 2387–2392.
30. (a) Rechfeld, F.; Gruber, P.; Kirchmair, J.; Boehler, M.; Hauser, N.; Hechenberger, G.; Garczarczyk, D.; Lapa, G. B.; Preobrazhenskaya, M. N.; Goekjian, P.; Langer, T.; Hofmann, J., Thienoquinolines as novel disruptors of the PKCepsilon/RACK2 protein-protein interaction. *J. Med. Chem.* **2014**, *57* (8), 3235–3246; (b) RohitKumar, H. G.; Asha, K. R.; Raghavan, S. C.; Advi Rao, G. M., DNA intercalative 4-butylaminopyrimido[4',5':4,5]thieno(2,3-b)quinoline induces cell cycle arrest and apoptosis in leukemia cells. *Cancer Chemother. Pharmacol.* **2015**, *75* (6), 1121–1133.
31. Teja, C.; Nawaz Khan, F. R., Facile synthesis of 2-acylthieno[2,3-b]quinolines via Cu-TEMPO-catalyzed dehydrogenation, sp(2)-C-H functionalization (nucleophilic thiolation by S8) of 2-haloquinolinyl ketones. *Org. Lett.* **2020**, *22* (5), 1726–1730.
32. Bansal, Y.; Silakari, O., The therapeutic journey of benzimidazoles: A review. *Bioorg. Med. Chem.* **2012**, *20* (21), 6208–6236.
33. Sofi, F. A.; Sharma, R.; Rawat, R.; Chakraborti, A. K.; Bharatam, P. V., Visible light promoted tandem dehydrogenation-deaminative cyclocondensation under aerobic conditions for the synthesis of 2-aryl benzimidazoles/quinoxalines from ortho-phenylenediamines and arylmethyl/ethyl amines. *New J. Chem.* **2021**, *45* (10), 4569–4573.
34. Olin, J.; Schneider, L., Galantamine for Alzheimer's disease. *Cochrane Database Syst. Rev.* **2001**, *1*, CD001747.
35. Zhang, Y.; Shen, S.; Fang, H.; Xu, T., Total synthesis of galanthamine and lycoramine featuring an early-stage C-C and a late-stage dehydrogenation via C-H activation. *Org. Lett.* **2020**, *22* (4), 1244–1248.
36. (a) Dai, H.; Ge, S.; Guo, J.; Chen, S.; Huang, M.; Yang, J.; Sun, S.; Ling, Y.; Shi, Y., Development of novel bis-pyrazole derivatives as antitumor agents with potent apoptosis induction effects and DNA damage. *Eur. J. Med. Chem.* **2018**, *143*, 1066–1076; (b) Cottineau, B.; Toto, P.; Marot, C.; Pipaud, A.; Chenault, J., Synthesis and hypoglycemic evaluation of substituted pyrazole-4-carboxylic acids. *Bioorg. Med. Chem. Lett.* **2002**, *12* (16), 2105–2108; (c) Stasi, R.; Evangelista, M. L.; Amadori, S., Novel thrombopoietic agents: A review of their use in idiopathic thrombocytopenic purpura. *Drugs* **2008**, *68* (7), 901–912.

37. Zhu, Y. F.; Wei, B. L.; Wei, J. J.; Wang, W. Q.; Song, W. B.; Xuana, L. J., Synthesis of pyrazolones and pyrazoles via Pd-catalyzed aerobic oxidative dehydrogenation. *Tetrahedron Lett.* **2019**, *60* (17), 1202–1205.
38. Bernadou, J.; Bonnafous, M.; Labat, G.; Loiseau, P.; Meunier, B., Model systems for metabolism studies. Biomimetic oxidation of acetaminophen and ellipticine derivatives with water-soluble metalloporphyrins associated to potassium monopersulfate. *Drug. Metab. Dispos.* **1991**, *19* (2), 360–365.
39. Johansson, T.; Weidolf, L.; Jurva, U., Mimicry of phase I drug metabolism—novel methods for metabolite characterization and synthesis. *Rapid. Commun. Mass. Spectrom.* **2007**, *21* (14), 2323–2331.
40. Klunk, L. J.; Ringel, S.; Neiss, E. S., The disposition of 14C-indapamide in man. *J. Clin. Pharmacol.* **1983**, *23* (8–9), 377–384.
41. Sun, H.; Ehlhardt, W. J.; Kulanthaivel, P.; Lanza, D. L.; Reilly, C. A.; Yost, G. S., Dehydrogenation of indoline by cytochrome P450 enzymes: A novel "aromatase" process. *J. Pharmacol. Exp. Ther.* **2007**, *322* (2), 843–851.
42. Que, L., Jr.; Tolman, W. B., Biologically inspired oxidation catalysis. *Nature* **2008**, *455* (7211), 333–340.
43. Balamurugan, K.; Perumal, S.; Reddy, A. S. K.; Perumal, Y.; Dharmarajan, S., A facile domino protocol for the regioselective synthesis and discovery of novel 2-amino-5-arylthieno-[2,3-b]thiophenes as antimycobacterial agents. *Tetrahedron Lett.* 50 (45), 6191–6195.
44. Marcos-Escribano, A.; Bermejo, F. A.; Bonde-Larsen, A. L.; Retuerto, J. I.; Sierra, I. H., 1,2-Dehydrogenation of steroidal 6-methylen derivatives. Synthesis of exemestane. *Tetrahedron* **2009**, *65* (36), 7587–7590.
45. Shi, F.; Tse, M. K.; Zhou, S.; Pohl, M. M.; Radnik, J.; Hubner, S.; Jahnisch, K.; Bruckner, A.; Beller, M., Green and efficient synthesis of sulfonamides catalyzed by nano-Ru/Fe(3)O(4). *J. Am. Chem. Soc.* **2009**, *131* (5), 1775–1779.
46. Zhou, L.-M.; Qu, R.-Y.; Yang, G.-F., An overview of spirooxindole as a promising scaffold for novel drug discovery. *Expert Opin. Drug Discov.* **2020**, *15* (5), 603–625.
47. Bhaskar, G.; Arun, Y.; Balachandran, C.; Saikumar, C.; Perumal, P. T., Synthesis of novel spirooxindole derivatives by one pot multicomponent reaction and their antimicrobial activity. *Eur. J. Med. Chem.* **2012**, *51*, 79–91.
48. Liang, L.; Yang, G.; Xu, F.; Niu, Y.; Sun, Q.; Xu, P., Copper-catalyzed aerobic dehydrogenation of C–C to C=C bonds in the synthesis of pyridazinones. *Eur. J. Org. Chem.* **2013**, *2013* (27), 6130–6136.
49. Jonnala, S.; Nameta, B.; Chavali, M.; Bantu, R.; Choudante, P.; Misra, S.; Sridhar, B.; Dilip, S.; Reddy, B. V. S., Design, synthesis, molecular docking and biological evaluation of 1-(benzo[d]thiazol-2-ylamino)(phenyl)methyl)naphthalen-2-ol derivatives as antiproliferative agents. *Lett. Org. Chem.* **2019**, *16* (10), 837–845.
50. Zhao, J.; Huang, H.; Wu, W.; Chen, H.; Jiang, H., Metal-free synthesis of 2-aminobenzothiazoles via aerobic oxidative cyclization/dehydrogenation of cyclohexanones and thioureas. *Org. Lett.* **2013**, *15* (11), 2604–2607.
51. Keri, R. S.; Chand, K.; Budagumpi, S.; Balappa Somappa, S.; Patil, S. A.; Nagaraja, B. M., An overview of benzo[b]thiophene-based medicinal chemistry. *Eur. J. Med. Chem.* **2017**, *138*, 1002–1033.
52. Zhang, X.; Zeng, W.; Yang, Y.; Huang, H.; Liang, Y., Transition-metal-free method for the synthesis of Benzo[b]thiophenes from o-Halovinylbenzenes and K2S via direct SNAr-type reaction, cyclization, and dehydrogenation process. *Synlett.* **2013**, *24* (13), 1687–1692.
53. Yang, Y.; Shu, W. M.; Yu, S. B.; Ni, F.; Gao, M.; Wu, A. X., Auto-tandem catalysis: Synthesis of 4H-pyrido[1,2-a]pyrimidin-4-ones via copper-catalyzed aza-Michael addition-aerobic dehydrogenation-intramolecular amidation. *Chem. Commun. (Camb)* **2013**, *49* (17), 1729–1731.
54. Wang, C.; Xiao, J., Iridacycles for hydrogenation and dehydrogenation reactions. *Chem. Commun. (Camb)* **2017**, *53* (24), 3399–3411.
55. Cheng, R.; Guo, T.; Zhang-Negrerie, D.; Du, Y.; Zhao, K., One-pot synthesis of quinazolinones from anthranilamides and aldehydes via p-toluenesulfonic acid catalyzed cyclocondensation and phenyliodine diacetate mediated oxidative dehydrogenation. *Synthesis* **2013**, *45*, 2998–3006.
56. Mascal, M., Chemicals from biobutanol: Technologies and markets. *Biofuel Bioprod. Biorefin.* **2012**, *6* (4), 483–493.
57. Yet, L., *Privileged Structures in Drug Discovery: Medicinal Chemistry and Synthesis*. New York: John Wiley and Sons, **2018**, 237–283.
58. Phan, N. H.; Kim, H.; Shin, H.; Lee, H. S.; Sohn, J. H., Dehydrosulfurative C-N cross-coupling and concomitant oxidative dehydrogenation for one-step synthesis of 2-Aryl(alkyl)aminopyrimidines from 3,4-dihydropyrimidin-1H-2-thiones. *Org. Lett.* **2016**, *18* (19), 5154–5157.

11 Industrial Applications of Dehydrogenation Reactions
Process Design of Reactors

Ravi Tejasvi

11.1 INTRODUCTION

A chemist creates a new chemical product or pathway to produce known materials efficiently. However, this happens either in a test tube or on a lab scale. Mass-production of said chemical or implementation of such a process to make material in bulk requires chemical engineers. Chemical engineers are responsible for the design and operation of process plants and the products' proper packing and safe transportation. Therefore, it seems prudent to consider the process entirely, from raw materials to finished products and from individual equipment parts to the whole plant.

A chemical engineer designs the equipment to execute the process at the specified reaction parameters based on the chemistry of the methods and products. Process equipment design starts with writing the underlying chemical reaction and calculating the entailing mass and energy balances. After this comes the real fun of chemical engineering. After bringing in the reactants and the required energy, the process is no longer theoretical. The practical challenges include pressure loss in the transportation pipelines, corrosion and toxic leakage prevention, the maintenance of reaction pressure, efficient heat transfer through various heat-exchange equipment, the homogenization of reactant composition in the reactor, the swift output of the process products, the separation of the intended outcome, conveying, and the safe packing of the final product.

Moreover, the reactants vary in composition, purity, and phase. In contrast, the chemical reactions vary regarding the intermediate steps, catalysts, end-products, selectivity, and yield. Therefore, each design exercise is unique and needs a thorough understanding of the founding elements of chemical engineering. A chemical engineer must study chemical reaction engineering, molecular transport processes, separation processes, process plant safety, risk analysis, plant and equipment design, process control, plant management, and plant and process economics in addition to the core courses in mathematics, physics, and chemistry. Practicing chemical engineers would tell you that the data they work with are often unreliable or incomplete. Hence, they must navigate at every step by making the correct assumptions backed by the founding principles of all of these courses of study. The purpose of discussing chemical engineering here is to make readers aware of the actions conducted by chemical engineers in bringing the brainchildren of chemistry to the point of commercial production and revenue generation. Knowing the chemical engineering involved in the design procedure will allow any reader with a chemistry background to understand the entire process better, re-invent, and help make the existing practices sustainable.

Dehydrogenation is a reaction of removing hydrogen from an organic reactant and forming a new chemical.[1–3] The dehydrogenation reaction variants include low-temperature catalytic dehydrogenation, catalytic cracking, and thermochemical cracking. However, from a chemical engineering perspective, all of these variants have the following commonalities[4]: (i) the product volume is larger than the reactant, (ii) swift separation of the product chemicals from the generated hydrogen is needed, and (iii) the reaction temperature and pressure must be sustainably maintained. This three-pronged problem exists for all dehydrogenation reactions, and the solution requires knowledgeable communication between chemists and chemical engineers. The chemist communicates the requirements of sustainable production based on the dehydrogenation reaction. However, the planning, commissioning, operation, and maintenance of the plant are carried out by chemical engineers.

The preamble presented here shows that the whole chemical engineering affair revolves around designing, operating, and maintaining the process reactors. Hence, the upcoming sections of the chapter discuss the most popular choices among various reactor forms, their distinctive designs, specific advantages, and performance weaknesses. Sometimes, obtaining a precise conversion may require more than one reactor. Therefore, the networking of reactors is also discussed. Finally, a case study of the process design of reactors for styrene production is presented. It aims to elaborate on the role of chemical engineering in the bulk production of a product initially envisaged by chemistry.

11.2 PROCESS DESIGN OF REACTORS

Process design refers to estimating the size, shape, and flow patterns calculated based on physicochemical laws, conservation equations, and empirical correlations. There are typically three stages of the process design of reactors. As mentioned earlier, the design exercise starts with building the total mass and energy conservation equations. It gives a *ballpark* estimate of the volume and shape of the potential reactor design. Afterwards, the effects of additional details are evaluated and amalgamated with the basic design. These extra considerations may include alteration in the reactor pressure, energy losses, and deployment of compensatory measures such as pumps, heat exchangers, and separators. It is to be ensured that no component is underutilized or overexerted. For example, the pipes should be filled with turbulent flow without the flow rate exceeding the maximum permissible limit. Finally, the entire design undergoes a holistic evaluation to estimate the efficiency and conversion expected from the reactor. A general outline of the design procedure is as follows:

Step-1 Collect all relevant physical and chemical property data and the kinetic and thermodynamic data over the various operating conditions, such as pressure, temperature, flow rate, and catalyst concentration from the literature, laboratory, or pilot plant studies.

Step-2 Identify the dominant rate-controlling mechanism, and choose the suitable reactor type based on the information gathered in step 1.

Step-3 Hypothesize the reactor conditions that can give the desired performance. The reactor performance is measured in terms of the attainable conversion and yield.

Step-4 Size the reactor and estimate its performance. Since the design relationships are rarely analytically soluble, semi-empirical methods are the best-practiced approaches.

Step-5 Choose suitable construction materials considering their compatibility with the reactants. Put a preliminary mechanical design for the reactor, including the vessel design, the regulation of heat transfers, and the arrangement of internals along with the general arrangement.

Step-6 Conduct a cost analysis of the stipulated design, including the capital, operating, and maintenance costs for a stipulated future period.

Step-7 Repeat steps 4 to 6, as necessary, to optimize the design.

Figure 11.1 shows the general protocol for a process design size estimate of ideal reactors.

Since the design is based on the *process* parameters and is aimed at achieving specific *process* targets, the exercise is appropriately named *process design*. The *process parameters* and the desired *conversion* are the standard design inputs, and they vary widely per production requirements. Hence, the reactor design templates, also known as the ideal reactors, are useful.

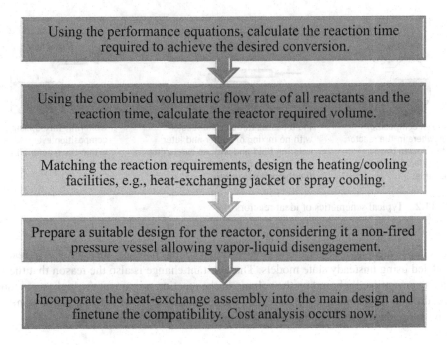

FIGURE 11.1 A general process design protocol of ideal reactors.

11.2.1 Ideal Reactors

Ideal reactors are design templates and are valuable in judging reactor performance, estimating the shapes and the dimensional proportions, and determining the reactor sequence. Since the input information varies widely, the ideal reactor designs are classified into three distinct types based on the reactant flow regime that exists within the reactor. These types are the batch reactor (BR), the mixed flow reactor (MFR), and the plug flow reactor (PFR). Other reactor designs exist, such as semi-batch, membrane, and falling-film reactors. Still, these ideal designs often serve as design guides or performance evaluation standards. There are also other ways to classify chemical reactors. For example, chemical reactors can be classified according to the phases processed within the reactor, for example, single-phase or multi-phase reactors. Multi-phase reactors are typically used to process gas-liquid, liquid-liquid, gas-solid, or liquid-solid catalytic reactors, and their performance analysis is more complex than single-phase reactors. Another classification of chemical reactors can be according to the mode of heat removal, including isothermal, non-isothermal, or adiabatic reactors. Let us start with ideal reactors. Figure 11.2 shows the typical schematics of ideal reactors.[5,6]

11.2.1.1. Batch Reactors (BR)

Batch reactors are the most common type and are typically used to process small amounts of reactants. The industrial involvement of batch reactors occurs in various processes such as the synthesis of dyes, production of specialty chemicals (including benzene sulphonic acid from benzene and oleum, nitrobenzene from benzene and mixed acid, or polyurethane from toluene di-isocyanate and polyglycol), batch fermentation (e.g., fermentation of molasses to ethanol), manufacture of active pharmaceutical ingredients (e.g., synthesis of cypermethrin), manufacture of amino and alkyd resins, etc.[7,8]

Fixed amounts of the material are put in these reactors, and desired temperatures are set while the materials are unceasingly stirred. Characteristically, there can be no input or output during the reaction. Our mouths, test tubes, flasks, and the earth are excellent examples of batch reactors. Accordingly, batch reactors' process design is the most straightforward. Here, only the reactor

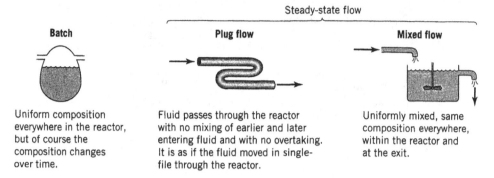

FIGURE 11.2 Typical schematics of ideal reactors.[6]

volume is to be calculated. The reactant concentration keeps changing; therefore, these reactors are best modeled using unsteady state models. The constant change is also the reason that the volume estimate is conventionally based on the volume of the reaction mixture at the highest reaction temperature, the maximum reaction pressure, and the specified degree of reactant conversion.

The performance equation of an ideal batch reactor can be given as follows:

$$V = F_{A_0} \int_0^{X_A} \frac{dX_A}{(-r_A)} \quad \{\text{for constant volume processes}\}$$

Equation 11.1

$$\theta_B = N_{A_0} \int_0^{X_A} \frac{dX_A}{(-r_A)V}$$

Equation 11.2

where N_{A_0} = Initial amount of limiting reactant, kmol

θ_B = Time required to achieve the conversion X_A, s

$F_{A_0} = \dfrac{N_{A_0}}{\theta_B}$ = Molar flow rate of limiting reactant A, kmol/s

X_A = Fractional conversion of limiting reactant A

$-r_A$ = Rate of reaction

V = Working volume of the reactor, m³

Batch reactors may be very advantageous in certain situations because they can obtain high conversions if a longer residence duration is permissible. Batch reactors are easy to clean and deemed suitable for producing small specimen amounts of products while the process is in its testing phases. However, the high labor cost, difficulty maintaining large-scale production, and extended downtime when cleaning are a few factors that limit their usage.

11.2.1.2. Mixed Flow Reactor (MFR)

The mixed flow reactor (MFR) is a non-batch, flow-type reactor. It has three characteristic features: (i) it is a continuous flow reactor, that is, reactants are continuously fed in, and products are continuously withdrawn, (ii) it is stirred unceasingly to keep the reaction mixture concentration and temperature homogenous everywhere, and (iii) it comes in the shape of a tank, specifically, its height is comparable to its diameter. Since the reactor enjoys a continuous flow, nothing accumulates within,

Industrial Applications of Dehydrogenation Reactions

so these reactors can be conveniently modeled using a steady-state mole balance. In various texts, this reactor has also been called the *back mix reactor*, the *continuous stirred tank reactor (CSTR)*, and the *constant flow stirred tank reactor*, all citing the characteristic flow pattern that maintains the compositional homogeneity.[7,9]

Industrial forms of MFRs include agitated vessel-type reactors, sparged vessels (including gas-induced agitator types), bubble column-type reactors, loop reactors (with recirculation, recirculating pump, and external heat exchanger), and a jet reactor with recirculation through an ejector.[7,10,11] The human stomach and our living rooms are the two closest examples of MFR. Industrially, MFR is deployed in various processes, including the liquid-phase carbonylation of methanol to acetic acid, oxidation of acetaldehyde to acetic acid, liquid-phase polymerization of styrene to polystyrene, solution polymerization of propylene to polypropylene, chlorination of benzene to chlorobenzene, oxidation of toluene to benzoic acid, and solution polymerization of vinyl chloride to polyvinyl chloride.

In an ideal MFR, the inflowing reactants immediately achieve the targeted composition due to the dilution of reactants by the products. The rate of reaction is minimum in an ideal MFR compared to all types of continuous flow reactors, as it provides a minimum concentration of reactants during reactions.[6,7,12] Like batch reactors, flow reactors, such as MFR or plug flow reactors, can also be modeled in terms of the *space-time* (τ), the time required to process one reactor volume of feed. The performance equation for an ideal MFR can be given as follows:

$$V = \frac{F_{A_0} X_A}{(-r_A)_{exit}} \qquad \text{Equation 11.3}$$

$$\tau_M = \frac{V C_{A_0}}{F_{A_0}} \qquad \text{Equation 11.4}$$

where, F_{A_0} = Molar flow rate of limiting reactant A, kmol/s

C_{A_0} = Molar concentration of limiting reactant A, kmol/m³

X_A = Fractional conversion of limiting reactant A

$(-r_A)_{exit}$ = Rate of reaction (based on the concentration measured at the exit)

V = Working volume of the reactor, m³

τ_M = Time velocity of the MFR, s

As explained earlier, MFR offers extensive mixing and a lower reaction rate. It not only lowers the rate of heat generation in case of exothermic reaction but also quickly disseminates the locally generated heat. It renders the temperature control more accessible vis-à-vis other ideal reactor forms. The tank shape requires less fixed cost for lower-order reactions with lower equilibrium conversion because the cost per unit volume of an MFR is low. In MFR, the fine catalyst particles can be effectively suspended throughout the liquid reaction system.

Moreover, an MFR can provide higher heat and mass transfer for slurry-like viscous reaction mixtures. Despite all of these benefits, an ideal MFR requires maximum volume among all types of continuous reactors for any given positive order reaction and required conversion. This is because of the *minimum rate of reaction* among all types of continuous flow reactors. Furthermore, unlike a stirred tank batch reactor, it is difficult to change over and process other products with MFR. Still, MFRs are the industry favorite simply because of their ease of design and operation.

11.2.1.3. Plug Flow Reactor (PFR)

The presumption of fully developed flow in a filled pipe is that the fluid flows with a flat velocity profile at any location and at any time. This presumption holds good; every cross-section can be seen moving like a plug. Since the same flow regime of a flat velocity profile occurs in a tube-shaped reactor, this reactor is uniquely and appropriately named a *plug flow reactor* (PFR). In PFR, only lateral or radial mixing is permitted. This means that the composition varies only axially and not radially or temporally. Hence, in an ideal PFR, all fluid particles get the same residence time. The local reactant concentration is only a function of the distance from the entry. MFR is known for the complete mixing of the reaction mixture, whereas ideal PFR is known for complete lateral mixing without a back mix or fluid overtake. Hence, PFR has also been named an unmixed, slug, or *piston flow reactor*.

Our intestines are the nearest example of PFRs. Industrially, PFRs are used as fixed bed catalytic reactors, packed tower-type reactors, tubular reactors, shell and tube heat exchangers, double pipe heat exchangers, sieve tray towers, trickle bed reactors, etc. These reactors are extensively used in the polymerization of ethylene to low-density polyethylene, hydrolysis of ethylene oxide to ethylene glycol, formation of NH_3 from Haber's process, methanol formation reaction, steam reforming reaction, air oxidation of methanol to formaldehyde, oxidation of SO_2 to SO_3 in sulphuric acid manufacturing, oxidation of ethylene to ethylene oxide, steam cracking-based olefins production, splitting of edible oil to fatty acids, hydrocracking of heavy gas oil, and hydrodenitrogenation of lube oil distillate.[7]

In ideal PFR, the fluid properties and the flow properties vary only axially. Any location always sees the exact composition of the reaction mixture. Therefore, the performance equations of the ideal PFR can be given as follows:

$$V = F_{A_0} \int_0^{X_A} \frac{dX_A}{(-r_A)} \qquad \text{Equation 11.5}$$

$$\tau_P = C_{A_0} \int_0^{X_A} \frac{dX_A}{(-r_A)} \qquad \text{Equation 11.6}$$

where, F_{A_0} = Molar flow rate of limiting reactant, kmol/s

C_{A_0} = Molar concentration of limiting reactant A, kmol/m³

X_A = Fractional conversion of limiting reactant A

$(-r_A)$ = Rate of reaction

V = Working volume of the reactor, m³

τ_P = Time velocity of the PFR, s

For a positive order reaction, plug flow reactors require the minimum volume among all types of continuous flow reactors. The volume required by an ideal PFR is much smaller than that required by an ideal MFR, particularly for a higher-order reaction with a higher desired conversion. PFR does not use moving mechanical parts; hence, it is suitable for reaction systems involving high pressures, corrosive materials, or toxic gases. These qualities render the cost per unit production volume from PFR higher than from MFR. Given the high investment and operational costs for lower conversions, PFR is unsuitable for lower-order reactions. Table 11.1 summarizes the pros and cons of various reactor designs.

TABLE 11.1
Advantages and Disadvantages of Various Reactor Designs[13]

Reactor geometry	Phases involved	Advantages	Disadvantages	Specific applications
Mixed flow or stirred tank reactor	Homogeneous L-L and heterogeneous G-L and L-S (finely suspended)	Continuous operation Good temperature control Easily adaptable for two-phase systems Simple construction Low operating (labor) cost Easy to clean	Lowest conversion/unit volume (requires large reactor size for high conversion) Bypassing and channeling problems for poor agitation	Any process requiring high heat and mass transfer rates When different concentration streams are required, CSTRs are used in a series
Plug flow or tubular reactor	Primarily gaseous and some L-L reactions	High conversion/unit (~75%) Low operating (labor) cost Good heat transfer	Low heat/mass transfer rates* Undesirable thermal gradient Temperature control is difficult (hot spot formation for exothermic reactions) Shutdown and cleaning are expensive	Continuous production Fast reaction on a large scale for both homogeneous and heterogeneous systems Thermal cracking of ethane Thermal decomposition of dichloroethane to vinyl chloride
Packed bed reactor	G-S, L-S, G-L-S (for gas-phase reaction using a solid catalyst)	Highest conversion (~95%) per unit catalyst weight for most reactions in the catalytic reactor	Poor heat transfer Poor temperature control Catalyst replacement difficult It cannot be used for very small catalyst particle sizes Channeling of fluid flow and ineffective use of reactor bed	Most common for high-temperature catalytic reactions and gas conversions - Ammonia synthesis - Oxidation of SO_2 - Isomerization of n-alkanes - Reaction of benzene with ethylene/propylene to form ethylbenzene/cumene
Fluidized bed reactor	G-S, L-S, G-L-S For gas-solid noncatalytic reactions, fluidized beds are always used Also used for gas-phase reactions using a solid catalyst	Good mixing High transfer rates Uniform temperature Good temperature control Suitable for a high rate of catalyst deactivation/frequent regeneration Allows continuous regeneration through use of an auxiliary loop Can handle large amounts of feed and solids	Low conversion (~32%) due to the severe bypassing of reactant gas Catalyst attrition and dust formation due to severe agitation Uncertain scale-up	Heterogeneous reactions such as coal combustion, coal gasification, biomass pyrolysis, fluidized coking/polymerization, catalytic cracking of heavy oil, and use of calcium carbonate to fix H_2S in the petroleum industry

G, L, and S refer to the gas, liquid, and solid phases, respectively.
*Reduced dimensions required for improved transfer rates and geometry such as tube bundles in shell and tube heat exchanger.

11.2.2 Non-Ideal Reactors

Non-ideal reactors are deviations from the ideal reactors' flow patterns. These non-ideal reactors exist by design and are used widely to serve various aims. For example, a dead volume occurs in an MFR where the flow-caused mixing cannot reach. Such a reactor is used in fermentation industries, for example, wineries, and the dead volume acts as the inbuilt bacteria-culture bank. Baffles in the MFR enhance mixing but present dead volumes in their vicinities.

Similarly, catalytic converters in automobiles are packed catalytic beds put in the path of an otherwise PFR design. This placement allows all the vehicular exhausts to undergo the catalytic process but causes lateral mixing. Hence, an ideal PFR model cannot correctly model the behavior there and invites the model to be considered non-ideal. *Tanks-in-series* and *dispersion* models are the most used models for analyzing non-ideal flow reactors. Batch reactors are also not deprived of non-idealities. All microbial and biochemical processes utilize the growth of microbial cultures in a controlled setup. These setups are incubated as batch reactors. However, these microbes are living organisms and therefore die upon aging. Accordingly, population replenishment is required and achieved through the periodical introduction of the new culture. This inflow of microbes, the reactants of the biochemical production process, causes the non-ideality in the otherwise batch reactor behavior. Therefore, such an arrangement is termed a semi-batch reactor.

Industry never uses a single reactor to achieve the target output. This is partially because no single reactor can perfectly model the actual reaction pattern and partially because every reactor can only partly execute the reaction. For example, our mouth, stomach, and intestine cannot on their own digest food completely. A network of reactors is preferred to complete the reaction to the farthest extent in the shortest duration.

11.3 REACTOR NETWORKING

The mouth chews food while mixing it with saliva to break the starches. Since we bite food and take time to chew it before taking another bite, the mouth can be deemed a batch reactor. However, the continuous addition of saliva makes the mouth a semi-batch reactor. The esophagus acts as a peristaltic pipeline to get mouth-chewed food to the stomach. The chewed food is mixed with various digestive juices in the stomach, and the food components are processed into simpler products.[14] The stomach resembles MFR because of the food movement's inherent inflow and outflow pattern. Then, the food reaches the intestines. The food moves linearly outwards while the nutrients are absorbed into the blood through the intestinal walls. Although the food propagation resembles PFR, given the nutrient absorption, the intestines are better modeled as non-ideal PFR. Overall, the entire digestive tract, including the mouth, stomach, and intestines, can be seen as a reactor network having a semi-batch, an MFR, and a non-ideal PFR arranged in a series. If any one of these malfunctions, then the whole network goes awry. The same happens to industrial reactor networks when any component reactor malfunctions. Hence, industries, as a rule, keep a parallel line on standby so that the production load can be transferred to one line while the components of the parallel line undergo periodic maintenance.

Each reactor model equation can be transformed to describe a relation between the reciprocal of the reaction rate $1/(-r_A)$ and the limiting reactant's conversion X_A. Each of these graphs presents a distinctive *area under the curve* that signifies the volume of the reactor required to achieve the target conversion. Table 11.2 lists the relations and their typical graphs. It is self-explanatory that any reactor in a series of reactors has the feed conversion the same as the exit conversion of the preceding unit, which in the case of the series starting reactor is zero. PFR requires a smaller volume for the exact conversion, as shown in Figure 11.3. The critical thing to remember here is that MFR has faster output delivery, but PFR has a higher degree of product conversion. Thus, based on the process requirements, various volumes and shapes of reactors can be arranged in a series, parallel, or a combination thereof to obtain the required conversion.

TABLE 11.2
Performance Equations of Ideal Reactors and their Graphical Representations

Ideal reactor type	Equation	Characteristic performance plot[6]
1. Batch Reactor (BR)	$\dfrac{V}{F_{A_0}} = \displaystyle\int_0^{X_A} \dfrac{dX_A}{(-r_A)}$	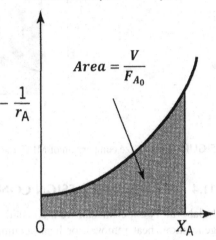
2. Mixed Flow Reactor (MFR)	$\dfrac{V}{F_{A_0}} = \dfrac{X_A}{(-r_A)_{exit}}$	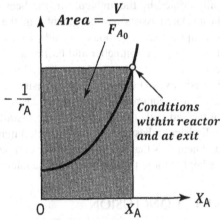
3. Plug Flow Reactor (PFR)	$\dfrac{V}{F_{A_0}} = \displaystyle\int_0^{X_A} \dfrac{dX_A}{(-r_A)}$	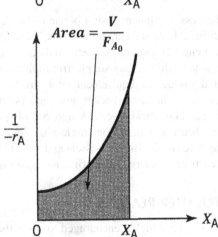

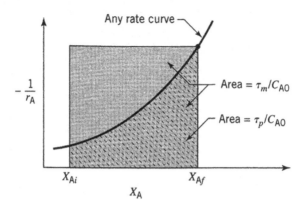

FIGURE 11.3 Size comparison of MFR and PFR.[6]

11.4 ADDITIONAL DESIGN CONSIDERATIONS FOR REAL REACTORS

In laboratory experiments, we are usually not concerned with heat-related problems such as vapor generation, heat removal, or heat accumulation. In a test tube or small setups, which are naturally cooled by the ambient air, the heat problems do not grow to become significant. However, heat-related issues can be gigantic at the industrial plant level. Therefore, as a standard design principle, some additional considerations are deemed prudent in the process design of reactors. For example, an impeller and baffles are required in MFR for proper mixing and temperature and concentration homogenization. Impeller design is based on the reaction mixture's required mixing intensity, vessel capacity, and viscosity. Extra volume (typically 40–50% of the reactor volume) is given to accommodate the generated vapors and for droplet disengagement and defoaming. Internal immersed coils or outer jackets are installed to provide the required heat exchange. In PFR, long cylindrical tubes can be put in a coiled manner to save space. However, this entails a pressure drop and heat loss that may need further compensation. In addition, the length-to-diameter ratio should be kept at more than 20 to minimize lateral mixing.

11.5 CONCLUSION

Process equipment design is one of the few white-collar jobs in the primarily blue-collar field of chemical engineering professions. To create a design, one needs extensive knowledge of all chemical engineering core subjects and an integrating approach toward their wholesome application. A reactor, which looks simple from afar, involves the diligent determination of process parameters and production requirements and careful consideration of the means. Hence, this chapter aimed to acquaint chemistry background readers with a basic understanding of chemical reaction engineering and chemical reactor design. Simply put, chemical reaction engineering is the superimposition of chemical kinetics on single-phase/multi-phase fluid flow. Similarly, process equipment design involves designing the vessels and facilities that can either carry out or help carry out the chemical reactions at a large scale efficiently and safely.

FURTHER READING

Readers are highly encouraged to study the referred texts for a holistic view of the presented chapter. References 2 and 8 are classic undergraduate-level textbooks of chemical reaction engineering. References 4, 6, and 7 are advanced books and offer broader coverage of reaction engineering details. References 3 and 9 are undergraduate-level textbooks of process equipment design and provide a wholesome course on the detailed design methods of various process plant components.

REFERENCES

1. Speight JG. 2017. Industrial organic chemistry. *Environ Org Chem Eng.*:87–151. https://doi.org/10.1016/B978-0-12-804492-6.00003-4
2. Khichar KK, Dangi SB, Dhayal V, Kumar U, Hashmi SZ, Sadhu V, Choudhary BL, Kumar S, Kaya S, Kuznetsov AE, et al. 2020. Structural, optical, and surface morphological studies of ethyl cellulose/grapheme oxide nanocomposites. *Polym Compos.* 10(7):1–11. https://doi.org/10.1002/pc.25576
3. Dhayal V, Hashmi SZ, Kumar U, Choudhary BL, Kuznetsov AE, Dalela S, Kumar S, Kaya S, Dolia SN, Alvi PA. 2020. Spectroscopic studies, molecular structure optimization and investigation of structural and electrical properties of novel and biodegradable Chitosan-GO polymer nanocomposites. *J Mater Sci* [Internet]. 55(30):14829–14847. https://doi.org/10.1007/S10853-020-05093-5/FIGURES/15 [accessed 2022 Nov 28].
4. Coker AK. 2007. Process planning, scheduling, and flowsheet design. In: *Ludwig's Applied Process Design for Chemical Petrochemical and Plants* [Internet]. 3rd ed. [place unknown]: Elsevier, pp. 1–68. https://doi.org/10.1016/B978-075067766-0/50008-7
5. Levenspiel O, Bischoff KB. 1964. Patterns of flow in chemical process vessels. *Adv Chem Eng.* 4(C):95–198. https://doi.org/10.1016/S0065-2377(08)60240-9
6. Levenspiel O. 1990. *Chemical Reaction Engineering.* 3rd ed. [place unknown]: John Wiley & Sons, Inc.
7. Thakore S, Bhatt B. 2015. *Introduction to Process Engineering and Design.* 2nd ed. [place unknown]: McGraw Hill.
8. Carberry JJ. 2020. *Chemical Reaction and Reactor Engineering.* 1st ed. Carberry JJ, Varma A, editors. New York: CRC Press. https://doi.org/10.1201/9781003065562
9. Manos G. 2008. Introduction to chemical reaction engineering. In: *Concepts of Chemical Engineering 4 Chemists* [Internet]. Vol. 45. Cambridge: Royal Society of Chemistry, pp. 21–54. https://doi.org/10.1039/9781847557674-00021
10. Froment G, DeWilde J, Bischoff K. 2011. *Chemical Reactor Analysis and Design.* 3rd ed. [place unknown]: John Wiley & Sons, Inc.
11. Salmi T, Mikkola J-P, Warna J. 2018. *Chemical Reaction Engineering and Reactor Technology.* 2nd ed. [place unknown]: CRC Press.
12. Fogler HS. 2019. *Elements of Chemical Reaction Engineering.* 5th ed. [place unknown]: Prentice Hall.
13. Ray S. 2020. *Process Equipment and Plant Design.* [place unknown]. https://doi.org/10.1016/c2017-0-02434-5
14. Your Digestive System & How It Works | NIDDK. 2022. *National Institutes of Health* [Internet]. https://www.niddk.nih.gov/health-information/digestive-diseases/digestive-system-how-it-works [accessed 2022 Oct 9].

12 Future Aspects of Dehydrogenative Reactions

Rama Gaur and Syed Shahabuddin

12.1 INTRODUCTION

Dehydrogenative reactions have been proven to be of great importance in a number of important applications, including the following.

1. Catalysis: Dehydrogenative reactions are commonly used in catalysis to remove hydrogen atoms from organic molecules, leading to the formation of new carbon-carbon bonds or carbon-heteroatom bonds. This approach has found applications in the production of pharmaceuticals, agrochemicals, and fine chemicals.
2. Energy storage: Dehydrogenative reactions can be used to store hydrogen for use in fuel cells, which are important for producing clean energy. This approach involves the use of materials such as metal hydrides, which can release hydrogen upon dehydrogenation.
3. Polymer synthesis: Dehydrogenative reactions are used in the synthesis of polymers, where hydrogen atoms are removed from monomers to form new bonds between carbon atoms. This approach is important for producing high-performance materials, such as conductive polymers, optical materials, and liquid crystals.
4. Materials science: Dehydrogenative reactions are important in the development of new materials, such as porous materials and metal-organic frameworks. These materials can be used for gas storage, separation, and catalysis.

Overall, dehydrogenative reactions play a vital role in a variety of important applications and are essential for the development of new materials, catalysts, and energy storage systems. Tremendous growth has been observed in the applications of dehydrogenative reactions toward a better and sustainable future. The development or progress of any process demands safety, sustainability, and improved efficiency. Environmental safety entails minimizing the use of hazardous chemicals and minimal hazardous manufacturing waste. The use of green chemicals and catalysts and eco-friendly methods not only aids in enhancing the effectiveness of the process but also makes it an environmentally friendly and sustainable approach. Several such perspectives are discussed as follows.

12.2 DEVELOPMENT OF THE EXISTING CATALYTIC TECHNOLOGIES VIA THE DESIGNING OF MORE SELECTIVE, ACTIVE CATALYSTS

A series of important organic reagents such as aldehydes, esters, ether, and acids, which are extensively used in the synthesis of organic compounds, are primarily derived from dehydrogenative reactions. For instance, organic esters are synthesized by the dehydrogenation of primary alcohols, which is a widely exploited, environmentally friendly method but suffers from the limitation of poor yield. The development of catalytic technologies by introducing catalysts to the reaction process will add value to the synthetic methodology. Onoda and Fujita reported the use of iridium-based catalyst with catalytic function switching to increase the yield up to 92% and 89% for both dehydrogenative esterification and dehydrative etherification, respectively (Onoda & Fujita, 2022).

12.3 DEVELOPING SUSTAINABLE, ECO-FRIENDLY, AND GREEN DEHYDROGENATION TECHNOLOGIES

Dehydrogenation reactions pose a challenge and are not thermodynamically favorable. Traditional methods for dehydrogenating different substrates involve the use of high pressure and temperature, considerable amounts of oxidizing agents, and sacrificial hydrogen acceptors. However, these methods have limitations such as harsh reaction environments, poor selectivity, and the formation of organic and inorganic waste material in significant amounts. To address these limitations, sustainable acceptorless approaches for dehydrogenation reactions have been introduced by utilizing light and photocatalysts. These photocatalytic acceptorless dehydrogenation (PAD) reactions have been successfully applied to various organic molecules, including amines, alcohols, acids, and passive substrates. The realization of PADs for activating molecules with low energy opens possibilities for further growth and development and promotes the use of sustainable, economical, and environmentally friendly methods (Verma, 2022).

PADs are viewed as a sustainable approach to meet future demands as hydrogen is considered one of the green by-products obtained through these reactions. According to the reports available about PADs, their use is presently limited to only a few types of substrates, indicating that there is considerable potential for future expansion and development. Steps toward the scale-up of PADs are highly desirable for sustainable growth and future advancements (Verma, 2022).

12.4 SELECTIVE HYDROGEN OXIDATION: DESIGN AND DEVELOPMENT OF A NOVEL CATALYST AND FACILE PROCESS

Catalysts play an important role in selective hydrogen oxidation by improving the selectivity and yield. Several catalyst materials based on transition metal compounds (TMCs) have been reported to be widely used because of their multiple oxidation states, stability, high activity, and ease of processing.

Despite incredible progress in the field, some challenges still need to be addressed. One of the important constraints is the development of a competent enantioselective heterogeneous catalyst for the synthesis of methods for the highly in-demand enantiomeric heterocyclic compound relevant to the pharmaceutical industry. Compared to other contemporaneous routes, the synthesis of enantiomeric heterocyclic compounds via the application of heterogeneous transition metal catalysts has been less explored to date. Normally, acceptorless dehydrogenation coupling (ADC) reactions are performed at high temperature and involve stoichiometric reagents. However, asymmetric reactions desire milder reaction conditions to obtain products possessing high chirality. Thus, a cautious chiral catalyst fabrication technique is highly desirable for the synthesis of highly enantioselective and pure heterocycle compounds. Furthermore, the synthesis of these heterocyclic compounds has been extensively studied using noble metal-based homogeneous and heterogeneous catalysts. However, investigations into earth-abundant heterogeneous catalysts belonging to the 3d-transition metal series that can give competent results to that of noble metal catalysts are highly desirable.

The potential of TMCs has still not been explored regarding their worth in dehydrogenative reactions such as alcohol oxidation to carbonyl compounds. According to the available reports, commonly used catalysts for organic transformations involve the use of noble metals and their complexes. These catalysts suffer from limitations such as they are expensive, have low abundance, have poor recyclability, etc. (Toyooka & Fujita, 2020; Awasthi & Singh, 2019; Cherepakhin & Williams, 2018; Sarbajna et al., 2017; Mittal & Awasthi, 2019). To overcome these limitations, efforts have been made to heterogenize the catalyst by incorporating transition metals such as Mn, Co, Ni, etc. to improve the conversion efficiency, regenerability, and reutility of the catalyst (Mittal et al., 2020; Chen et al., 2021; Mittal et al., 2020; Pradhan et al., 2020; Dai et al., 2017; Shao et al., 2018). Mixed TMCs with a spinel structure are recognized as having a varied chemical structure, a high surface area, and higher active sites in contrast to simple metal oxides. Hence, they offer better conversion for the reactions (Gawande et al., 2012; Yuan et al., 2014). For example, zinc catalysis is generally restricted to Lewis acid-mediated reactions because of its reduced redox performance, where only two oxidation states (0 and +2) are possible. Nevertheless, in recent times, oxidation reactions using hydroperoxides and catalyzed by zinc (II) have been increasingly explored. In contrast, cobalt has received considerable attention in catalysis over the years due to its earth-abundance, affordability, non-toxicity, and excellent redox behaviour (Mittal & Awasthi, 2021; Chen et al., 2021). Dehydrogenative oxidation reactions have not been investigated for their potential use with spinel-structured mixed TMCs (Mittal, R., & Awasthi, S. K. 2022).

Mittal and Awasthi showed the development of a bimetallic oxide catalyst for the dehydrogenative oxidation reactions of alcohols. The practical application of this method was demonstrated by using a ZCO-1 catalyst to carry out the gram-scale synthesis of valuable chemicals such as terephthalic acid, nicotinic acid, and succinic acid.

Likewise, García-Muñoz et al. documented the photocatalytic potential of mesoporous TiO_2 catalysts grafted with iron for generating hydrogen from water-ethanol mixtures in the gas phase (Garcia-Munoz et al., 2022).

In general, dehydrogenation reactions possess a high activation barrier and thus require an elevated temperature to operate. This challenge can be overcome by the use of metals with suitable ligand systems so that the process can be achieved under mild conditions. In addition, the use of chiral ancillary ligands with achiral/chiral metal complexes promotes asymmetric induction and results in enantioselective synthesis. Single-pot synthesis of N-heterocyclic compounds can easily be accomplished by activating alcohols or amines through transition metal catalysts. Hence, the development of suitable and efficient transition metal catalysts for the dehydrogenation of organic substrates has important implications in core-organic synthesis.

12.5 DEVELOPMENT OF MEMBRANE SEPARATION TECHNIQUES FOR REMOVING HYDROGEN FROM THE DEHYDROGENATION PRODUCT

Membrane technologies are extensively employed in several modern technological processes. In the chemical industry, membranes are utilized to address numerous issues ranging from improving the efficiency of specific processes to ensuring compliance with stringent environmental regulations. A unique synergistic effect is anticipated in the field of membrane catalysis when it becomes feasible to not only hasten the primary reaction through the use of a catalyst that influences the process kinetics but also alter the reaction equilibrium towards the products by separating one or more products from the primary reaction volume via membrane separation (Shelepova, & Vedyagin, 2022).

Pd alloy membranes exhibit great potential for H_2 separation in reactions involving hydrogen substrates. Gallucci et al. demonstrated the benefits of using Pd-based membranes in various devices (Gallucci et al., 2007). The statistics of hydrogen separation markets show that Pd membranes are the most comprehensively researched and used. However, the poisoning of the membrane due to other metals used is a matter of concern as it affects the separation efficiency. Rahimpour et al. reported the use of Pd-based membranes in membrane reactors for the production of pure hydrogen (Rahimpour et al., 2017). It was found that the use of a Pd-based membrane in the reactor increases the yield of the product and improves the selectivity, and the limitations associated with the thermal equilibrium are removed. Generally, the use of palladium in alloys with other metals, particularly silver, to enhance hydrogen permeability is highly recommended (Erdali et al., 2022). Dogan and coworkers accomplished the effective dehydrogenation of isobutane on a $CrOx/Al_2O_3$ catalyst in a membrane reactor (Erdali et al., 2022).

12.6 GREEN, SUSTAINABLE, SAFE, AND ENVIRONMENTALLY FRIENDLY MANUFACTURING TECHNIQUES

Catalytic dehydrogenative oxidation reactions are a promising approach to achieve carboxylic acid derivatives from alcohols in a green and sustainable way, as they do not require oxidants and generate hydrogen gas as the only by-product. This makes the process atom economic and environmentally friendly. By utilizing sustainable catalytic processes, it is possible to mitigate environmental concerns, yield value-added products, and capitalize on the value of preparatory reagents while reducing energy consumption and waste generation (Mittal & Awasthi, 2022). Additionally, this approach provides an opportunity to explore alcohols as potential hydrogen storage compounds in the future. Monochromates positively affect the isobutane dehydrogenation performance (Erdali et al., 2022).

A vast array of chemical products are currently available, and they play a significant role in improving our lives. However, the manufacturing processes used to create these products often result in the creation of large amounts of waste and a high consumption of energy. For instance, producing chemicals in bulk such as aniline and phenol requires multi-step, energy-intensive processes such as the benzene nitration/reduction and cumene process (Rappoport, 2004; Tyman, 1996; Fiege et al., 2000; Lawrence & Marshall, 1998). Moreover, the production of fine chemicals including electronic and pharmaceutical materials involves intricate and accurate preparation, which often requires the pre-functionalization of substrate by reactions such as boronation, halogenation, etc. and the use of stoichiometric reagents. However, the current manufacturing processes for these compounds frequently employ toxic procedures using stoichiometric organic- or inorganic-based oxidizing agents via liquid-phase oxidations. Accordingly, there is a pressing need to reduce or eliminate the involvement of oxidation processes, including dehydrogenation, oxygenation, and oxidative functionalization, which has resulted in significant research efforts to develop efficient catalytic oxidation methods (Anastas & Warner, 1998; Sheldon & Kochi, 1981; Schlögl & Mizuno, 2009). Efficient catalytic oxidation methods that use O_2 as an oxidizing agent or acceptorless dehydrogenation without any oxidants have the potential to be much more environmentally friendly than traditional processes (Yamaguchi et al., 2022).

12.7 DEVELOPMENT OF NEW CATALYSTS REPLACING NOXIOUS METALS AND METAL OXIDES

A new method has been developed for combining xanthene with β-keto molecules using MoS_2 Quantum Dots (QDs) as a photoredox catalyst under visible light. This approach has several advantages, including the ability to use non-functionalized starting materials and a wide range of substrates, mild reaction conditions, the use of water as a solvent, and the catalyst's ability to be recycled up to six times without a loss of yield or selectivity. A study also investigated the kinetic solvent isotope effect (KSIE) and endorsed the reductive quenching mechanism of QD photocatalysts through a cyclic-voltammetric study (Deore & De, 2022).

A promising method for producing carboxylic acids is the use of water and hydroxides for the catalytic ADC of alcohols. This protocol is gaining attention due to its abilities of producing only H_2 gas as a by-product, avoiding the use of external additives and oxidizing agents, and possessing eco-friendliness and high atom economy. Milstein et al. introduced a PNN-Ru complex that efficiently converted benzylic or aliphatic alcohols into carboxylic acids in alkaline solutions. This pioneering research has since inspired several scientists to develop various catalytic systems for this process (Wu et al., 2022). Because of their exceptional σ-electron donating properties, N-heterocyclic carbenes have been identified as the exclusive group of auxiliary ligands in a wide variety of metal complexes. Wu et al. reported the use of monodentate NHC-Ru complex with a miniscule loading of Ru (62.5 ppm) for widespread investigation, and the gram-scale production of numerous carboxylic acids was attained in open air (Wu et al., 2022).

12.8 DEVELOPMENT OF REFORMED HETEROGENEOUS CATALYSTS VIA SURFACE MODIFICATIONS

Typically, the dehydrogenative coupling reaction of silanes with alcohols employs precious metal complexes as catalysts and strong organic bases However, these catalysts are difficult to separate from the catalytic system and can lead to the self-polymerization of silanes in the presence of strong organic bases. To address these issues, researchers have developed a new catalyst called Pd/PDVB-Vim, which utilizes the coordination of imidazolium nitrogen with palladium nanoparticles (Pd NPs). This alkaline imidazole-functionalized porous organic polymer is designed to overcome the limitations of traditional catalysts. The Pd/PDVB-Vim catalyst has been characterized and applied in the dehydrogenative coupling reaction of silanes with alcohols under mild conditions (30°C). The results of the characterization show that the Pd NPs are uniformly dispersed on PDVB-Vim with an ultrasmall particle size (approximately 2.00 nm). Furthermore, the strong interaction between the nitrogen on PDVB-Vim and the Pd NPs is evident in Pd/PDVB-Vim, which leads to the catalyst's superior stability and reusability (Liu et al., 2022).

Using water as a reaction medium is difficult because organic molecules are generally insoluble in water. However, self-assembled structures such as micelles and vesicles can serve as nanoreactors to facilitate organic reactions in water. In pursuit of performing a chemical reaction in water, researchers have explored the potential of using these self-assembled structures as nanoreactors (Kulshrestha et al., 2022). Kulshrestha et al. developed environmentally friendly surfactants by synthesizing copper-based ionic liquids using amino acids such as phenylalanine or valine. They then used these surfactants to construct metallovesicles, which were employed as nano-catalytic reactors for the formation of carbon-carbon bonds in propargyl amines under oxidative conditions (Kulshrestha et al., 2022).

The researchers observed that the chirality of the amino acid is retained in $(AACl_2)_2(CuCl_4)$, which suggests that this approach may be a promising method for the one-pot synthesis of chiral compounds in a more environmentally friendly manner for future pharmaceutical manufacturing and other Cu(II) catalyzed reactions (Kulshrestha et al., 2022).

In this study, a series of biologically significant 2-amino-4H-chromenes with various substituents were synthesized using a one-pot multicomponent reaction. The reaction was catalyzed by p-cymene Ru(II) organometallic complexes that included N-O chelated carbazole-based hydrazone ligands (Tamilthendral et al., 2022).

Song and colleagues developed a new catalyst, called Pd/PDVB-Vim, for the improved catalytic dehydrogenative coupling of silanes with alcohols. The catalyst utilizes the strong interaction between imidazolium nitrogen and Pd NPs and is constructed using imidazole-functionalized porous organic polymer to stabilize the Pd NPs (Liu et al., 2022).

12.9 EFFICIENT HETEROGENIZED-HOMOGENEOUS CATALYST DEVELOPMENT FOR MODIFIED DEHYDROGENATION REACTIONS

The search for strategies to form carbon-heteroatom bonds is a constant focus of synthetic organic chemistry, as it creates new possibilities for the synthesis of functional organic molecules. Recently, the functionalization of the C-H bond has arisen as a favorable technique in modern organic synthesis due to its many inherent benefits. However, several reported approaches have significant drawbacks. For example, they often require oxidants, expensive and sensitive metal catalysts and additives, toxic organic reagents, complex purification processes, and harsh reaction environments. Consequently, synthetic chemists are increasingly turning to green approaches for the functionalization of C-H for the preparation of valorized chemicals and reactions. Although C(sp3)-H functionalization has been reported to be emerging among the C-H bond functionalizations (4), it remains a thought-provoking endeavour, mainly due to low selectivity and no activation (Brahmachari et al., 2022).

The ADC of alcohols and water/hydroxides is an emerging and elegant method for producing carboxylic acids. Consequently, there is high demand for the development of effective and useful catalysts or catalytic systems for these striking transformations. In this context, Wu et al. developed a series of cyclometallated N-heterocyclic carbene-Ru (NHC-Ru) complexes via the tuning of ligand (Ru-1), which was the better complex in the preceding work (Wu et al., 2022).

The advantages of a hydrogen economy are clear, although substantial investigation is necessary to achieve the necessary technological developments. Formic acid is considered an eco-friendly material for hydrogen storage due to its non-toxicity and facile storage. Formic acid is widely used for dehydrogenation reactions and results in only gaseous products (H_2/CO). The production of formic acid can release only gaseous products (H_2/CO) through dehydrogenation. CO can also be converted back into formic acid using catalysts under moderate conditions, resulting in a CO-neutral hydrogen storage cycle. This conversion process involves the use of nano-heterogeneous catalysts such as Au and Pd, which are used in aqueous formic acid under optimum temperature (20–50 °C). However, during the progress of the reactions, the chemical intermediates get adsorbed on the surface of nanoparticle surfaces and thus deactivate the Pd monometallic systems. Recent research has resulted in the development of high-performance catalysts, including bi- and tri-metallic Pd and Au combinations that produce high-grade H_2 with insignificant CO. The direct formic acid fuel cell (DFAFC) is an important advance towards model development, scale-up, and commercialization. Nonetheless, further research is necessary, particularly for mobile applications.

Yu and Pickup marked the significant prospects of developing DFAFC in their scale up and commercialization (Yu & Pickup, 2008). The authors demonstrated the use of high-performance nanocatalysts for the production of high-quality H_2 with minimal CO concentration. Furthermore, the CO generated can be converted into formic acid, resulting in a neutral CO hydrogen.

12.10 DEVELOPMENT OF DEHYDROGENATIVE TECHNOLOGIES FOR HYDROGEN ENERGY STORAGE AND UTILIZATION

The performance of the catalytic material for both the dehydrogenation and the hydrogenation steps can be ameliorated further by removing by-products from the equation. The advancement of dehydrogenative technology signifies a crucial achievement in the pursuit of effective storage and use of hydrogen energy. These state-of-the-art advancements provide an environmentally friendly and adaptable answer to the difficulties of storing and using hydrogen as a renewable energy resource. Dehydrogenation procedures entail the liberation of hydrogen from diverse hydrogen-rich substances, such as ammonia, alcohols, or hydrocarbons, often employing catalysts and heat. This innovative method allows for the development of small, densely packed hydrogen storage devices and simplifies the incorporation of hydrogen into many energy applications, including fuel cells for transportation and large-scale energy storage. Dehydrogenative technologies are playing a crucial role in the advancement of green energy by reducing energy losses and improving the hydrogen economy. These technologies provide a cleaner, more accessible, and effective method for storing and using hydrogen.

REFERENCES

Anastas, P. T., & Warner, J. C. (1998). Green chemistry. *Frontiers, 640.*

Awasthi, M. K., & Singh, S. K. (2019). Ruthenium catalyzed dehydrogenation of alcohols and mechanistic study. *Inorganic Chemistry, 58*(21), 14912–14923.

Brahmachari, G., Bhowmick, A., & Karmakar, I. (2022). Catalyst-and additive-free C (sp3)–H functionalization of (Thio) barbituric acids via C-5 dehydrogenative aza-coupling under ambient conditions. *ACS Omega, 7*(34), 30051–30063.

Chen, C., Wang, Z. Q., Gong, Y. Y., Wang, J. C., Yuan, Y., Cheng, H., ... & Verpoort, F. (2021). Cobalt embedded in nitrogen-doped porous carbon as a robust heterogeneous catalyst for the atom-economic alcohol dehydrogenation to carboxylic acids. *Carbon, 174,* 284–294.

Cherepakhin, V., & Williams, T. J. (2018). Iridium catalysts for acceptorless dehydrogenation of alcohols to carboxylic acids: Scope and mechanism. *ACS Catalysis, 8*(5), 3754–3763.

Dai, Z., Luo, Q., Jiang, H., Luo, Q., Li, H., Zhang, J., & Peng, T. (2017). Ni (ii)–N′ NN′ pincer complexes catalyzed dehydrogenation of primary alcohols to carboxylic acids and H 2 accompanied by alcohol etherification. *Catalysis Science & Technology, 7*(12), 2506–2511.

Deore, J. P., & De, M. (2022). Photoredox C (sp3)– C (sp3) cross-dehydrogenative coupling of xanthene with β-keto moiety using MoS2 Quantum Dot (QD) catalyst. *Advanced Synthesis & Catalysis, 364*(17), 3049–3058.

Erdali, A. D., Cetinyokus, S., & Dogan, M. (2022). Investigation of isobutane dehydrogenation on CrOx/Al2O3 catalyst in a membrane reactor. *Chemical Engineering and Processing-Process Intensification, 175*, 108904.

Fiege, H., Voges, H. W., Hamamoto, T., Umemura, S., Iwata, T., Miki, H., . . . & Paulus, W. (2000). *Phenol derivatives. Ullmann's encyclopedia of industrial chemistry*. Wiley Online Library.

Gallucci, F., Chiaravalloti, F., Tosti, S., Drioli, E., & Basile, A. (2007). The effect of mixture gas on hydrogen permeation through a palladium membrane: Experimental study and theoretical approach. *International Journal of Hydrogen Energy, 32*(12), 1837–1845.

Garcia-Munoz, P., Zussblatt, N. P., Chmelka, B. F., & Fresno, F. (2022). Production of hydrogen from gas-phase ethanol dehydrogenation over iron-grafted mesoporous Pt/TiO$_2$ photocatalysts. *Chemical Engineering Journal, 450*, 138450.

Gawande, M. B., Pandey, R. K., & Jayaram, R. V. (2012). Role of mixed metal oxides in catalysis science—versatile applications in organic synthesis. *Catalysis Science & Technology, 2*(6), 1113–1125.

Kulshrestha, A., Kumar, G., & Kumar, A. (2022). Cu (II)-amino acid ionic liquid surfactants: Metallovesicles as nano-catalytic reactors for cross dehydrogenative coupling reaction in water. *ChemistrySelect, 7*(16), e202200159.

Lawrence, F. R., & Marshall, W. J. (1998). *Aniline, Ullmann's encyclopedia of industrial chemistry*. Wiley Online Library.

Liu, S., Shi, S., Ding, S., Xiao, W., Wang, H., Zeng, R., . . . & Song, W. (2022). Imidazole functionalized porous organic polymer stabilizing palladium nanoparticles for the enhanced catalytic dehydrogenative coupling of silanes with alcohols. *ChemistrySelect, 7*(40), e202203056.

Mittal, R., & Awasthi, S. K. (2019). Recent advances in the synthesis of 5-substituted 1H-tetrazoles: A complete survey (2013-2018). *Synthesis, 51*(20), 3765–3783.

Mittal, R., & Awasthi, S. K. (2021). A synergistic magnetically retrievable inorganic-organic hybrid metal oxide catalyst for scalable selective oxidation of alcohols to aldehydes and ketones. *ChemCatChem, 13*(22), 4799–4813.

Mittal, R., & Awasthi, S. K. (2022). Bimetallic oxide catalyst for the dehydrogenative oxidation reaction of alcohols: Practical application in the synthesis of value-added chemicals. *ACS Sustainable Chemistry & Engineering, 10*(4), 1702–1713.

Mittal, R., Mishra, A., & Awasthi, S. K. (2020). A greener approach for the chemoselective Boc protection of amines using sulfonated reduced graphene oxide as a catalyst in metal-and solvent-free conditions. *Synthesis, 52*(4), 591–601.

Onoda, M., & Fujita, K. I. (2022). Dehydrogenative esterification and dehydrative etherification by coupling of primary alcohols based on catalytic function switching of an iridium complex. *ChemistrySelect, 7*(30), e202201135.

Pradhan, D. R., Pattanaik, S., Kishore, J., & Gunanathan, C. (2020). Cobalt-catalyzed acceptorless dehydrogenation of alcohols to carboxylate salts and hydrogen. *Organic Letters, 22*(5), 1852–1857.

Rahimpour, M. R., Samimi, F., Babapoor, A., Tohidian, T., & Mohebi, S. (2017). Palladium membranes applications in reaction systems for hydrogen separation and purification: A review. *Chemical Engineering and Processing: Process Intensification, 121*, 24–49.

Rappoport, Z. (2004). *The chemistry of phenols*. John Wiley & Sons.

Sarbajna, A., Dutta, I., Daw, P., Dinda, S., Rahaman, S. W., Sarkar, A., & Bera, J. K. (2017). Catalytic conversion of alcohols to carboxylic acid salts and hydrogen with alkaline water. *ACS Catalysis, 7*(4), 2786–2790.

Schlögl, R., & Mizuno, N. (2009). *Modern heterogenous oxidation catalysis*. Wiley Online Library.

Shao, Z., Wang, Y., Liu, Y., Wang, Q., Fu, X., & Liu, Q. (2018). A general and efficient Mn-catalyzed acceptorless dehydrogenative coupling of alcohols with hydroxides into carboxylates. *Organic Chemistry Frontiers, 5*(8), 1248–1256.

Sheldon, R. A., & Kochi, J. K. (1981). Oxidation with molecular oxygen. In *Metal-catalyzed oxidations of organic compounds*. Academic Press.

Shelepova, E. V., & Vedyagin, A. A. (2022). Comparative analysis of the dehydrogenation of hydrocarbons and alcohols in a membrane reactor. *Kinetics and Catalysis, 63*(1), 43–51.

Tamilthendral, V., Ramesh, R., & Malecki, J. G. (2022). New ruthenium (ii) catalysts enable the synthesis of 2-amino-4 H-chromenes using primary alcohols via acceptorless dehydrogenative coupling. *New Journal of Chemistry, 46*(45), 21568–21578.

Toyooka, G., & Fujita, K. I. (2020). Synthesis of dicarboxylic acids from aqueous solutions of diols with hydrogen evolution catalyzed by an iridium complex. *ChemSusChem, 13*(15), 3820–3824.

Tyman, J. H. P. (1996). *Synthetic and natural phenols*. Elsevier.

Verma, P. K. (2022). Advancement in photocatalytic acceptorless dehydrogenation reactions: Opportunity and challenges for sustainable catalysis. *Coordination Chemistry Reviews, 472*, 214805.

Wu, Z., Wang, Z. Q., Cheng, H., Zheng, Z. H., Yuan, Y., Chen, C., & Verpoort, F. (2022). Gram-scale synthesis of carboxylic acids via catalytic acceptorless dehydrogenative coupling of alcohols and hydroxides at an ultralow Ru loading. *Applied Catalysis A: General, 630*, 118443.

Yamaguchi, K., Jin, X., Yatabe, T., & Suzuki, K. (2022). Development of environmentally friendly dehydrogenative oxidation reactions using multifunctional heterogeneous catalysts. *Bulletin of the Chemical Society of Japan, 95*(9), 1332–1352.

Yu, X., & Pickup, P. G. (2008). Recent advances in direct formic acid fuel cells (DFAFC). *Journal of Power Sources, 182*(1), 124–132.

Yuan, C., Wu, H. B., Xie, Y., & Lou, X. W. (2014). Mixed transition-metal oxides: Design, synthesis, and energy-related applications. *Angewandte Chemie International Edition, 53*(6), 1488–1504.

13 Utilizing Ruthenium (Ru) Complexes in Dehydration Reactions of Saturated and Unsaturated Compounds

Khushbu G. Patel and Saami Ahmed

13.1 INTRODUCTION

Ru is a platinum group metal that along with Rh Pd, Os, and Ir, has an atomic number of 44. It is incredibly uncommon in the crust of the earth, occurring in parts per billion in ores that also include several other platinum group elements. Ruthenium is a strong, lustrous metal that is silvery-white and has a glossy surface. There are 7 stable isotopes in it. Ru and its complexes are mostly used to measure the levels of ferritin, calcitonin, cyclosporine, and folate in the human body for illness diagnosis. Immunosuppressant, antibacterial, and anticancer activities are the main focus of treatment. It is also used as a catalyst in various organic and inorganic reactions. Ru's coordination and organometallic chemistry has recently grown significantly.

Although abundant literature has been reported about iridium-based catalysts for alkane dehydrogenation, very few works have been published on ruthenium catalysts for the same. Indeed, catalytic complexes of ruthenium have already proven to be a great precursor and have significant applications in various organic transformations such as olefin metathesis and asymmetric hydrogenation. Ruthenium metal is approximately 10 times less expensive than iridium metal, and moreover, the inertness of ruthenium towards various polar functional groups and impurities makes it more a promising dehydrogenation catalyst. The advancement towards exploring the potential application of ruthenium-based dehydrogenation catalysts is a tremendous achievement. Ruthenium is photolytically active in $Pt/TiO_2/RuO_2$ and in complexes of the $[tris(2,2-bipyridine)ruthenium(II)]^{2+}$ kind. Due to the systems' photoactivity, they can produce hydrogen and oxygen from water [1, 2].

In this chapter, contemporary investigations on experimental and theoretical aspects were conducted to highlight the remarkable characteristics of ruthenium catalysts to dehydrogenate alcohols, acid, and alkenes in the aqueous phase.

13.2 ALCOHOL DEHYDROGENATION REACTIONS BASED ON RU

An investigation was carried out to study alcohol as a hydrogen-storing organic compound where an atom economic reaction of alcohol dehydrogenation was taken into account, which produces carboxylic acid and hydrogen gas and can be used as a fuel [3]. Reported in 2016, the catalytic dehydrogenation of alcohol to produce oxidatively synthesized carboxylic acid by employing Ir-, Rh-, and Ru-based metal complexes was investigated in contrast to ruthenium pincer complex-mediated acceptorless dehydrogenated alcohol [4]. Figure (Scheme) 13.1 shows alcohol dehydrogenation at high temperatures (110 °C) and a base concentration in both water and organic solvent.

Mahendraetal in 2019 formulated pyridylamine-ligated arene-Ru(II) complexes to catalyze a primary alcohol dehydrogenation reaction to form carboxylic acid. Several spectro-analytical

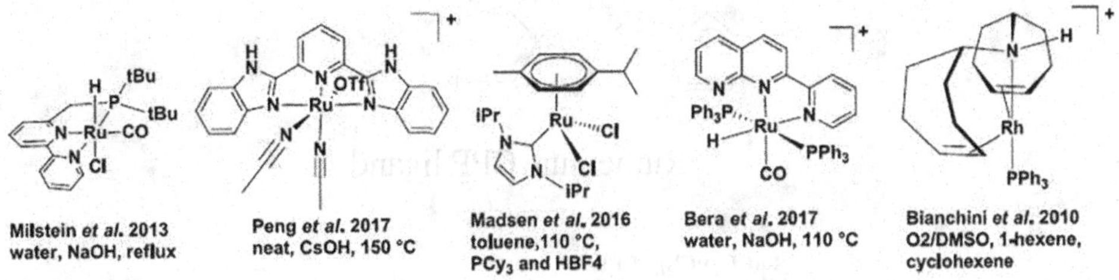

Utilizing Ruthenium (Ru) Complexes in Dehydration Reactions

FIGURE (SCHEME) 13.1 Alcohol dehydrogenation mediated by ruthenium complex to form carboxylic acid.

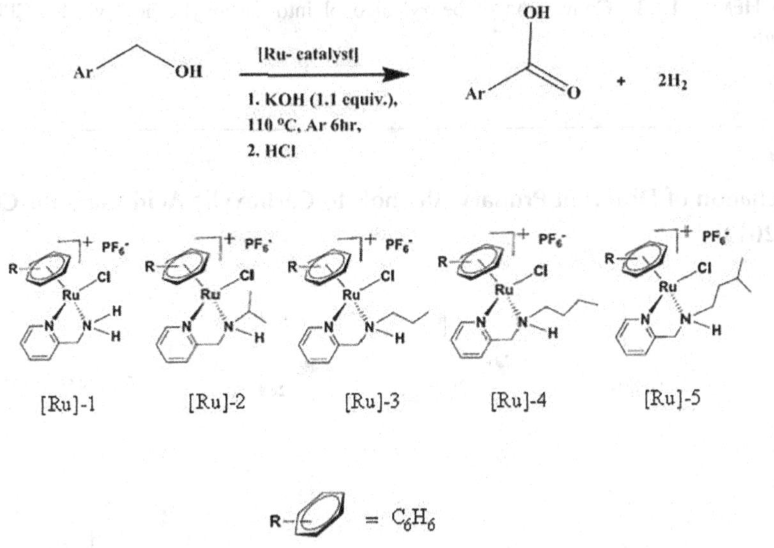

FIGURE (SCHEME) 13.2 Stages of Ru-coordinated intermediate in benzyl alcohol dehydrogenation.

techniques are used to describe all the produced Ru complexes, and single crystal X-ray crystallography is used to identify the arrangement of atoms in the structures of complexes [Ru]-1, [Ru]-2, and [Ru]-5. The conversion of primary alcohol to carboxylic acid or its salt along with the release of hydrogen gas in the presence of toluene as a solvent has been studied. To produce the desired carboxylic acid (approximately 86%) along with hydrogen gas production from diverse forms of alcohol such as aliphatic, aromatic, hetero aromatic alcohol, etc., a suitable method must be employed. Under catalytic and regulated reaction conditions, the intermediate coordinated with the Ru catalyst such as diol Ru species has been investigated to determine the clear route of the mechanism followed by the arene Ru catalyst [5].

According to Liu et al. 2019, a hybrid N-heterocyclic carbene (NHC)-phosphine-phosphine ligand (CPP) that is produced in situ from [Ru(COD)Cl$_2$] can be used to selectively catalyze the dehydrogenation reaction of turning primary alcohol into carboxylic acids.

By employing nuclear magnetic resonance spectroscopy (NMR) over a reaction mixture, the facial geometry of ruthenium complex was discovered. The fac-ruthenium catalyst exhibits a great turnover number of 20,000 with strong catalytic activity, high stability, and a wider substrate range at a very low catalyst concentration, that is, 0.002 mol%. Both the NHC's anchoring function and the hemi-lability of the phosphine were considered to be responsible for the extraordinarily high catalytic stability. The catalytic complex formed via facial coordination of the CPP ligand with

Ruthenium CPP ligand

FIGURE (SCHEME) 13.3 Conversion of benzyl alcohol into carboxylic acid via Ru-CPP catalytic dehydrogenation.

TABLE 13.1
Dehydrogenation of Different Primary Alcohols to Carboxylic Acid using Ru-CPP Catalyst[a] (Liu et al. 2019)

2a, 79%; 2b 95%; 2c 90%; 2d 55%; 2e 94%; 2f 92%; 2g 68%; 2h 90%; 2i 76%; 2j 25%

[a] Reaction conditionsw: [Ru(COD)Cl$_2$]n (1 mol%), ligand CPP (1 mol%), alcohol (1 mmol), and KOH (2.0 equiv) in toluene (4 mL) at 120 °C for 24 h under a nitrogen atmosphere; after the reaction, corresponding acid salts were converted into carboxylic acids by treatment with hydrochloric acid (4 mmol); the isolated yield is 99%.

Ru metal exhibits favorable conditions for the dehydrogenation of bulky adamantanyl-methanol, cholesterol, and sterically hindered alcohol, such as ortho-substituted benzylic alcohols, etc. [6].

In 2017, Wang et al. coordinated an unsymmetrical pyrazolyl-pyridylamino-pyridine ligand with Ru (II) metal. Furthermore, by employing X-ray crystallography, NMR, and an elemental analysis, its dimeric pincer-type ruthenium(II)-NNN complexes were determined. This dimeric Ru(II)

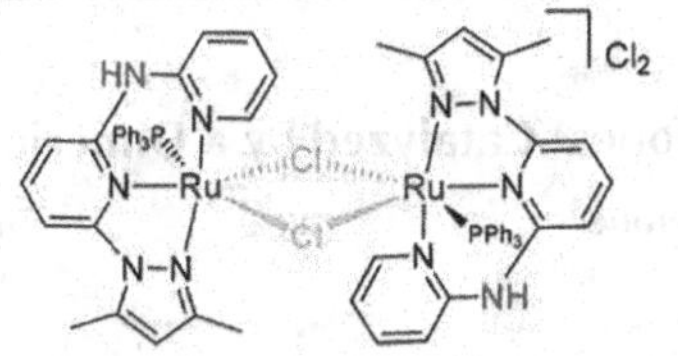

Dimeric Ru(II)-NNN complex catalyst bearing a pyrazolylpyridylamino-pyridine ligand Ru(II) catalyst

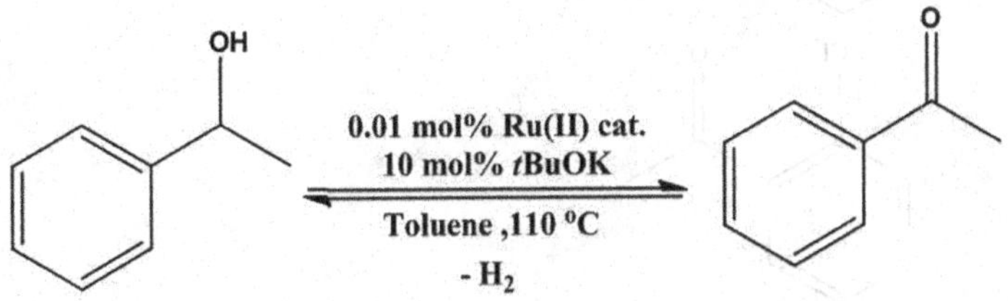

FIGURE (SCHEME) 13.4 Ruthenium-based catalytic dehydrogenation of alcohol.

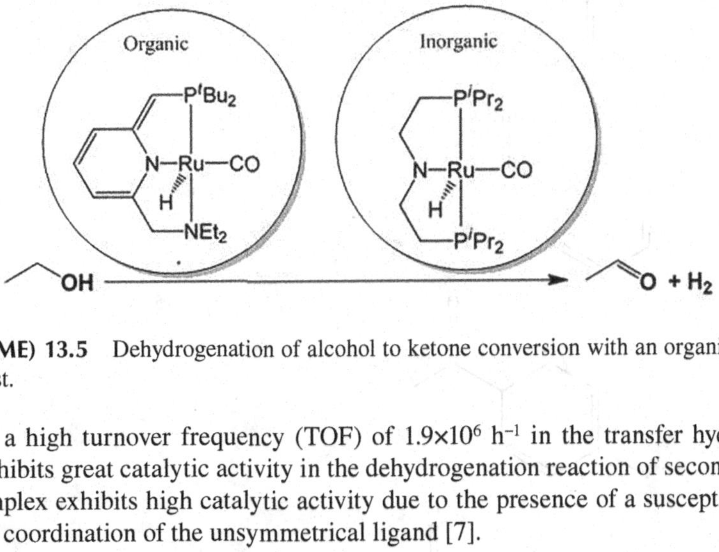

FIGURE (SCHEME) 13.5 Dehydrogenation of alcohol to ketone conversion with an organic and inorganic Ru-based catalyst.

catalyst shows a high turnover frequency (TOF) of 1.9×10^6 h^{-1} in the transfer hydrogenation of ketones and exhibits great catalytic activity in the dehydrogenation reaction of secondary alcohol.

The Ru complex exhibits high catalytic activity due to the presence of a susceptible NH group and hemilabile coordination of the unsymmetrical ligand [7].

13.2.1 ALIPHATIC VERSUS AROMATIC LIGANDS

Xinzheng Yang in 2013 discovered the catalytic dehydrogenation of ethanol via a pincer ruthenium and iron complex. By employing density functional theory, a self-promoted mechanism for an aliphatic PNP pincer ruthenium complex, a (PNP)Ru(H)CO 1Ru, PNP = N,N'-bis(diphenylphosphine)-2,6-diaminopyridine-based ethanol dehydrogenation, transferred the proton from the nitrogen of ligand to the In-Ru complexes by bridging ethanol molecule. The considerable contrast in the mechanism of Ru complex with the aromatic and aliphatic pincer ligands has also been examined [8].

To assess the effectiveness of the iron counterpart of 1Ru, (PNP)Fe(H)CO(1Fe) in the catalytic dehydrogenation reaction of ethanol, a computational analysis was explored. The activation energy required for the dehydrogenation of ethanol in the presence of an Fe-based catalyst is 0.7 kcal/mol, which is less than the amount of energy consumed in the presence of an Ru-based catalyst. Therefore, an economical Fe-based catalyst exhibits high catalytic activity in hydrogen gas

TABLE 13.2
Transfer Hydrogenation of Ketones[a] Catalyzed by a Dimeric Ru(II)-NNN Complex Catalyst

Entry	Ketone	Time (min.)	Yield [b] (%)
1.	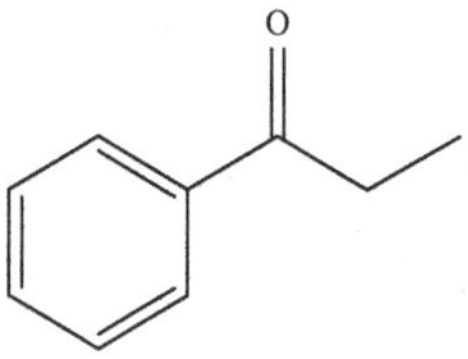 propiophenone	20	95–97
2.	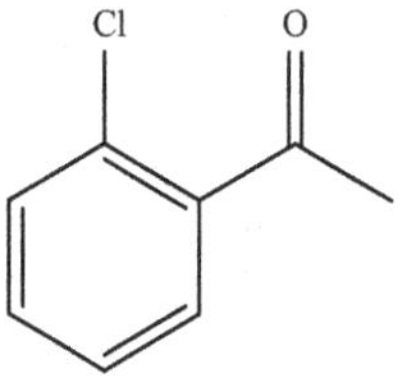 2'-chloroacetophenone	2	97–99
3.	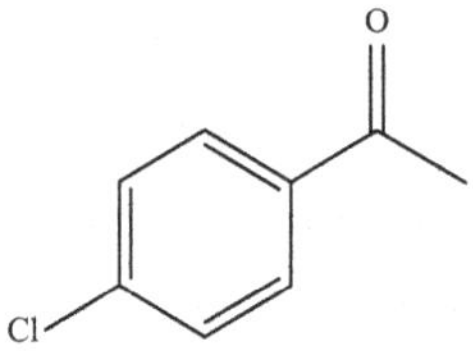 4'-chloroacetophenone	2	97–99
4.	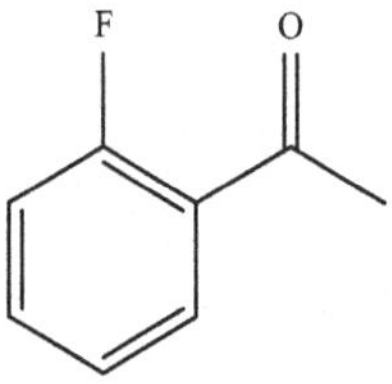 2'-fluoroacetophenone	2	97–99
5.	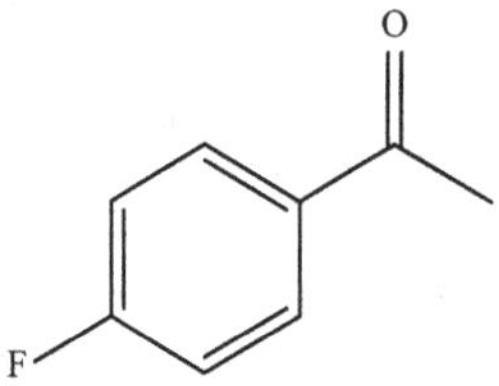 4'-fluoroacetophenone	5	97–99
6.	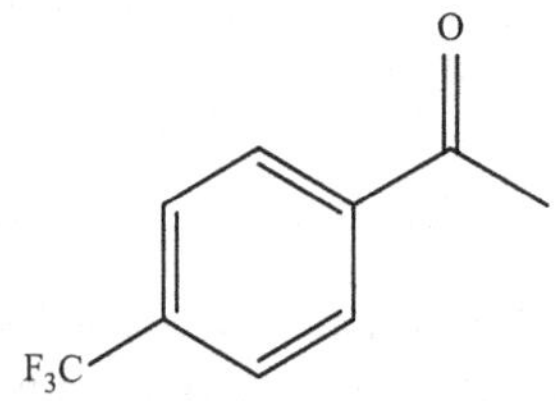 4'-(trifluoromethyl)acetophenone	2	97–99
7.	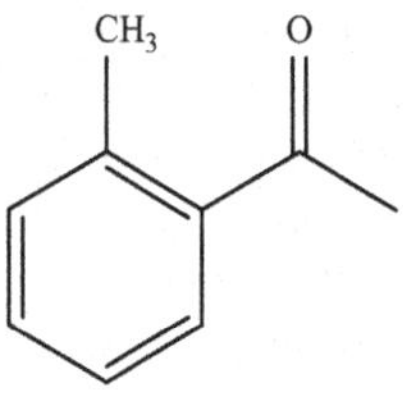 2'-methylacetophenone	3	97–99

Conditions: ketone, 2.0 mmol (0.1 M in 20 mL iPrOH); Ru catalyst, 0.01 mol%; ketone/iPrOK/Ru catalyst = 10000:20:1; 0.1 MPa N2, 82 °C. b Determined by GC analysis.

generated from ethanol. A potent oxidant-free oxidation was reported by Kim et al., 2006, where the dehydrogenation of an ample range of acceptor-free alcohol in the presence of a recyclable ruthenium-based catalyst was carried out. From readily available reagents via a nano synthesis pathway, the catalyst for dehydrogenation was synthesized [9].

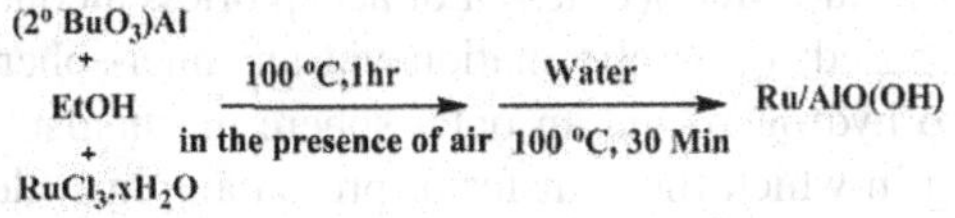

FIGURE (SCHEME) 13.6 Preparation of the ruthenium catalyst.

TABLE 13.3
Catalytic Dehydrogenation of Various Alcohols with RuAlO(OH) (Kim et al., 2006)

Entry	Substrate	Product	Mol % (Ru)	Time (hr)	Temperature (°C)	Yield[a] (%)
1	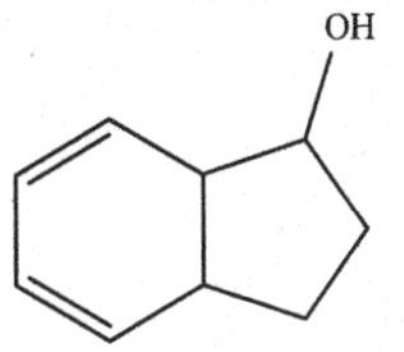 indanol	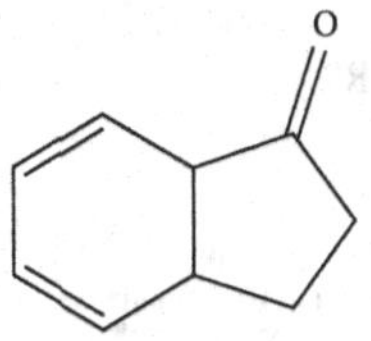 indanone	3.0	20	80	> 99
2	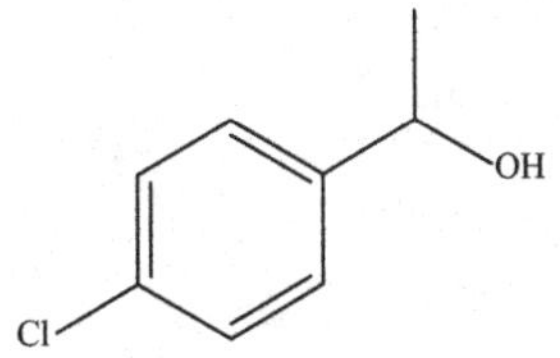 1-(4-chlorophenyl)ethanol	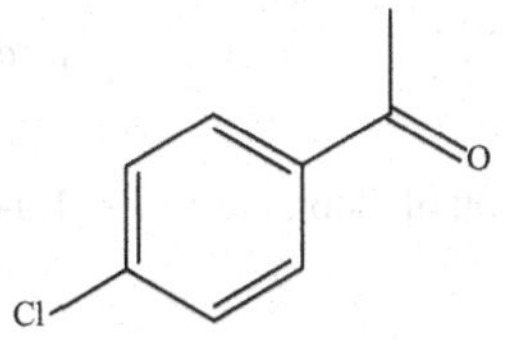 4'-chloroacetophenone	4.5	36	110	> 99
3	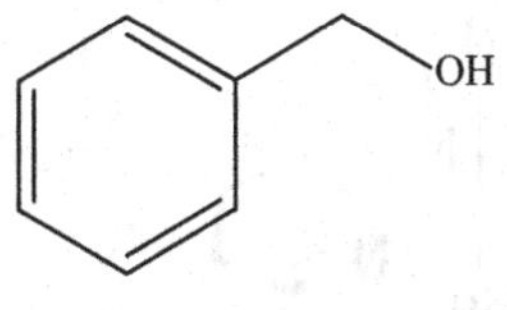 benzyl alcohol	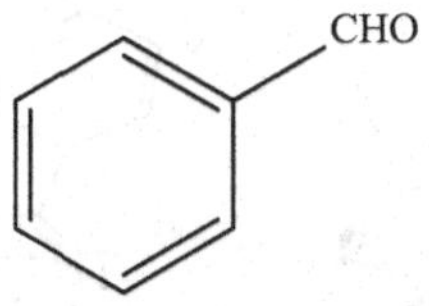 benzaldehyde (CHO)	3.0	5	80	> 99
4	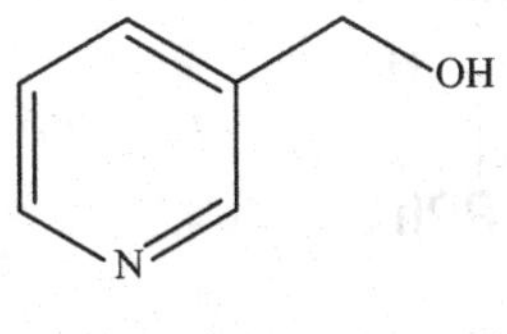 3-pyridylmethanol	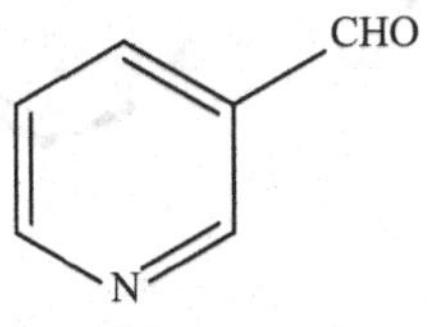 3-pyridinecarboxaldehyde (CHO)	6.0	36	110	80
5	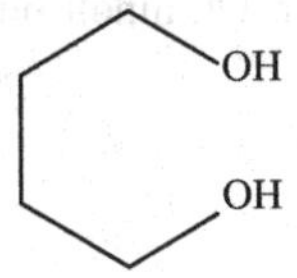 1,4-butanediol	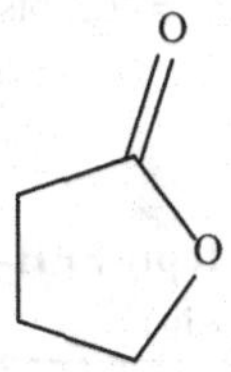 γ-butyrolactone	6.0	32	110	94
6	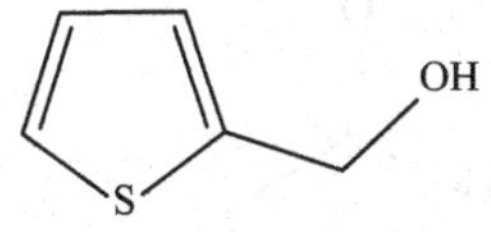 2-thiophenemethanol	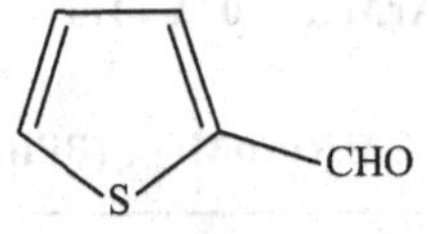 2-thiophenecarboxaldehyde (CHO)	3.0	5	80	97

a Isolated yield

Tseng et al., 2015, reported that reversible hydrogenation-dehydrogenation reactions between ketones and alcohols are skillfully accomplished by using a single pre-catalyst HRu(bMepi)(PPh$_3$)$_2$ (bMepi = 1,3-bis(6′-methyl-2′-pyridylimino)isoindolate). A series of kinetic and isotopic labelling studies, intermediate isolation, and assessment of Ru(b$_4$Rpi)(PPh$_3$)$_2$Cl (R = H, Me, Cl, OMe, OH) complexes were used to investigate the mechanism behind the acceptorless catalytic dehydrogenation of alcohol mediated by HRu(bMepi)(PPh$_3$)$_2$. For acceptorless alcohol dehydrogenation (AAD), two mechanisms were proposed: (i) H-elimination via an inner-sphere mechanism and (ii) the bifunctional transfer of two hydrogens via an outer-sphere mechanism. The process was accomplished by isotopic labeling in which the transfer of proton and hydride takes place in a series of distinct stages. The catalytic modification experiment indicates the significant role of the imine group present on the bMepi pincer scaffold in alcohol dehydrogenation.

In addition, this H-elimination via an inner-sphere mechanism involving HRu(bMepi)(PPh$_3$)$_2$ as a catalyst has been affirmed by kinetic catalyst modification experiments. After the dissociation of a ligand PPh$_3$, H-elimination takes place, which is a turnover-limiting step [10].

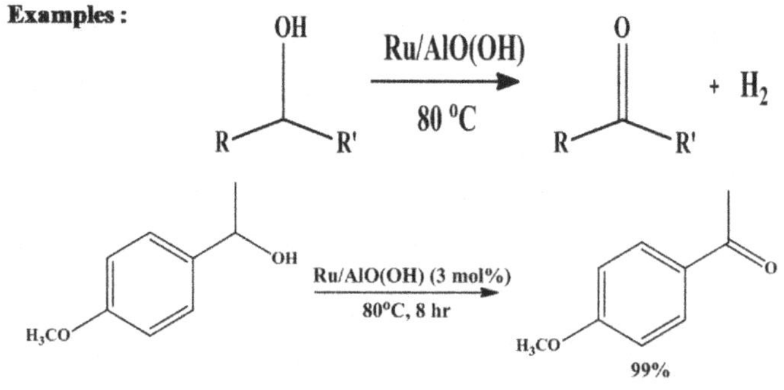

FIGURE (SCHEME) 13.7 Dehydrogenation of alcohol using an Ru-based catalyst.

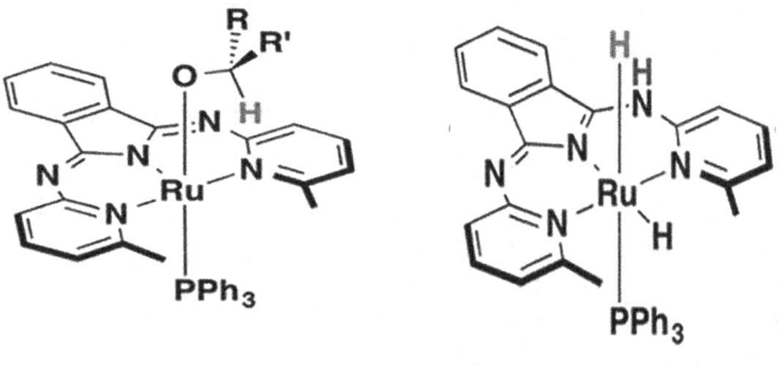

FIGURE (SCHEME) 13.8 HRu(bMepi)(PPh$_3$)$_2$ (bMepi = 1,3-bis(6-methyl- 2-pyridylimino)isoindolate) catalyst.

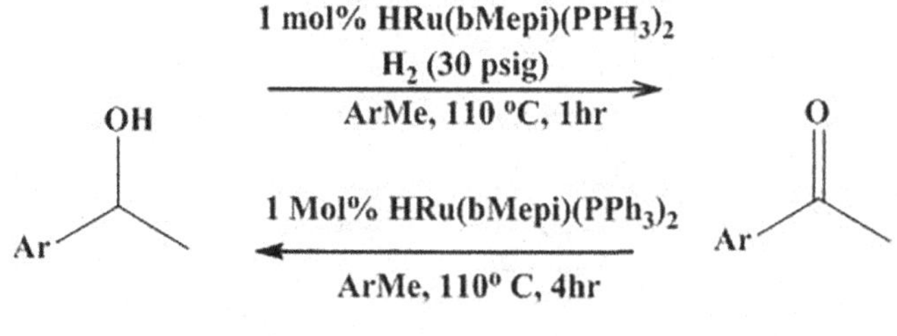

FIGURE (SCHEME) 13.9 Reversible dehydrogenation-hydrogenation reactions catalyzed by Ru complex.

Utilizing Ruthenium (Ru) Complexes in Dehydration Reactions

R= H, CH$_3$
Catalyst Ru(H)(Cl)(CO)(HN{CH$_2$CH$_2$P(iPr)$_2$}$_2$)

$$\text{ROH} \xrightarrow[\text{Ru catalyst}]{H_2} \underset{H}{\overset{O}{\underset{\|}{C}}}_H \xrightarrow{H_2O} HO\underset{}{\overset{OH}{\diagup}} \xrightarrow[\text{Ru catalyst}]{H_2} \underset{OH}{\overset{O}{\underset{\|}{C}}}_H \xrightarrow[\text{Ru catalyst}]{H_2} CO_2$$

FIGURE (SCHEME) 13.10 Methanol dehydrogenation by using an Ru-catalyst.

The Ru PNP pincer complexes were first employed as catalysts by Beller et al. [11], which was later adopted by Milstein and co-workers [12].

In 2019, a methanol reformation mechanism mediated by two ruthenium-based catalysts was investigated by Beller et al. [13]. The formic acid dehydrogenation was a rate-determining step in pincer-catalyzed methanol reformation. The accumulation of formic acid during the course of a reaction significantly alters the pH, which scales down the catalytic activity [14]. As a result, a collective effect was observed by effectively dehydrogenating formic acid below 100 °C. With 40 mmol KOH, a TON of 8,761 was observed, whereas a reduction in catalyst concentration to 10 mmol leads to a 13,386 TON.

Initially, methanol gets dehydrogenated to formaldehyde followed by a reaction with water to produce methylene glycol. Further dehydrogenation produces formic acid, and processing in a final step generates H$_2$ and CO$_2$. Three molecules of hydrogen gas and one molecule of carbon dioxide are released during the course of the reaction.

In 2017, Sun et al. reported that the Ru-based complex **1** mediated acceptorless dehydrogenation [15]. In the presence of 1 equiv. of KOtBu in p-xylene, comprehensively 17 secondary alcohols, including both aliphatic and aromatic alcohols, were oxidized to give 21% yield when catalytic loading was around 0.025 mol%. The formation of the corresponding esters was caused by the use of primary alcohols. After keeping benzyl alcohol at 50 °C for 48 hrs, 41% benzaldehyde was produced. In 2017, a dimeric Ru NNN pincer complex **4** was used by Yu and co-workers [16]. In addition to the hydrogenation of ketones, secondary alcohols's acceptorless dehydrogenation was also manifested. At 110 °C, a variety of alcohols were oxidized to their corresponding ketones in the presence of a 0.1 mol% catalyst and 10 mol% KotBu.

Mostly acetophenone derivatives with substituted electron withdrawing or electron releasing groups tend to exhibit high yield. Inflation by 0.5 mol% in catalytic loading during the conversion of aliphatic or cyclic alcohol results in a high conversion rate. Complex **2** was also discovered to be efficient for secondary alcohol dehydrogenation. Notably, a 0.5 mol% **2** catalyst in the absence of a base leads to excellent ketone yield (Figure (Scheme) 13.12). In 2017, Wang et al. reported the requisite of increased catalyst loading of 2 mol%, which was required for the dehydrogenation of tetrahydroquinolines and indolines to generate aromatic N-heterocyclic compounds.

Employing an Ru PNP pincer as a catalyst complex **5** in an acceptorless dehydrogenation reaction of primary and secondary alcohol was investigated by Muthaiah's group [17]. On heating for 24 hours at reflux, with the PNP ligand and a RuCl$_2$(PPh$_3$)$_3$ precursor, the water soluble complex was produced (Figure (Scheme) 13.13).

FIGURE (SCHEME) 13.11 Ruthenium-based pincer complexes applied to the acceptorless dehydrogenation of alcohols.

FIGURE (SCHEME) 13.12 Complex 2-mediated acceptorless dehydrogenation of alcohol.

Ru catalysts have been reported to govern the catalytic oxidation of carboxylic acid salt in a basic medium, in contrast to the report of alcohol acceptorless dehydrogenation being carried out in an aqueous medium to produce aldehyde and ketones [18–20]. The yield of aliphatic ketones and aldehyde were found to range from 35% to 75%, whereas the conversion of benzylic substrates to ketones give high yield. However, a comparatively considerable quantity of 5 mol% catalytic loading was required for the dehydrogenation of water.

FIGURE (SCHEME) 13.13 Ruthenium PNP complex synthesis **5**.

13.2.2 Dehydrogenation of Formic Acid (FA)

Formic acid having a 4.4 wt.% low hydrogen content is also considered suitable as an LOHC [21]. At 298 K, the standard free energy is −7.61 kcal mol^{-1} for dehydrogenation, which makes it exergonic from a thermodynamic standpoint [22]. Based on previous research, Ru-, Ir-, Fe-, and Co-based pincer catalytic complexes were developed for FA dehydrogenation. A list of NNN- and NCN-based ruthenium complexes composed of two pyrazol-3-ylbranches was compiled by Nakahara et al. in 2018.

The catalytic activity of the pyrazol rings differed greatly when its 5th position is substituted with bytrifluoromethyl or tert-butyl groups [23]. When the concentration of both KHMDS and complex **6a** is 0.25 mol%, the TON reaches 3,000 in 2 hours, which indicates high catalytic activity.

Syringe pump-continuous dosage of FA in the reaction mixture was given to maintain its concentration. The analogue **6b** (TON = 65) produced lower activity due to the significant role of acidity. Furthermore, the NCN ruthenium complex **7** shows reduced catalytic activity with TON = 660. The outstanding results were obtained with complex **8**, which were obtained by the dehydrochlorination of **6a**. The TON was 3,700 in 2 hours without the presence of a base because of the liberation of ammonia gas from complex **8**. The Beller group explored the catalytic activity of complexes **9a** and **9b**, which mark distinct features in methanol dehydrogenation [13]. The catalyst performance of complex **9b** can be enhanced by using a 1 M phosphoric acid/KH$_2$PO$_4$ buffer setup. This leads to a TOF of 9,219 h^{-1}, and the reaction completed in 40 minutes. Meanwhile, dehydrogenation with N-methylated complex **9b** along with the presence of a KOH base takes six hours to complete with a 26,388 TON. The solvent-dependent complex 10 that mediates formic acid dehydrogenation was studied by Kawanami et al. in 2019 [24]. At 80 °C, 1,4-dioxane under 20 Mpa H$_2$ gas pressure results in the finest yield. The Baratta group first synthesized another ruthenium complex (**11b**), which was later assessed by Beller in FA dehydrogenation [25]. The FA dehydrogenation with a cyclometalated ruthenium CNN catalytic complex yielded 96% product under a neutral reaction medium and was completed in 22 hours with an 11,910 TON. During NMR experiments, the Ru catalytic complex **11b** was visible in a basic medium. An evaluation was performed to confirm the role of Complex **11a** in the dehydrogenation reaction. Indeed, in the presence of **11b**, high catalytic activity with a TON of 9,085 was obtained in 3 hours under aqueous/triglyme optimal conditions.

13.2.3 Dehydrogenation of C-N Bond by Ru Catalyst

A brief description in the literature discusses amine dehydrogenation to produce imines or nitriles. Jensen et al., 2011, reported the secondary amines dehydrogenation to imines [26].

Meanwhile, the conversion of primary amines to nitriles via acceptorless dehydrogenation was reported by Szymczak et al. [27]. The double dehydrogenation was carried out by a ruthenium metal-based NNN hydride pincer complex **12**. In this way, 16% to 75% yields of respective nitriles were obtained from aliphatic and aromatic amines. A recent pathway for imine synthesis has been opened as this system withstands the oxidation of the thioether functional group.

FIGURE (SCHEME) 13.14 Role of some selected noble pincer complexes in the dehydrogenation of formic acid.

FIGURE (SCHEME) 13.15 Conversion of primary amines to nitriles via acceptorless dehydrogenation.

In addition, this N heterocycles dehydrogenation via acceptorless mechanism was investigated by Yu and Jones's group [28]. The acceptorless dehydrogenation of indolines and secondary alcohols was also possible using their ruthenium hydride complex.

13.2.4 RUTHENIUM-CATALYZED DEHYDROGENATION OF ALKENE

Alkenes are essential building blocks in a broad dimension of applications, from petrochemistry to natural product synthesis. The main constituents of petroleum and natural gas are alkanes, the most ample and economical hydrocarbons. By Fischer Tropsch synthesis, an extensive range of hydrocarbon can be produced from 'syngas' (a gas mixture of CO and H_2 generated during the gasification of natural gas, coal, and biomass). Despite the fact that alkanes are primarily used as fuel, their synthetic utility has been limited. This brings the functionalization of alkanes into trend to generate more advance and upgraded products.

Olefins are less abundant than alkanes but are exceptionally functional synthetic intermediates in organic reactions and are extensively explored as raw materials in the chemical industry. The most explicit and atom-efficient pathway for the conversion of low cost and plenteous alkane feedstock into olefin is catalytic alkane dehydrogenation (CAD). To break the simple C-H bond of alkanes during catalytic dehydrogenation, high activation energy is required. Moreover, the wide number of C-H bonds in alkane makes the selective dehydrogenation quiet challenging.

The dehydrogenation of higher alkanes results in low product selectivity, although some simple alkanes such as ethane, isobutenem, etc. can be converted into olefin in the presence of heterogeneous catalysts under a high temperature of around 500–900 °C, and they have little energy production. In contrast, the homogeneous catalyst-mediated dehydrogenation of alkane exhibits sufficient product selectivity and does not require a high temperature [29]. An endothermic acceptorless alkane dehydrogenation is a direct way to produce alkene and hydrogen gas from alkane. The high demand of energy for the dehydrogenation reaction could be remunerated by the choice of a proper sacrificial hydrogen acceptor moiety. The transfer dehydrogenation approach of adding a hydrogen acceptor allows the course of the reaction to proceed.

In 1984, Felkin et al. reported the transfer dehydrogenation at 150 °C of coenzyme (COA) with tert-butylethylene (TBE) in the presence of a ruthenium metal-based catalyst, specifically, [(p-F-$C_6H_4)_3$P]Ru(H)$_4$ yield 55 Tos [30]. Leitner and colleagues synthesized a range of advanced ruthenium bismethallyl catalytic complexes that had bidentate chelating phosphines ligand [Cy$_2$P(CH$_2$)nPCy$_2$]Ru(3-C$_4$H$_7$) (n = 1,2,3) and catalyzed the acceptorless dehydrogenation of COA [31]. However, the TOF and TON for the degradation reaction was lower by 1.9 h^{-1} and 5 h, respectively, after 2 days due to ligand degradation under dehydrogenation conditions.

In 2011, Roddick and coworkers developed a powerful ruthenium dehydrogenation catalyst. For COA transfer and acceptorless dehydrogenation, they synthesized a pincer ruthenium metal-based complex with ligand $^{(CF3)2}$PCP$^{(CF3)2}$)Ru(H)(COD) **14** (Figure (Scheme) 13.16) [32]. At 200 °C, the

FIGURE (SCHEME) 13.16 [Cy$_2$P(CH$_2$)nPCy$_2$]Ru catalyst for the transfer dehydrogenation of alkene.

thermal degradation of **14** in the presence of an equimolar mixture of COA and TBE yielded a TOF and total TON of 170 hr^{-1} and 230, respectively, which was equivalent to those of (^{tBu2}PCPt^{Bu2})Ir(H)$_2$. However, the complete degradation of the catalyst at 200 °C within half an hour resulted in a decline in TON. Also investigated was COA acceptorless dehydrogenation by **14**. By refluxing for 1 hour the COA solution of **14** at 140 °C and 590 Torr pressure, 10 Tos of cyclooctane (COE) was produced at a rate of 14 h^{-1}. The bulkiness of ligand and its electron-withdrawing nature could play a significant role in alkane dehydrogenation. Surprisingly, this catalytic-driven Figure (Scheme) is not affected by the addition of O$_2$, H$_2$O, or even N$_2$.

13.3 CONCLUSION

In the area of the coordination and organometallic chemistry of Ru, there has been a notable increase in research and assessment during recent years. Numerous articles have lately been published on the development of Ru-based compounds and their uses in a variety of fields, including medicine, catalysis, biology, nanoscience, redox, and photoactive materials. These advancements may be connected to Ru's exceptional capacity to exist in a variety of oxidation states. This chapter summarized the most recent applications of transition metal Ru-based complexes in catalytic dehydrogenation. Dehydrogenations of HC-CH, HC-NH, and HC-OH bonds were specifically unearthed in the beginning of 2020. Because they offer exceptional activity and good selectivity, heterogeneous ruthenium catalysts are among the most frequently used catalysts, particularly for hydrogenation and dehydrogenation reactions. Various advanced complexes have been recommended, with compelling improvements especially in the acceptorless dehydrogenation of acid, nitriles, alkane, alcohol, and alkenes.

13.4 ABBREVIATIONS

hr.:	Hour
Min.:	Minutes
°C:	Degree centigrade
Ru:	Ruthanium
Ir:	Iridium
Rh:	Rhodium
NHC:	Hybrid N-heterocyclic carbine
CPP:	Phosphine-Phosphine ligand
NMR:	Nuclear magnetic resonance spectroscopy
TOF:	Turnover frequency
PNP:	*N,N'*-bis(diphenylphosphine)-2,6-diaminopyridine
AAD:	Acceptorless alcohol dehydrogenation
PPh$_3$:	Triphenylphosphine
COE:	Cyclooctane
Pcy:	Tricyclohexylphosphine
TBE:	Tris/Borate/EDTA (buffer solution)
OsO$_4$:	Osmium Tetroxide
LOHC:	Liquid organic hydrogen carrier

REFERENCES

[1] Robbins, P., and Lee, D., 1979. *Guide to Precious Metals and Their Markets*. Van Nostrand-Reinhold, New York.

[2] Wilkinson, G., Stone, F.G.A., and Abel, E.W. (eds.), 1982. *Comprehensive Organometallic Chemistry*. Pergamon, Oxford.

[3] Li, H., and Hall, M.B., 2014. Mechanism of the formation of carboxylate from alcohols and water catalyzed by a bipyridine-based ruthenium complex: A computational study. *Journal of the American Chemical Society*, *136*(1), pp.383–395.

[4] Hou, C., Zhang, Z., Zhao, C., and Ke, Z., 2016. DFT study of acceptorless alcohol dehydrogenation mediated by ruthenium pincer complexes: Ligand tautomerization governing metal ligand cooperation. *Inorganic Chemistry*, *55*(13), pp.6539–6551.

[5] Awasthi, M., and Singh, S., 2019. Ruthenium catalyzed dehydrogenation of alcohols and mechanistic study. *Inorganic Chemistry*, *58*(21), pp.14912–14923.

[6] Liu, H., Jian, L., Li, C., Zhang, C., Fu, H., Zheng, X., Chen, H., and Li, R., 2019. Dehydrogenation of alcohols to carboxylic acid catalyzed by in situ-generated facial ruthenium-CPP complex. *The Journal of Organic Chemistry*, *84*(14), pp.9151–9160.

[7] Wang, Q., Chai, H., and Yu, Z., 2017. Dimeric ruthenium(II)-NNN complex catalysts bearing a pyrazolyl-pyridylamino-pyridine ligand for transfer hydrogenation of ketones and acceptorless dehydrogenation of alcohols. *Organometallics*, *36*(18), pp.3638–3644.

[8] Yang, X., 2013. A self-promotion mechanism for efficient dehydrogenation of ethanol catalyzed by pincer ruthenium and iron complexes: Aliphatic versus aromatic ligands. *ACS Catalysis*, *3*(12), pp.2684–2688.

[9] Kim, W., Park, I., and Park, J., 2006. Acceptor-free alcohol dehydrogenation by recyclable ruthenium catalyst. *Organic Letters*, *8*(12), pp.2543–2545.

[10] Tseng, K., Kampf, J., and Szymczak, N., 2015. Mechanism of N,N,N amide ruthenium(II) hydride mediated acceptorless alcohol dehydrogenation: Inner-sphere β-H elimination versus outer-sphere bifunctional metal–ligand cooperativity. *ACS Catalysis*, *5*(9), pp.5468–5485.

[11] Nielsen, M., Alberico, E., Baumann, W., Drexler, H.-J., Junge, H., Gladiali, S., and Beller, M., 2013. Low-temperature aqueous-phase methanol dehydrogenation to hydrogen and carbon dioxide. *Nature*, *495*, pp.85–89.

[12] Alberico, E., Lennox, A., Vogt, L., Jiao, H., Baumann, W., Drexler, H., Nielsen, M., Spannenberg, A., Checinski, M., Junge, H., and Beller, M., 2016. Unravelling the mechanism of basic aqueous methanol dehydrogenation catalyzed by ru–PNP pincer complexes. *Journal of the American Chemical Society*, *138*(45), pp.14890–14904.

[13] Agapova, A., Junge, H., and Beller, M., 2019. Developing bicatalytic cascade reactions: Ruthenium-catalyzed hydrogen generation from methanol. *Chemistry: A European Journal*, *25*(40), pp.9345–9349.

[14] Monney, A., Barsch, E., Sponholz, P., Junge, H., Ludwig, R., and Beller, M., 2014. Base-free hydrogen generation from methanol using a bi-catalytic system. *Chemical Communications*, *50*(6), pp.707–709.

[15] Wang, Z., Pan, B., Liu, Q., Yue, E., Solan, G., Ma, Y., and Sun, W., 2017. Efficient acceptorless dehydrogenation of secondary alcohols to ketones mediated by a PNN-Ru(II) catalyst. *Catalysis Science & Technology*, *7*(8), pp.1654–1661.

[16] Wang, Q., Chai, H., and Yu, Z., 2017. Dimeric ruthenium(II)-NNN complex catalysts bearing a pyrazolyl-pyridylamino-pyridine ligand for transfer hydrogenation of ketones and acceptorless dehydrogenation of alcohols. *Organometallics*, *36*(18), pp.3638–3644.

[17] Bhatia, A., and Muthaiah, S., 2018. Well-defined ruthenium complex for acceptorless alcohol dehydrogenation in aqueous medium. *ChemistrySelect*, *3*(13), pp.3737–3741.

[18] Balaraman, E., Khaskin, E., Leitus, G., and Milstein, D., 2013. Catalytic transformation of alcohols to carboxylic acid salts and H2 using water as the oxygen atom source. *Nature Chemistry*, *5*(2), pp.122–125.

[19] Choi, J., Heim, L., Ahrens, M., and Prechtl, M., 2014. Selective conversion of alcohols in water to carboxylic acids by in situ generated ruthenium trans dihydrido carbonyl PNP complexes. *Dalton Transactions*, *43*(46), pp.17248–17254.

[20] Zhang, L., Nguyen, D., Raffa, G., Trivelli, X., Capet, F., Desset, S., Paul, S., Dumeignil, F., and Gauvin, R., 2016. Catalytic conversion of alcohols into carboxylic acid salts in water: Scope, recycling, and mechanistic insights. *ChemSusChem*, *9*(12), pp.1413–1423.

[21] Sordakis, K., Tang, C., Vogt, L., Junge, H., Dyson, P., Beller, M., and Laurenczy, G., 2017. Homogeneous catalysis for sustainable hydrogen storage in formic acid and alcohols. *Chemical Reviews*, *118*(2), pp.372–433.

[22] Bernskoetter, W., and Hazari, N., 2017. Reversible hydrogenation of carbon dioxide to formic acid and methanol: Lewis acid enhancement of base metal catalysts. *Accounts of Chemical Research*, *50*(4), pp.1049–1058.

[23] Nakahara, Y., Toda, T., Matsunami, A., Kayaki, Y., and Kuwata, S., 2018. Protic NNN and NCN pincer-type ruthenium complexes featuring (trifluoromethyl)pyrazole arms: Synthesis and application to catalytic hydrogen evolution from formic acid. *Chemistry: An Asian Journal*, *13*, pp.73–80.

[24] Guan, C., Zhang, D., Pan, Y., Iguchi, M., Ajitha, M., Hu, J., Li, H., Yao, C., Huang, M., Min, S., Zheng, J., Himeda, Y., Kawanami, H., and Huang, K., 2016. Dehydrogenation of formic acid catalyzed by a ruthenium complex with an N,N'-diimine ligand. *Inorganic Chemistry, 56*(1), pp.438–445.

[25] Léval, A., Junge, H., and Beller, M., 2020. Formic acid dehydrogenation by a cyclometalated-CNN ruthenium complex. *European Journal of Inorganic Chemistry, 2020*(14), pp.1293–1299.

[26] Gu, X.-Q., Chen, W., Morales-Morales, D., and Jensen, C.M., 2002. *Journal of Molecular Catalysis A: Chemical, 189*, pp.119–124.

[27] Tseng, K., Rizzi, A., and Szymczak, N., 2013. Oxidant-free conversion of primary amines to nitriles. *Journal of the American Chemical Society, 135*(44), pp.16352–16355.

[28] Wang, Q., Chai, H., and Yu, Z., 2018. Acceptorless dehydrogenation of *N*-heterocycles and secondary alcohols by ru(II)-NNC complexes bearing a pyrazoyl-indolyl-pyridine ligand. *Organometallics, 37*(4), pp.584–591.

[29] Ren, T., Patel, M., and Blok, K., 2006. Olefins from conventional and heavy feedstocks: Energy use in steam cracking and alternative processes. *Energy, 31*(4), pp.425–451.

[30] Felkin, H., Fillebeen-Khan, T., Gault, Y., Holmes-Smith, R., and Zakrzewski, J., 1984. Activation of C-H bonds in saturated hydrocarbons. The catalytic functionalisation of cyclooctane by means of some soluble iridium and ruthenium polyhydride systems. *Tetrahedron Letters, 25*, pp.1279–1282.

[31] Six, C., Gabor, B., Görls, H., Mynott, R., Philipps, P., and Leitner, W., 1999. Inter- and intramolecular thermal activation of sp^{3}C–H bonds with ruthenium bisallyl complexes. *Organometallics, 18*(17), pp.3316–3326.

[32] Gruver, B.C., Adams, J.J., Warner, S.J., Arulsamy, N., and Roddick, D.M., 2011. Acceptor pincer chemistry of ruthenium: Catalytic alkane dehydrogenation by (^{CF3}PCP)Ru(cod)(H). *Organometallics, 30*, pp.5133–5140.

14 Dehydrogenation Reactions Incorporating Membrane Catalysis

Hiralkumar Morker, Pratik Saha, Bharti Saini, and Anirban Dey

14.1 INTRODUCTION

Growing the scale of energy production and enhancing the effectiveness of energy conversion processes are both necessary to meet the demands of ever-increasing energy consumption whilst minimizing its environmental effect. The hunt has driven the tremendous research effort over the past ten years for alternative energy sources with low pollution emissions and reasonable costs. Among these, hydrogen is considered an attractive source of energy generation. Due to its incredibly low atomic mass, high energy density, and environmentally friendly oxygen oxidation (which produces just water), hydrogen is a viable energy source. Hydrogen has lately often been stated as a renewable energy source [1]. However, nature does not contain pure hydrogen (H_2). Hydrogen exists in the form of hydrocarbons (such as natural gas, crude oil, etc.), synthetic products (such as alcohol, hydrides, etc.), and finally, water.

Metals or their hydrides, which produce pure hydrogen when exposed to heat or water, might become viable H_2 sources for laboratory and mobile units. Although these systems often have poor efficiency, their usage can mitigate one of hydrogen's most significant limitations, notably, its low density [2]. For instance, using hydrogen as a fuel for cars and airplanes is one area where this limitation is crucial. Some hydrides have hydrogen densities that are significantly greater than liquid H_2. A few alloys such as fullerenes, nanotubes, and magnesium hydrides must be mentioned among the hydrogen transporters. To assure the perceived security of hydrogen storage and usage, such alloys are used to create cylinders that store enormous volumes of H_2 at a certain pressure and release it instantly or with minimal energy. There are various commercial hydrogen-generation processes, and each has its drawbacks. The simplest and least-expensive processes are coal gasification, methane reforming, and other hydrocarbons to produce synthesis gas (a combination of H_2 and CO_2). Although the methane-water reaction is the most appealing, carbon dioxide creation involves significant energy absorption, whereas the generation of CO is only moderately exothermic. The synchronous creation of carbon oxides, which particularly causes certain environmental issues, is the main drawback of these procedures [3].

Additionally, difficulties with irreversible catalyst degradation arise when fuel cell operation takes place even in slight amounts of CO present with H_2. Thus, to mitigate this and achieve pure H_2, research institutions and commercial enterprises are becoming increasingly interested in multifunctional reactor designs that integrate reaction and extraction [4]. Along with the potential for easier segregation and heat regulation and the reduction in the costs of plant equipment and upkeep, the significant advantage of such an approach originates, first and foremost, from the achievement of increased conversion and product yield. This drives the primary justification for using membrane reactors for H_2 generation through the process of dehydrogenation (DH). The process can go on past equilibrium conversion because DH's elimination of hydrogen allows this [5].

However, the utilization of commercial DH procedures under typical demanding circumstances necessitates membranes with appropriate thermal, mechanical, and chemical resilience in addition to good permeability and selectivity [6]. To mitigate the problems of H_2 storage, a liquid organic H_2 carrier (LOHC) appears to be a viable option. Various types of research have been conducted on LOHCs since the 1980s, but they have become more popular in the last decade. The main advantage of LOHCs is that they are in a liquid state at room temperature, so their transportation and storage are easy and safe. These compounds are typically heterocyclic organic compounds that can undergo DH to generate H_2 and aromatic compounds. Their ability to ensure good thermodynamic and kinetic stability during the storage of H_2 signifies their potential merits. They have the capacity to provide up to (6–8 wt%) more storage than metal hydrides (3wt%). Among LOHCs, methylcyclohexane (MCH) is preferred due to its strong compatibility with the current traditional transportation system. Having a lower boiling point also ensures that MCH is less volatile and safer. However, due to the poor biodegradability and acute high aquatic toxicity of MCH, this type of LOHC poses health hazards to humans. Thus, to ameliorate unfavorable conditions, membranes are used in conjunction with the catalyst for the maximum production of pure H_2. A catalytic membrane reactor (CMR) integrates porous inorganic membranes, which implements molecular sieve mechanisms for the proper removal of H_2 even at high temperatures. CMRs have high mechanical and thermal stability, thus ensuring the effective removal of H_2 through the DH process. Moreover, the synthesis of such membranes is relatively easy. This process is not widely commercialized due to the enormous energy required to extract H_2 during the process [7–9]. Thus, this chapter provides in-depth knowledge pertaining to the application of catalytic membrane reactors for the DH of hydrocarbons to achieve pure H_2. H_2 will become a fuel in the upcoming years; accordingly, these processes are attractive to researchers.

14.2 THE HISTORY OF MEMBRANE CATALYSIS

A relatively new and quickly growing area of chemistry is membrane catalysis. Russian scientists contributed significantly to its creation. Its fundamental principles were created less than 50 years ago. The fast growth of membrane catalysis began towards the end of the 20th century after a phase of theoretical and experimental knowledge-building throughout the first 20 years. International conventions and symposiums that place particular emphasis on membrane catalytic reactions and membrane reactors are organized annually. There has been an exponential growth in the number of publications, articles, and patents published internationally about membrane reactors and the mechanism involved. Since 1994, an international conference about "Catalytic reactions in membrane reactors" has been conducted every three years under the sponsorship of the European Membrane Society; the tenth meeting took place in St. Petersburg in June 2011. The advancement and optimization of the production of membranes for membrane reactors for different applications were the focuses of the collective monograph, which was released in 2011 [6].

Palladium (Pd)-based composites have an extraordinary capacity to consume enormous quantities of H_2; they were initially identified by T. Graham in 1861 and served as the foundation for the first membrane reactors. Soviet researchers employed a cylindrical Pd membrane to examine the process of ethylene hydrogenation over a century later. However, membrane catalysis as a concept was not developed until the 1960s. Researcher V. M. Gryaznov first demonstrated the advantages of incorporating catalysts and improving properties in a single substance. He was known as the pioneer of catalysis in membrane reactors. He along with his colleagues first observed the impact of the integration of DH and hydrogenation processes on a Pd membrane in 1964 [10].

It has been experimentally demonstrated that when a substance that can emit hydrogen, such as cyclohexane, is introduced in a Pd tube (reaction zone) while a substance that can soak up hydrogen, such as o-xylene, is heated inside the reaction zone, benzene could form within the zone while toluene is produced by o-xylene hydro-alkylation outside the tube in the absence of H_2 gas. The result was found more in integrated processes than in individual ones [11]. After some time in the USA, another technique was identified to convert ethane into ethylene using a Pd membrane where the membrane was used to separate H_2. The ethylene production in this technique was much higher than the equilibrium output. Detailed research into reactions involving the absorption and expulsion of hydrogen on palladium membrane catalysts started with the discovery of these couplings, and subsequent studies focused on oxidation processes on silver membranes [10].

A significant part of the reactions that were first investigated that utilized membrane catalysts included the elimination of hydrogen. This is not unexpected given that the membrane-based catalysts would be more effective in these processes, where the rate of conversion in the desired reaction frequently surpasses the maximum yields, and the specificity of desired product creation frequently significantly rises [12]. Additionally, this offers the crucial chance to use ultrapure hydrogen that has gone through the membrane. The curiosity in membrane catalysis skyrocketed around the turn of the century, and articles in this area began to multiply significantly. The advent of novel materials, including composites of asymmetric (anisotropic) polymeric and ceramic barriers, which significantly increase the effectiveness of membrane catalysts, led to the broader practical use of membrane reactors. The operating temperature range of these membranes is increased, and the application's temperature can be raised because of the usage of high thermal stability ceramic membranes [4].

14.3 MAJOR MERITS OF H_2-ABSORPTIVE MEMBRANE CATALYSTS IN SUBSEQUENT REACTIONS WITH THE EXPULSION OF H_2

Due to the directed movement of reactants and energy, catalytic membrane devices can boost reaction speeds and selectivity if the subsequent process occurs on the membrane catalyst (MC).

$$A \rightarrow B+C \tag{1}$$

Conversion of A would be increased by the selective elimination of one of the by-products, such as product B, over the MC, which equilibrium conditions would constrain. This offers a potent tool to boost the reaction's selectivity and pace. The movement of H_2 from the region of its creation (B = H), which boosts both the equilibrium output and specificity of the DH, is indeed a specific instance and the most common illustration of such a process [6].

Accordingly, if by-product B is considered for further processing, then

$$D + B \rightarrow E \tag{2}$$

is fed in the process chamber via dispersal across the lamina, allowing autonomous authority of the interface concentrations of reacting species, D and B, and the inhibition of their competing surface assimilation. Thus, it is inevitable that just using a traditional modification slows the objective process. The ability to perpetuate the appropriate concentration with one stimulant, such as H_2, over the full extent of the catalyst is another benefit of utilizing membrane catalysts. A minimal surface hydrogen concentration makes things simpler to produce partial hydrogenation by-products without converting the initial molecule into less-valuable saturated products [8].

The orientation of chemical flows on the opposing side of the cell membrane can be changed as one technique of regulating its surface concentration anywhere along with the reactor's duration. Just on membrane catalysts, hydrogenation results in a drop inside the concentration of hydrogen along the membrane, whereas DH results in a rise. Because gradients of ionic strength across the barrier act as the driving force for hydrogen transmission, the velocity of hydrogen transfer with the identical or opposing passage path of the chemical to be void of H_2 and the H_2-abolishing gas from the different sides of the layer vary. As a result, distinct H_2 concentration configurations are created over the outside layer [2].

This subsequently impacts DH and hydrogenation kinetic models. For instance, it was demonstrated in several exploratory investigations that counter-flow input of H_2 and the chemical to be hydrolyzed on the separate border of the layer might boost outputs of intermediary hydrogenation by-products because the hydrogen congregation is maximal at the input of the hydrogenation region wherein the initial composite is stored, and unwanted hydrogenation of the pertinent outcomes are rare. H_2 congregation is at its lowest near the zone's exit; therefore, thorough hydrogenation might not be able to provide the requisite saturated result [13].

More significant quantities of hydrogen may get retrieved from the target area in the counter-flow input phase than in the co-current flow in procedures involving hydrogen discharge, including water-gas transfer when H_2 is removed by an inert gas. When a material that crosses the membrane is created on one surface and consumed on the other surface, the reaction coupling takes place. In this instance, a membrane catalyst efficiently carries out two processes in parallel within the same reactor without combining the reactants or end products [14]. The following section discusses the additional benefits of process coupling in membrane reactors.

14.4 DEHYDROGENATION ON PALLADIUM MEMBRANES

Almost 50 years of research have been spent on a wide range of hydrogen-involved processes in membrane catalysis. Especially in the case of hydrocarbon DH reactions, where the pace of the reaction is constrained by thermodynamic equilibrium, membrane catalysts have several benefits. These processes were initially conducted on palladium tubes that ranged in thickness from 50 to 100 mm, which are known as compact (non-porous) membranes. The circumstances of a few procedures used on such membranes showed promise for industrial utilization. For example, many investigations have been conducted regarding the DH processes of cyclohexane to benzene. Due to the comparatively cold temperatures where such a reaction takes place on a metal catalyst, there are no unfavorable side reactions that would otherwise cause catalyst carbonization and, eventually, deactivation. This reaction is thus used as a model [5].

Hydrogen extraction from the reaction can be improved by using palladium alloys. Cyclohexane conversion might be boosted from 18.7% to 99.5%. Similarly, in isopentane DH, by using Pd alloy in conjunction with 10% Ni, the yields comprising isoprene and isopentane at 860K are accordingly found to be 18.6% and 10.5%, respectively. Major merits of palladium composite membranes, which comprise a thin, thick metallic coating on the surface of a strong, porous substrate (preferably stainless steel), present themselves as the most appealing form of barrier for this purpose because of their high permeance and perm selectivity. They also ensure near 100% selectivity towards H_2 removal [15].

Compact membrane catalysts' apparent drawbacks include the high palladium price and the need to operate at above-average temperatures to obtain the necessary efficacy of H_2 extraction from the region. Additionally, using membranes that are as low as reasonably achievable is preferable because, per the Fick law, the hydrogen flux throughout a membrane is inversely related to its thickness for a given driving force [16]. However, the minimum thickness for compact membranes that are mechanically tolerable is 25 to 30 μm. Creating membranes with a robust reinforcement that is highly accessible to H_2 and a delicate uninterrupted coating of a Pd agglomerate is necessary for practical uses of catalytic membrane processes. These membrane catalysts are known as composite ones.

14.5 DEHYDROGENATION USING A COMPOSITE MEMBRANE CATALYST

These membranes are contrived of contrasting layers with distinct chemical or structural compositions. Among the layers, a particular layer delivers as the substrate, which has absorptive properties but is mechanically sturdy. A thermally secure and corrosion-resistant oxide is also possible. A thin catalyst layer, typically made of a palladium alloy, is applied to these macroporous materials. Typically, this layer ensures catalytic activity and specific hydrogen permeability. The substrate may occasionally be non-porous yet is extremely permeable to a particular molecule, most frequently H_2

in the processes supporting the study. In addition, it is feasible to employ mediums made from biological polymers, although this approach is unsuitable for DH procedures requiring relatively high temperatures since these materials often have poor thermostability [7].

A rather delicate transitional coating of the heat-contumacious compound is frequently put betwixt both the substratum and the interlayer to achieve a deformative-free surface of the metal and prevent the collective dispersion of the substratum and catalyst. Since the active layer's homogeneity and thin thickness are crucial to the excellent specificity and effectiveness of composite membranes, the techniques for depositing it have been the focus of countless studies that are continually refined. For the implementation of the active layer, many techniques, including deposition technique, electrochemical redox deposition, chemical vapor deposition (CVD), and plasma blasting, were suggested. The palladium alloy compacted coating cannot be thicker than 200 nm, which is within the finest composite membranes [17].

Together, all benefits of the compacted membrane catalyst are present in such membranes, notably selective permeability. However, relative to compact membrane catalysts, the hydrogen flow via a unit of area is significantly higher, and the quantity of the costlier metal upon the unit volume of the composites is significantly curtailed. Additionally, the composite membranes may have two or more pieces of different compact materials. Vanadium, niobium, tantalum, and titanium are reasonably cheap metals that are highly permeable to hydrogen. These metals are covered with diffusion limitations to limit the mutual migration of the metals from separate layers, followed by layers of biocompatible metal or an alloy that is porous to hydrogen. The vanadium membranes ($Pd/SiO_2/V$) are covered with silicon dioxide coatings (the resistance layer), and Pd yields the highest hydrogen flow. Such membranes' palladium layers serve various purposes, including catalysis and improved hydrogen transport specificity [18].

These membranes' substrate transition metals' surfaces are readily oxidized, and the oxidation causes a rapid permeability reduction. These transition metals are shielded from oxidation by the dense palladium covering. Complicated preparation requirements and a shorter operational life than compact membranes are two drawbacks of composite membrane catalysts. This can be caused by variations in the mechanical and chemical properties of materials and a greater propensity for the deterioration of the thin palladium layer. However, these membranes' catalytic capabilities remain the subject of active study [19]. Figure 14.1 shows the conventional DH reaction system and catalytic membrane DH system.

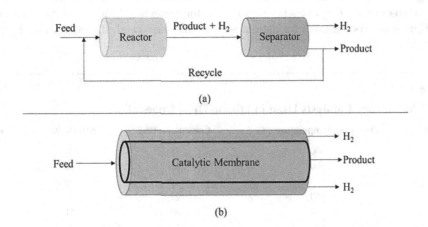

FIGURE 14.1 a) Conventional dehydrogenation reaction system and b) Catalytic membrane DH reaction system.

14.6 DEHYDROGENATION OF LOW MOLECULAR ALKANE, ALKENE, AND ALCOHOL

In recent years, much research has been conducted to illustrate the integration of membranes with catalytic systems for hydrogen production. The series of integrated cells were designed to verify the hydrogen production (rate, purity, etc.).

DH of lower molecular alkanes is the most viable process as it does not produce CO_2 as a by-product. Ethan, propane, ethanol, and methanol are the most widely used alkanes and alcohols for DH. Corresponding alkenes generated after DH can be used for other chemical processes or as fuel additives [20].

$$CH_3 \rightarrow CH_3 + CH_2 = CH_2 \tag{3}$$
$$2CH_3CH_2OH \rightarrow CH_3COOCH_2CH_3 + H_2 = CH_2 \tag{4}$$
$$2CH_3OH \rightarrow CH_3COOCH + 2H_2 \tag{5}$$

DH of formic acid is not desirable as it generates only CO_2 and H_2, which affects the H_2 purity and catalyst due to CO_2 coking on the membrane catalyst surface. The conversion reaction is shown as follows (equation 6). Most by-products of alcohol DH have been used as fuel or fuel additives [15].

$$HCOOCH \rightarrow H_2 + CO_2 \tag{6}$$

Usually, DH of alkanes and alkenes has been performed using alumina-supported Cr_2O_3 or Pt-based catalyst. Different additives were added to the Pt-based catalyst to enhance selectivity and prevent catalyst deactivation due to CO_2 build-up on the catalyst surface. Moreover, the DH of alkanes has several advantages over the DH of other hydrocarbons. Some significant factors affect the DH process, such as a higher conversion rate, no side reactions, namely, isomerization, no formation of carbon or CO_2, less catalyst deactivation, etc. In addition, these reactions are highly endothermic, so continuous heat source/input is required [21]. Table 14.1 shows the different membrane catalysts used in ethane DH.

These problems can be solved using membrane and catalyst system integration. In the DH process, if generated hydrogen is removed from the reaction zone, then hydrogen conversion increases. For this purpose, the membrane can be installed within the reaction zone. Catalytic membranes can be installed within the reactor to separate hydrogen from the reaction zone, enhancing the conversion rate. Furthermore, in this way, the reaction temperature can be lowered to a certain amount, leading to energy savings. As the amount of heat required decreases during the reaction, it lowers the coke formation, which can cause catalyst deactivation. By-products

TABLE 14.1
Different Membrane Catalysts Used in Ethane Dehydrogenation

Membrane	Operating Temperature (°C)	Conversion (%)	Selectivity	Reference
Pt-Sn/MgO	600	38	100	[22]
Pt-In/SiO$_2$	600	65	99	[23]
Cr$_2$O$_3$/SiO$_2$	650	49	98	[24]
Cr$_2$O$_3$	650	27	87	[25]
Cr/Si	650	62	81	[26]
Pt-Al$_2$O$_3$	600	75	96	[27]
Li-Mg-Sm	600	95	53	[28]

of alkanes (which are alkenes) were used for further DH. Reactions are shown in the following (equations 3, 7–9).

$$CH_3 - CH_3 \rightarrow CH_2 = CH_2 \tag{3}$$
$$CH_2 = CH_2 \rightarrow CH = CH + H_2 \tag{7}$$
$$CH_3OH \rightarrow CH_2O + H_2 \tag{8}$$
$$CH_2O + CH_3OH \rightarrow CH_3COOH + H_2 \tag{9}$$

Propane is another promising alkane for DH. Generally, propane DH occurs in a packed bed or fluidized bed reactor at 550–650 °C and in normal pressure conditions. Pt- or Cr_2O_3-based catalysts are mostly used. All possible DH reactions of propane are shown in Figure 14.2. To avoid coke formation, suitable catalysts and operating conditions should be used [1, 15, 19, 29].

14.7 DEHYDROGENATION OF CYCLOHEXANE AND METHYLCYCLOHEXANE

Cyclohexane and methylcyclohexane have 7.19% and 6.16% hydrogen, respectively (high hydrogen content). Membrane catalytic DH of these cyclo hydrocarbons is an emerging model as their DH occurs at relatively lower temperatures with no by-product formation during the process. Certain thermodynamic restrictions exist for the DH of other aromatic hydrocarbons, such as reduced conversion. The membrane catalytic method can increase exchange rates [7]. The DH of the methylcyclohexane catalytic membrane system is depicted in Figure 14.3.

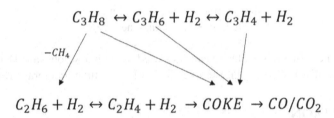

FIGURE 14.2 All possible propane dehydrogenation reactions [15].

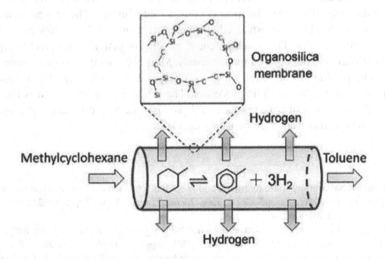

FIGURE 14.3 Membrane catalytic dehydrogenation of methylcyclohexane [7].

TABLE 14.2
Different Membrane Catalysts Used in Methylcyclohexane Dehydrogenation

Membrane Catalyst	Operating Conditions	Conversion (%)	Reference
Pt/Ga	Temp.: 450 °C; Pressure: 1 bar	17	[32]
Pt/Al$_2$O$_3$-TiO$_2$	Temp.: 450 °C; Pressure: 1 bar	95	[33]
Mo/SiO$_2$	Temp.: 400 °C; Pressure: 22 bar	90	[34]
Pt/SiO$_2$	Temp.: 350 °C; Pressure: 1 bar	65	[35]
Cu-Pt/SiO$_2$	Temp.: 400 °C; Pressure: 1 bar	92	[36]
Pt/Al$_2$O$_3$	Temp.: 250 °C; Pressure: 1 bar	60	[37]
Pt/C	Temp.: 300 °C; Pressure: 1 bar	95	[31]

Primarily, palladium-based catalytic membranes are used as thin films or as a tube. In addition, it has been investigated that coupling of the hydrogenation and DH has been shown to be energy-saving as the heat of hydrogenation can be used for DH. The advantages of a catalytic membrane system are that i) gas separation is easy and ii) gas separation occurs within the reaction zone. Currently, silica-based membranes are used for the membrane catalytic DH of methylcyclohexane and cyclohexane [7, 8, 30, 31].

$$Cycloxane \rightarrow Benzene + 3H_2 \qquad (10)$$
$$Methylcyclohexane \rightarrow Toluene + 3H_2 \qquad (11)$$
$$Decalin \rightarrow Naphthalene + 5H_2 \qquad (12)$$

Table 14.2 shows the different membrane catalysts used in methylcyclohexane DH and their operating conditions. Usually, DH takes place in fixed bed or hollow fiber membrane reactors.

14.8 CONCLUSION

The assimilation of various membrane reactors in systems for generating pure-quality H$_2$ has attracted scientific attention over the past ten years. Integrating membrane reactors to remove hydrogen from various resources has been proven to be far more effective than conventional methodologies. Simultaneously, the yield obtained is considerably higher. The synthesized membranes established on the Pd blend have satisfactorily provided pure-quality H$_2$ in various studies listed in this paper's subsequent topics. The development of novel film materials and technologies for producing H$_2$ are both actively underway simultaneously. Moreover, using such membrane reactors has helped curb CO by-product formation in various processes. Industrially significant monomers such as ethylene, propylene, etc. all use DH processes. The efficiency of film reactors has contributed significantly to the reduction of coke production. Likewise, the heat discharged during hydrogenation processes can be compensated for by using such systems.

REFERENCES

[1] M. Saidi and M. Safaripour, "Pure hydrogen and propylene coproduction in catalytic membrane reactor-assisted propane dehydrogenation," *Chem. Eng. Technol.*, vol. 43, no. 7, pp. 1402–1415, 2020, https://doi.org/10.1002/ceat.201900209.

[2] A. Zaluska, L. Zaluski, and J. O. Ström-Olsen, "Nanocrystalline magnesium for hydrogen storage," *J. Alloys Compd.*, vol. 288, no. 1–2, pp. 217–225, 1999, https://doi.org/10.1016/S0925-8388(99)00073-0.

[3] H. Abdallah, "A review on catalytic membranes production and applications," *Bull. Chem. React. Eng. Catal.*, vol. 12, no. 2, pp. 136–156, 2017, https://doi.org/10.9767/bcrec.12.2.462.136-156.

[4] S. P. Patil, A. B. Bindwal, Y. B. Pakade, and R. B. Biniwale, "On H2 supply through liquid organic hydrides – Effect of functional groups," *Int. J. Hydrog. Energy*, vol. 42, no. 25, pp. 16214–16224, 2017, https://doi.org/10.1016/j.ijhydene.2017.05.170.

[5] T. Peters and A. Caravella, "Pd-based membranes: Overview and perspectives," *Membranes (Basel).*, vol. 9, no. 2, pp. 1–5, 2019, https://doi.org/10.3390/membranes9020025.

[6] N. L. Basov, M. M. Ermilova, N. V Orekhova, and A. B. Yaroslavtsev, "Membrane catalysis in the dehydrogenation and hydrogen production processes," *Russ. Chem. Rev.*, vol. 82, no. 4, pp. 352–368, 2013, https://doi.org/10.1070/rc2013v082n04abeh004324.

[7] D. Acharya, D. Ng, and Z. Xie, "Recent advances in catalysts and membranes for mch dehydrogenation: A mini review," *Membranes (Basel).*, vol. 11, no. 12, pp. 1–20, 2021, https://doi.org/10.3390/membranes11120955.

[8] Y. Sekine and T. Higo, "Recent trends on the dehydrogenation catalysis of liquid organic hydrogen carrier (LOHC): A review," *Top. Catal.*, vol. 64, no. 7–8, pp. 470–480, 2021, https://doi.org/10.1007/s11244-021-01452-x.

[9] E. Gianotti, M. Taillades-Jacquin, J. Rozière, and D. J. Jones, "High-purity hydrogen generation via dehydrogenation of organic carriers: A review on the catalytic process," *ACS Catal.*, vol. 8, no. 5, pp. 4660–4680, 2018, https://doi.org/10.1021/acscatal.7b04278.

[10] J. K. Ali, E. J. Newson, and D. W. T. Rippin, "Deactivation and regeneration of PdAg membranes for dehydrogenation reactions," *J. Memb. Sci.*, vol. 89, no. 1–2, pp. 171–184, 1994, https://doi.org/10.1016/0376-7388(93)E0219-A.

[11] S. J. Ahn, G. N. Yun, A. Takagaki, R. Kikuchi, and S. T. Oyama, "Effects of pressure, contact time, permeance, and selectivity in membrane reactors: The case of the dehydrogenation of ethane," *Sep. Purif. Technol.*, vol. 194, pp. 197–206, 2018, https://doi.org/10.1016/j.seppur.2017.11.037.

[12] "Application of membrane separation processes in petrochemical industry: A review," *Desalination*, vol. 235, no. 1–3, pp. 199–244, 2009, https://doi.org/10.1016/j.desal.2007.10.042.

[13] A. Brune, T. Wolff, A. Seidel-Morgenstern, and C. Hamel, "Analysis of membrane reactors for integrated coupling of oxidative and thermal dehydrogenation of propane," *Chem. Ing. Tech.*, vol. 91, no. 5, pp. 645–650, 2019, https://doi.org/10.1002/cite.201800184.

[14] E. V. Shelepova, L. Y. Ilina, and A. A. Vedyagin, "Theoretical predictions on dehydrogenation of methanol over copper-silica catalyst in a membrane reactor," *Catal. Today*, pp. 35–42, 2019, https://doi.org/10.1016/j.cattod.2017.11.023.

[15] P. Quicker, V. Höllein, and R. Dittmeyer, "Catalytic dehydrogenation of hydrocarbons in palladium composite membrane reactors," *Catal. Today*, vol. 56, no. 1–3, pp. 21–34, 2000, https://doi.org/10.1016/S0920-5861(99)00259-X.

[16] J. P. Collins, R. W. Schwartz, R. Sehgal, T. L. Ward, C. J. Brinker, G. P. Hagen, and C. A. Udovich, "Catalytic dehydrogenation of propane in hydrogen permselective membrane reactors," *Ind. Eng. Chem. Res.*, vol. 35, no. 12, pp. 4398–4405, 1996, https://doi.org/10.1021/ie960133m.

[17] T. Shimbayashi and K. ichi Fujita, "Metal-catalyzed hydrogenation and dehydrogenation reactions for efficient hydrogen storage," *Tetrahedron*, vol. 76, no. 11, p. 130946, 2020, https://doi.org/10.1016/j.tet.2020.130946.

[18] H. Weyten, J. Luyten, K. Keizer, L. Willems, and R. Leysen, "Membrane performance: The key issues for dehydrogenation reactions in a catalytic membrane reactor," *Catal. Today*, vol. 56, no. 1–3, pp. 3–11, 2000, https://doi.org/10.1016/S0920-5861(99)00257-6.

[19] C. J. Dittrich, "The role of heat transfer on the feasibility of a packed-bed membrane reactor for propane dehydrogenation," *Chem. Eng. J.*, vol. 381, p. 122492, 2020, https://doi.org/10.1016/j.cej.2019.122492.

[20] J. Sheng et al., "Oxidative dehydrogenation of light alkanes to olefins on metal-free catalysts," *Chem. Soc. Rev.*, vol. 50, no. 2, pp. 1438–1468, 2021, https://doi.org/10.1039/d0cs01174f.

[21] E. V. Shelepova and A. A. Vedyagin, "Intensification of the dehydrogenation process of different hydrocarbons in a catalytic membrane reactor," *Chem. Eng. Process. Process Intensif.*, vol. 155, no. April, p. 108072, 2020, https://doi.org/10.1016/j.cep.2020.108072.

[22] V. Galvita, G. Siddiqi, P. Sun, and A. T. Bell, "Ethane dehydrogenation on Pt/Mg(Al)O and PtSn/Mg(Al)O catalysts," *J. Catal.*, vol. 271, no. 2, pp. 209–219, 2010, https://doi.org/10.1016/j.jcat.2010.01.016.

[23] E. C. Wegener et al., "Structure and reactivity of Pt–In intermetallic alloy nanoparticles: Highly selective catalysts for ethane dehydrogenation," *Catal. Today*, vol. 299, no. March, pp. 146–153, 2018, https://doi.org/10.1016/j.cattod.2017.03.054.

[24] T. V. M. Rao, E. M. Zahidi, and A. Sayari, "Ethane dehydrogenation over pore-expanded mesoporous silica-supported chromium oxide: 2. Catalytic properties and nature of active sites," *J. Mol. Catal. A Chem.*, vol. 301, no. 1–2, pp. 159–165, 2009, https://doi.org/10.1016/j.molcata.2008.12.027.

[25] K. Nakagawa et al., "The role of chemisorbed oxygen on diamond surfaces for the dehydrogenation of ethane in the presence of carbon dioxide," *J. Phys. Chem. B*, vol. 107, no. 17, pp. 4048–4056, 2003, https://doi.org/10.1021/jp022173b.

[26] X. Zhao and X. Wang, "Oxidative dehydrogenation of ethane to ethylene by carbon dioxide over Cr/TS-1 catalysts," *Catal. Commun.*, vol. 7, no. 9, pp. 633–638, 2006, https://doi.org/10.1016/j.catcom.2006.02.005.

[27] A. M. Champagnie, T. T. Tsotsis, R. G. Minet, and A. I. Webster, "A high temperature catalytic membrane reactor for ethane dehydrogenation," *Chem. Eng. Sci.*, vol. 45, no. 8, pp. 2423–2429, 1990, https://doi.org/10.1016/0009-2509(90)80124-W.

[28] A. L. Y. Tonkovich, J. L. Zilka, D. M. Jimenez, G. L. Roberts, and J. L. Cox, "Experimental investigations of inorganic membrane reactors: A distributed feed approach for partial oxidation reactions," *Chem. Eng. Sci.*, vol. 51, no. 5, pp. 789–806, 1996, https://doi.org/10.1016/0009-2509(95)00325-8.

[29] M. Sheintuch and O. Nekhamkina, "Architecture alternatives for propane dehydrogenation in a membrane reactor," *Chem. Eng. J.*, vol. 347, no. April, pp. 900–912, 2018, https://doi.org/10.1016/j.cej.2018.04.137.

[30] L. Meng and T. Tsuru, "Hydrogen production from energy carriers by silica-based catalytic membrane reactors," *Catal. Today*, vol. 268, pp. 3–11, 2016, https://doi.org/10.1016/j.cattod.2015.11.006.

[31] C. Zhang, X. Liang, and S. Liu, "Hydrogen production by catalytic dehydrogenation of methylcyclohexane over Pt catalysts supported on pyrolytic waste tire char," *Int. J. Hydrog. Energy*, vol. 36, no. 15, pp. 8902–8907, 2011, https://doi.org/10.1016/j.ijhydene.2011.04.175.

[32] O. Sebastian et al., "Stable and selective dehydrogenation of methylcyclohexane using supported catalytically active liquid metal solutions—Ga52Pt/SiO2 SCALMS," *ChemCatChem*, vol. 12, no. 18, pp. 4533–4537, 2020, https://doi.org/10.1002/cctc.202000671.

[33] X. Yang, Y. Song, T. Cao, L. Wang, H. Song, and W. Lin, "The double tuning effect of TiO_2 on Pt catalyzed dehydrogenation of methylcyclohexane," *Mol. Catal.*, vol. 492, no. March, 2020, https://doi.org/10.1016/j.mcat.2020.110971.

[34] N. Boufaden, R. Akkari, B. Pawelec, J. L. G. Fierro, M. Said Zina, and A. Ghorbel, "Dehydrogenation of methylcyclohexane to toluene over partially reduced Mo-SiO2 catalysts," *Appl. Catal. A Gen.*, vol. 502, pp. 329–339, 2015, https://doi.org/10.1016/j.apcata.2015.05.026.

[35] K. Mori, Y. Kanda, and Y. Uemichi, "Dehydrogenation of methylcyclohexane over zinc-containing platinum/alumina catalysts," *J. Japan Pet. Inst.*, vol. 61, no. 6, pp. 350–356, 2018, https://doi.org/10.1627/jpi.61.350.

[36] X. Zhang, N. He, L. Lin, Q. Zhu, G. Wang, and H. Guo, "Study of the carbon cycle of a hydrogen supply system over a supported Pt catalyst: Methylcyclohexane-toluene-hydrogen cycle," *Catal. Sci. Technol.*, vol. 10, no. 4, pp. 1171–1181, 2020, https://doi.org/10.1039/c9cy01999e.

[37] M. D. Irfan Hatim, M. A. Umi Fazara, A. Muhammad Syarhabil, and F. Riduwan, "Catalytic dehydrogenation of methylcyclohexane (MCH) to toluene in a palladium/alumina hollow fibre membrane reactor," *Procedia Eng.*, vol. 53, pp. 71–80, 2013, https://doi.org/10.1016/j.proeng.2013.02.012.

15 A Greener Dehydrogenation
Environmentally Benign Reactions

Ankita Saini, Monalisa Bourah and Sunil Kumar Saini

15.1 INTRODUCTION

The petrochemical sector relies heavily on dehydrogenation reactions. Various side reactions in a traditional cracking or pyrolysis process prevent high selectivity towards olefins. When planning a reaction as an environmentally friendly procedure, it is critical to take into account the unique thermodynamic, environmental and economic characteristics of each dehydrogenation reaction. In recent years, development of a greener process in synthetic organic chemistry has drawn attention worldwide as a sensitive issue to prevent environmental pollution. Direct dehydrogenation (DDH) has been the only economically viable reaction to date. Dehydrogenation cannot be viewed as a long- or even short-term sustainable process without the use of a green chemistry involving a stable recyclable catalyst, acceptorless reagents, nanotechnology, water-mediated reactions or photocatalysis. DDH can be conducted with or without the presence of an oxidant such as O_2, CO_2, and N_2O or nanoparticles. The most typically reported oxidants in the oxidative dehydrogenation (ODH) of alkane are oxygen and carbon dioxide [1, 2]. Other oxidants, however, have been demonstrated to be viable.

With the fast diminishing rate of fossil feedstock and the increasing demand for environmentally friendly renewable energy, the research and development of alternative methods for the regeneration of fuels are becoming unavoidably significant. Both clean chemical synthesis and hydrogen storage systems could benefit from catalytic dehydrogenation and considering them as future applications [3]. Agarwal *et al.* [4] have explored several intensification and integration tactics to identify pathways for enhancing the sustainability of the dehydrogenation method. It was demonstrated that measures such as heat recovery, additional waste and recycling of off gas could reduce CO_2 emissions by up to 70%. In this chapter, we discuss the sustainable and greener route for carrying out dehydrogenation reactions using oxidative catalytic dehydrogenation, acceptorless dehydrogenation, nanoparticles-supported reactions, photocatalysis and water-mediated reactions, as depicted in Figure 15.1.

15.1.1 Oxidative Catalytic Dehydrogenation

Platinum is a key component in numerous catalyst-based dehydrogenation processes. Because of the strong ability of Pt to activate C-H bonds and its low activity in cleaving C-C bonds, it is prominently used. However, the most active catalysts known are chromium oxide (CrO_3)-based or Pt-assisted with tin oxide (SnO_2). Even these need to be regenerated on a regular basis due to the process of coking. As a result, the future for designing a DDH catalyst must pivot to decreasing profound dehydrogenation reactions that result in the coking of the reaction. Regardless of catalyst, certain motifs have emerged in efforts to increase C-H bond activation while minimizing substrate adsorption on surface sites. In addition, supports with the help of weak to moderate superficial acidic-coupled stimulating components such as B, Ga, Cu and Ni have a lower utilization of Pt, which is ideal for possible industrially usable catalysts. Furthermore, a number of noble- and non-noble-based inter-metallic alloys and several mixed metal-oxides are evolving as suitable substitutions for traditional catalysts that are both economically and environmentally beneficial. Because of

FIGURE 15.1 Various dehydrogenative methods for carrying out greener dehydrogenation reactions.

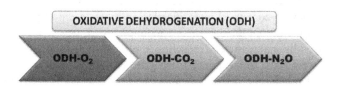

FIGURE 15.2 Types of ODH using O_2, CO_2 and N_2O.

the high cost of noble-metal catalysts, Ni-Ga alloys with equivalent catalytic dehydrogenation performance to noble metal catalysts were discovered to be the ideal candidates to replace expensive catalysts. CrO_3 materials are well-known for being highly active dehydrogenation catalysts for light alkanes. The conversion of lower alkanes utilizes gallium-oxide based on zeolites. The regeneration frequency will then be decreased, resulting in fewer regeneration cycles that produce CO_2, which lowers the carbon output of alkane. To overcome these issues, ODH can be carried out [5–7]. The ODH reactions can be explored in the presence of easily available oxidants such as O_2, CO_2 and N_2O. Herein, we discuss each ODH using O_2, CO_2 and N_2O (Figure 15.2).

15.1.2 Oxygen-Based Oxidative Dehydrogenation

Since the oxidative dehydrogenation of alkane using oxygen (ODH-O_2) is potentially limitless, exothermic and possesses favourable thermodynamics, it is a desirable pathway to produce alkene.

A Greener Dehydrogenation

This is very advantageous for scaling up reactions at the industrial level [8]. Although oxygen is inexpensive and non-toxic, special precautions must be taken to prevent a combustible oxygen/alkane ratio. Compared to DDH and ODH-CO_2, ODH-O_2 reaction pathways are substantially more straightforward and have less byproducts and side reactions than conventional catalysts. The key restriction on ODH-O_2 is excessive oxidation. Gaseous oxygen is used with catalysts made of metal oxides. The catalysts utilized in this process often contain Pt, Sn, V, Cu, B, Ni and Mo [9–12]. When the temperature is between 400 and 700 °C, the reaction takes place through a partial oxidation that produces alkene and water. When compared with a conventional steam cracking process, this exothermic ODH reaction can save energy because it requires lower operating temperatures. ODH-O_2 provides a route that completely prevents the creation of coke. In this section, we briefly highlight a few ODH-O_2 catalyzed reactions.

Sam et al. explored $Mg_2V_2O_7$ as a catalyst for ODH-O_2 reaction [13]. At 550 °C, the catalyst demonstrated a conversion of 6.95% and a selectivity for propene of 53.5%, making it significantly more productive than $Mg_2V_2O_8$ and $Mg_2V_2O_6$. This emphasized the significant role of a V-double bond in the production of ODH-O_2 catalysts. MoOx-based catalysts were shown to function slower than vanadia-based catalysts for ODH-O_2 reactions [14]. In this review by Cavari et al., the key elements influencing the catalytic capabilities of supported vanadium oxide and molybdenum oxide systems and the traits of catalysts achieving exceptional olefin yields are examined.

Høj et al. depicted that Mo loading up to 10 wt% boosts selectivity, which highlights that the polymeric form of MoOx provides optimum selectivity [15]. However, selectivity was marginally lower at the same conversion as 2wt% V-catalysts. An increment on the overall yield is observed as a catalyst with Mo only with a loading of 10 wt% on Al_2O_3 produced 9.4% yield when compared to a catalyst with a combined loading of 4wt% Mo and 2 wt% V that produced a yield of 10%. Barman et al. used surface organometallic chemistry to graft monomeric V species onto SiO_2 to illustrate how to alter preparation methods to improve dispersion [16]. It was found that monomeric V-species are inherently more selective. Below 500 °C, for ODH-O_2-catalyzed reactions, the incipient wetness impregnation (IWI) approach-produced catalysts are more selective than monomeric V-based catalysts. But once this controlled temperature was reached, monomeric V-based catalysts showed an equivalent or higher selectivity.

Recently, Matam et al. [17] discovered that in an ODH of propane followed by DDH and a second ODH run reaction cycle, MoOx/Al_2O_3 as catalyst was tested for both oxidative and non-ODH of propane and showed effective results. The catalyst has highly scattered Mo-oxide species in +6 oxidation state as $[MoVIO_4]^{-2}$ possessing tetrahedral coordination moieties. Interestingly, at a similar propene conversion for the second ODH-O_2 phase, propene selectivity is enhanced by 15%, resulting in a 5% improvement in overall alkene yield. Kumar et al. investigated how effective oxygen functionalized hexagonal boron nitride performs catalysis for the ODH of light alkanes, namely, ethane, propane, butane and isobutane [18]. It was found that the ODH reaction usually takes place at the same temperature for isomers, indicating that it is independent of the isomer's form.

Because of the exothermic nature of its reactions, its thermodynamic potential and its ability to prevent coke formation, ODH-O_2 has numerous intrinsic advantages. Metal oxides still struggle with profound oxidation. However, it is evident that newly developed metal-based catalysts, despite their early development and limited structural and mechanism understanding, have definite commercialization promise because of their great selectivity at high alkane conversion.

15.1.3 Carbon Dioxide-Based Oxidative Dehydrogenation

To reduce the over-oxidation of products such as CO and CO_2, the soft-oxidation of alkane using CO_2 has been suggested as an alternative method to utilize O_2. The ODH-CO_2 reaction has regained interest in the context of combined efforts to reduce human CO_2 emissions [19–20]. Technologies that use CO_2 are generally sought after to create a circular economy. Due to the expensive equipment allocated to gaseous O_2, which necessitates high-tech reactors and sufficient diluent to assure operational safety, the ODH processes with gaseous O_2 cannot compete with conventional techniques of steam cracking and pyrolysis. The alternative strategy is to use carbon dioxide in place of oxygen because it has a lower oxidative capacity than oxygen that hinders the complete oxidation of olefin. The use of CO_2 in the ODH of alkane can also be more appealing from an economic and environmental perspective than the steam used in cracking procedures. The benefit of ODH-CO_2 is that it uses the reverse water-gas shift (RWGS) reaction to consume the by-product of alkane dehydrogenation, namely, H_2. As a result, there may be a greater conversion of alkane to alkene and a change in the thermodynamic equilibrium for the dehydrogenation of alkane. ODH-CO_2 catalysts do not typically over-oxidize alkane. As could be expected in an ODH-O_2 process, there is also a higher cost involved in getting concentrated CO_2 than in getting air. Herein, we briefly discuss a few ODH-CO_2-catalyzed reactions.

Deng et al. investigated the ODH of ethane (ODE) to ethylene using CO_2 as an oxidant in the catalytic performance of Fe-Cr/ZrO_2 catalysts [21]. Higher ethylene selectivity and reduced CO_2 conversion are provided by the catalyst made by the coprecipitation approach. The results show that the active sites for the dehydrogenation of ethane are Cr(III) species, and during the reaction, Fe_2O_3 is reduced to Fe_3O_4, which may facilitate the RWGS process. Moreover, Oliveira et al. studied the catalytic characteristics of Cr-ZrO_2 catalysts for the CO_2-ODH of propane, which was found to be affected by a change not only in the content of metal from 2.5 to 15 wt% but also in the hydrothermal production process [22]. It was found that when Cr^{+6} species are reduced at temperatures higher than 500 °C, the CO_2 on Cr-ZrO_2 desorbs. The accumulation of carbon, which is only oxidized with O_2 is linked to the activity that decreases over time.

As an alternative to the dehydrogenation process of producing propene, Michorczyk et al. carried out greener dehydrogenation of propane in the presence of CO_2 [23]. With a molar ratio of CO_2:C_3H_8 (5:1) at 873 K, the catalytic selectivity and performance of chromium oxide-based catalysts in combination with SiO_2 and Al_2O_3 showed 25.2% and 24% yields of C_6H_6, respectively. Zhang et al. highlighted the use of CO_2 as an oxidant in a photocatalytic ODH process over Pd-TiO_2 catalysts at moderate temperature, 308 K, for ethylene and syngas production [24]. A 1% Pd-TiO_2 catalyst demonstrated ethylene production, and the presence of CO_2 considerably increased the production. Mamoori et al. described the in-situ collection and use of CO_2 in the ODH of ethane over an adsorbent catalyst made of double salt KCa and Cr-impregnated H-ZSM-5 [25]. Due to its higher turnover frequency (TOF) and lower surface density, Cr10/HZSM-5 along with a 1:1 wt ratio of K-Ca under 5,000 mLg^{-1}h^{-1} and 5 vol% of ethane feed concentration demonstrated the maximum ethane conversion at 25% with ethylene selectivity of 88%. Wang et al. described that by applying up to 5 wt% of Cr over silicalite-1, a variety of CrOx/silicalite-1 catalysts with defined composition of CrOx species can be produced for CO_2-ODH [26]. It was discovered that the polymeric Cr^{6+} oxides for the CO_2-ODH of propane are much more active but less selective than isolated Cr^{6+} oxides. The activity increases as the degree of polymerization for the polymeric Cr^{6+} oxides increases with strong adsorption properties.

One might contend that a commercial ODH-CO_2 process would greatly lessen the carbon footprint of a single chemical or industrial usage. However, CO_2 has since been reported to be used for various alkane substrates. The dehydrogenation reaction performed by the addition of CO_2 has a number of outcomes. It helps the RWGS reaction, which eliminates H_2 from the system and moves the athermodynamic equilibrium limit to enhance alkene synthesis.

Furthermore, CO_2 can dissociate on a catalyst's surface and supply O-species, which can directly dehydrogenate alkane. Contrary to alkane, which adsorbs on the acidic sites, CO_2 will predominantly adsorb on the basic sites because it is an acidic molecule. Therefore, when designing a catalyst, the proper balance of both acidic and basic sites is crucial. Due to the potential for utilizing CO_2 and creating neutral carbon, ODH-CO_2 has generated interest, mostly in academia.

A Greener Dehydrogenation

15.1.4 NITROUS OXIDE-BASED OXIDATIVE DEHYDROGENATION

N_2O is another important soft oxidant. The global warming potential of N_2O is around 310 times more than that of CO_2, and both are greenhouse gases. As a result, reactions that eliminate N_2O gas from the atmosphere are highly favoured. The combustion of fossil fuels results in producing N_2O as a by-product in the industrial sector. ODH-O_2 and ODH-N_2O normally occur at temperatures lower than DDH, but ODH-CO_2 typically occurs between or somewhat closer to DDH reaction temperatures [27]. Herein, we highlight several ODH-N_2O-catalyzed works reported in the literature.

According to Bulánek et al., the presence of an extra framework, that is, Fe-oxo complexes could be the cause of Fe-catalytic ZSM-5's activity in the ODH of propane (ODHP) with nitrous oxide [28]. In addition, Pérez-Ramírez et al. depicted that small amounts of clearly isolated Fe species in Fe-ZSM-5 has proved to be active in ODHP with N_2O, but massive Fe clusters facilitated the profound oxidation of significant reaction intermediates to COx [29]. Zhou et al. synthesized a series of Ni-Al mixed oxides for the ODH of ethane with the N_2O oxidant [30]. These compounds showed good activity in N_2O decomposition. It was found that compared to other catalysts at 460 °C, the Ni_3Al-MO catalyst has a TOF value of 99 h^{-1}, which is much higher than that of Fe-, Cr-, Mo- and V-based catalysts. In the ODH of ethane with N_2O, the Ni_3Al-MO catalyst performs better than other catalysts. Moreover, Wu et al. examined the feasibility of employing Fe-MFI produced by reductive ion-exchange as a catalyst for the ODH of propane with nitrous oxide [31]. Acid posttreatments on Fe-MFI clearly have a good impact on how well the ODH of propane with nitrous oxide functions. The maximum propylene output progressively rises to around 25% with cumulative acid post-treatments.

The ODH-N_2O reaction has received the least amount of research, and recent developments have been little. This might be because employing N_2O as an oxidant has high economic cost, yet doing so might be advantageous in an integrated process where N_2O is produced as a by-product.

15.1.5 COMPARISON OF CATALYTIC ODH OVER DDH

Table 15.1 summarizes both chemical and catalytic properties linked with DDH and ODH reactions using O_2, CO_2 and N_2O.

TABLE 15.1
Chemical and Catalytic Summary of Reaction Conditions in ODH Reactions.

Type of Reaction	DDH	ODH-O_2	ODH-CO_2	ODH-N_2O
Operating Temperature Range, °C	550–630 °C	400–600 °C	400–600 °C	400–600 °C
Thermodynamics	Endothermic	Exothermic	Endothermic	Exothermic
Side Reactions Encountered	Cracking and Coking	Cracking and Over-Oxidation	Cracking and Coking and Dry Reforming	Cracking and Coking and Over-Oxidation
Role of Redox Side	Not Directly Involved	Oxygen Activation	Carbon Dioxide Activation	Nitrous Oxide Oxidation
Role of Acidic Sites	Alkane Activation, Coking	Alkane Activation	Alkane Activation, Coking	Alkane Activation, Coking
Role of Basic Sites	Alkene Desorption	Alkene Desorption	Alkene Desorption and Carbon Dioxide Activation	Alkene Desorption
Deactivation Process	Coking	Sintering	Over-Reduction and Coking	Coking

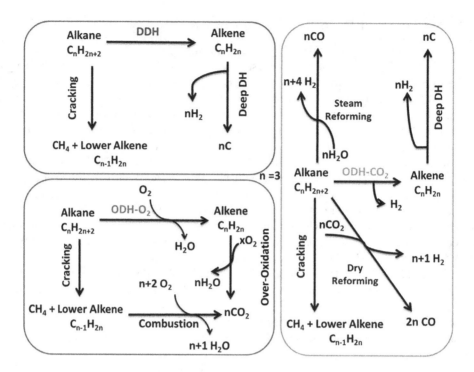

FIGURE 15.3 Typical pathway of DDH reactions and ODH-O_2 and ODH-CO_2 catalyzed reactions.

The temperature ranges of each reaction overlap, whereas ODH-O_2 and ODH-N_2O are normally conducted at lower temperatures than DDH, and ODH-CO_2 is often carried out at temperatures between or close to DDH temperatures. Using the proper catalysts, ODH-CO_2 typically reacts at temperatures between 550 and 850 °C and at atmospheric pressure. As a result, new high-activity catalysts are continually being developed. Since ODH with CO_2 is extremely endothermic and ODH with O_2 is strongly exothermic, the restrictions can be lessened by combining the heat exchange properties of these two processes. This method might offer an efficient technique to make alkene while using little energy. Although over-oxidation is frequently seen in the ODH-O_2 and ODH-N_2O processes, coking is still a significant concern in DDH, ODH-CO_2 and ODH-N_2O. Figure 15.3 depicts the typical pathway of DDH reactions and ODH-O_2 and ODH-CO_2 catalyzed reactions.

Each ODH process involves the active participation of redox sites, whereas DDH does not. Alkanes are activated by acid sites. However, in the case of ODH-N_2O and ODH-O_2 reactions where acid sites are strong, alkane will be held on the surface and cause over-oxidation or coking. Meanwhile, basic sites are recognized for facilitating alkene desorption. The most common deactivation mechanisms also differ amongst these reactions. Coking is a major reason for the loss of on-stream activity in DDH, ODH-CO_2 and ODH-N_2O. Despite relatively quick deactivation of DDH catalysts, it is interesting that the activity can be restored following an oxidative process, greatly increasing the catalyst's lifespan.

Accordingly, ODH processes successfully utilize CO_2 as an attractive, soft oxidant, and CO_2 can be in-situ reduced to value added CO, a crucial molecule in subsequent processes. Although ODH-CO_2 is not a commercially viable catalytic technology, a number of catalysts including oxide, metallic, bi-metallic, nitride and carbide catalysts have been thoroughly investigated with excellent catalytic performance, bringing this process closer to being viable. However, the ODH-CO_2 process still needs to make considerable development before it can be used in the commercial sector, despite the substantial advancements made in recent years.

15.2 ACCEPTORLESS DEHYDROGENATION

One of the most crucial processes in the production of organic compounds is carrying out the oxidation of alcohols to carbonyl compounds including aldehydes, ketones and acids. Alcohols are thermally dehydrogenated catalytically [32]. Alcohols are essential building blocks in the synthesis of organic compounds because they act as precursors to carbonyl compounds when they react with stoichiometric amounts of inorganic and high molecular weight oxidants. Alcohols can be derived from a wide variety of sources [33–34]. The acceptorless dehydrogenation (AD) [35] of alcohols under thermal conditions using metal catalysts is one of the most researched methods for improving the viability and atom efficiency of the process [36, 37]. The direct synthesis of carbonyl compounds without the use of H_2 acceptors such as O_2, H_2O_2 or sacrificial olefin is made possible by the catalytic AD reaction of alcohol. Instead, molecular H_2 is released, which may subsequently be utilized in hydrogen economy. For the AD process, a number of homogeneous catalyst systems, including the Rh(II) complex and Cp*Ir complex catalysts, have been reported [38]. Low catalytic activity, hostile environments, the requirement for additives and challenges with synthesis nevertheless make these systems problematic. Both heterogeneous noble metal catalysts—namely, those made of Ru, Ag, Au and Pt—and non-noble metal catalysts, such as those made of Ni, Co and Cu, have recently been created to address these issues [39]. These catalysts must be heavily protected from air exposure due to their air sensitivity to achieve high activity.

Fuse et al. [40] achieved the first AD conversion of aliphatic secondary alcohols to corresponding ketones by developing a ternary hybrid catalyst made up of photoredox catalyst, thiophosphate organocatalyst and nickel catalyst under visible light irradiation at ambient temperature. Catalytic AD of aliphatic and aromatic alcohols using a ternary hybrid catalyst system with visible light irradiation at ambient temperature was carried out. Three primary steps were involved in the reaction: the transfer of a hydrogen atom from an alcohol substrate's -C-H bond to the thiyl radical produced by the photooxidizedorgano catalyst, the nickel catalyst's interception of the resulting carbon-centered radical and the elimination of the -hydride. High yields of the reaction took place under benign conditions, and it did not result in the production of any unwanted by-products, except for the H_2 gas from a variety of alcohols. Furthermore, Gangwar et al. [41] optimized stable Cp*Co(CO)I_2 complex for the efficient dehydrogenation of secondary alcohols into ketones and ethers under mild conditions. Acetophenone is produced with a 48% yield when 1-phenylethanol is combined with 3 mol% catalyst and base tBuOK in acetone at 80 °C for 20 hours. An improved yield of the anticipated ketone, specifically, 57%, is obtained by subsequently treating the substrate and catalyst with the ligand 8-hydroxyquinoline.

Putro et al. [42] conducted reactions using a Cu-Fe(III) catalyst obtained from a Cu-Fe layered double hydroxide (LDH) precursor. It showed the most efficient catalytic activity compared to Cu-Fe catalysts synthesized by various treatments and another conventional supported Cu-Fe$_3$O$_4$ catalyst made using co-precipitation and physical methods. Under base- and oxidant-free conditions, these Cu-Fe catalysts have shown exceptional efficiency for the dehydrogenation of several alcohols. The LDH structure can be quickly and efficiently converted into a Cu-Fe catalyst for the efficient AD of alcohols.

In 2019, Sahoo et al. [43] optimized a process for catalytic non-ODH of partially saturated N-heterocycles that produces N-aromatics and the development of H_2 gas at room temperature because water is used as the reaction medium. They developed visible-light active [Ru(bpy)$_3$]$^{2+}$ as a photoredox catalyst for the dehydrogenation of partially saturated N-heterocycles to N-heterocyclic arenes in water media as discussed in Figure (scheme) 15.1. This method produces large yields of N-heterocyclic arenes. In the presence of two catalytic conditions, namely, cobalt complex as a proton-reduction catalyst and [Ru(bpy)$_3$]$^{2+}$ as a photoredox catalyst, unsubstituted indoline produced 90% isolated yields of indole when subjected to a dehydrogenation process under optimal conditions, that is, N-heterocycles, 0.25 mmol, [Ru(bpy)$_3$]$^{2+}$, 1.0 mol%, proton reduction catalyst and 2.5 mol% at room temperature for 12 hours. Additionally, an indoline derivative substitute produced

FIGURE (SCHEME) 15.1 Ru complex catalyzed the ADH of partially saturated N-heterocycles to N-heterocyclic arenes [43].

products with a good yield. The reaction was performed under neutral and mild conditions with the liberation of H_2 at room temperature.

The AD-based reaction also offers noteworthy benefits in a number of organic synthesis processes, such as the borrowing hydrogen method for the synthesis of imines, amides and esters.

15.3 NANOPARTICLE-BASED CATALYZED DEHYDROGENATION

The use of nanoparticles as catalysts in organic synthesis has drawn attention in recent decades due to its unique properties and efficiency. Nanocatalysts are recyclable and highly efficient compared to other catalysts [44–46]. Nanocatalyzed reactions and the preparation of nanoparticles can be performed in environmentally benign solvents at low temperature. From this point of view, a nanocatalyzed reaction is a safer and greener reaction with minimal waste of environmentally polluted material. Numerous nanoparticles have been synthesized, some of them are water soluble and numerous nanoparticles are successfully used in dehydrogenation reactions. Here, we briefly describe a particular dehydrogenation reaction carried out in the presence of a nanocatalyst.

Because of their unique electronic interactions with formic acid, supported PdNPs have been extensively utilized as catalysts for the selective dehydrogenation of formic acid [47–48]. Gold (Au) is viewed as a superior cocatalyst to increase the overall catalytic efficiency of bi-and trimetallic Pdnanocatalysts. Furthermore, Pd/N-MSC-30 nanocatalyst was used for the dehydrogenation of FA with extremely high catalytic activity and recyclability, affording the highest TOF of 8414 h^{-1} at 333 K [49]. A wet chemical reduction method was used to create palladium nanoclusters (PdNCs) immobilized by an N-functionalized porous carbon support (N-MSC-30). This was prepared by a tandem low temperature heat treatment approach serving as a support for stabilizing PdNCs. Au nanoparticles supported by protonated titanate nanotubes (TiNTs) were used as a composite catalyst to the additive-free dehydrogenation of formic acid at low temperature [50]. The multiwalled open-ended structure of TiNTs depicts effective electronic interaction with transitional metals. However, the optimal catalytic hydrogen production efficiency with other catalysts occurs at 300 °C. The charge distribution between both Au and Ti nanotubes was discovered to be the most important factor in the enhancement of the rate of the dehydrogenation reaction.

Monodispersed AgPd alloy NPs, assembled with mpg-C_3N_4, that is, mesoporous graphitic carbon-nitride for unique catalysis by mpg-C_3N_4@AgPd, was employed for the dehydrogenation of ammonia borane hydrolytically at ambient temperature [51]. These monodispersed AgPd alloy NPs were prepared by a high-temperature organic-phase surfactant-assisted process through the co-reduction of Ag(I)acetate and Pd(II)acetylacetonate. The AgPd alloy NPs that synthesized and then assembled on mesoporous graphitic carbon nitride helped to reduce graphene oxide using a liquid-phase self-assembly method by following the subsequent chemical reaction.

$$NH_3BH_3 + 2H_2O \rightarrow NH_4^+ + BO_2^- + 3H_2$$

AgPd nanoparticles fixed on $WO_{2.72}$ nanorods (NRs) were used as catalysts for the dehydrogenation of formic acid, and they transferred hydrogenation from nitro-aromatic (Ar-NO_2) to aromatic-amines (Ar-NH_2), which further reacted with aldehydes to generate benzene-fused hetero-cyclic moieties [52]. The catalysis of AgPd/WO$_{2.72}$ is found to be Ag/Pd-dependent. The high activity of the appropriate ratio of AgPd/WO$_{2.72}$ is the result of a strong interfacial contact between the two, resulting in the lattice expansion of AgPd and polarization of the charges from AgPd to WO$_{2.72}$. Monodispersed AgPd and AuPd alloy nanoparticles supported on carbon material were also highly efficient formic acid dehydrogenation catalysts [53].

Carbon nanotubes (CNTs) were involved as catalysts in the catalytic dehydrogenation of formaldehyde in alkaline solution using O$_2$ [54]. Compared to other carbon-based catalysts, CNTs displayed superior catalytic performance due to an sp^2 carbon-rich surface, hydrophilicity and numerous surface defects. The peroxide species generated by the activation of O$_2$ adsorbed on CNTs are responsible for C-H activation, leading to H$_2$ generation. Graphene and sp^{2}C-nanomaterials such as CNTs and carbon nanofibers (CNFs) have attracted attention as effective catalysts for various reactions attributed to their special structures and properties [55]. CNFs and CNTs as catalysts have been extensively reported in ODH reactions because of their high activity and stability [56]. Thus, carbon as a catalytic material has numerous advantages over conventional metal-supported catalysts due to its unique control of surface acidity and basicity and the π-electron density through surface functionalization. In carbon materials, the short- and long-range order regulates the macroscopic properties and overall performance in industrial processes. The nanostructure of CNFs has a great influence on catalytic behavior. Schlögl and coworkers [57] initiated the applications of nanocarbons to the ODH of ethylbenzene. The major benefit of using carbon nanomaterials is their high crystallinity causing their high resistance against oxidation.

Accordingly, numerous nanoparticle catalysts are employed in dehydrogenation reactions. In this section, we mentioned some recent works that explored the nanocatalyst dehydrogenation reaction.

15.4 PHOTOCATALYSIS-BASED DEHYDROGENATION

An alternate method that uses clean solar energy to activate CO$_2$ and lower alkane molecules under benign conditions is called photocatalysis [58, 59]. Moreover, photoredox catalysts can be called green catalysts. Photocatalysts are used in small amounts and do not require high temperatures, pressures or harsh conditions [60]. The ODH of alkane in the presence of CO$_2$ as a mild oxidant is one of the potential methods for creating alkene, as was previously indicated. Photo-oxidation of alkane with CO$_2$ may be a potential technique to enhance the process because ODH with CO$_2$ occurs at high temperatures. The Pd on TiO$_2$ materials has been shown to be the potential catalyst for photocatalytic ODH with CO$_2$ at ambient temperature. In addition, alcohol/water combinations can undergo photocatalytic dehydrogenation. In recent years, considerable research has been conducted on the photocatalytic synthesis of H$_2$ by a water splitting process coupled with the oxidation of organic substrate. In general, mild circumstances are usually sufficient for photocatalytic conversion of alcohol/water combinations to H$_2$. As a result, compared to equivalent electrochemical or thermal dehydrogenations, the selectivity of the photochemical process is typically higher. The most popular photocatalysts for the hydrogen oxidation process are those based on the semiconductor TiO$_2$. Most catalyst modifications involve the loading of noble metal nanoparticles to a semiconductor's surface to increase the rate at which catalytic hydrogen is produced when exposed to visible light [61]. The lifetime of the electron and hole pair can be extended by the metal. It can serve as an electron scavenger by encouraging proton-reduction and reducing electron/hole recombination. Furthermore, combining these materials with organometallic complexes that operate as co-catalysts is another effective tactic to support their performance in visible light.

Metallacarboranes system as a photocatalyst was employed by Teixidor *et al.* [62] They reported that with low loadings of [Co(C$_2$B$_9$H$_{11}$)$_2$], specifically, 0.1 to 0.01 mol%, metallacarboranes could promote the oxidation of aliphatic and aromatic alcohols in aqueous media in a quantitative conversion attributed to their high solubility. Additionally, the recovery of the photocatalyst is easy through precipitation with [NMe$_4$]Cl [62]. Moreover, metallacarborane could covalently link to magnetic nanoparticles layered with silica and be applied in a heterogeneous catalytic system.

The metallacarboranes were efficiently photooxidized alcohols by loading 0.1 and 0.01mol% [63]. Ruthenium cobaltbis(dicarbollide) photoredox catalytic system photooxidized alcohols to aldehydes and acids. This was prompted by an H+-coupled electron-transfer process under UV light irradiation with 0.005 mol% of catalyst. In the catalytic system, the photoredoxmetallacarborane and ruthenium oxidation catalysts were linked by noncovalent interactions that persisted their activity even after water dissolution [64]. Giustiniano et al. in 2022 also reviewed the visible light photoredox catalyst in water. They extensively discussed the photoredox catalyst dehydrogenation reaction in aqueous media [65].

The selective ODH of gas-phase ethanol was carried out in 2009 in a fluidized bed photoreactor by using photocatalyst VO_x/TiO_2 [66]. Photocatalytic dehydrogenation of alcohols under anaerobic conditions in aqueous solution has been thoroughly studied with the aim of generating H_2 from alcohol and water [67–69]. In anaerobic dehydrogenation, TiO_2 is used as a photocatalyst, and the alcohol acts as an active sacrificial reagent or a hydrogen donor by decomposing to CO_2. Lower reactivity TiO_2 photocatalysts, such as rutile TiO_2, are useful for selective coupling, while highly active TiO_2 photocatalysts tend to oxidize excessive ethanol to CO_2 and acetic acid.

The dehydrogenation of ethanol without an acceptor can undergo C-C coupling in aqueous solution and yield 2,3-butanediol by selectively releasing H_2 in the presence of the TiO_2 photocatalyst [70]. Primary aliphatic-alcohols can be directly converted into acetals with H_2 generation through TiO_2-photocatalytic dehydrogenation reaction coupling at room temperature, without any hydrogen acceptors (Figure (scheme) 15.2) [71]. The reaction followed photocatalytic alcohol dehydrogenation and H^+ catalytic acetylation, in which H^+ catalysts were generated by the dehydrogenation process. The approach exhibited a fast rate of reaction and good product selectivity, which represents a novel green process for the conversion from primary alkyl alcohols to bio-renewable ethanol and 1-butanol.

Visible-light organo-photoredox catalyst was used for the catalytic ODH of N-heterocycles by Balaraman et al. in 2017 [72]. The reaction proceeded significantly under base- and additive-free conditions with ambient atmosphere at room temperature and a low catalyst loading of 0.1–1 mol%. 2-methylquinoline was aromatized by the irradiation of visible-light upon a catalytic amount of rose bengal (1 mol%) in the presence of ambient air as an oxidant at room temperature. In a biphasic medium, another ODH of tetrahydro-β-carbolines, indolines and iso-quinolines was preceded by the use of reusable and homogeneous cobalt-phthalocyanine used as a photoredox catalyst. The major advantages of using a biphasic medium system is attributed to its ease in the separation of product and reusability of the photoredox catalyst [73].

Cyclohexane is also selectively oxidized to benzene over $MoOx/TiO_2$ photocatalysts in the presence of gaseous oxygen at 35°C under UV irradiation [74]. Using a fluidized bed reactor improves the oxidation of cyclohexane to benzene over MoOx-based catalysts. Moreover, the photoactivity enhanced in a fluidized photoreactor is found to be associated with greater light absorption, because of the utilization of scattered light by the catalyst [75, 76].

Photocatalysts Au/TiO_2 were also used in the ODH of ethanol. The catalysts with Au loading between 0.5 and 2 wt% were synthesized by photo-deposition using different deposition times over

FIGURE (SCHEME) 15.2 TiO_2-photocatalytic ADC reaction of primary alcohols into acetals [71].

TiO$_2$, which was prepared by the sol-gel method [77]. The catalytic activity largely depends on Au loading and its particle size and metal distribution on the TiO$_2$ surface. The catalyst with 0.5 wt.% Au prepared within 120 min of the deposition time showed high TiO$_2$ photoactivity in ethanol ODH. The acetaldehyde selectivity and maximum conversion of 95% and 82%, respectively, were obtained at a concentration of 0.2 vol% ethanol at 60 °C. These results were significantly higher than those obtained over pure TiO$_2$ and a commercial catalyst. There is currently great interest in exploring typical nanostructures based on a metal-organic framework (MOF), graphene oxide, SiO$_2$ and TiO$_2$ for the effective improvement of the photocatalytic dehydrogenation of FA. A newer strategy of a Mott–Schottky heterojunction has been utilized between a nanocatalyst and its support to adjust the catalytic property and surface charge density of a metal-based nanocatalyst for the desired photocatalytic dehydrogenation of formic acid [78].

Through the previous examples and reactions, it is clearly highlighted that photocatalysis represents one of the most significant fields in research that addresses organic synthesis through which the possibility of introducing open-shell species under mild reaction conditions will emerge. Taking advantage of photoredox catalysis in numerous dehydrogenation reaction transformations that are not always attainable through ground-state reactivities can be achieved.

15.5 WATER-MEDIATED DEHYDROGENATION REACTIONS

Water is an environmentally benign solvent, and the use of water in organic synthesis is a new revolution in green chemistry. Moreover, water has played an important role in the mechanism of many catalytic reactions. Among different alternative energy sources, H$_2$ is an appealing, environmentally benign, safe and clean energy source for enabling a more sustainable future as its combustion generates water as a by-product [79–81]. However, the storage of hydrogen is one of the difficult issues due to hydrogen being an exceptionally highly volatile gas that consequently possesses a lower volumetric density. This can be overcome effectively if H$_2$ is stored reversibly as various chemical energy transporters. In this way, liquid organic compounds, such as formic acid, methanol, methylamine, etc., are an alluring choice. In this section, we discuss some important water-mediated dehydrogenation reactions.

Because of sustainability, low toxicity, biodegradability, non-flammability and stability, formic acid has proved to be a promising liquid organic hydrogen transporter. It can undergo a dehydrogenation reaction producing H$_2$, which is significant for hydrogen-based energy applications [82]. Formic acid is effortlessly acquired from either carbon dioxide and hydrogen or non-edible biomass. Considered as a significant industrial by-product, formic acid has been broadly used to produce H$_2$ in both gas and transfer dehydrogenation reactions under mild conditions in the liquid phase and in the presence of catalysts. Eqs. 1 and 2 represent formic acid decomposition that involves a dehydrogenation process and dehydration process, respectively, that produces CO as a by-product [83].

$$HCOOH \rightarrow CO_2 + H_2 \quad \Delta G_{298K} = -35.0 \text{ kJ mol}^{-1} \quad \text{(Eq. 1)}$$
$$HCOOH \rightarrow H_2O + CO \quad \Delta G_{298K} = -14.9 \text{ kJmol}^{-1} \quad \text{(Eq. 2)}$$

The efficient and selective dehydrogenation of aqueous-phase formic acid is a great challenge to the production of hydrogen with non-noble metal-based heterogeneous catalysts. In recent years, aqueous phase FA dehydrogenation using non-noble metal-based catalysts has been extensively studied. Here, we describe several cases of the aqueous phase catalytic dehydrogenation of FA.

Non-noble metal complexes are widely used as catalysts in synthetic chemistry. Iron(II) complexes along with hydrophilic multi-dentate water soluble ligand as m-Trisulfonated-tris[2-(diphenylphosphino)ethyl] phosphine sodium salt (PP$_3$TS) are active in FA dehydrogenation reactions in aqueous solutions [84]. Employing this as a catalyst resulted in the production of a 1:1 ratio of H$_2$:CO$_2$. Cobalt pincer complex was utilized in the dehydrogenation of formic acid in aqueous media under mild conditions and achieved excellent results of hydrogen production (Figure (scheme) 15.3) [85].

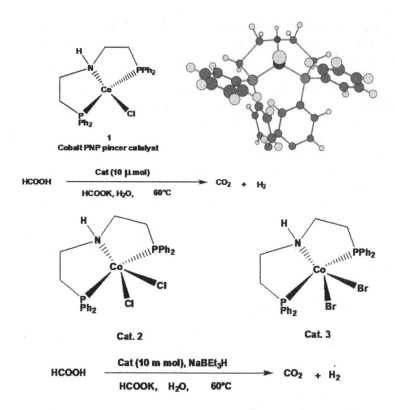

FIGURE (SCHEME) 15.3 Cobalt PNP pincer-catalyzed dehydrogenation of FA [85].

In an aqueous medium, catalytic FA dehydrogenation reactions by using phenyl that substituted a cobalt PNP pincer complex exhibited higher activity, wherein formic acid was completely decomposed into H_2 and CO_2 in less than an hour in the presence of potassium formate and reached a catalyst TON of 2,260. The activation of the catalyst was increased by adding sodium triethylborohydrate. Figure (scheme) 15.4 depicts that the initial step of the reaction is the replacement of chloro-ligand in the complex by formate, which results in the formation of complex A1. Next, the corresponding transition state for the CH activation and dissociation, TS(A2/B), leads to CO_2 release and the formation of parent amine complex B. The amine complex B releases H_2 [85].

One more technique was developed for FA dehydrogenation in water by utilizing an iridium complex containing conjugated N,N′-diimine ligands that is free of the addition of bases or additives (Figure (scheme) 15.5) [86]. Using [Cp*Ir(L1)Cl]Cl (L1=2,2′-bi-2-imidazoline), a TOF of 4,87,500 h^{-1} at 90 °C and a TON of 2,400,000 with an in-situ prepared catalyst from [IrCp*Cl$_2$]$_2$ and 2,2′-bi-1,4,5,6-tetrahydropyrimidine (L2) at 80 °C were obtained [86].

Furthermore, the complete dehydrogenation of water-mediated formic acid by using γ-Mo$_2$N catalyst derived from biomass with a concentration of 40 vol% achieved a high gas production rate [87]. The main advantage for carrying out the reaction of this catalyst was the ease in the synthesis of γ-Mo$_2$N catalyst with high yield by an easy pyrolysis reaction. Using this non-noble metal-based heterogeneous catalyst in a water-mediated reaction enhanced the dehydrogenation performance of FA.

In the aqueous phase, FA dehydrogenation by using nanoparticles has also been studied. Ag Pd core-shell NPs have shown potential in catalyzing the dehydrogenation of formic acid in aqueous media at a low temperature of 50 °C without the use of an additive. The initial TOFs determined were in the range of 125–252h^{-1} at a 25–50 °C temperature range [88]. This clearly shows that bimetallic nanoparticle catalysts are more effective than their single component counterparts for the

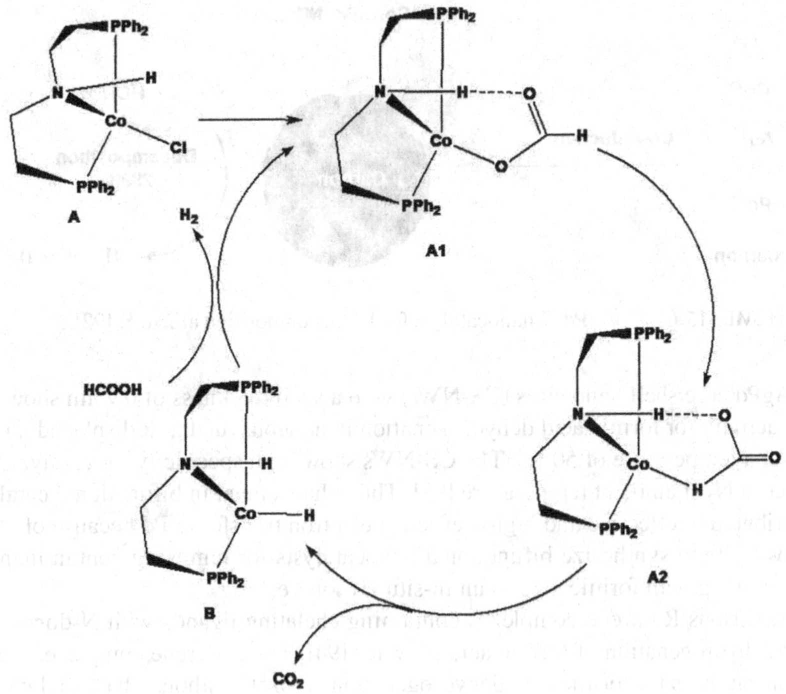

FIGURE (SCHEME) 15.4 Reaction pathway of the cobalt catalyzed dehydrogenation of FA [85].

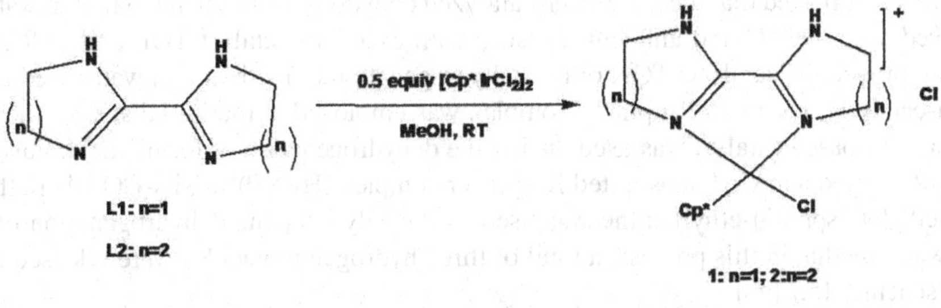

FIGURE (SCHEME) 15.5 Synthetic pathway for [Cp*Ir(L)Cl]Cl complex [86].

active dehydrogenation of FA [89]. In addition, monodispersed AgPd alloy nanoparticles with a 2.2-nm diameter were proved to be highly active heterogeneous catalysts for the dehydrogenation of FA in water [90]. The $Ag_{42}Pd_{58}$ nanoparticles revealed the highest catalytic activity with an initial TOF of 382 h^{-1} at 50 °C. This showed that the high potential of binary alloy nanoparticles makes them more convenient catalysts for the dehydrogenation of FA and H_2 generation. Monodispersed AuPd NPs of 4 nm were more active in catalyzing the dehydrogenation of FA in water at 50 °C without additives and displayed a TOF of 230 h^{-1} [91].

Another work was carried out by Q. Jiang in 2013 using CoAuPdnano-alloy supported on carbon (CoAuPd/C) as an active catalyst for the dehydrogenation of FA as shown in Figure (scheme) 15.6 [92]. The high stability of Co^0 in the nanoalloy structure was first used in catalytic FA dehydrogenation in water. Moreover, the CoAuPd over C has a lower consumption of noble metals and exhibits 100% production of H_2, which is the highest selectivity and stability toward H_2 generation from FA in water media without additives at 298 K.

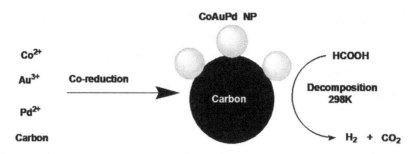

FIGURE (SCHEME) 15.6 CoAuPd/C nanocatalyst for FA decomposition at 298 K [92].

The Ag/AgPd core/shell nanowires (CS-NWs) with a shell thickness of 0.9 nm showed the highest catalytic activity for formic acid dehydrogenation in aqueous media. It displayed an initial TOF of 1,400 h^{-1} at a temperature of 50 °C. The CS-NWs show high specificity for catalyzing NO_2^- and NO_3^- in water to N_2 at ambient temperature [93]. The enhancement in bifunctional catalytic performance is attributed to effective and highly efficient electron transfer to Pd because of nanowires. It reveals a new route to synthesize bifunctional nanocatalysts for removing contaminants from the polluted water along with formic acid as an in-situ H_2 source.

Moreover, various Ru-arene complexes containing chelating ligands with N-donor were developed for the dehydrogenation of formic acid in water [94]. The Ru-arene complexes were found to show efficient catalysis for formic acid dehydrogenation at 90 °C without a base and with a TON of 2,248. Rodríguez et al. [95] synthesized an Ru-based catalytic complex carrying a chelating ligand as bis(olefin) diazadiene. It was used for the catalysis of the dehydrogenation of methanol in water leading to the production of three molecules of H_2 at a temperature below 100 °C. Additionally, Nielsen et al. [96] showed that the Ru-pincer-catalyzed dehydrogenation of methanol in water media was achieved at 65–90 °C and ambient pressure with excellent catalyst TOF and TON. Monney et al. [97] improved the catalytic TON of the dehydrogenation of methanol in water media with the use of a bi-catalytic, where an Ru-pincer complex was employed at the initial step of the reaction, and a second Ru-based catalyst was used during the dehydrogenation of formic acid. Furthermore, A. K. Das et al. used a methyl-substituted Ru-pincer complex (HPNP)Ru(H)$_2$-CO where the HPNP bis[2-(dimethylphosphino)-ethyl]amine was used as a catalyst for the dehydrogenation of methylamine in water media; in this process, a total of three hydrogen molecules were released as shown in Figure (scheme) 15.7 [98].

In Figure (scheme) 15.8, a highly efficient and water-soluble Ir-complex containing 2-(2-benzimidazolyl)-6-hydroxypyridine as a ligand was used for the dehydrogenation of alcohols and the synthesis of N-hetercycle in aqueous media [99]. The oxidative cyclization of 2-amino/nitroaryl alcohols with alcohol/water is carried out with double dehydrogenation.

From these discussed reactions, it is clear that water-mediated dehydrogenation reactions have also received much attention from chemists due to their green chemistry features, unique properties and exceptional reactivity.

15.6 CONCLUSION AND FUTURE PROSPECTS

Herein, we summarized the recent approaches on greener dehydrogenation for sustainable and cleaner reactions. Greener and more efficient routes for carrying out dehydrogenation reactions are highlighted using oxidative catalysts, photocatalysis, water-mediated reactions, nanoparticle-supported reactions and AD. Among these, catalytic dehydrogenation based on mild oxidants, nanoparticle-supported reactions and photocatalysts presents an attractive approach for clean synthesis and its future applications as a hydrogen storage system. In addition, it is shown that CO_2 has proved to be an attractive oxidant that could be sufficiently explored in ODH reactions compared

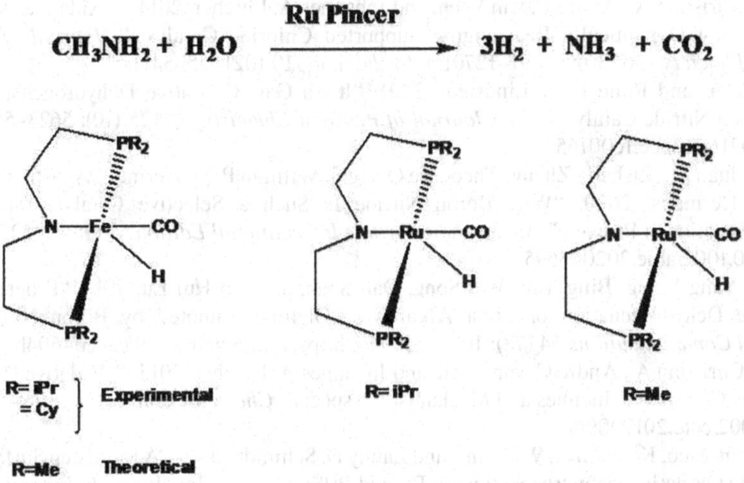

$$CH_3NH_2 + H_2O \xrightarrow{\text{Ru Pincer}} 3H_2 + NH_3 + CO_2$$

FIGURE (SCHEME) 15.7 Ru-pincer complex-catalyzed dehydrogenation reaction of methyl amine [98].

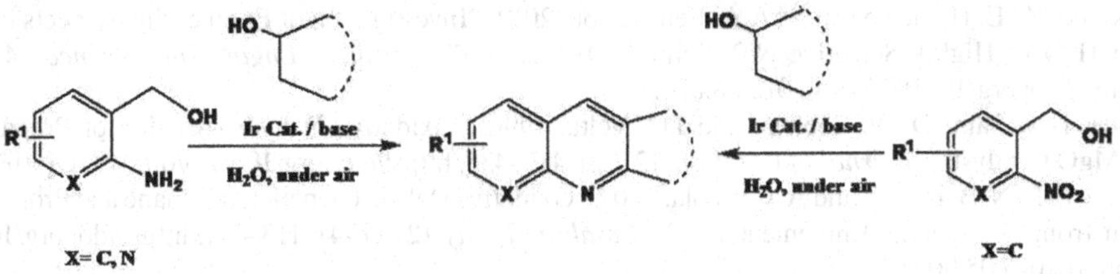

FIGURE (SCHEME) 15.8 Water soluble Ir-complex-catalyzed dehydrogenation reaction of alcohols [99].

to O_2 and N_2O. However, these methods have not been commercially viable until now, and various types of methodologies can be extensively explored for their commercialization at an industrial scale. Accordingly, going by a greener route is much more sustainable and presents great significance for the future prospects of dehydrogenation reactions.

15.7 CONFLICT OF INTEREST

The authors declare no conflict of interest.

REFERENCES

1. Hosseini, Seyed Ehsan, and Mazlan Abdul Wahid. 2016. "Hydrogen Production from Renewable and Sustainable Energy Resources: Promising Green Energy Carrier for Clean Development". *Renewable and Sustainable Energy Reviews* 57: 850–866. https://doi.org/10.1016/j.rser.2015.12.112.
2. Li, Guomin, Ce Liu, Xinjiang Cui, Yanhui Yang, and Feng Shi. 2021. "Oxidative Dehydrogenation of Light Alkanes with Carbon Dioxide". *Green Chemistry* 23 (2): 689–707. https://doi.org/10.1039/d0gc03705b.
3. Fairuzov, Danis, Ilias Gerzeliev, Anton Maximov, and Evgeny Naranov. 2021. "Catalytic Dehydrogenation of Ethane: A Mini Review of Recent Advances and Perspective of Chemical Looping Technology". *Catalysts* 11 (7): 833. https://doi.org/10.3390/catal11070833.
4. Agarwal, Ashwin, Debalina Sengupta, and Mahmoud El-Halwagi. 2018. "Sustainable Process Design Approach for On-Purpose Propylene Production and Intensification". *ACS Sustainable Chemistry & Engineering* 6 (2): 2407–2421. https://doi.org/10.1021/acssuschemeng.7b03854.

5. Gärtner, Christian A., André C. van Veen, and Johannes A. Lercher. 2014. "Oxidative Dehydrogenation of Ethane on Dynamically Rearranging Supported Chloride Catalysts". *Journal of the American Chemical Society* 136 (36): 12691–12701. https://doi.org/10.1021/ja505411s.
6. Kraus, Peter, and Rune Peter Lindstedt. 2021. "It's a Gas: Oxidative Dehydrogenation of Propane Over Boron Nitride Catalysts". *The Journal of Physical Chemistry C* 125 (10): 5623–5634. https://doi.org/10.1021/acs.jpcc.1c00165.
7. Venegas, Juan M., Zisheng Zhang, Theodore O. Agbi, William P. McDermott, Anastassia Alexandrova, and Ive Hermans. 2020. "Why Boron Nitride Is Such a Selective Catalyst for the Oxidative Dehydrogenation of Propane". *Angewandte Chemie International Edition* 59 (38): 16527–16535. https://doi.org/10.1002/anie.202003695.
8. Shi, Lei, Yang Wang, Bing Yan, Wei Song, Dan Shao, and An-Hui Lu. 2018. "Progress in Selective Oxidative Dehydrogenation of Light Alkanes to Olefins Promoted by Boron Nitride Catalysts". *Chemical Communications* 54 (78): 10936–10946. https://doi.org/10.1039/c8cc04604b.
9. Gärtner, Christian A., André C. van Veen, and Johannes A. Lercher. 2013. "Oxidative Dehydrogenation of Ethane: Common Principles and Mechanistic Aspects". *Chemcatchem* 5 (11): 3196–3217. https://doi.org/10.1002/cctc.201200966.
10. Donsì, Francesco, Kenneth A. Williams, and Lanny D. Schmidt. 2005. "A Multistep Surface Mechanism for Ethane Oxidative Dehydrogenation on Pt- and Pt/Sn-Coated Monoliths". *Industrial & Engineering Chemistry Research* 44 (10): 3453–3470. https://doi.org/10.1021/ie0493356.
11. Solsona, B., A. Dejoz, T. Garcia, P. Concepcion, J. Nieto, M. Vazquez, and M. Navarro. 2006. "Molybdenum–Vanadium Supported on Mesoporous Alumina Catalysts for the Oxidative Dehydrogenation of Ethane". *Catalysis Today* 117 (1–3): 228–233. https://doi.org/10.1016/j.cattod.2006.05.025.
12. Skoufa, Z., E. Heracleous, and A.A. Lemonidou. 2012. "Investigation of Engineering Aspects in Ethane ODH Over Highly Selective Ni0.85Nb0.15Ox Catalyst". *Chemical Engineering Science* 84: 48–56. https://doi.org/10.1016/j.ces.2012.08.007.
13. Siew Hew Sam, D., V. Soenen, and J.C. Volta. 1990. "Oxidative Dehydrogenation of Propane Over VMgO Catalysts". *Journal of Catalysis* 123 (2): 417–435. https://doi.org/10.1016/0021-9517(90)90139-b.
14. Cavani, F., N. Ballarini, and A. Cericola. 2007. "Oxidative Dehydrogenation of Ethane and Propane: How Far from Commercial Implementation?". *Catalysis Today* 127 (1–4): 113–131. https://doi.org/10.1016/j.cattod.2007.05.009.
15. Høj, Martin, Thomas Kessler, Pablo Beato, Anker D. Jensen, and Jan-Dierk Grunwaldt. 2014. "Structure, Activity and Kinetics of Supported Molybdenum Oxide and Mixed Molybdenum–Vanadium Oxide Catalysts Prepared by Flame Spray Pyrolysis for Propane OHD". *Applied Catalysis A: General* 472: 29–38. https://doi.org/10.1016/j.apcata.2013.11.027.
16. Barman, Samir, Niladri Maity, Kushal Bhatte, Samy Ould-Chikh, Oliver Dachwald, Carmen Haeßner, and Youssef Saih et al. 2016. "Single-Site Voxmoieties Generated on Silica by Surface Organometallic Chemistry: A Way to Enhance the Catalytic Activity in the Oxidative Dehydrogenation of Propane". *ACS Catalysis* 6 (9): 5908–5921. https://doi.org/10.1021/acscatal.6b01263.
17. Matam, Santhosh K., Caitlin Moffat, Pip Hellier, Michael Bowker, Ian P. Silverwood, C. Richard A. Catlow, and S. David Jackson et al. 2020. "Investigation OfMoox/Al2o3 Under Cyclic Operation for Oxidative and Non-Oxidative Dehydrogenation of Propane". *Catalysts* 10 (12): 1370. https://doi.org/10.3390/catal10121370.
18. Kumar, Sonu, Andrey Lyalin, Zhenguo Huang, and Tetsuya Taketsugu. 2022. "Catalytic Oxidative Dehydrogenation of Light Alkanes Over Oxygen Functionalized Hexagonal Boron Nitride". *Chemistryselect* 7 (1). https://doi.org/10.1002/slct.202103795.
19. Wang, Shaobin, and Z. H. Zhu. 2004. "Catalytic Conversion of Alkanes to Olefins by Carbon Dioxide Oxidative Dehydrogenationa Review". *Energy & Fuels* 18 (4): 1126–1139. https://doi.org/10.1021/ef0340716.
20. Mukherjee, Deboshree, Sang-Eon Park, and Benjaram M. Reddy. 2016. "CO2 as a Soft Oxidant for Oxidative Dehydrogenation Reaction: An Eco Benign Process for Industry". *Journal of CO2 Utilization* 16: 301–312. https://doi.org/10.1016/j.jcou.2016.08.005.
21. Deng, Shuang, Songgeng Li, Huiquan Li, and Yi Zhang. 2009. "Oxidative Dehydrogenation of Ethane to Ethylene with CO2 Over Fe–Cr/Zro2 Catalysts". *Industrial & Engineering Chemistry Research* 48 (16): 7561–7566. https://doi.org/10.1021/ie9007387.
22. de Oliveira, João F.S., Diogo P. Volanti, José M.C. Bueno, and Adriana P. Ferreira. 2018. "Effect of CO2 in the Oxidative Dehydrogenation Reaction of Propane Over Cr/Zro2 Catalysts". *Applied Catalysis A: General* 558: 55–66. https://doi.org/10.1016/j.apcata.2018.03.020.

23. Michorczyk, Piotr, Kamila Zeńczak, Rafał Niekurzak, and Jan Ogonowski. 2012. "Dehydrogenation of Propane with CO2—A New Green Process for Propene and Synthesis Gas Production". *PJCT* 14 (4): 77–82. https://doi.org/10.2478/v10026-012-0106-1.
24. Zhang, Ronghao, Hong Wang, Siyang Tang, Changjun Liu, Fan Dong, Hairong Yue, and Bin Liang. 2018. "Photocatalytic Oxidative Dehydrogenation of Ethane Using CO2 as a Soft Oxidant Over Pd/ TiO_2 Catalysts to C2H4 and Syngas". *ACS Catalysis* 8 (10): 9280–9286. https://doi.org/10.1021/acscatal.8b02441.
25. Al-Mamoori, Ahmed, Shane Lawson, Ali A. Rownaghi, and Fateme Rezaei. 2020. "Oxidative Dehydrogenation of Ethane to Ethylene in an Integrated CO2 Capture-Utilization Process". *Applied Catalysis B: Environmental* 278: 119329. https://doi.org/10.1016/j.apcatb.2020.119329.
26. Wang, Jian, Yong-Hong Song, Zhao-Tie Liu, and Zhong-Wen Liu. 2021. "Active and Selective Nature of Supported Crox for the Oxidative Dehydrogenation of Propane with Carbon Dioxide". *Applied Catalysis B: Environmental* 297: 120400. https://doi.org/10.1016/j.apcatb.2021.120400.
27. Nagieva, I. T. 2015. "Kinetics and Mechanism of Homogeneous Oxidative Dehydrogenation of Cyclohexane with Nitrous Oxide". *Russian Journal of Physical Chemistry A* 89 (10): 1762–1766. https://doi.org/10.1134/s0036024415100246.
28. Bulánek, Roman, Blanka Wichterlová, Kateřina Novoveská, and Viktor Kreibich. 2004. "Oxidation of Propane with Oxygen and/Or Nitrous Oxide Over Fe-ZSM-5 with Low Iron Concentrations". *Applied Catalysis A: General* 264 (1): 13–22. https://doi.org/10.1016/j.apcata.2003.12.020.
29. Pérez-Ramírez, J., and A. Gallardo-Llamas. 2005. "Impact of the Preparation Method and Iron Impurities in Fe-ZSM-5 Zeolites for Propylene Production Via Propane Oxidative Dehydrogenation with N2O". *Applied Catalysis A* 279: 117–123. https://doi.org/10.1016/j.apcata.2004.10.020
30. Zhou, Yanliang, Jian Lin, Lin Li, Ming Tian, Xiaoyu Li, Xiaoli Pan, Yang Chen, and Xiaodong Wang. 2019. "Improving the Selectivity of Ni-Al Mixed Oxides with Isolated Oxygen Species for Oxidative Dehydrogenation of Ethane with Nitrous Oxide". *Journal of Catalysis* 377: 438–448. https://doi.org/10.1016/j.jcat.2019.07.050.
31. Wu, Guangjun, Fei Hei, Naijia Guan, and Landong Li. 2013. "Oxidative Dehydrogenation of Propane with Nitrous Oxide Over Fe–MFI Prepared by Ion-Exchange: Effect of Acid Post-Treatments". *Catalysis Science & Technology* 3 (5): 1333. https://doi.org/10.1039/c3cy20782j.
32. Nicolau, Guillermo, Giulia Tarantino, and Ceri Hammond. 2019. "Acceptorless Alcohol Dehydrogenation Catalysed By Pd/C". *Chemsuschem* 12 (22): 4953–4961. https://doi.org/10.1002/cssc.201901313.
33. Nielsen, Martin, Anja Kammer, Daniela Cozzula, Henrik Junge, Serafino Gladiali, and Matthias Beller. 2011. "Efficient Hydrogen Production from Alcohols Under Mild Reaction Conditions". *Angewandte Chemie International Edition* 50 (41): 9593–9597. https://doi.org/10.1002/anie.201104722.
34. Crabtree, Robert H. 2017. "Homogeneous Transition Metal Catalysis of Acceptorless Dehydrogenative Alcohol Oxidation: Applications in Hydrogen Storage and to Heterocycle Synthesis". *Chemical Reviews* 117 (13): 9228–9246. https://doi.org/10.1021/acs.chemrev.6b00556.
35. Trincado, Monica, Jonas Bösken, and HansjörgGrützmacher. 2021. "Homogeneously Catalyzed Acceptorless Dehydrogenation of Alcohols: A Progress Report". *Coordination Chemistry Reviews* 443: 213967. https://doi.org/10.1016/j.ccr.2021.213967.
36. Wang, Tao, Jin Sha, Maarten Sabbe, Philippe Sautet, Marc Pera-Titus, and Carine Michel. 2021. "Identification of Active Catalysts for the Acceptorless Dehydrogenation of Alcohols to Carbonyls". *Nature Communications* 12 (1). https://doi.org/10.1038/s41467-021-25214-1.
37. Friedrich, Anja, and Sven Schneider. 2009. "Acceptorless Dehydrogenation of Alcohols: Perspectives for Synthesis and H_2 Storage". *Chemcatchem* 1 (1): 72–73. https://doi.org/10.1002/cctc.200900124.
38. Krylov, O. V., A. Kh. Mamedov, and S. R. Mirzabekova. 1995. "Oxidation of Hydrocarbons and Alcohols by Carbon Dioxide on Oxide Catalysts". *Industrial & Engineering Chemistry Research* 34 (2): 474–482. https://doi.org/10.1021/ie00041a007.
39. Nielsen, Martin, Henrik Junge, Anja Kammer, and Matthias Beller. 2012. "Towards a Green Process for Bulk-Scale Synthesis of Ethyl Acetate: Efficient Acceptorless Dehydrogenation of Ethanol". *Angewandte Chemie International Edition* 51 (23): 5711–5713. https://doi.org/10.1002/anie.201200625.
40. Nielsen, Martin, Henrik Junge, Anja Kammer, and Matthias Beller. 2012. "Towards a Green Process for Bulk-Scale Synthesis of Ethyl Acetate: Efficient Acceptorless Dehydrogenation of Ethanol". *Angewandte Chemie International Edition* 51 (23): 5711–5713. https://doi.org/10.1002/anie.201200625.
41. Gangwar, Manoj Kumar, Pardeep Dahiya, Balakumar Emayavaramban, and Basker Sundararaju. 2018. "Cp*CoIII-Catalyzed Efficient Dehydrogenation of Secondary Alcohols". *Chemistry: An Asian Journal* 13 (17): 2445–2448. https://doi.org/10.1002/asia.201800697.

42. Putro, Wahyu S., Takashi Kojima, Takayoshi Hara, Nobuyuki Ichikuni, and Shogo Shimazu. 2018. "Acceptorless Dehydrogenation of Alcohols Using Cu–Fe Catalysts Prepared from Cu–Fe Layered Double Hydroxides as Precursors". *Catalysis Science & Technology* 8 (12): 3010–3014. https://doi.org/10.1039/c8cy00655e.
43. Sahoo, Manoj K., and Ekambaram Balaraman. 2019. "Room Temperature Catalytic Dehydrogenation of Cyclic Amines with the Liberation of H_2 Using Water as a Solvent". *Green Chemistry* 21 (8): 2119–2128. https://doi.org/10.1039/c9gc00201d.
44. Bayat, Ahmad, Mehdi Shakourian-Fard, Nona Ehyaei, and Mohammad Mahmoodi Hashemi. 2015. "Silver Nanoparticles Supported on Silica-Coated Ferrite as Magnetic and Reusable Catalysts for Oxidant-Free Alcohol Dehydrogenation". *RSC Advances* 5 (29): 22503–22509. https://doi.org/10.1039/c4ra15498c.
45. Kon, Kenichi, S.M.A. Hakim Siddiki, and Ken-ichi Shimizu. 2013. "Size- and Support-Dependent Pt Nanocluster Catalysis for Oxidant-Free Dehydrogenation of Alcohols". *Journal of Catalysis* 304: 63–71. https://doi.org/10.1016/j.jcat.2013.04.003.
46. Wang, Fei, Ruijuan Shi, Zhi-Quan Liu, Pan-Ju Shang, Xueyong Pang, Shuai Shen, Zhaochi Feng, Can Li, and Wenjie Shen. 2013. "Highly Efficient Dehydrogenation of Primary Aliphatic Alcohols Catalyzed by Cu Nanoparticles Dispersed on Rod-Shaped La2o2co3". *ACS Catalysis* 3 (5): 890–894. https://doi.org/10.1021/cs400255r.
47. Gu, Xiaojun, Zhang-Hui Lu, Hai-Long Jiang, Tomoki Akita, and Qiang Xu. 2011. "Synergistic Catalysis of Metal–Organic Framework-Immobilized Au–Pd Nanoparticles in Dehydrogenation of Formic Acid for Chemical Hydrogen Storage". *Journal of the American Chemical Society* 133 (31): 11822–11825. https://doi.org/10.1021/ja200122f.
48. Wang, Zhi-Li, Jun-Min Yan, Yun Ping, Hong-Li Wang, Wei-Tao Zheng, and Qing Jiang. 2013. "An Efficient Coaupd/C Catalyst for Hydrogen Generation from Formic Acid at Room Temperature". *Angewandte Chemie International Edition* 52 (16): 4406–4409. https://doi.org/10.1002/anie.201301009.
49. Li, Zhangpeng, Xinchun Yang, Nobuko Tsumori, Zheng Liu, Yuichiro Himeda, Tom Autrey, and Qiang Xu. 2017. "Tandem Nitrogen Functionalization of Porous Carbon: Toward Immobilizing Highly Active Palladium Nanoclusters for Dehydrogenation of Formic Acid". *ACS Catalysis* 7 (4): 2720–2724. https://doi.org/10.1021/acscatal.7b00053.
50. Li, Renhong, Xiaohui Zhu, Xiaoqing Yan, Donghai Shou, Xin Zhou, and Wenxing Chen. 2016. "Single Component Gold on Protonated Titanate Nanotubes for Surface-Charge-Mediated, Additive-Free Dehydrogenation of Formic Acid Into Hydrogen". *RSC Advances* 6 (102): 100103–100107. https://doi.org/10.1039/c6ra19703e.
51. Kahri, Hamza, Melike Sevim, and Önder Metin. 2016. "Enhanced Catalytic Activity of Monodispersed Agpd Alloy Nanoparticles Assembled on Mesoporous Graphitic Carbon Nitride for the Hydrolytic Dehydrogenation of Ammonia Borane Under Sunlight". *Nano Research* 10 (5): 1627–1640. https://doi.org/10.1007/s12274-016-1345-x.
52. Yu, Chao, Xuefeng Guo, Bo Shen, Zheng Xi, Qing Li, Zhouyang Yin, Hu Liu, et al. 2018. "One-Pot Formic Acid Dehydrogenation and Synthesis of Benzene-Fused Heterocycles Over Reusable Agpd/WO2.72 Nanocatalyst". *Journal of Materials Chemistry A* 6 (46): 23766–23772. https://doi.org/10.1039/c8ta09342c.
53. Zhang, Sen, Önder Metin, Dong Su, and Shouheng Sun. 2013. "Monodisperse Agpd Alloy Nanoparticles and Their Superior Catalysis for the Dehydrogenation of Formic Acid". *Angewandte Chemie International Edition* 52 (13): 3681–3684. https://doi.org/10.1002/anie.201300276.
54. Lu, Nan, Xiaoqing Yan, Hui Ling Tan, Hisayoshi Kobayashi, Xuehan Yu, Yuezhou Li, Jiemei Zhang, et al. 2022. "Carbon-Catalyzed Oxygen-Mediated Dehydrogenation of Formaldehyde in Alkaline Solution for Efficient Hydrogen Production". *International Journal of Hydrogen Energy*. https://doi.org/10.1016/j.ijhydene.2022.06.134.
55. Su, Dang Sheng, Siglinda Perathoner, and Gabriele Centi. 2013. "Nanocarbons for the Development of Advanced Catalysts". *Chemical Reviews* 113 (8): 5782–5816. https://doi.org/10.1021/cr300367d.
56. Chen, De, Anders Holmen, Zhijun Sui, and Xinggui Zhou. 2014. "Carbon Mediated Catalysis: A Review on Oxidative Dehydrogenation". *Chinese Journal of Catalysis* 35 (6): 824–841. https://doi.org/10.1016/s1872-2067(14)60120-0.
57. Maciá-Agulló, J.A., D. Cazorla-Amorós, A. Linares-Solano, U. Wild, D.S. Su, and R. Schlögl. 2005. "Oxygen Functional Groups Involved in the Styrene Production Reaction Detected by Quasi in Situ XPS". *Catalysis Today* 102–103: 248–253. https://doi.org/10.1016/j.cattod.2005.02.023.
58. Zhang, Guoting, Chao Liu, Hong Yi, Qingyuan Meng, Changliang Bian, Hong Chen, Jing-Xin Jian, Li-Zhu Wu, and Aiwen Lei. 2015. "External Oxidant-Free Oxidative Cross-Coupling: A Photoredox Cobalt-Catalyzed Aromatic C–H Thiolation for Constructing C–S Bonds". *Journal of the American Chemical Society* 137 (29): 9273–9280. https://doi.org/10.1021/jacs.5b05665.

59. Niu, Linbin, Hong Yi, Shengchun Wang, Tianyi Liu, Jiamei Liu, and Aiwen Lei. 2017. "Photo-Induced Oxidant-Free Oxidative C–H/N–H Cross-Coupling Between Arenes and Azoles". *Nature Communications* 8 (1). https://doi.org/10.1038/ncomms14226.
60. Crisenza, Giacomo E. M., Adriana Faraone, Eugenio Gandolfo, Daniele Mazzarella, and Paolo Melchiorre. 2021. "Catalytic Asymmetric C–C Cross-Couplings Enabled by Photoexcitation". *Nature Chemistry* 13 (6): 575–580. https://doi.org/10.1038/s41557-021-00683-5.
61. He, Ke-Han, Fang-Fang Tan, Chao-Zheng Zhou, Gui-Jiang Zhou, Xiao-Long Yang, and Yang Li. 2017. "Acceptorless Dehydrogenation of N-Heterocycles by Merging Visible-Light Photoredox Catalysis and Cobalt Catalysis". *Angewandte Chemie International Edition* 56 (11): 3080–3084. https://doi.org/10.1002/anie.201612486.
62. Guerrero, Isabel, Zsolt Kelemen, Clara Viñas, Isabel Romero, and Francesc Teixidor. 2020. "Metalla carboranes as Photoredox Catalysts in Water". *Chemistry: A European Journal* 26 (22): 5027–5036. https://doi.org/10.1002/chem.201905395.
63. Guerrero, Isabel, Arpita Saha, Jewel Ann Maria Xavier, Clara Viñas, Isabel Romero, and Francesc Teixidor. 2020. "Noncovalently Linked Metallacarboranes on Functionalized Magnetic Nanoparticles as Highly Efficient, Robust, and Reusable Photocatalysts in Aqueous Medium". *ACS Applied Materials & Interfaces* 12 (50): 56372–56384. https://doi.org/10.1021/acsami.0c17847.
64. Guerrero, Isabel, Clara Viñas, Xavier Fontrodona, Isabel Romero, and Francesc Teixidor. 2021. "Aqueous Persistent Noncovalent Ion-Pair Cooperative Coupling in a Ruthenium Cobaltabis(Dicarbollide) System as a Highly Efficient Photoredox Oxidation Catalyst". *Inorganic Chemistry* 60 (12): 8898–8907. https://doi.org/10.1021/acs.inorgchem.1c00751.
65. Crisenza, Giacomo E. M., and Paolo Melchiorre. 2020. "Chemistry Glows Green with Photoredox Catalysis". *Nature Communications* 11 (1). https://doi.org/10.1038/s41467-019-13887-8.
66. Ciambelli, P., D. Sannino, V. Palma, V. Vaiano, and R. S. Mazzei. 2009. "A Step Forwards in Ethanol Selective Photo-Oxidation". *Photochemical & Photobiological Sciences* 8 (5): 699. https://doi.org/10.1039/b818053a.
67. Ni, Meng, Michael K.H. Leung, Dennis Y.C. Leung, and K. Sumathy. 2007. "A Review and Recent Developments in Photocatalytic Water-Splitting Using TiO_2 for Hydrogen Production". *Renewable and Sustainable Energy Reviews* 11 (3): 401–425. https://doi.org/10.1016/j.rser.2005.01.009.
68. Esswein, Arthur J., and Daniel G. Nocera. 2007. "Hydrogen Production by Molecular Photocatalysis". *Chemical Reviews* 107 (10): 4022–4047. https://doi.org/10.1021/cr050193e.
69. Leung, Dennis Y. C., Xianliang Fu, Cuifang Wang, Meng Ni, Michael K. H. Leung, Xuxu Wang, and Xianzhi Fu. 2010. "Hydrogen Production Over Titania-Based Photocatalysts". *Chemsuschem* 3 (6): 681–694. https://doi.org/10.1002/cssc.201000014.
70. Lu, Haiqiang, Jianghong Zhao, Li Li, Liming Gong, Jianfeng Zheng, Lexi Zhang, Zhijian Wang, Jian Zhang, and Zhenping Zhu. 2011. "Selective Oxidation of Sacrificial Ethanol Over TiO_2-Based Photocatalysts During Water Splitting". *Energy Environmental Science* 4 (9): 3384. https://doi.org/10.1039/c1ee01476e.
71. Zhang, Hongxia, Zhenping Zhu, Yupeng Wu, Tianjian Zhao, and Li Li. 2014. "TiO_2-Photocatalytic Acceptorless Dehydrogenation Coupling of Primary Alkyl Alcohols Into Acetals". *Green Chemistry* 16 (9): 4076–4080. https://doi.org/10.1039/c4gc00413b.
72. Sahoo, Manoj K., Garima Jaiswal, Jagannath Rana, and Ekambaram Balaraman. 2017. "Organo-Photoredox Catalyzed Oxidative Dehydrogenation of N-Heterocycles". *Chemistry: A European Journal* 23 (57): 14167–14172. https://doi.org/10.1002/chem.201703642.
73. Srinath, S., R. Abinaya, Arun Prasanth, M. Mariappan, R. Sridhar, and B. Baskar. 2020. "Reusable, Homogeneous Water Soluble Photoredox Catalyzed Oxidative Dehydrogenation of N-Heterocycles in a Biphasic System: Application to the Synthesis of Biologically Active Natural Products". *Green Chemistry* 22 (8): 2575–2587. https://doi.org/10.1039/d0gc00569j.
74. Ciambelli, P., D. Sannino, V. Palma, and V. Vaiano. 2005. "Photocatalysed Selective Oxidation of Cyclohexane to Benzene on Moox/TiO_2". *Catalysis Today* 99 (1–2): 143–149. https://doi.org/10.1016/j.cattod.2004.09.034.
75. Sannino, D., V. Vaiano, P. Ciambelli, P. Eloy, and E.M. Gaigneaux. 2011. "Avoiding the Deactivation of SulphatedMoox/TiO_2 Catalysts in the Photocatalytic Cyclohexane Oxidative Dehydrogenation by a Fluidized Bed Photoreactor". *Applied Catalysis A: General* 394 (1–2): 71–78. https://doi.org/10.1016/j.apcata.2010.12.025.
76. Ciambelli, P., D. Sannino, V. Palma, V. Vaiano, and R. S. Mazzei. 2011. "Intensification of Gas-Phase Photoxidative Dehydrogenation of Ethanol to Acetaldehyde by Using Phosphors as Light Carriers". *Photochemical and Photobiological Sciences* 10 (3): 414–418. https://doi.org/10.1039/c0pp00186d.

77. Sannino, Diana, Vincenzo Vaiano, Paolo Ciambelli, M. Carmen Hidalgo, Julie J. Murcia, and J. Antonio Navío. 2012. "Oxidative Dehydrogenation of Ethanol Over Au/TiO$_2$ Photocatalysts". *Journal of Advanced Oxidation Technologies* 15 (2). https://doi.org/10.1515/jaots-2012-0206.
78. Liu, Hu, Xinyang Liu, Weiwei Yang, Mengqi Shen, Shuo Geng, Chao Yu, Bo Shen, and Yongsheng Yu. 2019. "Photocatalytic Dehydrogenation of Formic Acid Promoted by a Superior Pdag@G-C3N4 Mott–Schottky Heterojunction". *Journal of Materials Chemistry A* 7 (5): 2022–2026. https://doi.org/10.1039/c8ta11172c.
79. Bockris, J. O.'M. 1972. "A Hydrogen Economy". *Science* 176 (4041): 1323–1323. https://doi.org/10.1126/science.176.4041.1323.
80. Bockris, J. 2002. "The Origin of Ideas on a Hydrogen Economy and Its Solution to the Decay of the Environment". *International Journal of Hydrogen Energy* 27 (7–8): 731–740. https://doi.org/10.1016/s0360-3199(01)00154-9.
81. Midilli, A., M. Ay, I. Dincer, and M.A. Rosen. 2005. "On Hydrogen and Hydrogen Energy Strategies". *Renewable and Sustainable Energy Reviews* 9 (3): 255–271. https://doi.org/10.1016/j.rser.2004.05.003.
82. Loges, Björn, Albert Boddien, Felix Gärtner, Henrik Junge, and Matthias Beller. 2010. "Catalytic Generation of Hydrogen from Formic Acid and Its Derivatives: Useful Hydrogen Storage Materials". *Topics in Catalysis* 53 (13–14): 902–914. https://doi.org/10.1007/s11244-010-9522-8.
83. Navlani-García, Miriam, Kohsuke Mori, David Salinas-Torres, Yasutaka Kuwahara, and Hiromi Yamashita. 2019. "New Approaches Toward the Hydrogen Production from Formic Acid Dehydrogenation Over Pd-Based Heterogeneous Catalysts". *Frontiers in Materials* 6. https://doi.org/10.3389/fmats.2019.00044.
84. Montandon-Clerc, Mickael, Andrew F. Dalebrook, and Gábor Laurenczy. 2016. "Quantitative Aqueous Phase Formic Acid Dehydrogenation Using Iron(II) Based Catalysts". *Journal of Catalysis* 343: 62–67. https://doi.org/10.1016/j.jcat.2015.11.012.
85. Zhou, Wei, Zhihong Wei, Anke Spannenberg, Haijun Jiao, Kathrin Junge, Henrik Junge, and Matthias Beller. 2019. "Cobalt-catalyzed aqueous dehydrogenation of formic acid". *Chemistry: A European Journal* 25 (36): 8459–8464. https://doi.org/10.1002/chem.201805612.
86. Wang, Zhijun, Sheng-Mei Lu, Jun Li, Jijie Wang, and Can Li. 2015. "Unprecedentedly High Formic Acid Dehydrogenation Activity on an Iridium Complex with Ann, N'-Diimine Ligand in Water". *Chemistry: A European Journal* 21 (36): 12592–12595. https://doi.org/10.1002/chem.201502086.
87. Yu, Zhongliang, Yanyan Yang, Song Yang, Jie Zheng, Xiaogang Hao, Guoqiang Wei, Hongcun Bai, Abuliti Abudula, and Guoqing Guan. 2022. "Selective Dehydrogenation of Aqueous Formic Acid Over Multifunctional Γ-Mo2n Catalysts at a Temperature Lower Than 100°C". *Applied Catalysis B: Environmental* 313: 121445. https://doi.org/10.1016/j.apcatb.2022.121445.
88. Tedsree, Karaked, Tong Li, Simon Jones, Chun Wong Aaron Chan, Kai Man Kerry Yu, Paul A. J. Bagot, Emmanuelle A. Marquis, George D. W. Smith, and Shik Chi Edman Tsang. 2011. "Hydrogen Production from Formic Acid Decomposition at Room Temperature Using a Ag–Pd Core–Shell Nanocatalyst". *Nature Nanotechnology* 6 (5): 302–307. https://doi.org/10.1038/nnano.2011.42.
89. Huang, Yunjie, Xiaochun Zhou, Min Yin, Changpeng Liu, and Wei Xing. 2010. "Novel Pdau@Au/C Core–Shell Catalyst: Superior Activity and Selectivity in Formic Acid Decomposition for Hydrogen Generation". *Chemistry of Materials* 22 (18): 5122–5128. https://doi.org/10.1021/cm101285f.
90. Zhang, Sen, Önder Metin, Dong Su, and Shouheng Sun. 2013. "Monodisperse Agpd Alloy Nanoparticles and Their Superior Catalysis for the Dehydrogenation of Formic Acid". *Angewandte Chemie International Edition* 52 (13): 3681–3684. https://doi.org/10.1002/anie.201300276.
91. Metin, Önder, Xiaolian Sun, and Shouheng Sun. 2013. "Monodisperse Gold–Palladium Alloy Nanoparticles and Their Composition-Controlled Catalysis in Formic Acid Dehydrogenation Under Mild Conditions". *Nanoscale* 5 (3): 910–912. https://doi.org/10.1039/c2nr33637e.
92. Wang, Zhi-Li, Jun-Min Yan, Yun Ping, Hong-Li Wang, Wei-Tao Zheng, and Qing Jiang. 2013. "An Efficient Coaupd/C Catalyst for Hydrogen Generation From Formic Acid at Room Temperature". *Angewandte Chemie International Edition* 52 (16): 4406–4409. https://doi.org/10.1002/anie.201301009.
93. Liu, Hu, Xinyang Liu, Yongsheng Yu, Weiwei Yang, Ji Li, Ming Feng, and Haibo Li. 2018. "Bifunctional Networked Ag/Agpd Core/Shell Nanowires for the Highly Efficient Dehydrogenation of Formic Acid and Subsequent Reduction of Nitrate and Nitrite in Water". *Journal of Materials Chemistry A* 6 (11): 4611–4616. https://doi.org/10.1039/c8ta00600h.
94. Patra, Soumyadip, Mahendra K. Awasthi, Rohit K. Rai, Hemanta Deka, Shaikh M. Mobin, and Sanjay K. Singh. 2019. "Dehydrogenation of Formic Acid Catalyzed by Water-Soluble Ruthenium Complexes: X-Ray Crystal Structure of a Diruthenium Complex". *European Journal of Inorganic Chemistry* 2019 (7): 1046–1053. https://doi.org/10.1002/ejic.201801501.

95. Rodríguez-Lugo, Rafael E., Mónica Trincado, Matthias Vogt, Friederike Tewes, Gustavo Santiso-Quinones, and Hansjörg Grützmacher. 2013. "A Homogeneous Transition Metal Complex for Clean Hydrogen Production from Methanol–Water Mixtures". *Nature Chemistry* 5 (4): 342–347. https://doi.org/10.1038/nchem.1595.
96. Nielsen, Martin, Elisabetta Alberico, Wolfgang Baumann, Hans-Joachim Drexler, Henrik Junge, Serafino Gladiali, and Matthias Beller. 2013. "Low-Temperature Aqueous-Phase Methanol Dehydrogenation to Hydrogen and Carbon Dioxide". *Nature* 495 (7439): 85–89. https://doi.org/10.1038/nature11891.
97. Monney, Angèle, Enrico Barsch, Peter Sponholz, Henrik Junge, Ralf Ludwig, and Matthias Beller. 2014. "Base-Free Hydrogen Generation from Methanol Using a Bi-Catalytic System". *Chemical Communications* 50 (6): 707–709. https://doi.org/10.1039/c3cc47306f.
98. Debnath, Tanay, Tamalika Ash, Avik Ghosh, Subhendu Sarkar, and Abhijit Kr. Das. 2018. "Exploration of Unprecedented Catalytic Dehydrogenation Mechanism of Methylamine-Water Mixture in Presence of Ru-Pincer Complex: A Systematic DFT Study". *Journal of Catalysis* 363: 164–182. https://doi.org/10.1016/j.jcat.2018.04.021.
99. Maji, Milan, Kaushik Chakrabarti, Dibyajyoti Panja, and Sabuj Kundu. 2019. "Sustainable Synthesis of N-Heterocycles in Water Using Alcohols Following the Double Dehydrogenation Strategy". *Journal of Catalysis* 373: 93–102. https://doi.org/10.1016/j.jcat.2019.03.028.

16 Application of Pt- and Non Pt-Based Zeolitic Catalysts for the Dehydrogenation of Light Alkanes

Hardik Koshti and Rajib Bandyopadhyay

16.1 INTRODUCTION

Since 1930, the insistence for saturated acyclic hydrocarbons has continued to rise. This is mostly due to their higher reactivity and conventional usage as a construction constituent in the manufacture of commodities, polymers, and specialised chemicals for routine use. Remarkable socioeconomic growth in the last decade of the twentieth century was largely fuelled by the accessibility of energy in the form of gas and oil. Natural fossil resources are primarily a collection of saturated or aromatic hydrocarbons with no olefins. As a result, on-purpose and conventional technologies are on trend for producing light alkenes. After C_2H_4, C_3H_6 is the world's second-largest producer of acyclic hydrocarbons, but it is largely generated as a by-product of fluid catalytic cracking and steam crackers. Propylene demand is steadily increasing because of global economic expansion. Alternative propylene production systems using new, cheap, and on-purpose propylene are gaining popularity (e.g., dehydrogenation of C_3H_8). Likewise, straight dehydrogenation of isobutane to isobutene, a transitional alkene for the synthesis of high octane oxygenates (i.e., methyl tert-butyl ether [MTBE]), is commonly used. Because of the increased demand for refinery fuel, on-purpose olefin mechanisms, namely, straight dehydrogenation, have become more important than traditional techniques such as fluid catalytic cracking (FCC) and steam cracking. Direct dehydrogenation produces propylene and isobutylene in a variety of petrochemical complexes (Hamid and Ali 2004). On-purpose reactions, such as propane dehydrogenation, methane to olefins, methane to propylene, syngas to alkenes, and double decomposition can also yield olefins. Dehydrogenation is an industrially important chemical transition that is used to make a variety of plastic products and chemicals. Moreover, it was one of the first industrial techniques for the manufacture of petrochemicals. C_3H_6 is a low boiling, colourless, highly flammable, and volatile gas used in the industry. C_3H_6 is sold in three different ranks: polymer grade with a purity of 99.5 % or higher (PG), chemical grade with a purity of 90–96 % (CG), and refinery grade with a purity of 50–70 % (RG). Propylene oxide and polypropylene are both made from PG. Impurities such as carbonyl sulphide, which can poison catalysts, are rare in PG propylene. Polypropylene, propylene oxide oxo-alcohols, acrylonitrile, butanal, cumene, and propene oligomers are the most common chemical uses of propylene. Ethylene-propene rubbers and acrylic acid derivatives are two other applications. Most chemical derivatives use CG propylene extensively (e.g., acrylonitrile and oxo-alcohols). Polypropylene production dominates worldwide propylene demand, accounting for roughly 2/3 of overall propylene demand. Moreover, it is valuable in some chemical reactions (e.g., isopropanol or cumene). The conversion of RG propylene to CG or PG propylene for consumption in the manufacturing of propylene oxide, oxo-alcohols, acrylonitrile, and polypropyleneis is the most important market for RG propylene. Mechanical components, containers, fibres, ABS polymer, propylene oxide, propylene glycol, phenolic resin acrylic acid coatings, antifreeze, plasticizer cumene polycarbonates polyurethane, oxo-alcohol coatings, adhesives, and superabsorbent polymers are some of the major propylene consumers.

During World War II, the Germans used dimerization, hydrogenation, and catalytic dehydrogenation of C_4H_{10} to produce high-octane aviation fuel (Bhasin et al. 2001). The use of oxygenated hydrocarbons as gasoline fuels resulted in a surge in MTBE production. For the large-scale manufacture of light olefins, particularly isobutylene for MTBE, direct dehydrogenation of paraffins is used. n-butane can be changed into its isomer iso-butane, which can then be dehydrogenated to yield corresponding isobutylene. Later, iso-butylene and methanol are combined in a liquid phase reaction on a strong acidic resin to produce MTBE (Brockwell et al. 1991). Olah and Molnar in 2003 (Olah and Molnár 2003), Fei et al. in 2008 (Fei et al. 2008), and Nawaz et al. in 2009 (Nawaz and Wei 2010, Tang et al. 2010; Nawaz, Baksh et al. 2013; Nawaz, Chu et al. 2010) described the dehydrogenation of iso-butane as an equilibrium-restricted and extremely endothermic reaction carried out at 555°C–649°C using Cr- or Pt-based catalysts. With a lower pressure and lower temperature, the reaction rate and equilibrium conversion rise. At the same interval, a greater temperature promotes coking and cracking processes, which diminish the specificity of alkenes. Coking also poisons the catalyst, which is then routinely burned in an oxygen-diluents solution to revive it and produce heat. Dehydrogenation of saturated acyclic hydrocarbons is one of the utmost difficult chemical procedures to execute on a large scale because the thermodynamic stability restricts the greatest conversion per pass; it is also very endothermic and requires considerable heat. The complexation of the dehydrogenation reaction shows that a greater pressure has a negative impact on the reaction. Regardless, light alkane dehydrogenation processes are typically sluggish (Miracca and Piovesan 1999). Aromatization, cracking, and other side reactions are kinetically and thermodynamically preferred under specific reaction conditions; therefore, the problems are not confined to this sluggishness. To make a commercial process further profitable, a good idea is to enhance the conversion per pass at the point where the intended product's selectivity is highest so that downstream treatment and recycling are less expensive. Coke is generated and is an unavoidable by-product of the dehydrogenation reaction, which reduces catalyst activity over time (Miracca and Piovesan 1999; Nawaz, Zhu et al. 2009; Nawaz, Shu et al. 2009; Nawaz, Tang et al. 2009; Nawaz, Xiaoping et al. 2009; Nawaz, Chu et al. 2010; Nawaz, Qing et al. 2010; Nawaz and Wei 2013).

Zeolites and zeolite supports are environmentally benign, shape selective, non-corrosive, stable, and hydrothermally solid acid catalysts and have gained prominence in the last three decades for their use in dehydrogenation processes. Zeolites feature well-defined pore architectures with Bronsted and/or Lewis acid locates (sites). Direct protonation of alkane by Bronsted acid sites is the well-known starting step (Miracca and Piovesan 1999; Nawaz, Zhu et al. 2009; Nawaz, Shu et al. 2009; Nawaz, Tang et al. 2009; Nawaz, Xiaoping et al. 2009; Nawaz, Chu et al. 2010; Nawaz, Qing et al. 2010; Nawaz and Wei 2013; Nawaz et al. 2013; Vaezifar et al. 2011). Zeolite acid sites have been seen to be insufficient for moderate alkane protonation to produce an alkoxy species mechanism, as demonstrated by experiments (Miracca and Piovesan 1999; Nawaz, Zhu et al. 2009; Nawaz, Shu et al. 2009; Nawaz, Tang et al. 2009; Nawaz, Xiaoping et al. 2009; Nawaz, Chu et al. 2010; Nawaz, Qing et al. 2010; Nawaz and Wei 2013; Nawaz et al. 2013). Lewis acid sites of zeolite provide another route for the start of cracking and dehydrogenation. This is somehow connected to the production of carbenium ion-activated complexes and hydride abstraction from alkanes. Bimolecular hydride transfer and agglomeration reactions involve the original carbenium ion complexes and mono molecular processes.

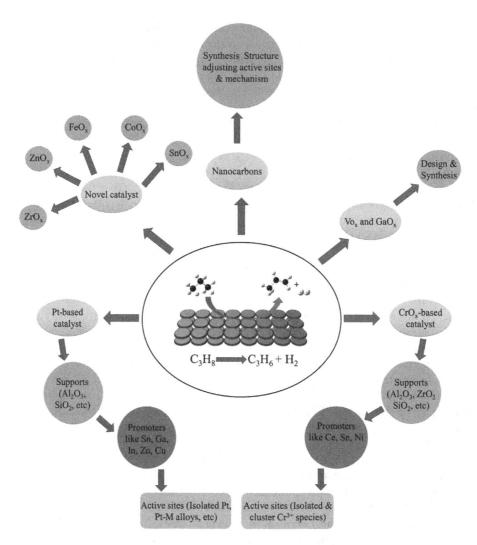

FIGURE 16.1 Various catalysts for the dehydrogenation of propane to propylene (Hu et al. 2019).

Zeolite's framework, structure, particle size, and acidity play major roles in determining how well it performs catalysis (Rownaghi et al. 2011; Rownaghi and Hedlund 2011; Rownaghi et al. 2012a, 2012b; Rownaghi, Rezaei, Stante et al. 2012; Li et al. 2016). Because of the typically slow mass transfer into and out of the active locates in zeolite's micropores, there is the potential for increased propylene oligomerisation and coke production due to a prolonged residence time (Li et al. 2014; Haw and Marcus 2005). There is general agreement that propane dehydrogenation reactions take place on zeolite's acid sites. Additionally, the zeolite's physicochemical and reactivity characteristics are significantly influenced by the SiO_2/Al_2O_3 ratio (Armaroli et al. 2006; Thakkar, Issa et al. 2017; Thakkar et al. 2016; Thakkar, Eastman et al. 2017). A 2-D channel system interconnecting micro networks with a 5–6 Å size is provided by upturned mordenite (MFI) zeolites such as H-ZSM-5 and silica solids such as MCM-41 (Michorczyk et al. 2008), MSU-x (Liu et al. 2007; Michorczyk et al. 2012), SBA-1 (Michorczyk et al. 2012), and SBA-15 (Li et al. 2016). In the oxidative dehydrogenation (ODH) of alkanes with CO_2, these materials have received extensive evaluation.

16.2 CHEMISTRY OF DEHYDROGENATION

The non-oxidative method to light olefins, known as direct dehydrogenation, is discussed here. The activation energy of the dehydrogenation of isobutane and propane to equivalent olefin transformations is 121–143 kJ/mol, as shown in (Airaksinen et al. 2002).

$$C_3H_8 \rightleftharpoons C_3H_6 + H_2$$
$$i\text{-}C_4H_{10} \rightleftharpoons i\text{-}C_4H_8 + H_2$$

Because the processes are reversible, strongly endothermic, and likely lead to volume expansion, a lower pressure (under vacuum) and higher temperature favour forward reactions. The energy demand for endothermic reactions is the most critical feature of light alkane dehydrogenation. In addition to catalysts, one of the key technological hurdles that distinguishes commercial reactor designs and processes is heat input to the reactor (Sanfilippo 2011). Conversely, high temperature stimulates secondary reactions and the generation of coke while also deactivating the catalyst (Miracca and Piovesan 1999; Nawaz, Zhu et al. 2009; Nawaz, Shu et al. 2009; Nawaz, Tang et al. 2009; Nawaz, Xiaoping et al. 2009; Nawaz, Chu et al. 2010; Nawaz, Qing et al. 2010; Nawaz and Wei 2013).

Because C-H bonds are much more reactive than C-C bonds in paraffins and olefins, a catalyst that favours C-H bond rupture over C-C bond cleavage to avoid side reactions is a favourable option for paraffin dehydrogenation. Undesirable side and secondary reactions, are a problematic concern for the more reactive olefins than their respective paraffins. Alkane hydrogenolysis as catalysed by platinum sites requires C_xH_y adspecies and/or hydrogen atoms (Nawaz and Wei 2013; Cecilia et al. 2015).

For the direct dehydrogenation reaction, catalysis is widely used. Many materials have been investigated as light alkane catalysts; however, two forms of combinations are used in practise: catalysts based on metal oxide and catalysts based on noble metals (e.g., Pt-based and promoted Cr-based catalysts) (Airaksinen et al. 2002; Sanfilippo 2011). As a result of the harsh reaction conditions, catalysts are aimed to restrict unpreventable termination because of morphological transformations, volatilization, sintering, and hydrothermal restoration with oxidants resulting in a long active life (Airaksinen et al. 2002; Sanfilippo 2011). The only metal that has been thoroughly researched is platinum. CrO_x, however, is the most well-known metal oxide, with commercial applications in

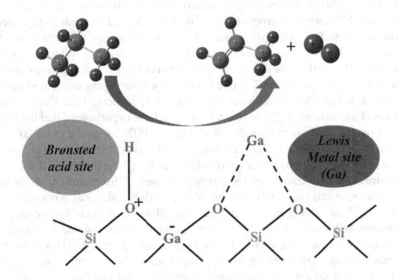

FIGURE 16.2 Dehydrogenation reaction catalysed by gallosilicate MFI zeolite (Choi et al. 2017).

mild alkane dehydrogenation. The usage of Ga, V, In, Zn, and Mo oxides has produced promising results (Miracca and Piovesan 1999; Nawaz, Shu et al. 2009; Nawaz, Tang et al. 2009; Nawaz, Xiaoping et al. 2009; Nawaz, Chu et al. 2010; Nawaz, Zhu et al. 2009; Nawaz, Qing et al. 2010; Nawaz and Wei 2013; Rownaghi, Rezaei, Stante et al. 2012; Li et al. 2016; Li et al. 2014; Haw and Marcus 2005; Armaroli et al. 2006; Thakkar, Issa et al. 2017; Thakkar et al. 2016; Thakkar, Eastman et al. 2017; Michorczyk et al. 2008; Liu et al. 2007; Michorczyk et al. 2012; Airaksinen et al. 2002; Sanfilippo 2011). The inevitable build-up of coke on the catalyst surface causes a gradual loss in catalytic activity and necessitates routine renewal. Because some catalysts also utilise the redispersion of active metal, each catalyst undergoes a sequence of introductions to changing atmospheres: dropping during reaction and oxidising during regeneration. An appropriate catalyst is one that has longer stability, permits high reaction speeds, and has a higher olefin selectivity (Sattler, Ruiz-Martinez et al. 2014; Sattler, Gonzalez-Jimenez et al. 2014). The following are the primary categories of dehydrogenation catalysts described in patents and academic publications.

16.3 NON PT-BASED ZEOLITIC CATALYST

16.3.1 Vanadium Oxide-Based Zeolitic Materials

Vanadium-based materials are the most effective, selective, and thoroughly studied catalysts for the ODH of alkanes (Nakagawa et al. 2001; Liu et al. 2001; Chen et al. 2002; Sattler, Gonzalez-Jimenez et al. 2014). It is usually found that reinforced V_2O_5 catalysts are more specific than unassisted bulk V_2O_5 catalysts because the VO_x species reacts with oxide supports (such as ZrO_2, Nb_2O_5, Al_2O_3, TiO_2, SiO_2, etc.) (Pak et al. 2002; Chen et al. 2002; Liu et al. 2016). The interaction amongst the deposited V_2O_5 and the oxide support is thought to determine the structure of subsequent surface VO_x species (Kondratenko et al. 2005; Rownaghi and Taufiq-Yap 2010; Rownaghi et al. 2009). It is believed that factors such as the calcination temperature, loading, additives, proportion of different vanadium oxide species, and specific support used affect vanadium oxide's catalytic performance in the oxidative dehydrogenation of propane (ODP) and its specificity for propylene. Both unsupported and supported vanadium catalysts react according to the Mars-Van Krevelen redox mechanism, which consists of the activation of C-H bonds by lattice oxygen (Chen et al. 2002; Sugiyama et al. 2006). Lattice oxygen atoms are involved in the C-H bond activation stage, and a detailed mechanistic and kinetic investigation has demonstrated that every reaction turnover needs a reduction (2 electron) of high-valent M^{n+} cations to produce 1 M^{n-2} or 2 M^{n-1} ions. The created alkyl species desorb into propylene, while residual −OH groups interact with nearby groups to create reduced vanadium centres and water molecules (Sugiyama et al. 2006). The formation of water may be encouraged by the addition of CO_2. By the dissociative chemisorption of O_2, the reduced vanadium centres can be reoxidised (Sugiyama et al. 2006). The rate constant for re-oxidising condensed centres with O_2 on MoO_x- and VO_x-based catalysts is higher than those for re-reducing them through the C-H bond activation stages (Langanke et al. 2014). Several scientists have hypothesised that the ODP rate rises as active metal oxides grow more reducible (Chen et al. 2002; Langanke et al. 2014; Grabowski 2006; Takehira et al. 2004). In the ODP reaction over VO_x provinces and oxygen vacancies, O-H groups and surface O_2 are the primary reactive intermediates (Botavina et al. 2008).

V_2O_5 is commonly well-known for oxidation processes of hydrocarbon, while at the same time, considerable research into ODH utilising V-Mg mixed oxides was reported by Wachs and Weckhuysen in 1997 and Zhang et al. in 2011 (Zhang, Zhou, Tang et al. 2012; Zhang, Zhou et al. 2011; Weckhuysen and Wachs 1996). Monolayer coverage causes isolated vanadium crystallites (species) to develop, such as when V-O-V, VO, or V-O-hosted (diverse dioxides with Zr, Mg, Ti, or Al) are combined, an anchoring site (hydroxyl groups) should be present on the support's surface to produce a well-scattered catalyst. According to studies, V species that are straight attached to the γ/θ-alumina support are active positions for dehydrogenation, while remote active sites result in deactivation because they produce coke. Reduction of vanadium oxide with CO, CH_4, and H_2 was seen to never finish, and oxide species such V^{3+}, V^{4+}, and V^{5+} was seen to stay on the catalyst surface (Zhang, Zhou et al. 2011). According to Zhao et al. (Michorczyk et al. 2010), the intrinsic activity for propane dehydrogenation and coke deposition is closely connected to the presence of –OH groups on the surface of the VO_x/Al_2O_3 catalyst. The surface –OH groups on the pre-reduced catalyst significantly reduced coke deposition compared to the material without pre-reduction. However, the valence state transition of the active vanadium species generated by these hydroxyl groups resulted in an unavoidable reduction in the intrinsic activity of the pre-reduced sample. As the support, Baek et al. (2012) developed mesoporous alumina with a constrained pore size distribution and a substantial specific surface area, which outperformed the conventional alumina-supported catalyst in terms of activity and stability. Alkali or alkali earth metals have traditionally been added to VO_x/Al_2O_3 as promoters to change the acidic characteristics of alumina and increase its selectivity to olefins. In contrast to the nanoparticles of V_2O_5, Liu et al. (Liu et al. 2016) discovered that a moderate quantity of Mg promoted the synthesis of two-dimensional VO_x species. As a result, the selectivity to propene dramatically enhanced from 82 to 93 %, and the catalyst stability was improved because of the significantly decreased coke deposition rate.

Zhang et al. (Zhang, Jia et al. 2011) investigated a variety of V-Cr/SBA-15/Al_2O_3/FeCrAl metal massive material for the ODH reaction including V values of 10 wt. % and Cr matters ranging from 0 to 12.5 wt. %. Catalysts were assessed in fixed bed reactor (FBR) at 450–700 °C, 1 atm, GHSV = 14,400 ml/g.h. for the C_3H_8/CO_2 space velocity and for a molar ratio of 3 for the V_{CO2}/V_{C3H8}. X-ray photoelectron spectroscopy (XPS) analysis was used to determine the oxidation states of vanadium and were found to be +5 and +4. The comparative content and overall quantity of the V^{5+} species rose with the incorporation of the Cr species and peaked at 10 wt. % Cr content. With even more of an upsurge in the Cr level, to 12.5 %, V^{5+} species sharply decreased. More Cr^{3+} species are developing, which prevents the emergence of distributed tetrahedral V^{5+} species as a result. The catalysts also contained Cr^{3+} and Cr^{6+} species, with the quantity of Cr^{6+} reaching a peak at 10 wt. % Cr. By showing C_3H_8 conversion of 49.9 %, a specificity of 86.5 % for C_3H_6, and a yield of 42.5 % at 650 °C, the V-Cr/SBA-15/Al_2O_3/FeCrAl catalyst having 10 wt. % Cr and 10 wt. % validated the best efficiency. Above these loadings, the selectivity to secondary products such as methane and ethane rose, while the efficiency of propylene declined. The primary active component in the ODP reaction was the tetrahedral vanadium (V^{5+}) species. It was thought that the altered redox characteristics and increased catalytic activity of the catalyst were caused by their significant contact with chromium oxide and vanadium oxide.

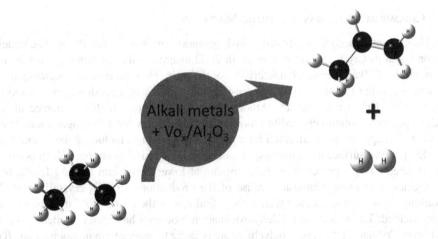

FIGURE 16.3 Catalytic dehydrogenation using vanadium species (Liu et al. 2016).

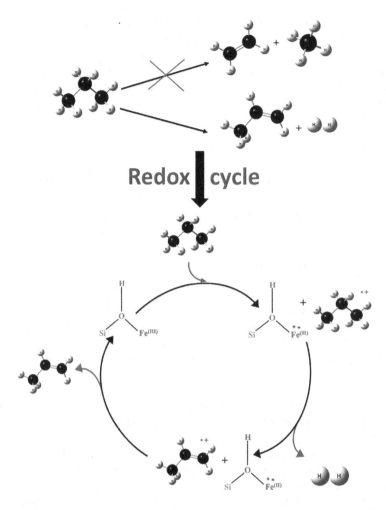

FIGURE 16.4 Possible reaction mechanism of the dehydrogenation of propane over Fe-silicate zeolites (Yun and Lobo 2014).

16.3.2 Chromium Oxide-Based Zeolitic Materials

Frey and Huppke introduced Cr_2O_3-based dehydrogenation catalysts in the 1930s. The widely used formulation contains Cr_2O_3. Airaksinen et al. in 2002 examined the assembly at various reaction conditions, effect of alkali metals, characteristics of the active sites, promoters, mechanism of dehydrogenation, and role of support. The surface-detected species of a fresh chromium oxide catalyst included Cr^{6+}, Cr^{5+}, Cr^{3+}, and Cr^{2+}. Among these surface types, highly dispersed monomeric Cr^{3+} or Cr^{2+} species are often believed to represent the active sites for dehydrogenation. Their relative concentration appears to be affected by a number of variables, including the amount of chromium loaded per unit surface area, method of calcination, and kind of support. With poorly bound supports, isolated CrO_4^{2-} species often predominate at lower chromium loading levels; however, Cr-O-Cr species were more common because of the production of Cr_2O_3 crystallites at elevated concentrations. According to recent spectroscopic findings, at the cost of Cr^{6+} species, surface Cr^{3+} species are created. The active site for dehydrogenation processes has been considered to contain a combination of Cr^{2+} and Cr^{3+} because polychromate is easier to reduce than monochromate (Grünert et al. 1986; Cimino et al. 1991; Weckhuysen et al. 2000).

In spectroscopic studies conducted by Shi et al. in 2008 (Shi et al. 2008) and Fang et al. in 2015 (Fang et al. 2015), it was discovered that three distinct forms of Cr^{3+} ions could be identified, specifically, three types of Cr^{3+} species: (i) isolated Cr^{3+} ion centres stabilised by the alumina surface; (ii) tiny amorphous Cr^{3+} clusters; and (iii) a reduction of Cr^{6+}/Cr^{5+} surface ions generates Cr^{3+} species. Induction-phase reduction of highly oxidised Cr species to active Cr species with a minor oxidation state and insignificant selectivity has been found to be irresponsive to the Cr loading. However, high selectivity of 90 % can be achieved at a conversion rate of 55 %. With the intention of up-surging the activity of chromium dehydrogenation catalysts, Puurunen et al. in 2001 (Puurunen et al. 2001) and Michorczyk et al. in 2010 (Michorczyk et al. 2010) suggested the accumulation of CO_2 as a mild oxidant. This removes evolved H_2 through a reaction known as RWGS reaction with no harm to Cr^{3+} active sites, efficiently shifting the equilibrium toward the product. Using MCM-41 as support, Kilicarslan et al. in 2013 attempted to boost the performance of chromium-based catalysts and observed that isobutane dehydrogenation deactivates at a slower rate. The outcome of the characterization further showed that the Cr^{6+} percentage concentration was greater than before, and the chromium was evenly distributed throughout the MCM-41 (Kilicarslan et al. 2013). The following are some crucial characteristics of chromium-based catalysts (Chaar et al. 1987).

1. Even with excess O_2, Cr_2O_3 is thermodynamically stable (regeneration environment).
2. On a new catalyst, different chromium species, including crystalline Cr_2O_3, soluble Cr^{6+}, and dispersed Cr^{3+}, develop.
3. During reduction, all Cr^{6+} species are totally reduced.
4. Coordinative unsaturated Cr^{2+}, both Cr^{2+} and Cr^{3+}, contribute to the catalytic activity.
5. The solid-state reaction that produces α-Cr_2O_3-$4Al_2O_3$ results in the everlasting deactivation of chromia-alumina catalysts.

Regardless of the support, Takehira and his colleagues (Takehira et al. 2004) confirmed that the electrochemical cycle among Cr^{6+} and $Cr(VI)O_4$ plays a major role in the dehydrogenation of C_3H_8 over both Cr/SiO_2 and Cr/Al_2O_3 catalysts. According to their research, under the reaction conditions, CO_2 oxidises a portion of Cr^{3+} to Cr^{6+} species. As a consequence, in the case of Cr/SiO_2, in addition to the straightforward dehydrogenation over Cr^{3+} species, the ODH of C_3H_8 over Cr^{6+} species also occurs. The hypothesised process for dehydrogenation over Cr-based catalysts is depicted in Figure 16.5. (Michorczyk et al. 2012; Atanga et al. 2018) According to this mechanism, the reduction of Cr^{6+} species at the primary phase of the dehydrogenation reaction creates scattered Cr^{2+} and Cr^{3+} sites that take part in the propylene formation via a non-oxidative route. Cr^{3+} and Cr^{2+} locates, in the presence of CO_2, may also participate in the production of H_2 in the ODP by a RWGS reaction and in a different oxidative route for the creation of propylene.

The hydrothermal technique at 180 °C was used by Wu et al. (Wu et al. 2013) to prepare a sequence of Cr_2O_3-ZrO_2 mixed oxides for ODH with CO_2 (ODPC). The reaction was conducted with a balance of N_2, 2.5 vol. % C_3H_8, and 5 vol. % CO_2 at 550 °C. A hydrothermal treatment catalyst produced a propane conversion of 53.3 % (C_3H_8), which was comparatively 1.6 times higher than the standard Cr_2O_3-ZrO_2. Because Cr^{6+} species are more active in reactions, increased efficacy of the thermally produced catalysts was ascribed to their increased surface area and Cr^{6+} concentration over Cr^{3+}. With the lack of CO_2, a conversion of 76.5 %, specificity of 60.3 %, and turnover frequency (TOF) value of 14.8 x 10^4 s^{-1} were attained in less than 10 minutes. However, TOF values and the conversion rate fell to 0.6 x 10^4 s^{-1} and 3 %, respectively, and specificity touched 97.4 % in 6 hrs. with coke sedimentation of 9 %. However, similar behaviour was shown in the presence of CO_2, where preliminary TOF and conversion declines from 10.3 x 10^4 s^{-1} and 53.3 % to 5.4 x 10^4 s^{-1} and 27.7 %, respectively, while an increase in specificity is observed from 79 % to 90.8 % in 6 hrs. with 2.7 wt. % coking (Wu et al. 2013).

Zhang et al. (Zhang, Wu et al. 2011) synthesized Na-H-ZSM-5 and employed it in propane's ODPC process at 550 °C with crystal sizes of 400 nm (H-ZSM-5-S) and 2 μm (H-ZSM-5-L).

Compared to H-ZSM-5, which has a larger crystal size (about 2 μm), the sub-micro size catalyst (about 400 nm) demonstrated noticeably higher activity. The authors also investigated the influence of promoters such as Cr^{6+} on the ODP reaction over H-ZSM-5 catalysts with both the existence and lack of CO_2. The primary propane conversion of 48.3 %, 96.0 %, and selectivity towards propylene were 11.3 %, 92.8 % over Cr/H-ZSM-5-S and Cr/H-ZSM-5-L, respectively. These values fell to 30.1 % and 91.8 % for the earlier catalyst after 8 hours on stream and to 8.4 % and 94.1 % for the subsequent catalyst. The sub-micro size catalyst had a substantially larger proportion of surface Cr^{6+} species than that of the micro size catalyst, which contributed to its enhanced catalytic efficiency, according to diffuse reflectance UV-Vis, H_2-TPR, XPS, and combined Raman data. Deactivation of the catalyst was impacted by a reduction of monochromate Cr^{6+} to polychromate Cr^{3+} in addition to the creation of coke, and the catalyst could be regenerated by oxidising Cr^{3+} once more and burning carbon deposits.

In a different study, Botavina and colleagues (Botavina et al. 2016) observed positive outcomes with a Cr deposition of 5 wt. % on SiO_2, but the catalyst rapidly lost its activity as a result of the chromium species clumping together. The same researchers employed direct hydrothermal synthesis in a subsequent work to enhance the distribution of Cr atoms on the substrate for the ODP reaction with CO_2 or $CO_2 + O_2$ atmosphere in the environment that was similar to an industrial environment. When CO_2 was used as the oxidant in ODP, the initial C_3H_8 conversion rose from 40 % to 70 % for 0.25-Cr/SiO_2 and stayed at 50 % after 6 hrs. for 2.0 wt. % Cr/SiO_2.

The effectiveness of a Si-Beta catalyst containing Cr in catalyzing propane dehydrogenation to propene with CO_2 assistance was examined by Michorczyk, Piotr, et al. (Michorczyk et al. 2020). Comparative catalytic studies show that $Cr_{2.0}$Si-beta was positively affected by CO_2 propylene production. Dealumination with acid nitrate was used, and the result was a decline in the Si/Al ratio from 17 to 1,000. Their research shows that the presence of CO_2 increases selectivity and yield to propene in in the dealuminated form of beta zeolite. Cr_2O_3-based catalysts were hosted on various pure siliceous supports, including ordered SBA-15, MCM-41, SBA-1, and MSU-x or non-ordered SiO_2, and a similar promoting effect was seen (Takahara et al. 1998; Michorczyk et al. 2011; Michorczyk et al. 2008; Liu et al. 2007; Michorczyk et al. 2012). To the contrary, CO_2 has an adverse effect on the $Cr_{2.0}$Al-beta catalyst that is comparable to the poisoning effect seen with chromium oxide catalysts based on γ-Al_2O_3.

Xie, Linjun, et al. (Xie, Wang et al. 2021) prepared sequences of Zn-enclosing silicate-1 zeolite (namely, Zn@S-1) with various Zn loadings via a ligand-protected direct hydrothermal route. The

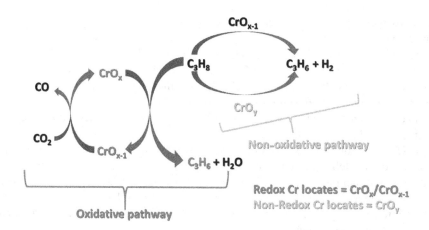

FIGURE 16.5 Probable mechanism of Cr-based catalyst on the ODPC reaction (Michorczyk et al. 2012; Atanga et al. 2018).

catalyst Zn@S-1 zeolite was used for the prospective propane dehydrogenation process, with zinc atoms serving as key constituents. They discovered in-situ restricted reduction of Zn species, with partly reduced Zn cations acting as active sites for the propane dehydrogenation (PDH) process. The Zn-4@S-1 catalyst showed high propylene selectivity, good dehydrogenation activity, and redevelopment proficiency in the PDH process, with a propylene selectivity of 97% attained at 580 °C.

Cheng, Yanhu, and colleagues used an incipient wetness approach to manufacture catalysts made of ultrafine ZSM-5-supported Cr_2O_3 particles and examined their catalytic efficiency for the dehydrogenation of C_2H_6 in the presence of CO_2 (Figure 16.6) (Cheng et al. 2015). Their findings show that ultrafine Na-type ZSM-5-supported Cr_2O_3 particles with a high Si/Al ratio turn out to be outstanding catalysts for the ODH of C_2H_6 with CO_2. With a C_2H_6 conversion of 65 % and C_2H_4 production of 49 % after 50 hours, the catalyst demonstrated both excellent stability and activity. The catalyst characterization results demonstrate that the greater extent of distribution of CrOx particles on the ultrafine ZSM-5 surface is responsible for the remarkable catalytic activity.

Recent studies have shown that mesoporous sieves containing Cr have remarkable catalytic activity in ODPC reactions (Baek et al. 2012; Michorczyk et al. 2012; Michorczyk et al. 2010). Michorczyk et al. (Michorczyk et al. 2010; Puurunen et al. 2001) created a sequence of Cr/SBA-1 catalysts with a well-organized cubic assembly having Cr/Si = 0.06 (atomic ratio) by integrating 14 different quantities of Cr (1–15 wt. %) into SBA-1 and a large definite surface area (S-BET > 900 m^2 g^{-1}). The findings of Michorczyk et al. were supported by those of Baek et al. (Baek et al. 2012) that the activity of the catalyst rose with Cr concentration, reaching an ideal loading of 4.3 wt. %. The integration method improved the dispersion of the Cr species inside the framework and network of the catalyst, thereby elevating the Brunauer-Emmett-Teller surface area. This contrasts with the standard wet impregnation route, where Cr rests on the pore mouths of support and on the surface. The amount of redox chromium species was found to be inversely correlated with the dehydrogenation activity. Expanding an in-situ UV-Vis DR analysis, Michorczyk's group (Michorczyk et al. 2010) looked into the properties of Cr/SBA-1 in more detail. The Cr^{6+} species was being quickly decreased at the initial ODPC reaction, according to the results. Grounded on their findings, the authors concluded that the resulting Cr^{3+} and Cr^{2+} species start redox cycles that are accountable for the increased catalytic efficiency of the Cr/SBA-1. Takehira et al. (Takehira et al. 2004) reported in a different experiment the impact of CO_2 on the oxidation of propane over a Cr/MCM-41

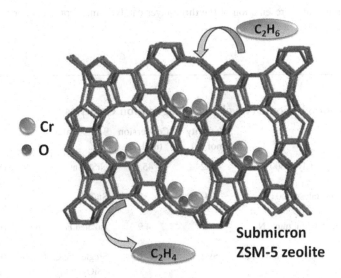

FIGURE 16.6 Dehydrogenation of ethane with CO_2 above Cr sustained on submicron ZSM-5 zeolite (the image of the zeolite framework is reprinted from (Xu et al. 2017; Cheng et al. 2015)).

catalyst. They discovered that the addition of CO_2 doubled the yield of C_3H_6 and C_3H_8. With both the existence and lack of CO_2 during the process over the Cr-MCM-41, the catalyst activity steadily reduced, indicating catalyst deactivation.

To separate hydrogen in the existence of lighter hydrocarbons and for PDH, Pati, Subhasis, et al. (Pati et al. 2020) successfully synthesised and tested a three-layer coating with LTA zeolite on the periphery and an ultrathin (~1μm) Pd membrane on the underside of an Al_2O_3 cylindrical fibre support (Figure 16.7) (Pati et al. 2020). Their findings show that related to the only internal layered Pd membrane, the three-layer coating exhibited superior performance for milder hydrocarbons (C_3H_8 and C_3H_6). They reported that the three-layer coating showed outstanding thermal strength in the presence of propane and propene at 600 °C for 72 hrs. According to their findings, the consistent nano-sized holes (0.4 nm) in the LTA zeolite were effective at blocking hydrocarbon molecules. They demonstrated the membrane for the dehydrogenation of propane by using 7.5 wt. % Cr/Al_2O_3 as the PDH catalyst at 600 °C. They claimed that the catalytic-membrane reactor produced 15 % more propane than a traditional FBR while maintaining equivalent propene specificity. The simultaneous elimination of H_2 from the product mixture through the dehydrogenation of propane is primarily accountable for the increase in the activity of the catalyst.

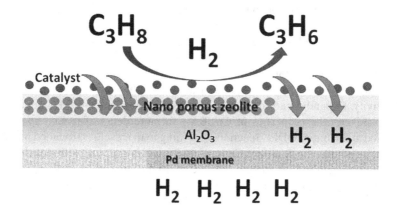

FIGURE 16.7 Graphical representation of the three-layer catalytic membrane reactor for dehydrogenation (Pati et al. 2020).

TABLE 16.1
Catalytic Application of Some Novel Dehydrogenation Catalysts

Reaction Condition	Selectivity (mol %)	Conversion (mol %)	Catalyst	Reference
5 ml min^{-1} isobutane, 4 g catalyst, 600 °C	90.2	45.6	$NiSn/SnO_2$	(Wang, Wang et al. 2016)
8 ml min^{-1} propane, 4 g catalyst, 600 °C	94	35	Sn/SiO_2	(Wang, Zhang et al. 2016)
1.65 ml min^{-1} propane + 53.4 ml min^{-1} N_2, 0.5 g catalyst, 650 °C	>99	4.9	Isolated Fe(II) on SiO_2	(Hu, Schweitzer, Zhang et al. 2015)
1.65 ml min^{-1} propane + 53.35 ml min^{-1} N_2, 1 g of catalyst, 560 °C	>90	4–10	Single-site Co (II) on SiO_2	(Hu, Schweitzer, Das et al. 2015)
1.65 ml min^{-1} propane + 53.35 ml min^{-1} N_2, 1 g of catalyst, 650 °C	86–95	60–40	Single-site Co (II) on SiO_2	(Hu, Schweitzer, Das et al. 2015)

TABLE 16.1 (Continued)
Catalytic Application of Some Novel Dehydrogenation Catalysts

Reaction Condition	Selectivity (mol %)	Conversion (mol %)	Catalyst	Reference
2.4 ml min^{-1} isobutane + 3.6 ml min^{-1} N$_2$, 0.3 g of catalyst, 550 °C	93–88 (27 hrs.)	14–36 (27 hrs.)	2.3V/Al$_2$O$_3$	(Rodemerck et al. 2016)
1 ml min^{-1} propane + 4 ml min^{-1} Ar, 0.05–0.1 g of catalyst, 550 °C	80–90 (25 hrs.)	24–8 (25 hrs.)	Ga/Al$_2$O$_3$	(Szeto et al. 2018)
10 ml min^{-1} isobutane, 0.5 g of catalyst, 600 °C	44–66 (0–100 mins.)	26–22 (0–100 mins.)	Mo/Al$_2$O$_3$	(Zhao et al. 2017)
7 ml min^{-1} propane + 7 ml min^{-1} H$_2$ + 11 ml min^{-1} N$_2$, 0.25 g of catalyst, 600 °C	83–90 (4 hrs.)	33–22 (4 hrs.)	12V1Mg/Al$_2$O$_3$	(Wu et al. 2017)
1 ml min^{-1} propane + 19 ml min^{-1} N$_2$, 0.1 g of catalyst, 530 °C	33–55 (6 hrs.)	18.5–2 (6 hrs.)	In-Ga-O	(Tan et al. 2015)
1 ml min^{-1} propane + 19 ml min^{-1} N$_2$, 0.2 g of catalyst, 600 °C	92	41.3–24.7	Hierarchical porous carbon	(Hu et al. 2018)
2 ml min^{-1} isobutane + 12 ml min^{-1} N2, 4 g of catalyst, 560 °C	82–90	63–71	Metal sulphide	(Wang, Li et al. 2014)
0.3 ml min^{-1} propane + 14.7 ml min^{-1} He, 0.25 g of catalyst, 550 °C	92	16–12	Hybrid nanocarbon	(Wang et al. 2014)

16.4 PT-BASED ZEOLITIC CATALYST

The first source of inspiration was the employment of bifunctional supported metal catalysts for catalytic remodelling as active locates for hydrogenation, isomerization, hydrocracking, dehydrogenation, and cyclization processes. For dehydrogenation catalysts, however, the acidic function must be minimised to prevent adverse effects. All of the Group VIIIB noble metals are typically involved in the hydrogenation or dehydrogenation of alkanes. It has been proven that platinum has excellent C-H bond activation but also poor C-C cleavage activity. As a result, Universal Oil Products (UOP) commercialised it in the Pacol process in 1968 and then in Oleflex in the early 1970s. However, Pt has a low efficacy for hydrogenolysis but is deactivated due to carbon accumulation. Extensive research has been conducted to reverse the association between the hydrogenolysis rate (i.e., coking) and the particle size (Monzón et al. 2003; Goodwin et al. 2004). It is thought that two processes—(i) side reactions that result in coke covering the active surface area and (ii) the thermal gradient during the restoration process—cause the clumping or sintering of Pt-based dehydrogenation catalysts. As a result, the Pt-support interaction is critical and is determined by the robustness of the Pt-O-M bond (where M stands for the cation in the host). However around 500 °C, when the relations between the host and the dynamic surface complexes of PtO$_x$ and PtO$_x$Cl$_y$ are created, by the addition of modest amounts of oxygen and chlorine, Pt can be redispersed (Damyanova et al. 1997). The traditional support for the majority of catalysts, including platinum-based dehydrogenation catalysts, is high surface area alumina, but it is acidic. Promoters are thus employed to reduce acidity. Utilizing more steady derivatives of alumina, such as α- or θ-alumina, which have lower surface areas and comparatively fewer Bronsted acid locates than γ-alumina, is another tactic.

16.4.1 Sn Metal as Promoter

It is thought that active sites of zeolites have a wide range of crystal shapes and strengths, which affects how accessible they are. Numerous studies have examined Pt-Sn-based catalysts reinforced on zeolite and/or amorphous materials for mild alkane dehydrogenation (Sattler, Ruiz-Martinez et al. 2014; Sattler, Gonzalez-Jimenez et al. 2014). According to favourable geometrical and/or electrical effects, tin was chosen as a promoter (Faro Jr et al. 2003). The size of platinum ensembles shrinks with geometric effects. Tin covers the 5d band of platinum in a Pt-Sn alloy or solid solution, minimizing hydrocarbon interaction with platinum, and as a result, the amount of hydrogenolysis and coke production also decreases. Through the positive charge transfer from Sn^+ ions, Sn also alters the electronic density of platinum (Nawaz and Wei 2013). Tin aids in the accessibility of adequate platinum locates for dehydrogenation at reaction temperature and slows the creation of the Pt-Sn alloy (Miracca and Piovesan 1999; Nawaz, Zhu et al. 2009; Nawaz, Shu et al. 2009; Nawaz, Tang and Wei 2009; Nawaz, Xiaoping et al. 2009; Nawaz, Chu et al. 2010; Nawaz, Qing et al. 2010; Nawaz and Wei 2013; Airaksinen et al. 2002; Weckhuysen and Wachs 1996).

According to Nawaz et al. (2010) and Zhang et al. (2012), tin additionally stabilises the surface area and lowers the acidity of the supports. Therefore, we can conclude that interactions among platinum, tin, and the support affect how well Pt-Sn-based catalysts function in the dehydrogenation process (Zhang, Zhou, Shi et al. 2012; Nawaz, Qing et al. 2010). According to Weckhuysen et al. (1998) and Faro et al. (2003), the aggregation of platinum particles and carbon deposition is a key cause of the platinum-based catalyst's deactivation during PDH (Rownaghi, Rezaei, Hedlund et al. 2012; Faro Jr et al. 2003). Tin promoters make it easier for coke to be transferred from active platinum sites. Nawaz et al. (Nawaz, Wei et al. 2010) investigated a variety of additional promoters, including La, Mg, Cu, Zn, Ru, Na, and Ca, although their role is disputed (Nawaz, Wei et al. 2010).

According to reports by Miracca and Piovesan (1999), Airaksinen et al. (2002), and Nawaz, Zeeshan et al. (2010), the following properties of Pt-Sn-based catalysts are significant (Miracca and Piovesan 1999; Nawaz, Zhu et al. 2009; Nawaz, Shu et al. 2009; Nawaz, Tang et al. 2009; Nawaz, Xiaoping et al. 2009; Nawaz, Chu et al. 2010; Nawaz, Qing et al. 2010).

1. Through sintering and coke fouling, catalysts age through a "double mechanism."
2. The catalyst can withstand various percentages of coke accumulation while maintaining sufficient catalytic activity and permitting several hours of reaction time on stream without regeneration.
3. Because hydrogen suppresses coking processes, its existence increases the efficacy of the catalyst.
4. It is possible to regenerate using oxygen, steam, or air. Additionally, chlorine's redistribution of platinum (through the interaction between platinum and chloride on the surface) lessens sintering.

16.4.2 Ce Metal as Promoter

According to research by Mostafa Aly et al. (Aly et al. 2020), using Pt/Al_2O_3 catalysts with the right quantity of boron decreases carbon build up and improves propylene selectivity during PDH, which increases productivity. Interestingly, only boron added prior to platinum improves selectivity and carbon resistance, while boron added after platinum just lowers activity. In particular, advancement with 1 wt. % boron diminishes the carbon content by three times and boosts propylene selectivity from 90 to 98 %. Naseri et al. (2019) showed that, in the dehydrogenation of propane, Ce demonstrates enhanced aiding effects on the Pt-Sn-K/γ-Al_2O_3 catalyst, as demonstrated by the general evaluation of Zn, Ce, and Co as the promoter. The Pt-Ce contacts improve the ability of Pt atoms to counterattack the agglomeration, increasing selectivity. The catalysts Pt-Sn-K-$Co_{0.3}$/γ-Al_2O_3, Pt-Sn-K-$Zn_{0.7}$/γ-Al_2O_3, and Pt-Sn-K-$Ce_{0.5}$/γ-Al_2O_3 had the ultimate PDH performance among those made with various loadings of Ce, Zn, and Co. Additionally, the best conversion and selectivity were demonstrated by Pt-Sn-K-$Co_{0.3}$-$Zn_{0.7}$/γ-Al_2O_3.

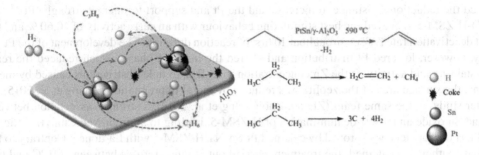

FIGURE 16.8 Dehydrogenation of C_3H_8 to C_3H_6 over a Pt-Sn-Na/γ-Al_2O_3 catalyst (reprinted with permission from Zhao) (Zhao et al. 2018).

16.4.3 Gallium (III) Oxide as Promoter

Numerous research teams have examined how zeolites behave in the ODPC reaction in relation to surface acidity. As reported, the efficacy of the catalyst reduced, but the specificity upgraded correspondingly as the Si/Al ratio of H-ZSM-5-supported Ga_2O_3 increased (Ren et al. 2012). Because Bronsted acidity is related to H-ZSM-5 rather than to Lewis acidity, which is associated with Ga_2O_3, removing the aluminium from H-ZSM-5 resulted in a substantially larger reduction in Bronsted acidity than in Lewis acidity. The catalytic efficacy of the Ga_2O_3-assisted H-ZSM-48 catalyst for the ODPC process was reported by Ren et al. (Ren et al. 2012) in the same work. However, Ga_2O_3/H-ZSM-48 with an Si/Al ratio of 130 demonstrated the greatest propylene output of 22.2 %. The effectiveness with many Si/Al ratios was equivalent to the previous observations. Due to its weaker acidity, Ga_2O_3/H-ZSM-48 had a greater selectivity for propylene than Ga_2O_3/H-ZSM-5, especially at minimal propane conversion. However, Ga_2O_3/HZSM-48's stability was poorer than Ga_2O_3/H-ZSM-5's stability, which may be due to the material's unidimensional pore structure and numerous weak acid sites.

16.4.4 Alkaline Earth Metals as Promoter

By reducing side reactions, Sn-aided Pt-H-ZSM-5 lowers coke production, according to studies concerning the impact of metal promoters on the catalytic performance of zeolites (Liu et al. 2021). As a result, it improves catalytic activity, stability, and Pt dispersion because of its geometrical effect. As shown in Table 16.2, adding Na, La, Zn, and Mg can enhance the synergy between the zeolite support and the metals by neutralising the strong acid locates on the catalyst surface (Tasbihi et al. 2007). These assertions were supported by Zhang et al. (Zhang et al. 2010), who co-impregnated H-ZSM-5 with Pt and Sn and then impregnated it with various K concentrations in the absence of CO_2 for PDH. Characterization findings suggested that adding K lowered the acid locates in the catalyst while improving Pt dispersion and the relationship between Pt and Sn. This increased the catalyst's activity, selectivity, and stability, but the performance grades down past an optimal composition (about 0.8 wt. % K). At this K composition, a higher conversion of 33 % was achieved, along with a yield of 30 % and specificity of 91.43 %. These results are better than those of Pt-Sn/H-ZSM-5, which had conversion as high as 27.5 %, yield as low as 12 %, and selectivity as low as 37.3 % after 8 hours on stream.

16.4.5 Transition Metals as Promoters

Zhang et al. (Zhang et al. 2015) employed hydrothermal synthesis to introduce Zn into the H-ZSM-5 framework and utilized it as a platform for Pt catalysts. Interestingly, using this Zn/H-ZSM-5 caused the reduction resistance to increase and the Pt and support to interact strongly. Pt-Na/Zn (1 %)-H-ZSM-5 displayed the best stimulating behaviour with an early activity of 40.60 % and the least deactivation rate (7.3%) throughout 10 hrs. of reaction duration. The development of the Pt-Zn alloy, however, lowered Pt distribution and altered the metallic phase, which reduced the reaction stability and activity as the Zn concentration rose. Due to hole obstructions caused by metal atoms, the surface area of the zeolite was reduced by metal impregnation (Zhang et al. 2015). In a later study by the same team (Zhang, Zhou, Tang et al. 2012), a stronger association between Pt and Sn oxide and lower dealumination of H-ZSM-5, leading to an improved catalytic efficacy for PDH to propylene was proved by calcined Pt-Sn-Na/H-ZSM-5 with La-doped. Contrary to the catalyst without La calcined, the treatment carried out at temperatures between 500 °C and 650 °C resulted in a strong linkage between the La atoms and the host, which, in turn, suppressed the dealumination of the host and increased bonding between the Sn atom and the host and between Sn and the Pt oxides. The catalyst from La, calcined at 650 °C, had the maximum reaction efficiency and durability in a PDH test. Additionally, the creation of tiny crystal sizes caused by the addition of a metal to the zeolite structure enriched the surface area of the zeolite support. The same group (Razavian et al. 2015) had previously described a similar characteristic of the isomorphous replacement of tin in H-ZSM-5 and employed it as a host for Pt-Na metal catalyst dispersal. The H-ZSM-5-based Pt catalyst improved by Sn showed higher propane efficiency and durability compared to the traditional impregnated Pt-Na/H-ZSM-5 and Pt-Sn-Na/H-ZSM-5. Additionally, it displayed a minimum deactivation rate (6.2 %), revealed selectivity > 98 % (propylene), and had a comparable propane conversion of roughly 39.0 % after 9 hours. Selectivity of 98.6 % and productivity of 30 % were reported after 175 hours on the stream (Zhang, Zhou et al. 2011).

TABLE 16.2
Metal-Supported Zeolites for the Dehydrogenation of Propane.

Catalyst	Si/Al	Conversion (C_3H_8) (%)	Selectivity (C_3H_6) (%)	Yield (C_3H_6) (%)	Ref.
Pt-Sn/H-ZSM-5	0.24	27	34	11	(Zhang et al. 2010)
Pt-Sn-K(0.2 %)/H-ZSM-5	1.6	22	58	13	(Zhang et al. 2010)
Pt-Sn-K(0.5 %)/H-ZSM-5	5.5	25	89	23	(Zhang et al. 2010)
Pt-Sn-K(0.8 %)/H-ZSM-5	20.2	31	92	30	(Zhang et al. 2010)
Pt-Sn-K(1 %)/H-ZSM-5	-	21	94	21	(Zhang et al. 2010)
Pt-Sn-K(1.2 %)/H-ZSM-5	60	22	92	24	(Zhang et al. 2010)
5 wt. % Cr/SBA-1	-	33	86	28	(Sattler, Ruiz-Martinez et al. 2014)
H-ZSM-5	20.2	28	71	20	(Razavian et al. 2015)
H-ZSM-5	140	29	21	6	(Kim et al. 2016)
Pt/H-ZSM-5	140	27	35	10	(Kim et al. 2016)
Pt-Sn/H-ZSM-5	140	27	40	9	(Kim et al. 2016)
Pt-Na/H-ZSM-5	400	35	92	32	(Zhang et al. 2015)
Pt-Na-Zn(0.5 %)/H-ZSM-5	400	36	95	34	(Zhang et al. 2015)
Pt-Na-Zn(1 %)/H-ZSM-5	400	36	98	36	(Zhang et al. 2015)
Pt-Na-Zn(1.5 %)/H-ZSM-5	400	38	99	37	(Zhang et al. 2015)
Pt-Sn-Na/H-ZSM-5 (500[a])	120	37	98	36	(Zhang et al. 2007)
Pt-Sn-Na/H-ZSM-5 (400[b])	120	32	99	32	(Zhang et al. 2006)
Pt-Sn-Na-La/H-ZSM-5	333–400	31	98	30	(Zhang, Zhou, Tang et al. 2012)

a = calcination temperature, b = hydrothermal treatment temperature.

Kim et al. (Kim et al. 2016) recently developed micro meter-thin SAPO-34 zeolite skins on alumina cylindrical supports and applied them in a 600 °C PDH reaction. At W.H.S.V. of 0.1–0.5 h^{-1}, propane selectivities of more than 80 % and conversions of 65–75 % were reported. The remarkable performance was caused by the membrane's effective elimination of generated hydrogen during the ODP reaction.

By enclosing subnanometer Pt-Zn groups in completely siliceous zeolites, Qiming Sun et al. (Sun et al. 2020) unveiled a novel method for producing catalysts with elevated performance for PDH reactions. For the easy encapsulation of atomically distributed and subnanometer bimetallic Pt-Zn atoms inside silicalite-1 zeolites, they created a ligand-protected direct hydrogen reduction technique and showed how it dramatically reduced the size of the metal atoms that were trapped inside the zeolites. Due to the creation of ultrasmall Pt-Zn alloy groups, the addition of Zn species enhances the catalytic efficiency and durability of Pt clusters. The specific activity of the Pt-Zn4@S-1-H in the production of propylene was extraordinarily high. The most stable Pt-based heterogeneous catalyst for the PDH reaction to date was reported, and it is significant that no deactivation was seen. Additionally, they discovered that adding Cs$^+$ ions considerably enhanced regeneration efficiency.

Zhikang Xu et al. (2019) constructed a new method for creating very stable Pt/Sn-zeolite catalysts. Their approach consisted of three fundamental steps. To produce the lattice defects, the zeolite must first undergo the following stages: (i) exclusion of Al atoms from the crystal structure; (ii) insertion of Sn into the defects to yield Sn-containing zeolite; and (iii) easy deposition of Pt on Sn-zeolite superficial to produce Pt/Sn-zeolite. Pt/Sn-Y and Pt/Sn-Beta catalysts were effectively produced using this method. They achieved the greatest propylene specificity of ≥ 99 %, 50 % propane conversion, a great TOF of 114 s^{-1} (0.006 h^{-1}) a lower rate of deactivation, and excellent regenerability in the Pt-Sn$_{2.00}$/Sn-Beta catalyst for PDH at 570 °C. Nawaz Z et al. (Nawaz, Shu et al. 2009) examined the process challenges and comprehended the kinetics of the reaction in mixed alkane feed dehydrogenation. They observed decreased conversion at a higher time-on-stream and higher propylene selectivity. They demonstrated that zeolite-supported bi-metallic material (Pt-Sn/ZSM-5) promotes dehydrogenation more effectively than zeolite- and mono-metallic catalysts alone. Bi-metallic zeolite-based catalytic alteration was discovered to be significantly more enhanced than mono-metallic ZSM-5 and ZSM-5. There was unacceptably low dehydrogenation capacity of ZSM-5 with a SAR of 140, whereas a monometallic catalyst (Pt/ZSM-5) showed a modest dehydrogenation capacity but activity better than ZSM-5. By using bimetallic catalysts, an initial conversion of 57 % and final conversion of 30 % were achieved.

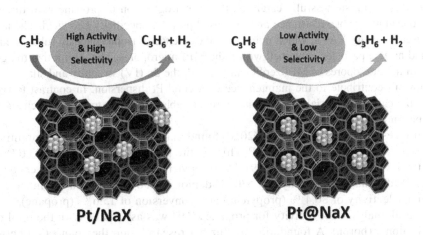

FIGURE 16.9 High activity and selectivity by an enclosed Pt-cluster in the zeolite cavity (reprinted with permission from Zhang ziyang) (Zhang et al. 2021).

Through a two-step post-synthesis process, Xie et al. (Xie, Chai et al. 2021) created zeolite-stabilised Pt-Zn catalysts and researched their potential for direct propane to propylene dehydrogenation. They outlined how the extremely distributed Pt-Zn species are in close proximity on the zeolite support, encouraging electronic interaction. The reaction parameters and catalyst compositions were both perfectly tuned. With 0.1Pt-2Zn/Si-Beta at 550 °C, 98 % propylene selectivity and a viable deactivation rate of 0.02 h^{-1} were established. They concluded that the 0.1Pt-2Zn/Si-Beta PDH catalyst is a viable PDH catalyst for future applications due to its outstanding functioning, low cost, and ease of scalability. Compared to θ-Al_2O_3 and amorphous silica, Lee, Su-Un, et al.'s (Lee et al. 2020) work reveals the effects of using zeolite Beta as a host on the physicochemical properties and catalytic effectiveness of supported Pt-Sn materials for the PDH process. According to their research, the Pt-Sn catalysts supported by Beta zeolite converted propane more efficiently, and due to the presence of Bronsted acid sites, light gases (C_{1-2}) and C_{5+} were also produced. Before the catalyst deactivated, the propylene selectivity rose as these fractured products covered the acid sites after a certain reaction time and suppressed the side reaction. Based on their comparison of Pt-Sn catalysts supported by Beta zeolites with SiO_2/Al_2O_3 ratios of 38 and 300, they found that, although it had less Pt dispersion, the Pt-Sn/Beta (300) had the finest PDH catalytic efficacy. With Bronsted acid locates participating in the cyclization and oligomerization stages of alkane aromatization, the enhanced surface area permits Pt scattering as functional catalytic sites, and acidity is essential to cut coke production.

To create a composite material that can be used as a host for Pt-Sn-based catalysts, Li, Hongda, et al. (2019) wrapped titanium-based zeolite TS-1 in SBA-16 silica. According to their report, the composite 10 % TS-1 content produced the best results in the PDH reaction compared to the effects of other TS-1 percentages (0–20 %). The electrical impact of the Ti species, which stabilises the SnO_x species, and the reciprocal mass transfer over the hierarchical porous construction of the optimal acidity were assumed to be the causes of the catalytic operation of Pt-SnO_x/TS-1@SBA-16.

Ivanushkin, G. G. et al. (2019) suggested that the dealumination of zeolites enables the reduction of the support's acidity and assimilation of Sn atoms into the zeolite framework (BEA, SiO_2/Al_2O_3 = 25). The associated Pt-Sn/zeolite material exhibit a Pt-Sn interaction, which results in great specificity and an improvement in the active phase's forbearance to agglomeration. Their research shows that the most effective sample produces a conversion of 26–28 % (propane) at a selectivity of up to 98.8 % (propylene) at 550 °C and contains 0.5 % Pt and 0.5–1.0 % Sn.

The coke-resistant catalyst materials developed by Lezcano-González, Inés, et al. (2022) are an imperfect silicalite-1 with Sn as a promotor having tiny platinum particles dispersed on it. Lezcano-González, Inés, et al. successfully catalysed the dehydrogenation of propane with this material, revealing outstanding coke resistance and increased propane specificity and yield. The material is characterised by using XAFS, which points to the existence of isomorphously substituted Sn (IV) species and minute particles contained within the MFI micropores. They think that the geometric restraints in the micropores of MFI zeolite, along with the Sn (IV) structure and silanolim perfection groups, all contribute to the maintenance of a good Pt dispersion. In contrast to typical Pt/Sn-Al_2O_3, the catalyst material demonstrates a remarkable resistance to coke, which significantly reduces the amount of coke generated.

In their investigation, Qiu Bin et al. (2020) found that MWW-type zeolite containing Boron atoms was an effective catalyst for ODP. This zeolite showed a selectivity of 83.6 % (propylene) and total mild olefins selectivity of 91.2 % (propylene and ethylene) at a conversion of 3.9 % (propane). Additionally, the catalyst at 530 °C demonstrated a total mild olefin specificity of 91.6 % and selectivity of 80.4 % (propylene) at a conversion of 15.6 % (propane). According to a structural analysis, the activity for propane ODH was associated with the total of faulty trigonal B atoms (boron). A foundation for further research into the connection between the boron atom environment and ODH behaviour was provided by this clearly defined catalyst system.

16.5 CONCLUSION

Accordingly, it has been demonstrated that the kinds of promoter and support/template, which influence the catalyst's redox and surface acidic properties, have a considerable impact on the functioning of key parts in a vanadium-based catalyst. Acidic sites encourage the cracking of olefin compounds or propyl molecules and the production of coke because these centres are susceptible to the ODH reaction since they have many electrons. Additionally, the catalytic efficacy and specificity of nano-sized particles tend to be higher than those of large catalysts, suggesting that propane moiety has a better possibility of interacting with the active locates that extract hydrogen. Therefore, to improve the activity of V catalysts, the design strategy should concurrently take into account each of these aspects. The potential for the marketing utilization of CO_2 as a trivial oxidant beyond this class of materials in the ODH industry is present given the great future possibilities for vanadium-based catalysts.

Commercial Pt and CrO_x-based catalysts offer satisfactory dehydrogenation capabilities, but their widespread use has been constrained by the cost of Pt and the ecological issues connected with CrO_x. Therefore, there is still a need for the creation of innovative dehydrogenation catalysts. In recent years, single atom catalysts, carbon-based catalysts, metal sulphide-based catalysts, and metal oxide-based catalysts (VO_x, MoO_x, GaO_x, In-Ga-O, and Zn-Nb-O) have all received the most attention. The results of dehydrogenation processes by these catalyst are listed in Table 16.1.

The catalytic efficacy of chromium-based catalysts depends upon the catalyst's structure, promoter-support interaction, synthesis technique, reaction conditions, and active locates. For the ODP with CO_2, various publications have reported a variety of chromium catalytic systems. Despite the positive results from chromium-based catalysts, substitute catalysts should be created and established to permit more viable process equipment due to significant ecological and physiological problems related to chromium use.

However, this course of porous substances can be utilized as supports (metal oxide) in the advancement of bi-functional catalysts (zeolite composite and metals) for ODPC processes. It is accepted that the admittance to dynamic sites of zeolite differs significantly due to several crystal assemblies. The bi-functional catalyst's acidity and oxidising capacity should be maximised to prevent side reactions. The need to tailor redox and acid-based characteristics to specific uses and mechanical characteristics has fuelled an interest in replacing aluminium in MFI or CHA zeolite frameworks with other essentials; however, this type of composite material has not been thoroughly investigated for dehydrogenation reactions. Moreover, it is still unclear how the kinematics of non-framework entities, which alters the reaction mechanism and, therefore, the way that they contribute to catalytic reactions, behave under reaction circumstances that involve electrical conductivity variations, redox processes, and the adsorption of reactants/products. The zeolitic constituents seem to offer a tremendous possibility for creating new bi-functional materials that solve the issues with present constituents and exhibit high efficiency and durability in the ODPC process due to the capacity to smooth the surface, structural, and topological properties. In fact, this should be one of the main concerns for further study in the area of bi-functional materials and zeolitic composites.

REFERENCES

Airaksinen, Sanna MK, M Elina Harlin, and A Outi Krause. 2002. "Kinetic modeling of dehydrogenation of isobutane on chromia/alumina catalyst." *Industrial & Engineering Chemistry Research* 41 (23):5619–5626.

Aly, Mostafa, Esteban L Fornero, Andres R Leon-Garzon, Vladimir V Galvita, and Mark Saeys. 2020. "Effect of boron promotion on coke formation during propane dehydrogenation over Pt/γ-Al_2O_3 catalysts." *ACS Catalysis* 10 (9):5208–5216.

Armaroli, T, LJ Simon, M Digne, T Montanari, M Bevilacqua, V Valtchev, J Patarin, and G Busca. 2006. "Effects of crystal size and Si/Al ratio on the surface properties of H-ZSM-5 zeolites." *Applied Catalysis A: General* 306:78–84.

Atanga, Marktus A, Fateme Rezaei, Abbas Jawad, Mark Fitch, and Ali A Rownaghi. 2018. "Oxidative dehydrogenation of propane to propylene with carbon dioxide." *Applied Catalysis B: Environmental* 220:429–445.

Baek, Jayeon, Hyeong Jin Yun, Danim Yun, Youngbo Choi, and Jongheop Yi. 2012. "Preparation of highly dispersed chromium oxide catalysts supported on mesoporous silica for the oxidative dehydrogenation of propane using CO_2: Insight into the nature of catalytically active chromium sites." *ACS Catalysis* 2 (9):1893–1903.

Bhasin, MM, JH McCain, BV Vora, T Imai, and PR Pujado. 2001. "Dehydrogenation and oxydehydrogenation of paraffins to olefins." *Applied Catalysis A: General* 221 (1–2):397–419.

Botavina, MA, Yu A Agafonov, NA Gaidai, E Groppo, V Cortés Corberán, AL Lapidus, and G Martra. 2016. "Towards efficient catalysts for the oxidative dehydrogenation of propane in the presence of CO_2: Cr/SiO_2 systems prepared by direct hydrothermal synthesis." *Catalysis Science & Technology* 6 (3):840–850.

Botavina, MA, Gianmario Martra, Yu A Agafonov, NA Gaidai, NV Nekrasov, DV Trushin, S Coluccia, and AL Lapidus. 2008. "Oxidative dehydrogenation of C3–C4 paraffins in the presence of CO_2 over CrO_x/SiO_2 catalysts." *Applied Catalysis A: General* 347 (2):126–132.

Brockwell, HL, PR Sarathy, and R Trotta. 1991. "Synthesize ethers." *Hydrocarbon Processing* 70 (9):133–141.

Cecilia, JA, C García-Sancho, JM Mérida-Robles, J Santamaría-González, R Moreno-Tost, and P Maireles-Torres. 2015. "V and V–P containing Zr-SBA-15 catalysts for dehydration of glycerol to acrolein." *Catalysis Today* 254:43–52.

Chaar, MA, D Patel, MC Kung, and HH Kung. 1987. "Selective oxidative dehydrogenation of butane over VMgO catalysts." *Journal of Catalysis* 105 (2):483–498.

Chen, Kaidong, Alexis T Bell, and Enrique Iglesia. 2002. "The relationship between the electronic and redox properties of dispersed metal oxides and their turnover rates in oxidative dehydrogenation reactions." *Journal of Catalysis* 209 (1):35–42.

Cheng, Yanhu, Fan Zhang, Yi Zhang, Changxi Miao, Weiming Hua, Yinghong Yue, and Zi Gao. 2015. "Oxidative dehydrogenation of ethane with CO_2 over Cr supported on submicron ZSM-5 zeolite." *Chinese Journal of Catalysis* 36 (8):1242–1248.

Choi, Seung-Won, Wun-Gwi Kim, Jung-Seob So, Jason S Moore, Yujun Liu, Ravindra S Dixit, John G Pendergast, Carsten Sievers, David S Sholl, and Sankar Nair. 2017. "Propane dehydrogenation catalyzed by gallosilicate MFI zeolites with perturbed acidity." *Journal of Catalysis* 345:113–123.

Cimino, A, D Cordischi, S De Rossi, G Ferraris, D Gazzoli, V Indovina, G Minelli, M Occhiuzzi, and M Valigi. 1991. "Studies on chromia/zirconia catalysts I. Preparation and characterization of the system." *Journal of Catalysis* 127 (2):744–760.

Damyanova, S, Paul Grange, and Bernard Delmon. 1997. "Surface characterization of zirconia-coated alumina and silica carriers." *Journal of Catalysis* 168 (2):421–430.

Fang, Deren, Jinbo Zhao, Shang Liu, Limei Zhang, Wanzhong Ren, and Huimin Zhang. 2015. "Relationship between Cr-Al interaction and the performance of $Cr-Al_2O_3$ catalysts for isobutane dehydrogenation." *Modern Research in Catalysis* 4 (2):50.

Faro Jr, Arnaldo C, Kátia Regina Souza, Vera Lúcia DL Camorim, and Mauri B Cardoso. 2003. "Zirconia-alumina mixing in alumina-supported zirconia prepared by impregnation with solutions of zirconium acetylacetonate." *Physical Chemistry Chemical Physics* 5 (9):1932–1940.

Fei, Wei, Tang Xiaoping, Zhou Huaqun, and Zeeshan Nawaz. 2008. "Development of propylene production enhancement technology." *Petrochemical Technology* 37 (10):979.

Goodwin, JRJG, S Kim, and WD Rhodes. 2004. "Meanings, functionalities and relationships." *Catalysis* 17:320.

Grabowski, R. 2006. "Kinetics of oxidative dehydrogenation of C_2-C_3 alkanes on oxide catalysts." *Catalysis Reviews* 48 (2):199–268.

Grünert, W, W Saffert, R Feldhaus, and K Anders. 1986. "Reduction and aromatization activity of chromia-alumina catalysts: I. Reduction and break-in behavior of a potassium-promoted chromia-alumina catalyst." *Journal of Catalysis* 99 (1):149–158.

Hamid, Halim, and Mohammed Ashraf Ali. 2004. "Kinetics of tertiary-alkyl ether synthesis." In *Handbook of MTBE and Other Gasoline Oxygenates*, 230–263. CRC Press.

Haw, James F, and David M Marcus. 2005. "Well-defined (supra) molecular structures in zeolite methanol-to-olefin catalysis." *Topics in Catalysis* 34 (1):41–48.

Hu, Bo, Neil M Schweitzer, Ujjal Das, HackSung Kim, Jens Niklas, Oleg Poluektov, Larry A Curtiss, Peter C Stair, Jeffrey T Miller, and Adam S Hock. 2015. "Selective propane dehydrogenation with single-site Co^{II} on SiO_2 by a non-redox mechanism." *Journal of Catalysis* 322:24–37.

Hu, Bo, Neil M Schweitzer, Guanghui Zhang, Steven J Kraft, David J Childers, Michael P Lanci, Jeffrey T Miller, and Adam S Hock. 2015. "Isolated Fe^{II} on silica as a selective propane dehydrogenation catalyst." *ACS Catalysis* 5 (6):3494–3503.

Hu, Zhong-Pan, Dandan Yang, Zheng Wang, and Zhong-Yong Yuan. 2019. "State-of-the-art catalysts for direct dehydrogenation of propane to propylene." *Chinese Journal of Catalysis* 40 (9):1233–1254.

Hu, Zhong-Pan, Hui Zhao, Chong Chen, and Zhong-Yong Yuan. 2018. "Castanea mollissima shell-derived porous carbons as metal-free catalysts for highly efficient dehydrogenation of propane to propylene." *Catalysis Today* 316:214–222.

Ivanushkin, GG, AV Smirnov, PA Kots, and II Ivanova. 2019. "Modification of acidic properties of the support for Pt–Sn/BEA propane dehydrogenation catalysts." *Petroleum Chemistry* 59 (7):733–738.

Kilicarslan, Saliha, Meltem Dogan, and Timur Dogu. 2013. "Cr incorporated MCM-41 type catalysts for isobutane dehydrogenation and deactivation mechanism." *Industrial & Engineering Chemistry Research* 52 (10):3674–3682.

Kim, Seok-Jhin, Yujun Liu, Jason S Moore, Ravindra S Dixit, John G Pendergast Jr, David Sholl, Christopher W Jones, and Sankar Nair. 2016. "Thin hydrogen-selective SAPO-34 zeolite membranes for enhanced conversion and selectivity in propane dehydrogenation membrane reactors." *Chemistry of Materials* 28 (12):4397–4402.

Kondratenko, Evgueni V, Maymol Cherian, Manfred Baerns, Dangsheng Su, Robert Schlögl, Xiang Wang, and Israel E Wachs. 2005. "Oxidative dehydrogenation of propane over V/MCM-41 catalysts: Comparison of O_2 and N_2O as oxidants." *Journal of Catalysis* 234 (1):131–142.

Langanke, Jens, Aurel Wolf, Jorg Hofmann, Katrin Böhm, Muhammad A Subhani, Thomas E Müller, Walter Leitner, and Christoph Gürtler. 2014. "Carbon dioxide (CO_2) as sustainable feedstock for polyurethane production." *Green Chemistry* 16 (4):1865–1870.

Lee, Su-Un, You-Jin Lee, Soo-Jin Kwon, Jeong-Rang Kim, and Soon-Yong Jeong. 2020. "Pt-Sn supported on beta zeolite with enhanced activity and stability for propane dehydrogenation." *Catalysts* 11 (1):25.

Lezcano-González, Inés, Peixi Cong, Emma Campbell, Monik Panchal, Miren Agote-Arán, Verónica Celorrio, Qian He, Ramon Oord, Bert M Weckhuysen, and Andrew M Beale. 2022. "Structure-activity relationships in highly active platinum-tin MFI-type zeolite catalysts for propane dehydrogenation." *ChemCatChem* 14 (7):e202101828.

Li, Hongda, Zhen Zhao, Jiacheng Li, Jianmei Li, Linlin Zhao, Jiachen Sun, and Xiaoqiang Fan. 2019. "Synthesis of Pt-SnO_x/TS$_{-1}$@ SBA$_{-16}$ composites and their high catalytic performance for propane dehydrogenation." *Chemical Research in Chinese Universities* 35 (5):866–873.

Li, Xin, Amit Kant, Yingxin He, Harshul V Thakkar, Marktus A Atanga, Fateme Rezaei, Douglas K Ludlow, and Ali A Rownaghi. 2016. "Light olefins from renewable resources: Selective catalytic dehydration of bioethanol to propylene over zeolite and transition metal oxide catalysts." *Catalysis Today* 276:62–77.

Li, Yuxin, Yanghuan Huang, Juhua Guo, Mingye Zhang, Dezheng Wang, Fei Wei, and Yao Wang. 2014. "Hierarchical SAPO-34/18 zeolite with low acid site density for converting methanol to olefins." *Catalysis Today* 233:2–7.

Liu, Dongyang, Liyuan Cao, Guohao Zhang, Liang Zhao, Jinsen Gao, and Chunming Xu. 2021. "Catalytic conversion of light alkanes to aromatics by metal-containing HZSM-5 zeolite catalysts—A review." *Fuel Processing Technology* 216:106770.

Liu, Gang, Zhi-Jian Zhao, Tengfang Wu, Liang Zeng, and Jinlong Gong. 2016. "Nature of the active sites of VO$_x$/Al_2O_3 catalysts for propane dehydrogenation." *ACS Catalysis* 6 (8):5207–5214.

Liu, Licheng, Huiquan Li, and Yi Zhang. 2007. "Mesoporous silica-supported chromium catalyst: Characterization and excellent performance in dehydrogenation of propane to propylene with carbon dioxide." *Catalysis Communications* 8 (3):565–570.

Liu, Yan, Junxia Wang, Guangdong Zhou, Mo Xian, Yingli Bi, and Kaiji Zhen. 2001. "Oxidative dehydrogenation of propane to propene over barium promoted Ni-Mo-O catalyst." *Reaction Kinetics and Catalysis Letters* 73 (2):199–208.

Michorczyk, Piotr, Jan Ogonowski, Piotr Kuśtrowski, and Lucjan Chmielarz. 2008. "Chromium oxide supported on MCM-41 as a highly active and selective catalyst for dehydrogenation of propane with CO_2." *Applied Catalysis A: General* 349 (1–2):62–69.

Michorczyk, Piotr, Jan Ogonowski, and Marta Niemczyk. 2010. "Investigation of catalytic activity of CrSBA-1 materials obtained by direct method in the dehydrogenation of propane with CO_2." *Applied Catalysis A: General* 374 (1–2):142–149.

Michorczyk, Piotr, Jan Ogonowski, and Kamila Zeńczak. 2011. "Activity of chromium oxide deposited on different silica supports in the dehydrogenation of propane with CO_2–A comparative study." *Journal of Molecular Catalysis A: Chemical* 349 (1–2):1–12.

Michorczyk, Piotr, Piotr Pietrzyk, and Jan Ogonowski. 2012. "Preparation and characterization of SBA-1-supported chromium oxide catalysts for CO_2 assisted dehydrogenation of propane." *Microporous and Mesoporous Materials* 161:56–66.

Michorczyk, Piotr, Kamila Zeńczak-Tomera, Barbara Michorczyk, Adam Węgrzyniak, Marcelina Basta, Yannick Millot, Laetitia Valentin, and Stanislaw Dzwigaj. 2020. "Effect of dealumination on the catalytic performance of Cr-containing Beta zeolite in carbon dioxide assisted propane dehydrogenation." *Journal of CO$_2$ Utilization* 36:54–63.

Miracca, Ivano, and Laura Piovesan. 1999. "Light paraffins dehydrogenation in a fluidized bed reactor." *Catalysis Today* 52 (2–3):259–269.

Monzón, A, Teresita Francisca Garetto, and Armando Borgna. 2003. "Sintering and redispersion of Pt/γ-Al$_2$O$_3$ catalysts: A kinetic model." *Applied Catalysis A: General* 248 (1–2):279–289.

Nakagawa, Kiyoharu, Chiaki Kajita, Kimito Okumura, Na-oki Ikenaga, Mikka Nishitani-Gamo, Toshihiro Ando, Tetsuhiko Kobayashi, and Toshimitsu Suzuki. 2001. "Role of carbon dioxide in the dehydrogenation of ethane over gallium-loaded catalysts." *Journal of Catalysis* 203 (1):87–93.

Naseri, Maryam, Farnaz Tahriri Zangeneh, and Abbas Taeb. 2019. "The effect of Ce, Zn and Co on Pt-based catalysts in propane dehydrogenation." *Reaction Kinetics, Mechanisms, and Catalysis* 126 (1):477–495.

Nawaz, Zeeshan, Faisal Baksh, Jie Zhu, and Fei Wei. 2013. "Dehydrogenation of C3–C4 paraffin's to corresponding olefins over slit-SAPO-34 supported Pt-Sn-based novel catalyst." *Journal of Industrial, and Engineering Chemistry* 19 (2):540–546.

Nawaz, Zeeshan, Xiaoping Tang, Yao Wang, and Fei Wei. 2010. "Parametric characterization and influence of tin on the performance of Pt– Sn/SAPO-34 catalyst for selective propane dehydrogenation to propylene." *Industrial and Engineering Chemistry Research* 49 (3):1274–1280.

Nawaz, Zeeshan, Yue Chu, Wei Yang, Xiaoping Tang, Yao Wang, and Fei Wei. 2010. "Study of propane dehydrogenation to propylene in an integrated fluidized bed reactor using Pt-Sn/Al-SAPO-34 novel catalyst." *Industrial, and Engineering Chemistry Research* 49 (10):4614–4619.

Nawaz, Zeeshan, Shu Qing, Gao Jixian, Xiaoping Tang, and Fei Wei. 2010. "Effect of Si/Al ratio on performance of Pt–Sn-based catalyst supported on ZSM-5 zeolite for n-butane conversion to light olefins." *Journal of Industrial, and Engineering Chemistry* 16 (1):57–62.

Nawaz, Zeeshan, Qing Shu, Shahid Naveed, and Fei Wei. 2009. "Light alkane (mixed feed) selective dehydrogenation using bi-metallic zeolite supported catalyst." *Bulletin of the Chemical Society of Ethiopia* 23 (3).

Nawaz, Zeeshan, Xiaoping Tang, Yao Wang, and Fei Wei. 2010. "Parametric characterization and influence of tin on the performance of Pt– Sn/SAPO-34 catalyst for selective propane dehydrogenation to propylene." *Industrial and Engineering Chemistry Research* 49 (3):1274–1280.

Nawaz, Z, X Tang, and F Wei. 2009. "Hexene catalytic cracking over 30% sapo-34 catalyst for propylene maximization: Influence of reaction conditions and reaction pathway exploration." *Brazilian Journal of Chemical Engineering* 26:705–712.

Nawaz, Zeeshan, and Fei Wei. 2010. "Hydrothermal study of Pt–Sn-based SAPO-34 supported novel catalyst used for selective propane dehydrogenation to propylene." *Journal of Industrial and Engineering Chemistry* 16 (5):774–784.

Nawaz, Zeeshan, and Fei Wei. 2013. "Light-alkane oxidative dehydrogenation to light olefins over platinum-based SAPO-34 zeolite-supported catalyst." *Industrial and Engineering Chemistry Research* 52 (1):346–352.

Nawaz, Zeeshan, Tang Xiaoping, and Wei Fei. 2009. "Influence of operating conditions, Si/Al ratio and doping of zinc on Pt-Sn/ZSM-5 catalyst for propane dehydrogenation to propene." *Korean Journal of Chemical Engineering* 26 (6):1528–1532.

Nawaz, Zeeshan, J Zhu, and F Wei. 2009. "Drastic enhancement of propene yield from 1-hexene catalytic cracking using a shape intensified Meso-SAPO-34 catalyst." *Journal of Engineering Science and Technology* 4 (4):409–418.

Olah, George A, and Árpád Molnár. 2003. *Hydrocarbon Chemistry*. John Wiley & Sons.

Pak, Chanho, Alexis T Bell, and T Don Tilley. 2002. "Oxidative dehydrogenation of propane over vanadia–magnesia catalysts prepared by thermolysis of OV (OtBu) 3 in the presence of nanocrystalline MgO." *Journal of Catalysis* 206 (1):49–59.

Pati, Subhasis, Nikita Dewangan, Zhigang Wang, Ashok Jangam, and Sibudjing Kawi. 2020. "Nanoporous zeolite-A sheltered Pd-hollow fiber catalytic membrane reactor for propane dehydrogenation." *ACS Applied Nano Materials* 3 (7):6675–6683.

Puurunen, Riikka L, Bram G Beheydt, and Bert M Weckhuysen. 2001. "Monitoring chromia/alumina catalysts in situ during propane dehydrogenation by optical fiber UV–visible diffuse reflectance spectroscopy." *Journal of Catalysis* 204 (1):253–257.

Qiu, Bin, Fan Jiang, Wen-Duo Lu, Bing Yan, Wen-Cui Li, Zhen-Chao Zhao, and An-Hui Lu. 2020. "Oxidative dehydrogenation of propane using layered borosilicate zeolite as the active and selective catalyst." *Journal of Catalysis* 385:176–182.

Razavian, Marjan, and Shohreh Fatemi. 2015. "Synthesis and application of ZSM-5/SAPO-34 and SAPO-34/ZSM-5 composite systems for propylene yield enhancement in propane dehydrogenation process." *Microporous and Mesoporous Materials* 201:176–189.

Ren, Yingjie, Jie Wang, Weiming Hua, Yinghong Yue, and Zi Gao. 2012. "Ga_2O_3/HZSM-48 for dehydrogenation of propane: Effect of acidity and pore geometry of support." *Journal of Industrial and Engineering Chemistry* 18 (2):731–736.

Rodemerck, Uwe, Sergey Sokolov, Mariana Stoyanova, Ursula Bentrup, David Linke, and Evgenii V Kondratenko. 2016. "Influence of support and kind of VOx species on isobutene selectivity and coke deposition in non-oxidative dehydrogenation of isobutane." *Journal of Catalysis* 338: 174–183.

Rownaghi, Ali A, and Jonas Hedlund. 2011. "Methanol to gasoline-range hydrocarbons: Influence of nanocrystal size and mesoporosity on catalytic performance and product distribution of ZSM-5." *Industrial and Engineering Chemistry Research* 50 (21):11872–11878.

Rownaghi, Ali A, Fateme Rezaei, and Jonas Hedlund. 2011. "Yield of gasoline-range hydrocarbons as a function of uniform ZSM-5 crystal size." *Catalysis Communications* 14 (1):37–41.

Rownaghi, Ali A, Fateme Rezaei, and Jonas Hedlund. 2012a. "Selective formation of light olefin by n-hexane cracking over HZSM-5: Influence of crystal size and acid sites of nano-and micrometer-sized crystals." *Chemical Engineering Journal* 191:528–533.

Rownaghi, Ali A, Fateme Rezaei, and Jonas Hedlund. 2012b. "Uniform mesoporous ZSM-5 single crystals catalyst with high resistance to coke formation for methanol deoxygenation." *Microporous and Mesoporous Materials* 151:26–33.

Rownaghi, Ali A, Fateme Rezaei, Matteo Stante, and Jonas Hedlund. 2012. "Selective dehydration of methanol to dimethyl ether on ZSM-5 nanocrystals." *Applied Catalysis B: Environmental* 119:56–61.

Rownaghi, Ali Asghar, and Yun Hin Taufiq-Yap. 2010. "Novel synthesis techniques for preparation of ultrahigh-crystalline vanadyl pyrophosphate as a highly selective catalyst for n-butane oxidation." *Industrial and Engineering Chemistry Research* 49 (5):2135–2143.

Rownaghi, Ali Asghar, Taufiq-Yap Yun Hin, and Tang W Jiunn. 2009. "Influence of the ethylene glycol, water treatment and microwave irradiation on the characteristics and performance of VPO catalysts for n-butane oxidation to maleic anhydride." *Catalysis Letters* 130 (3):593–603.

Sanfilippo, Domenico. 2011. "Dehydrogenations in fluidized bed: Catalysis and reactor engineering." *Catalysis Today* 178 (1):142–150.

Sattler, Jesper JHB, Ines D Gonzalez-Jimenez, Lin Luo, Brien A Stears, Andrzej Malek, David G Barton, Beata A Kilos, Mark P Kaminsky, Tiny WGM Verhoeven, and Eline J Koers. 2014. "Platinum-promoted Ga/Al_2O_3 as highly active, selective, and stable catalyst for the dehydrogenation of propane." *Angewandte Chemie* 126 (35):9405–9410.

Sattler, Jesper JHB, Javier Ruiz-Martinez, Eduardo Santillan-Jimenez, and Bert M Weckhuysen. 2014. "Catalytic dehydrogenation of light alkanes on metals and metal oxides." *Chemical Reviews* 114 (20):10613–10653.

Shi, Xuejun, Shengfu Ji, and Kai Wang. 2008. "Oxidative dehydrogenation of ethane to ethylene with carbon dioxide over Cr–Ce/SBA-15 catalysts." *Catalysis Letters* 125 (3):331–339.

Sugiyama, Shigeru, Takeshi Osaka, Yuuki Hirata, and Ken-Ichiro Sotowa. 2006. "Enhancement of the activity for oxidative dehydrogenation of propane on calcium hydroxyapatite substituted with vanadate." *Applied Catalysis A: General* 312:52–58.

Sun, Qiming, Ning Wang, Qiyuan Fan, Lei Zeng, Alvaro Mayoral, Shu Miao, Ruoou Yang, Zheng Jiang, Wei Zhou, and Jichao Zhang. 2020. "Subnanometer bimetallic platinum–zinc clusters in zeolites for propane dehydrogenation." *Angewandte Chemie* 132 (44):19618–19627.

Szeto, Kai C, Zachary R Jones, Nicolas Merle, César Rios, Alessandro Gallo, Frederic Le Quemener, Laurent Delevoye, Régis M Gauvin, Susannah L Scott, and Mostafa Taoufik. 2018. "A strong support effect in selective propane dehydrogenation catalyzed by Ga (i-Bu) 3 grafted onto γ-alumina and silica." *ACS Catalysis* 8 (8):7566–7577.

Takahara, I, W-C Chang, N Mimura, and M Saito. 1998. "Promoting effects of CO_2 on dehydrogenation of propane over a SiO_2-supported Cr_2O_3 catalyst." *Catalysis Today* 45 (1–4):55–59.

Takehira, Katsuomi, Yoshihiko Ohishi, Tetsuya Shishido, Tomonori Kawabata, Ken Takaki, Qinghong Zhang, and Ye Wang. 2004. "Behavior of active sites on Cr-MCM-41 catalysts during the dehydrogenation of propane with CO_2." *Journal of Catalysis* 224 (2):404–416.

Tan, Shuai, Laura Briones Gil, Nachal Subramanian, David S Sholl, Sankar Nair, Christopher W Jones, Jason S Moore, Yujun Liu, Ravindra S Dixit, and John G Pendergast. 2015. "Catalytic propane dehydrogenation over In_2O_3–Ga_2O_3 mixed oxides." *Applied Catalysis A: General* 498:167–175.

Tasbihi, M, F Feyzi, MA Amlashi, AZ Abdullah, and AR Mohamed. 2007. "Effect of the addition of potassium and lithium in Pt–Sn/Al_2O_3 catalysts for the dehydrogenation of isobutane." *Fuel Processing Technology* 88 (9):883–889.

Thakkar, Harshul, Stephen Eastman, Ahmed Al-Mamoori, Amit Hajari, Ali A Rownaghi, and Fateme Rezaei. 2017. "Formulation of aminosilica adsorbents into 3D-printed monoliths and evaluation of their CO_2 capture performance." *ACS Applied Materials and Interfaces* 9 (8):7489–7498.

Thakkar, Harshul, Stephen Eastman, Amit Hajari, Ali A Rownaghi, James C Knox, and Fateme Rezaei. 2016. "3D-printed zeolite monoliths for CO_2 removal from enclosed environments." *ACS Applied Materials and Interfaces* 8 (41):27753–27761.

Thakkar, Harshul, Ahlam Issa, Ali A Rownaghi, and Fateme Rezaei. 2017. "CO_2 capture from air using amine-functionalized kaolin-based zeolites." *Chemical Engineering and Technology* 40 (11):1999–2007.

Vaezifar, Sedigheh, Hossein Faghihian, and Mahdi Kamali. 2011. "Dehydrogenation of isobutane over Sn/Pt/Na-ZSM-5 catalysts: The effect of SiO_2/Al_2O_3 ratio, amount and distribution of Pt nanoparticles on the catalytic behavior." *Korean Journal of Chemical Engineering* 28 (2):370–377.

Wang, Guowei, Chunyi Li, and Honghong Shan. 2014. "Highly efficient metal sulfide catalysts for selective dehydrogenation of isobutane to isobutene." *ACS Catalysis* 4 (4):1139–1143.

Wang, Guowei, Haoren Wang, Huanling Zhang, Qingqing Zhu, Chunyi Li, and Honghong Shan. 2016. "Highly selective and stable NiSn/SiO_2 catalyst for isobutane dehydrogenation: Effects of Sn addition." *ChemCatChem* 8 (19):3137–3145.

Wang, Guowei, Huanling Zhang, Haoren Wang, Qingqing Zhu, Chunyi Li, and Honghong Shan. 2016. "The role of metallic Sn species in catalytic dehydrogenation of propane: Active component rather than only promoter." *Journal of Catalysis* 344:606–608.

Wang, Rui, Xiaoyan Sun, Bingsen Zhang, Xiaoying Sun, and Dangsheng Su. 2014. "Hybrid nanocarbon as a catalyst for direct dehydrogenation of propane: Formation of an active and selective core–shell sp2/sp3 nanocomposite structure." *Chemistry: A European Journal* 20 (21):6324–6331.

Weckhuysen, Bert M, Abdelhamid Bensalem, and Robert A Schoonheydt. 1998. "In situ UV–VIS diffuse reflectance spectroscopy–on-line activity measurements: Significance of Cr n+ species (n= 2, 3 and 6) in n-butane dehydrogenation catalyzed by supported chromium oxide catalysts." *Journal of the Chemical Society, Faraday Transactions* 94 (14):2011–2014.

Weckhuysen, Bert M, An A Verberckmoes, Jan Debaere, Kristine Ooms, Ivan Langhans, and Robert A Schoonheydt. 2000. "In situ UV–Vis diffuse reflectance spectroscopy—on line activity measurements of supported chromium oxide catalysts: Relating isobutane dehydrogenation activity with Cr-speciation via experimental design." *Journal of Molecular Catalysis A: Chemical* 151 (1–2):115–131.

Weckhuysen, Bert M, and Israel E Wachs. 1996. "Raman spectroscopy of supported chromium oxide catalysts. Determination of chromium—oxygen bond distances and bond orders." *Journal of the Chemical Society, Faraday Transactions* 92 (11):1969–1973.

Wu, Runxia, Pengfei Xie, Yanhu Cheng, Yinghong Yue, Songyuan Gu, Weimin Yang, Changxi Miao, Weiming Hua, and Zi Gao. 2013. "Hydrothermally prepared Cr_2O_3–ZrO_2 as a novel efficient catalyst for dehydrogenation of propane with CO_2." *Catalysis Communications* 39:20–23.

Wu, Tengfang, Gang Liu, Liang Zeng, Guodong Sun, Sai Chen, Rentao Mu, Sika Agbotse Gbonfoun, Zhi-Jian Zhao, and Jinlong Gong. 2017. "Structure and catalytic consequence of Mg-modified VOx/Al_2O_3 catalysts for propane dehydrogenation." *AIChE Journal* 63 (11):4911–4919.

Xie, Linjun, Yuchao Chai, Lanlan Sun, Weili Dai, Guangjun Wu, Naijia Guan, and Landong Li. 2021. "Optimizing zeolite stabilized Pt-Zn catalysts for propane dehydrogenation." *Journal of Energy Chemistry* 57:92–98.

Xie, Linjun, Rui Wang, Yuchao Chai, Xuefei Weng, Naijia Guan, and Landong Li. 2021. "Propane dehydrogenation catalyzed by in-situ partially reduced zinc cations confined in zeolites." *Journal of Energy Chemistry* 63:262–269.

Xu, Shutao, Yuchun Zhi, Jingfeng Han, Wenna Zhang, Xinqiang Wu, Tantan Sun, Yingxu Wei, and Zhongmin Liu. 2017. "Advances in catalysis for methanol-to-olefins conversion." In *Advances in Catalysis*, 37–122. Elsevier.

Xu, Zhikang, Yuanyuan Yue, Xiaojun Bao, Zailai Xie, and Haibo Zhu. 2019. "Propane dehydrogenation over Pt clusters localized at the Sn single-site in zeolite framework." *ACS Catalysis* 10 (1):818–828.

Yun, Jang Ho, and Raul F Lobo. 2014. "Catalytic dehydrogenation of propane over iron-silicate zeolites." *Journal of Catalysis* 312:263–270.

Zhang, Fan, Runxia Wu, Yinghong Yue, Weimin Yang, Songyuan Gu, Changxi Miao, Weiming Hua, and Zi Gao. 2011. "Chromium oxide supported on ZSM-5 as a novel efficient catalyst for dehydrogenation of propane with CO_2." *Microporous and Mesoporous Materials* 145 (1–3):194–199.

Zhang, Hong Jiang, Zhi Gang Jia, and Sheng Fu Ji. 2011. "Preparation, characterization, and Oxidative dehydrogenation of C_3H_8 with CO_2 of V-Cr/SBA-15/Al_2O_3/FeCrAl metal monolithic catalysts." *Advanced Materials Research* 287:1671–1674.

Zhang, Shaobo, Yuming Zhou, Yiwei Zhang, and Li Huang. 2010. "Effect of K addition on catalytic performance of PtSn/ZSM-5 catalyst for propane dehydrogenation." *Catalysis Letters* 135 (1):76–82.

Zhang, Yiwei, Yuming Zhou, Li Huang, Mengwei Xue, and Shaobo Zhang. 2011. "Sn-modified ZSM-5 as support for platinum catalyst in propane dehydrogenation." *Industrial and Engineering Chemistry Research* 50 (13):7896–7902.

Zhang, Yiwei, Yuming Zhou, Li Huang, Shijian Zhou, Xiaoli Sheng, Qianli Wang, and Chao Zhang. 2015. "Structure and catalytic properties of the Zn-modified ZSM-5 supported platinum catalyst for propane dehydrogenation." *Chemical Engineering Journal* 270:352–361.

Zhang, Yiwei, Yuming Zhou, Junjun Shi, Xiaoli Sheng, Yongzheng Duan, Shijian Zhou, and Zewu Zhang. 2012. "Effect of zinc addition on catalytic properties of PtSnK/γ-Al_2O_3 catalyst for isobutane dehydrogenation." *Fuel Processing Technology* 96:220–227.

Zhang, Yiwei, Yuming Zhou, Menghan Tang, Xuan Liu, and Yongzheng Duan. 2012. "Effect of La calcination temperature on catalytic performance of PtSnNaLa/ZSM-5 catalyst for propane dehydrogenation." *Chemical Engineering Journal* 181:530–537.

Zhang, Yiwei, Yuming Zhou, Yu Wang, Yi Xu, and Peicheng Wu. 2007. "Effect of calcination temperature on catalytic properties of PtSnNa/ZSM-5 catalyst for propane dehydrogenation." *Catalysis Communications* 8 (7):1009–1016.

Zhang, Yiwei, Yuming Zhou, Kangzhen Yang, Yian Li, Yu Wang, Yi Xu, and Peicheng Wu. 2006. "Effect of hydrothermal treatment on catalytic properties of PtSnNa/ZSM-5 catalyst for propane dehydrogenation." *Microporous and Mesoporous Materials* 96 (1–3):245–254.

Zhang, Zhiyang, Wenlong Xu, Xiaomei Ye, Yonglan Xi, Cunpu Qiu, Liping Ding, Gui Liu, and Qingbo Xiao. 2021. "Enormous passivation effects of a surrounding zeolitic framework on Pt clusters for the catalytic dehydrogenation of propane." *Catalysis Science and Technology* 11 (15):5250–5259.

Zhao, Huahua, Huanling Song, Lingjun Chou, Jun Zhao, Jian Yang, and Liang Yan. 2017. "Insight into the structure and molybdenum species in mesoporous molybdena–alumina catalysts for isobutane dehydrogenation." *Catalysis Science and Technology* 7 (15):3258–3267.

Zhao, Shiyong, Bolian Xu, Lei Yu, and Yining Fan. 2018. "Catalytic dehydrogenation of propane to propylene over highly active PtSnNa/γ-Al_2O_3 catalyst." *Chinese Chemical Letters* 29 (3):475–478.

17 Porous Inorganic Nanomaterials as Heterogeneous Catalysts for the Dehydrogenation of Paraffin

Sauvik Chatterjee and Nandini Mukherjee

17.1 INTRODUCTION

Since 1930, the demand for pure olefins has increased due to their reactivity and ability to be used as building blocks for the production of various commodity and speciality chemicals and polymers.[1] Light olefins, particularly ethylene, propylene and butene, are amongst the most important compounds in the chemical industry. These compounds are the feedstock for the production of polymers, namely, polyethelene, polypropylene and oxygenates, such as acetaldehyde, acetone, ethylene glycol, propylene oxide and glycerine, and a vast array of industrially important chemical intermediates, specifically, ethylbenzene, propanal, etc.[2-4] With the singular exception of the Eurozone, the market has witnessed a steady increase in the demand for and production of these compounds in recent decades.[5] The most popular ways of obtaining light olefins are as by-products in the steam cracking and fluid catalytic cracking of paraffin.[6] Paraffins are saturated hydrocarbons with the general chemical formula C_nH_{2n+2}. The dehydrogenation process involves paraffin analogs with $n=2$ or more where one or more H_2 is eliminated to form unsaturated hydrocarbons or olefins. These processes are not widely known due to their thermodynamic restraints. Dehydrogenation by cracking requires the cleavage of two C-H bonds and the formation of one hydrogen molecule and a C-C double bond, which is quite an energy-demanding process due to these reactions' endothermic nature. To address this high energy demand, several catalytic processes have been investigated where the catalysts are designed to tune the energy requirement and the kinetics. Oleflex and Catofin are two catalysts that were commercialized by industries around 1985-90, which has addressed these issues to a certain extent. The low cost and very high Lewis acidic property of Catofin, a CrO_x catalyst-based process, have made it very successful. However, the environmental concerns due to the acute toxicity and quick deactivation because of the high activity of the catalyst leading to various side-reactions are difficult challenges to the sustainability of the process.[7] In more recent times, shell gas cracking, methanol to olefin-olefin cracking processes (MTO-OCP), etc. have been popularized. Propane to olefin (PTO) conversion has been one of the most important techniques in this venture. The combination of dehydrogenation and a metathesis catalyst in PTO involves opportunities and challenges with respect to chemistry, physicochemical properties and operations design. Exposure of the catalyst to different gases and temperatures during the time of catalyst activation, operation and regeneration and the interaction of different components of the catalyst are two key points to understand and remember at the time of catalyst development. There have been reports of other metal oxides, such as VO_x, GaO_x, FeO_x, etc., carbon nanotube and single atom catalysts, and the progress observed has been incremental. The thermal dehydrogenation of ethane and propane has been investigated using a bimetallic alloy catalyst with Pt nanoparticles.[6] An Sn-Ga alloy with Pt is the most prominent among the catalysts used for paraffin dehydrogenation. Although Ga is more appropriate for propane dehydrogenation, Sn/Pt has been quite efficient for the conversion of ethane to ethylene.

In recent times, focus has been directed to the development of smart catalysts that can show efficiency in terms of the adsorption of paraffin on the surface, C-H bond activation, desorption of olefin and hydrogen generation.[8] Along with the active sites, the physicochemical properties of the catalyst support have emerged as a crucial factor for the success of the catalyst. Porous nanomaterials such as K-L zeolite, alumina and hydrotalcite are some of the popular catalyst supports due to their selectivity towards alkane adsorption, which, in turn, increases the life cycle of the catalyst.[9] Deposition of coke on the catalyst is one of the major challenges because of the poisoning effect. There have been reports of catalysts that have shown enhanced selectivity after the deposition of a trace amount of coke. The acidity of the supports has been reported to influence the activity of the catalyst. Mesoporous γ-Al_2O_3 and MCM-41 are the two most popular supports for V-based catalysts. This is due to their acidic nature, customized porosity and high surface area, and they have shown improved life cycles and selectivity.

In this chapter, we deeply examine the process of dehydrogenation, the key factors controlling the efficiency and the progress that has been made in this field by using sophisticated catalysts for better yield in terms of the turnover number (TON) and selectivity. The process of dehydrogenation involves research on many aspects other than catalysts such as reactor design, operations, engineering, etc., which are beyond the scope of this discussion.

17.2 PRINCIPLE OF DEHYDROGENATION

The principle of the dehydrogenation reaction has been well-discussed in various literature.[10,11] In this section, we touch upon the key factors that influence the outcome of the reaction.

The direct non-oxidative dehydrogenation of ethane (EDH), propane (PDH) and isobutane to their corresponding olefin has activation energy of $ca.$ 121–143 kJ/mol. Empirically, they can be presented as

$$C_2H_6 \rightarrow C_2H_4 + H_2$$
$$C_3H_8 \rightarrow C_3H_6 + H_2$$
$$i\text{-}C_4H_{10} \rightarrow C_4H_8 + H_2$$

These reactions, due to their exothermic nature and overall increase in volume, are favoured under high temperature and low pressure. This requirement of high energy, however, leads to the generation of undesired by-products that creates a drop in selectivity along with coking, which reduces the catalyst life cycle through poisoning. Under the same pressure and temperature conditions, an increase in the number of carbon atoms and the branching of paraffin lead to a rise in the dehydrogenation equilibrium level.[12,13] The dehydrogenation process is primarily governed by the C-H activation of paraffin, whereas the majority of catalysts (transition metal-based) are responsive to both C-H and C-C bond activation to a certain extent due to their abundance of d-orbitals. The activity towards the C-C bond, in turn, causes a further reaction with olefins and formations of undesired products including coke. This happens since the C-C bonds of olefins are more reactive than the C-H bonds in the paraffin. Therefore, to develop an efficient catalyst, one of the major requirements has been the higher selectivity towards the C-H bond activation of paraffin. Cracking also leads to the cleavage of paraffin into smaller hydrocarbons under similar conditions. Complicating the situation is the Lewis and Bronsted acidic properties of the catalyst that proceed via the formation of carbocation and end up rearranging the hydrocarbons. This produces isomerization and further dehydrogenation of the final products. For example, butane and isobutane will isomerize and dehydrogenate to 1-butene or 2-butene and undergo further dehydrogenation to produce 1,3-butadiene.[14]

17.3 HETEROGENEOUS CATALYSTS FOR PARAFFIN DEHYDROGENATION

In principle, most catalysts for dehydrogenation consist of two major parts, namely, active materials and support materials. Due to their reactive nature and site sensitivity, both parts are important and are discussed in this section. Amongst the active catalysts, there are single-atom catalysts, both mono-metallic and bimetallic. Metal oxide catalysts are also extensively studied and commercialized in many cases. In recent times, there are explorations of various metal organic frameworks used for dehydrogenation, both oxidative and non-oxidative. Different types of hetereogeneous catalytic nanomaterials are discussed in the context of paraffin dehydrogenation in the following section.

17.3.1 Metal Oxide Catalyst

Metal oxides are the most well-studied catalysts for paraffin dehydrogenation (PDH) due to their stability, good life cycle and ease of regeneration. Various nanoporous support-based metal oxide catalysts have been reported (Table 17.1). As mentioned earlier, the Catofin process uses a mixed oxide of chromium on alumina. There have been reports of oxides of Sn and V. VO_x has been discussed as an alternative competent of a widely acclaimed Pt-based single-atom (SA) catalyst. VO_x on mesoporous SiO_2 such as MCM-41 was studied compared with γ-Al_2O_3, ZrO_2, etc. MCM-41-supported VO_x catalyst was found to have stable catalytic properties for PDH in the absence of a gas phase O_2 and can be fully recovered with oxidative regeneration, which is attributed to the high thermal stability of VO_x.[15,16] Furthermore, Bai et al. reported a mesoporous γ-Al_2O_3-supported V-based catalyst to achieve high catalytic activity and stability.[17] Hu and co-workers examined a one-step process for the preparation of V-doped porous silica monolith VO_x-SiO_2 in an environmentally friendly way for non-oxidative propane dehydrogenation, which shows a constant propene

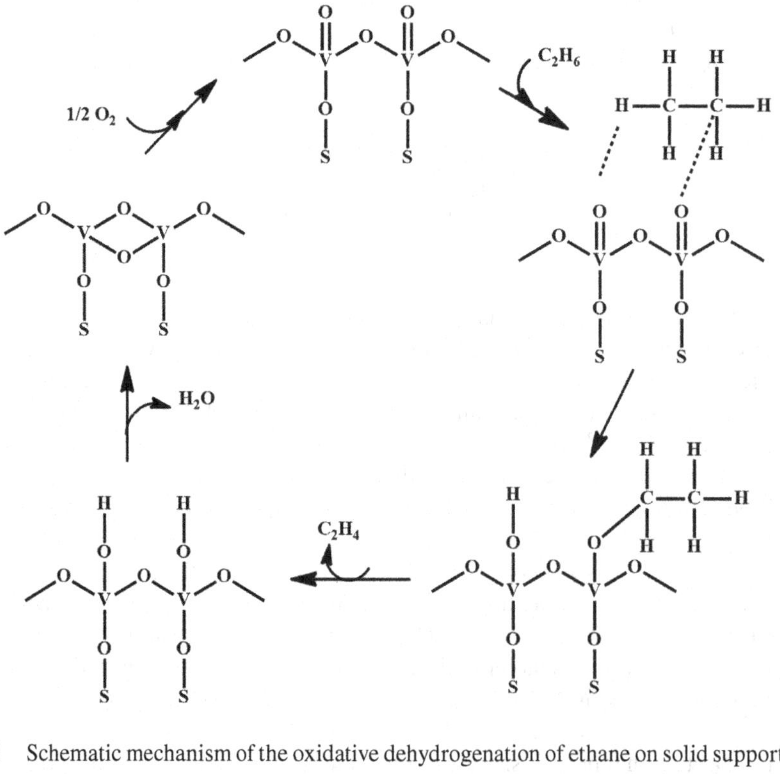

FIGURE 17.1 Schematic mechanism of the oxidative dehydrogenation of ethane on solid supported V_2O_5 (S = Support).

Source: (Adapted from ref. 19)

selectivity of around 91% over 8 reaction-regeneration cycles.[18] Similarly, V_2O_5-based catalyst supported on Al_2O_3 has been investigated for the oxidative dehydrogenation of ethane. These types of catalysts are believed to operate through a redox mechanism where paired metal centres undergo single electron reduction from V(V) to V(IV). Two neighbouring metal atoms participate in the dehydrogenation process, whereas only one of the two participates in C-H cleavage (Figure. 17.1).[19]

Recently, gallium oxide-based materials have been reported as highly promising catalysts for their ability to activate hydrocarbon species. Nakagawa et al. indicated that β-Ga_2O_3 has shown the highest activity among all the polymorphs of this oxide due to its abundant surface acid sites.[20] Various materials such as ZSM-5, HZSM-48, Al_2O_3, MgO, and TiO_2 have been used for the dispersion of Ga_2O_3. Shen et al. showed that Ga_2O_3/HZSM-5 has better selectivity and that the stability is enhanced when the Si/Al ratio is increased.[21] However, with an increased Si/Al ratio the initial activity of the catalyst drops. At an Si/Al ratio > 90, the catalytic stability is best amongst this class of catalysts. Xu et al. identifies Ga_2O_3/TiO_2 and Ga_2O_3/ZrO_2 to have better activity due to the presence of a greater number of acid sites with medium to high strength.[22] 5Ga_2O_3/SBA-15 has shown a propene selectivity of 92% after 30 h PDH at 620 °C.[23]

ZnO nanoparticles have also been reported in different forms and with various support materials as robust catalysts for PDH. Highly dispersed nanosized ZnO assemblies in dealuminated beta zeolite show almost 93% propylene selectivity with 53.3% conversion at 600 °C.[24] An increase in the Si/Al ratio has proved beneficial for higher propene selectivity and stability due to the selective acid sites.[25] ZnO nanospheres of 2–4 nm encapsulated in an N-doped cluster grown *in situ* from the carbonization of ZIF-8 on silicalite-1 (ZnO@NC/S-1) indicates more than 90% propylene selectivity and 44% conversion.[26] It has been observed that ZnO remains protected from volatilization and sintering at higher temperatures because of the presence of an NC layer. A mixed oxide of Zn and Nb was shown to have very high catalytic ability for PDH.[27] ZnO_x supported on rutile TiO_2 indicates a small $Zn_{1-3}O_x$ cluster with a 1–3 Zn atom as an active site.[28] A modification with ZrO_2 additive protects the catalyst from the formation of the detrimental $ZnTiO_3$ phase. It has also been reported that the presence of even 0.1% Pt in ZnO/Al_2O_3 improves the C-H activation ability and causes better hydrogen desorption due to the increase in the Lewis acidity of Zn^{+2}.[29]

Mesoporous cobalt-aluminate spinel prepared by the evaporation-induced self-assembly (EISA) method has been described for a conversion of 8% with more than 80% propane selectivity at a comparatively lower temperature of 550 °C. The co-centres, which occupy the tetrahedral sites, are identified as the active sites that show an estimated turnover frequency of 5.1 per hour.[30] Co-oxide grafted in HZSM-5 catalyst for EDH under optimal conditions presents a conversion efficiency of 54% with ethylene selectivity of 87.8%.[31]

Sarazen and Jones disclosed a Fe-containing MOF, Fe-BDC, to be used as a scaffold to develop PDH catalysis. Carbon-supported FeC catalysts have been developed by the calcination of Fe-BDC, which exhibits superior stability for propane formation for more than 48 hours of reaction time.[32] Co(II)-containing N-doped carbon material has been developed from a well-known metal organic framework, ZIF-67. This Co/NC-700 has shown 84% propene selectivity with over 16% conversion.

TABLE 17.1
Metal Oxide-Based Catalysts for the Dehydrogenation of Paraffin to Olefin

Sl No	Name of Catalyst	Process	Conversion (%)	Specificity (%)	TOF(/h)	Ref
1	Pt/SiO_2	Isobutane to isobutene	19	19.5	0.08	8
2	1:1 Pt/Sn/SiO_2	Isobutane to isobutene	3.4	78.9	0.052	8
3	1:3 Pt/Sn/SiO_2	Isobutane to isobutene	4.2	97.9	0.043	8
4	Pt/L	Isobutane to isobutene	22.5	14.1	0.64	8
5	Pt-Sn/Al_2O_3	Propane to propene	19.8	~100		15

(Continued)

TABLE 17.1 (Continued)
Metal Oxide-Based Catalysts for the Dehydrogenation of Paraffin to Olefin

Sl No	Name of Catalyst	Process	Conversion (%)	Specificity (%)	TOF(/h)	Ref
6	VO_x/MCM-41	Propane to propene	21.5	>90		15
7	CrO_x/MCM-41	Propane to propene	24.8	>75		15
8	VO_x/meso-Al_2O_3-373K	Propane to propene	12	15		17
9	VO_x-MCM-41	Propane to propene		79	0.00515 (/s)	18
10	60-VO_x-SiO_2-580	Propane to propene	57	91		18
11	Ga_2O_3	Ethane to ethylene (in the presence of CO_2)	19.6	95		20
12	Ga_2O_3	Ethane to ethylene (in the presence of Ar)	9.6	94		20
13	Cr_2O_3	Ethane to ethylene (in the presence of CO_2)	12.1	93.8		20
14	V_2O_5	Ethane to ethylene (in the presence of Ar)	12.5	91.7		20
15	β-Ga_2O_3	Ethane to ethylene	57.3	96.1		21
16	Ga_2O_3/HZSM-5(15)	Ethane to ethylene	34.6	97.7		21
17	Ga_2O_3/HZSM-5(25)	Ethane to ethylene	32.5	96.7		21
18	Ga_2O_3/TiO_2	Propane to propylene (in the presence of CO_2)	32	73		22
19	Ga_2O_3/Al_2O_3	Propane to propylene (in the presence of CO_2)	26	94		22
20	Ga_2O_3/ZrO_2	Propane to propylene (in the presence of CO_2)	30	65		22
21	Ga_2O_3/SiO_2	Propane to propylene (in the presence of CO_2)	6.4	92		22
22	Ga_2O_3/SBA-15	Propane to propylene (in the presence of Ar)	28.5	92.5	0.029	23
23	Ga_2O_3/ZSM-5	Propane to propylene (in the presence of Ar)	78.1	60.2	0.018	23
24	ZnHβ-10	Propane to propylene	6.5	7.6		24
25	Znβ-20	Propane to propylene	36.1	94.4		24
26	Znβ-10	Propane to propylene	53.3	92.9		24
27	Znβ-7	Propane to propylene	46.3	95.0		24
28	10%Zn/30HZSM-5	Propane to propylene	14.51	84.0		25
29	10%Zn/50HZSM-5	Propane to propylene	30.52	72.5		25
30	10%Zn/80HZSM-5	Propane to propylene	50.53	65.6		25
31	10%Zn/250HZSM-5	Propane to propylene	42.51	92.5		25
32	ZnO/S-1	Propane to propylene (in the presence of N_2)	13.7	57		26
33	ZnO@NC/S-1(0.0)	Propane to propylene (in the presence of N_2)	52.2	87		26
34	ZnO@NC/S-1(1.0)	Propane to propylene (in the presence of N_2)	56.4	85		26
35	2ZnTi	Propane to propylene (in the presence of N_2)	18	94		28
36	2Zn1.4ZrTi	Propane to propylene (in the presence of N_2)	21	92		28
37	2Zn5.6ZrTi	Propane to propylene (in the presence of N_2)	23	95		28
38	15Zn0.1Pt	Propane to propylene (in the presence of N_2)	35	97		29

Porous Inorganic Nanomaterials as Catalyst for Dehydrogenation

17.3.2 Nanoporous Materials-Based Single Atom (SA) and Single Atom Alloy (SAA) Catalysts

The other variants of catalyst that have been used for the conversion of paraffin to olefin are single atom catalysts (Table 17.2). These single atom catalysts or single atom alloy catalysts are generally supported upon porous nanomaterials with a high surface area and tunable porosity to achieve better reactivity and specificity.

The most common single atom catalyst for PDH was identified to be Pt almost half a century ago. Yang et al. presented a detailed mechanism of PDH on different Pt surfaces (e.g., Pt (111), Pt (100), Pt (211), etc.) with 17 dehydrogenation steps (Figure 17.2).

However, to use Pt as an efficient catalyst—by reducing the separation and regeneration costs of the catalyst—the authors were required to bind to or graft onto solid supports. These solid supports have a considerable role to play. The support not only modifies the effective area of the catalyst but also determines the life cycle of the catalyst by modulating the adsorption selectivity between the paraffin and olefin by virtue of the surface acidity/basicity and the pore size. Pt SA supported on various zeolite/silica supports is a well-reported and -studied catalyst for EDH and PDH reactions.

The dependencies of the particle size and atomic dispersion of Pt on the support was studied extensively by Zhang et al. who showed that the turnover frequency increases with a decrease in the particle size of the Pt nanoparticals, and atomically dispersed Pt indicates a 7-fold higher turnover frequency compared to Pt nanoparticles.[34] Pt on a commercially available porous ceria support, with strong metal support interaction (SMSI), displays a higher ethane to olefin conversion rate than Pt/SiO$_2$ that does not show SMSI. The reduction temperature also influences the propane selectivity. For the same Pt-ceria catalyst with SMSI, the reduction at 950 °C shows lower selectivity compared to 550 °C.[35] Recently, the Furukawa group developed a Pd-modified

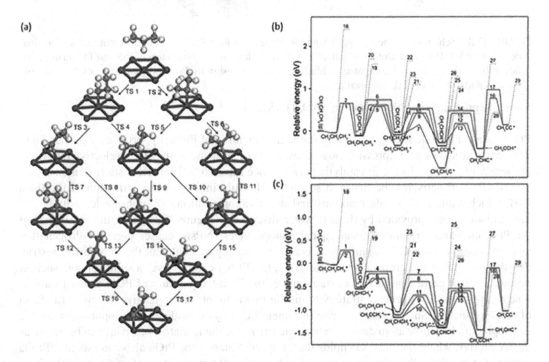

FIGURE 17.2 (a) Reaction network of propane dehydrogenation on Pt(111); (b and c) energy profile for propane dehydrogenation on Pt(111) (b) and Pt(211) (c) including both the dehydrogenation steps (solid lines) and the C–C cracking steps (dotted lines).[33] (Permission requested for reproduction.)

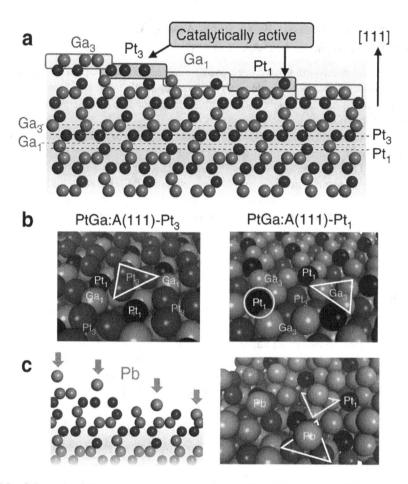

FIGURE 17.3 Schematic of the catalyst design for single atom Pt in PtGa: a) Four different surface terminations of PtGa:A(111) viewed along [101] direction (ball model). b) Diagonal view of Pt3 and Pt1 termination (space-filling model). Pt1 is highlighted in black. c) Catalyst design by Pb deposition to block the Pt3 (and Ga3) sites and to keep the Pt1 sites available.

Source: Reproduced from Nat. Commun. 2020, 11, 2838

Pt SA catalyst included in the thermally stable intermetallic of PtGa supported over porous SiO_2. This catalyst shows an impressive 30% conversion of propane with 99.6% selectivity towards propene without any change in catalytic performance after 96 h. Their DFT study demonstrated that the Pt well activates the first and second C-H while inhibiting the activation of the third C-H, which leads to less side reactions and deactivation resulting in higher selectivity.[36] The mechanistic insight provided by them is worth discussion (Figure 17.3). Since the semblance of the Pt atom—and formation of subsequent nanoparticles (NPs)—causes over-dehydrogenation and the scission of the C-H(C) moiety, it is a good strategy to single out the Pt atoms to restrict dehydrogenation at the propene stage. However, the Pt SA catalyst over a solid support such as porous silica or ceria is relatively unstable under the high temperature of PDH as the Pt atoms tend to agglomerate. A general strategy to inhibit this is to introduce inactive metals in the form of an alloy such as Sn. Thermally stable intermetallic alloys provide a good opportunity for the development of a single atom catalyst where the Pt1 in the cubic matrix of $PtGa_3$ can be active as an SA catalyst where the three Ga holds the Pt apart. Notably, the PtGa also consists of a Pt_3Ga matrix that has three Pt atoms that can lead to the over-dehydrogenation of paraffin. Since these Pt_3 parts are more prone towards Pd adsorption, the Pt_3 parts are blocked by Pd deposition and the formation of Pt single atom sites. Other intermetallic alloys such as $PtZn_1$, Pt_3V, Pt_3Co, Pt_3Cr, Pt_1Bi_1 and Pt_3Mn are reported to have good catalytic activity for the conversion of light alkane with high olefin selectivity.[35]

The effect of the support of the Pt SA or SAA has also been demonstrated by various groups. Jang et al. showed direct evidence of the suppression of metal sintering and coke formation with the reduction in the Lewis acid site on the support.[37] The Lewis acid sites of beta zeolites can improve the conversion by increasing the metal dispersion, while Brønsted acid sites can catalyze the cracking reactions, oligomerization and aromatization. When compared, the Pt-Sn/Beta catalysts (SiO_2/Al_2O_3 = 300) indicated better catalytic activity for PDH irrespective of the fact that the dispersion of Pt was lower than the Pt-Sn/Beta catalyst (SiO_2/Al_2O_3 = 38).[38] The use of ZSM-5 as support for the Pt SA or Pt-Sn alloy has been investigated due to the well-defined ten-membered ring structure of the support, which prevents coking by inhibiting the formation of larger hydrocarbons, thus increasing the life cycle of the catalyst. The Tsikiyiannis group reported 0.7% Pt-Sn-ZSM-5 as a PDH catalyst in the DH →SHC→DH microreactor process with 25% yield at 550 °C.[39] 'The Blekkan group proved that the Sn in Pt-Sn has an interaction with the γ-Al_2O_3 support, which stabilizes the oxidation state of Sn, unlike with SiO_2 where Sn was more readily reduced and alloy formation was possible.[40] Pt-Zn supported on zeolite L has shown superior activity for the dehydrogenation of butane and isobutane with 42% selectivity towards olefin formation.[41] Zn-containing ZSM-5 catalysts have been shown to be very attractive for the dehydrogenation of lower alkanes. However, the Zn incorporation method into the support plays a role in the catalytic activity. The volatilization of metallic Zn, sintering of the Pt-Zn phase and coking leads to rapid deactivation for both the Zn over-exchanged zeolites and the Zn impregnated Pt-Zn catalyst. Zn incorporation performed by an aqueous Zn(II) solution ion-exchange can limit the Zn loss and pore blocking at the time of catalyst activation. Moreover, several other Zn sites such as $(Zn-O-Zn)^{2+}$, $(Zn-(o-Zn)_n)^{2+}$, ZnO and Bronsted acid site can be formed during the ion exchange method. These isolated Zn(II) moieties in ZSM-5 and ZSM-11 cause aromatization in the ten-membered ring of the support and coke formation for the zeolite L and Y (12-membered ring). The Pt-Zn-loaded zeolite Na-Beta (nSi/nAl = 26, mPt = 0.5 wt.%, mZn = 1.6–2.6 wt.%) indicates propene selectivity exceeding 90% for at least 63 hrs. time-on-stream.[42]

Very recently, a Ga(III) catalyst was reported on zeolite BEA support, which shows 19% conversion with 82% propene selectivity.[43] The study demonstrates that the kinetics of the reaction were determined by the first and second C-H cleavages at low propane partial pressure, whereas at a higher propane pressure, the kinetics are governed by the H_2 desorption rate from the catalyst. The active Ga(III) sites were identified to be dehydrated and tetrahedrally coordinated with zeolite BEA. The low selectivity towards the formation of aromatic side products is attributed to the low Bronsted acidity and Lewis acidity of the catalyst support. Lobo et al. showed Ga(I) to be an active centre in a chabazite (Ga-CHA) framework to achieve 96% propylene selectivity.[44] They demonstrated by in-situ FTIR spectroscopy that Ga(I) originated from the precursor Ga_2O_3 by a reduction followed by diffusion into the CHA framework. The position of the metal in the framework's structure has a vital role to play in the reaction mechanism and kinetics of the dehydrogenation process. For example, a study by Yun and Lobo explained the differences in the pattern of selectivity, entropy of formation of the transition states, the estimated enthalpy and the activation energy in the case of H-[Fe]-ZSM-5 and H-[Al]-ZSM-5, which are iso-structural by nature.[45] This indicates, despite the similarity in the framework and structure, that the mechanisms of the reaction for different catalysts are different.

17.3.3 Metal Organic Frameworks (MOFs)-Based Catalyst

Metal organic frameworks (MOFs) are the modern miracle of materials science. These are crystalline porous materials made of organic entities connected through metal centres. MOFs have tunable porosity, a periodic structure and a very high surface area with an ample scope of surface

TABLE 17.2
SA- and SAA-Based Catalysts for the Dehydrogenation of Paraffin to Olefin

Sl No	Name of Catalyst	Process	Initial Conversion (%)	Initial Specificity (%)	TOF(/s)	Ref
1	0.05% Pt/Al$_2$O$_3$	Propane to propylene	5.5	95		31
2	0.1% Pt/Al$_2$O$_3$	Propane to propylene	12	91		31
3	0.3% Pt/Al$_2$O$_3$	Propane to propylene	15	85		31
4	1% Pt/Al$_2$O$_3$	Propane to propylene	22	70		31
5	3% Pt/Al$_2$O$_3$	Propane to propylene	26	52		31
6	5% Pt/Al$_2$O$_3$	Propane to propylene	32	61		31
7	2 Pt/SiO$_2$	Propane to propylene	20	75	0.17	35
8	Pt/CeO$_2$-red550 °C	Propane to propylene	21	78	0.33	35
9	Pt/CeO$_2$-red975 °C	Propane to propylene	15	95	0.013	35
10	Pt3Sn/SiO$_2$	Propane to propylene	25	98.6		36
11	PtGa/SiO$_2$	Propane to propylene	39	99.1		36
12	PtGa-Pb/SiO$_2$	Propane to propylene	27	99.6		36
13	Pt-Sn/Beta(300)	Propane to propylene	63.8	65.5		35
14	Pt-Sn/Beta(38)	Propane to propylene	65.7	55.8		35
15	Pt-Sn/θ-Al$_2$O$_3$	Propane to propylene	60.1	71.7		35
16	Pt-Sn/Q6	Propane to propylene	23.6	48.8		35
17	Pt/γ-Al2O3	Propane to propylene			0.8	37
18	PtSn/γ-Al2O3	Propane to propylene			1.2	37
19	Pt/SiO2	Propane to propylene			0.9	37
20	PtSn/SiO2	Propane to propylene			0.1	37
21	0.5Pt/Na-ZSM-5	Propane to propylene	44.8	55.4		39
22	0.5Pt/2.6Zn, Na-ZSM-5 (n$_{Si}$/n$_{Al}$=55)	Propane to propylene	17.7	68.7		39
23	0.5Pt/0.6Zn, Na-ZSM-5(n$_{Si}$/n$_{Al}$=55)	Propane to propylene	47.8	18.1		39
24	0.5Pt/1.3Zn, Na-ZSM-5(n$_{Si}$/n$_{Al}$=55)	Propane to propylene	19.2	30.0		39
25	0.5Pt/4.4Zn, Na-ZSM-5(n$_{Si}$/n$_{Al}$=55)	Propane to propylene	33.4	38.8		39
26	0.5Pt/2.5Zn, Na-ZSM-5(n$_{Si}$/n$_{Al}$=21)	Propane to propylene	23.9	67.5		39
27	0.5Pt/1.9Zn, Na-ZSM-5(n$_{Si}$/n$_{Al}$=21)	Propane to propylene	76.0	4.6		39
28	0.5Pt/2.5Zn, Na-ZSM-5(n$_{Si}$/n$_{Al}$=10)	Propane to propylene	78.9	5.1		39
29	0.5Pt/4.4Zn, Na-ZSM-5(n$_{Si}$/n$_{Al}$=10)	Propane to propylene	55.4	10.5		39

functionalization. These classes of materials are excellent supports of metal NPs and metal oxides. The robust nature of MOFs is very useful for operation as a heterogeneous catalyst. The tunable porosity and surface functionality make these catalysts selective towards both reactants and products. In addition to being used as support for metal NP or metal oxide catalysts, these materials can be highly effective catalysts in their own right.

The Farah group used a Co(II)-anchored NU-1000 MOF as a catalyst for the oxidative dehydrogenation of propane. Co(II) is deposited by either solvothermal deposition on MOF or atomic layer deposition in MOF—both of which were found to be active for propane oxidative dehydrogenation at a relatively lower temperature of 200 °C. They showed with mechanistic computations that the second abstraction of the C-H of propane is less energy-demanding than isopropyl migration. This study indicates the possibility of controlling propene selectivity depending upon the temperature.[46]

Very recently, Ghosh et al. reported N, S and P heteroatom-doped Fe-carbon nano-spheres for the CO_2-assisted dehydrogenation of ethylbenzene to produce 1-phenylethylene.[47] It was demonstrated that the presence of each dopant viz. N, S and P has a prominent role in high-temperature dehydrogenation while low-temperature dehydrogenation is majorly controlled by the iron content. The N-doped hierarchical Fe-hollow sphere carbon (Fe-N-C) catalyst shows the highest catalytic performance among these sets of catalysts due to the cooperative action of pyridinic N species and the $Fe3C$ phase for CO_2 activation.

17.4 OUTLOOK AND CONCLUSION

It has been a fascinating journey of catalyst development for a process as industrially important and chemically intriguing as the dehydrogenation of paraffin. The daily improvement on the front lines of mechanistic understanding and kinetic studies and their dependencies on various parameters show promising prospects in this field in the near future. Despite these innovations in various directions, it should be mentioned in this context that very few alternatives have been presented as industrially viable options apart from the classic Oleflex and Catofin. Multiple challenges such as stability, the cost-optimized synthesis of the catalyst and the thermodynamic viability for in situ dehydrogenation need to be addressed with these innovative catalysts. Further study of the structure-property correlation is essential to make an industrially viable catalyst. Because of the ever-growing requirement of light olefins, tremendous attention and investments are being given by both corporate industries and several federal agencies in the USA, the EU, Saudi Arabia, Iran, India, etc. Keeping in mind that reactor design and process control are the complementary components of dehydrogenation, smart catalyst development is of the utmost importance. Aspects such as the usage of CO_2 as a promoter for dehydrogenation to improve carbon utilization and the exploration of the oxidative pathways of dehydrogenation to stabilize the catalyst need to be major focuses of the research fraternity. Accordingly, enhanced coordination between industry and academia is solicited for the achievement of the best outcome of developmental research in this field.

REFERENCES

[1] D. Sanfilippo, I. Miracca, Dehydrogenation of paraffins: Synergies between catalyst design and reactor engineering. *Catal. Today* 111 (2006) 133–139.

[2] S. Budavari, M. O'Neil, A. Smith, P. Heckelman, J. Obenchain, *The Merck Index*, 12th ed.; Budavari, S., Ed., Merck & Co.: NJ, 1996; pp. 1348–1349.

[3] M. McCoy, M. Reisch, A. H. Tullo, P. L. Short, J.-F. Tremblay, The case for Saltigo. *Chem. Eng. News* 84 (2006) 59.

[4] *Ethylene Uses and Market Data*, htp://www.icis.com/Articles/2007/11/05/9075777/ethylene-uses-and-market-data.html (accessed Oct 25, 2013).

[5] *The Global Olefins and Polyolefins Markets in 2011—Slow Growth in Demand Amid Political and Economical Crisis*, http://www.researchandmarkets.com/research/bdd2c0/the_global_olefins (accessed Oct 25, 2013).

[6] Z. Nawaz, Light alkane dehydrogenation to light olefin technologies: A comprehensive review. *Rev. Chem. Eng.* (2015) http://doi.org/10.1515/revce-2015-0012.

[7] Z. Lian, S. Ali, T. Liu, C. Si, B. Li, D. S. Su, Revealing the Janus character of the coke precursor in the propane direct dehydrogenation on Pt catalysts from a kMC simulation. *ACS Catal.* 8 (2018) 4694–4704.

[8] M. L. Yang, Y. A. Zhu, X. G. Zhou, Z. J. Sui, D. Chen, First-principles calculations of propane dehydrogenation over PtSn catalysts. *ACS Catal.* 2 (2012) 1247–1258.

[9] R.D. Cortright, J.M. Hill, J.A. Dumesic, Selective dehydrogenation of isobutane over supported Pt/Sn catalysis. *Catal. Today* 55 (2000) 213–223.

[10] A. Alamdari, R. Karimzadeh, S. Abbasizadeh, Present state of the art of and outlook on oxidative dehydrogenation of ethane: Catalysts and mechanisms. *Rev. Chem. Eng.* 37 (2021) 481–532.

[11] J. A. Lercher, F. N. Naraschewski, *Nanostructured Catalysts—Selective Oxidations*; Hess, C., Schlçgl, R., Eds., Royal Society of Chemistry, Cambridge, 2011, p. 5.

[12] Z. Nawaz, F. Wei, Light alkane oxidative dehydrogenation to light olefins over Pt-based SAPO-34 zeolite supported catalyst. *Ind. Eng. Chem. Res.* 52 (2013) 346–352.

[13] Y. Chu, Q. Zhang, T. Wu, Z. Nawaz, Y. Wang, F. Wei, Ultra-dispersed Pt nanoparticles on SAPO-34/γ-Al2 O3 support for efficient propane dehydrogenation. *J. Nanosci. Nanotechnol.* 14 (2014) 6900–6906.

[14] H. Song, R. M. Rioux, J. D. Hoefelmeyer, R. Komor, K. Niesz, M. Grass, P. Yang, G. A. Somorjai, Hydrothermal growth of mesoporous SBA-15 silica in the presence of PVP-stabilized Pt nanoparticles: Synthesis, characterization, and catalytic properties. *J. Am. Chem. Soc.* 128 (2006) 3027–3037.

[15] S. Sokolov, M. Stoyanova, U. Rodemerck, D. Linke, E.V. Kondratenko, Comparative study of propane dehydrogenation over V-, Cr-, and Pt-based catalysts: Time on-stream behavior and origins of deactivation. *J. Catal.* 293 (2012) 67–75.

[16] U. Rodemerck, M. Stoyanova, E.V. Kondratenko, D. Linke, Influence of the kind of VOx structures in VOx/MCM-41 on activity, selectivity and stability in dehydrogenation of propane and isobutene. *J. Catal.* 352 (2017) 256–263.

[17] P. Bai, Z. Ma, T. Li, Y. Tian, Z. Zhang, Z. Zhong, W. Xing, P. Wu, X. Liu, Z. Yan, Relationship between surface chemistry and catalytic performance of mesoporous γ-Al_2O_3 supported VOX catalyst in catalytic dehydrogenation of propane ACS. *Appl. Mater. Inter.* 8 (2016) 25979–25990.

[18] P. Hu, W.-Z. Lang, X. Yan, L.-F. Chu, Y.-J. Guo, Influence of gelation and calcination temperature on the structure-performance of porous VO_x-SiO_2 solids in non-oxidative propane dehydrogenation. *J. Catal.* 358 (2018) 108–117.

[19] C. A. Gärtner, A. C. van Veen, J. A. Lercher, Oxidative dehydrogenation of ethane: Common principles and mechanistic aspects. *ChemCatChem* 5 (2013) 3196–3217.

[20]]K. Nakagawa, M. Okamura, N. Ikenaga, T. Suzuki, T. Kobayashi, Dehydrogenation of ethane over gallium oxide in the presence of carbon dioxide. *Chem. Commun.* (1998) 1025–1026.

[21] Z. Shen, J. Liu, H. Xu, Y. Yue, W. Hua, W. Shen, Dehydrogenation of ethane to ethylene over a highly efficient Ga_2O_3/HZSM-5 catalyst in the presence of CO_2. *Appl. Catal. A* 356 (2009) 148–153.

[22] B. Xu, B. Zheng, W. Hua, Y. Yue, Z. Gao, Support effect in dehydrogenation of propane in the presence of CO_2 over supported gallium oxide catalysts. *J. Catal.* 239 (2006) 470–477.

[23] C.-T. Shao, W.-Z. Lang, X. Yan, Y.-J. Guo, Catalytic performance of gallium oxide based-catalysts for the propane dehydrogenation reaction: Effects of support and loading amount. *RSC Adv.* 7 (2017) 4710–4723.

[24] C. Chen, Z. Hu, J. Ren, S. Zhang, Z. Wang, Z.-Y. Yuan, ZnO nanoclusters supported on dealuminated zeolite as a novel catalyst for direct dehydrogenation of propane to propylene. *ChemCatChem* 11 (2019) 868–877.

[25] C. Chen, Z.-P. Hu, J.-T. Ren, S. Zhang, Z. Wang, Z.-Y. Yuan, ZnO supported on high silica HZSM-5 as efficient catalysts for direct dehydrogenation of propane to propylene. *Mol. Catal.* 476 (2019) 110508.

[26] D. Zhao, Y. Li, S. Han, Y. Zhang, G. Jiang, Y. Wang, K. Guo, Z. Zhao, C. Xu, R. Li, C. Yu, J. Zhang, B. Ge, E. V. Kondratenko, ZnO nanoparticles encapsulated in nitrogen doped carbon nanoparticle and silicalite-1 composite for efficient propane dehydrogenation. *iScience* 13 (2019) 269–276.

[27] Y.-N. Sun, C. Gao, L. Tao, G. Wang, D. Han, C. Li, H. Shan, ZnNbO catalysts for propylene production via catalytic dehydrogenation of propane. *Catal. Commun.* 50 (2014) 73–77.

[28] S. Han, D. Zhao, H. Lund, N. Rockstroh, S. Bartling, D. E. Doronkin, J.-D. Grunwaldt, M. Gao, G. Jiang, E. V. Kondratenko, TiO_2 supported catalysts with ZnO and ZrO_2 for non-oxidative dehydrogenation of propane: Mechanistic analysis and application potential. *Catal. Sci. Technol.* 10 (2020) 7046–7055.

[29] G. Liu, L. Zeng, Z.-J. Zhao, H. Tian, T. Wu, J. Gong, Platinum-modified ZnO/Al_2O_3 for propane dehydrogenation: Minimized platinum usage and improved catalytic stability. *ACS Catal.* 6 (2016) 2158–2162.

[30] B. Hu, W.-G. Kim, T. P. Sulmonetti, M. L. Sarazen, S. Tan, J. So, Y. Liu, R. S. Dixit, S. Nair, C. W. Jones, A mesoporous cobalt aluminate spinel catalyst for nonoxidative propane dehydrogenation. *ChemCatChem* 9 (2017) 3330.

[31] H. Guo, C. Miao, W. Hua, Y. Yue, Z. Gao, Cobaltous oxide supported on MFI zeolite as an efficient ethane dehydrogenation catalyst. *Microporous Mesoporous Mater.* 312 (2021) 110791.

[32] M. L. Sarazen, C. W. Jones, MOF-derived iron catalysts for nonoxidative propane dehydrogenation. *J. Phys. Chem. C* 122 (2018) 28637–28644.

[33] M.-L. Yang, Y.-A. Zhu, C. Fan, Z.-J. Sui, D. Chen, X.-G. Zhou, DFT study of propane dehydrogenation on Pt catalyst: Effects of step sites. *Phys. Chem. Chem. Phys.* 13 (2011) 3257–3267.

[34] W. Zhang, H. Wang, J. Jiang, Z. Sui, Y. Zhu, D. Chen, X. Zhou, Size dependence of Pt catalysts for propane dehydrogenation: From atomically dispersed to nanoparticles. *ACS Catal.* 10 (2020) 12932–12942.

[35] J. Z. Chen, A. Talpade, G. A. Canning, P. R. Probus, F. H. Ribeiro, A. K. Datye, J. T. Miller, Strong metal-support interaction (SMSI) of Pt/CeO2 and its effect on propane dehydrogenation. *Catal. Today* 371 (2021) 4–10.

[36] Y. Nakaya, J. Hirayama, S. Yamazoe, K. Shimizu, S. Furukawa, Single-atom Pt in intermetallics as an ultrastable and selective catalyst for propane dehydrogenation. *Nat. Commun.* 11 (2020) 2838.

[37] E. J. Jang, J. Lee, H. Y. Jeong, J. H. Kwak, Controlling the acid-base properties of alumina for stable PtSn-based propane dehydrogenation catalysts. *Appl. Catal. A-Gen.* 572 (2019) 1–8.

[38] S.-U. Lee, Y.-J. Lee, S.-J. Kwon, J.-R. Kim, S.-Y. Jeong, Pt-Sn supported on beta zeolite with enhanced activity and stability for propane dehydrogenation. *Catalysts* 11 (2021) 25.

[39] R. K. Grasselli, D. L. Stern, J. G. Tsikoyiannis, Catalytic dehydrogenation (DH) of light paraffins combined with selective hydrogen combustion (SCH): II. DH + SHC catalysts physically mixed (redox process mode). *Appl. Catal. A* 189 (1999) 9.

[40] O. A. Barias, A. Holmen, E. A. Blekkan, Propane dehydrogenation over supported platinum catalysts: Effect of tin as promoter. *Catal. Today* 24 (1995) 361.

[41] B. D. Alexander, G. A. Huff Jr., *US Patent 5,453,558* (1995), to Amoco Corporation.

[42] P. L. De Cola, R. Gläser, J. Weitkamp, Non-oxidative propane dehydrogenation over Pt–Zn-containing zeolites. *Appl Catal A: General* 306 (2006) 85–97.

[43] L. Ni, R. Khare, R. Bermejo-Deval, R. Zhao, L. Tao, Y. Liu, J. A. Lercher, Highly active and selective sites for propane dehydrogenation in zeolite ga-BEA. *J. Am. Chem. Soc.* 144 (2022) 12347–12356.

[44] Y. Yuan, J. S. Lee, R. F. Lobo, Ga+-chabazite zeolite: A highly selective catalyst for nonoxidative propane dehydrogenation. *J. Am. Chem. Soc.* 144 (2022) 15079–15092.

[45] J. H. Yun, F. Raul, Lobo radical cation intermediates in propane dehydrogenation and propene hydrogenation over H-[Fe] zeolites. *J. Phys. Chem. C* 118 (2014) 27292–27300.

[46] Z. Li, A. W. Peters, V. Bernales, M. A. Ortuño, N. M. Schweitzer, M. R. DeStefano, L. C. Gallington, A. E. Platero-Prats, K. W. Chapman, C. J. Cramer, L. Gagliardi, J. T. Hupp, O. K. Farha, Metal–organic framework supported cobalt catalysts for the oxidative dehydrogenation of propane at low temperature. *ACS Cent. Sci.* 3 (2017) 31–38.

[47] A. Ghosh, A. Singha, R. Chatterjee, T. E. Müller, A. Bhaumik, B. Chowdhury, Influence of heteroatom-doped Fe-carbon sphere catalysts on CO_2- mediated oxidative dehydrogenation of ethylbenzene. *Mol. Catal.* 535 (2023) 112836.

Index

A

ABS polymer, 239
abstraction, 2, 58, 91, 102, 239, 271
acetaldehyde, 56, 62, 87, 153, 177, 227, 235, 262
acrylate, 114
acrylonitrile, 7, 61, 169, 238, 239
acylating agent, 78, 79
adhesive, 239
aerobic oxidation, 107, 108, 111
aggregation, 62, 250
albaconazole, 169
alcohol dehydrogenase, 152, 153, 154, 170
alkylating agent, 13, 20, 26, 35, 52, 54
allyl-palladium, 114, 121
Alzheimer, 154, 163, 171
aminoarenes, 15, 48
aminoarenes, 15, 48
aminophenol, 127, 164
aminothiophenols, 127, 128
ancillary ligand, 20, 28, 47, 186
anti-arrhythmic, 153
antifreeze, 239
auxiliary ligands, 187

B

benzaimidazole, 123
benzodiazepine, 123
benzothiazoles, 24, 53, 127, 128, 133, 172
bidentate, 16, 117, 121, 203
biorefinery, 124
boronation, 187
butanal, 238

C

caprolactam, 124
caprolactone, 124
carbanion, 80
carbon based catalyst, 91
catalytic membranes, 212, 214
catalytic oxidation, 187, 200
chemical hydrogen storage, 134, 234
chemical vapor deposition, 211
chemoselective, 19, 49, 171, 190
chloroquine, 154
cholesterol, 158, 169, 194
chromatography, 46
circular economy, 219
CO_2 emissions, 217, 219
combustion, 179, 221, 227, 273
commonalities, 172
contemporaneous, 132, 136, 185
conventional catalysts, 219

conversion efficiency, 265
crotonate, 114
cumene, 96, 97, 179, 187, 238, 239
cyclohexane conversion, 137, 210
cyclooctane, 92, 93, 100, 154, 204, 206
cyclopentane, 92, 93
cyclosporine, 192

D

dealkylation, 104, 111
dearomatized, 152
decarbonylation, 34, 69
dehydrated, 59, 269
dehydrocyclization, 66, 88, 90, 99, 111
dehydrogenative technologies, 10, 189
design procedure, 173, 175
detergents, 1
diethlyamine, 104
dimerization, 40, 42, 94, 99, 239

E

ecological, 54, 255
economically, 8, 62, 101, 217
electrocatalysis, 102
electrophile, 11, 47, 110
enamine, 101
enantiomeric, 131, 132, 185
encapsulated, 60, 65, 91, 265, 272
enolates, 4, 113
epoxidation, 55
etching agent, 55
exergonic, 201

F

fluid catalytic cracking, 87
fluidized bed reactor, 179

G

gallium oxide, 90, 91, 265, 272
gallium oxide-based catalyst, 90

H

hemiaminal, 14, 43
heteroarenes, 4, 136, 149
histidine, 153
homogenization, 173, 182
hydroacylation, 37, 38, 52
hydrodenitrification, 141
hydrogen extraction, 210
hydroperoxides, 186

Index

hydrotalcite, 64, 65, 124, 263
hygroscopic, 61, 141

I

impeller design, 182
indolines, 150, 199, 203, 226
industrial reactor networks, 180
inexpensive, 29, 63, 126, 162, 219
inorganic hydrides, 134, 135
intermediate isolation, 198
intermetallic, 215, 268, 269, 273
isolation, 102, 198
isomers, 4, 219
isotope, 102, 187, 192

K

Knoevenagel, 34, 58, 63, 66

L

lactams, 4, 54, 74
larock synthesis, 156, 170
light-assisted, 84

M

malignancies, 169
materials science, 10, 62, 184, 269
metallacarboranes, 225, 226, 235
metallovesicles, 188, 190
Micelle, 55, 62, 66, 188
microchannel reactor, 140
microwave-assisted, 50, 138
mixed flow reactor, 176, 177, 179, 181
molecular sieves, 56, 57, 66, 78
molybdenum oxide-based catalysts, 90
monoalkylation, 15, 16, 20, 35, 48, 50
monodentate, 187
monomers, 90, 184, 214
morpholine, 48, 123

N

nanoalloy, 229
nanocrystalline, 56, 66, 214, 249
nanoporous, 258, 264, 267
nanotechnology, 55, 217, 236
neratinib, 154, 170
N-ethyl carbazole, 136, 140
Ni-based catalyst systems, 91
non-ideal reactors, 180
nucleophile, 34, 35, 38, 39

O

oleflex, 89, 249, 262, 271
Oppenauer, 2, 19, 48, 68, 84
organohalide, 54
oxo-alcohols, 238, 239

oxygenates, 1, 238, 256, 262
oxygenation, 99, 111, 187

P

packed bed reactor, 179
Pacol process, 87, 88, 249
perhydrocarbazole, 140
phenelynediamine, 127, 128
phenylcyclohexylpiperidine, 87, 93
photoactivity, 192, 226, 227
photocatalytic potential, 186
photochemical, 104, 133, 225, 235
photoirradiation, 128
photooxidized, 223, 226
photoreactor, 226, 235
photoredoxmetallacarborane, 226
physicochemical properties, 254, 262, 263
pincer catalyst, 11, 72, 100, 150, 152, 160
pincer iridium complexes, 93, 94
platinum-based catalyst, 9, 88, 250
plug flow reactor, 178, 179, 181
polyamidation, 76
polyaromatic system, 136
polycarbonates, 239
polycyclic heteroarenes, 136
polycyclic hydroaromatic, 98, 100
polyol techniques, 138
Povarov reaction, 155, 156, 170
Prazosin, 169
pre-functionalization, 187
protonolysis, 59
pyrazine, 63, 74, 75, 85, 123
pyridinic N species, 271
pyrimidine, 123, 169, 170, 172, 228
pyrolysis, 137, 179, 217, 220, 228, 232
pyrrolidine, 17, 18, 63, 104, 123, 150

Q

quantum dots, 187
quinolone, 123, 132
quinoxaline, 63, 123, 132, 171

R

racemization, 39
reactor networking, 180
real reactors, 182
recrystallization, 46
recyclability, 58, 110, 130, 145, 184, 224
reformation, 66, 199
regeneration mode, 89
rejuvenation, 88
reusability, 110, 188, 226
Robinson catalyst, 69

S

Schrock's catalyst, 152
scyclooctane, 154

selenoxide, 4, 12, 113
single-pot synthesis, 186
skyrocketed, 87, 209
sol-gel, 55, 92, 227
spray pulse reactor, 137
stoichiometric reagents, 29, 113, 132, 151, 185, 187
substrate transition metals, 211
sustainability, 10, 184, 217, 227, 262
synergistic effect, 186

T

template, 55, 60, 174, 175, 255

V

valeric acid, 118

X

Xantphos ligand system, 160